The Finite Element Method
Displayed

DHATT, GOURI
THE FINITE ELEMENT METHOD DISP
000455049

517.95 DS3

The Finite Element Method Displayed

Gouri Dhatt
*Professor of Civil Engineering
Laval University
Quebec City, Canada*

and

Gilbert Touzot
Universite de Technologie, Compiègne, France

translated by
Gilles Cantin
*Professor of Mechanical Engineering,
Naval Postgraduate School, Monterey, California, USA*

A Wiley–Interscience Publication

JOHN WILEY & SONS
Chichester · New York · Brisbane · Toronto · Singapore

Copyright © 1984 by Gouri Dhatt and Gilbert Touzot

Reprinted August, 1985.

All rights reserved.

No part of this book may be reproduced by any means, nor transmitted, nor translated into a machine language without the written permission of the publisher.

Library of Congress Cataloging in Publication Data:

Dhatt, G.
   The finite element method displayed.
   Translation of: Une prèsentation de la mèthode des èlèments finis.
   "A Wiley-Interscience publication."
   Includes bibliographical references.
   1. Finite element method. I. Touzot, Gilbert. II. Title.
TA347.F5D513   1983   620'.001'515353   82 24843
ISBN 0 471 90110 5

British Library Cataloguing in Publication Data:

Dhatt, Gouri
   The finite element method displayed.
   1. Finite element methods
   I. Title   II. Touzot Gilbert
   515.3'53   TA347.F5

   ISBN 0 471 90110 5 (cloth)

Photosetting by Thomson Press (India) Limited, New Delhi and printed by Page Brothers, Norwich

FOREWORD

This book by my two colleagues of Compiègne is a welcome addition to the finite element literature. It is the result of many years of teaching the finite element methods in France and also in America. As a collaborator in the pedagogical experiments of Compiègne, the undersigned has seen early manuscripts of this book; it agrees well with his own views on what should be taught in such a course. It contains very concisely all that is needed to understand finite element methods in general.

The mathematical treatment is rigorously correct without being too sophisticated; this aspect should appeal to a vast readership. The choices of numerical algorithms for the solution of the discretized problems follow standard practice and are up to date. Implementation of the methods into computer codes is very well done and the reader should acquire valuable skills in writing engineering analysis software after studying the numerous examples in the text.

Finally it should be said that this book is written for those potential users of finite element methods who are not steeped in the theories of structural mechanics. It contains some structural mechanics applications but even the casual reader should have no problems with the concepts used in these examples.

Gilles Cantin
Monterey, June 1982

CONTENTS

Introduction 1

   0.1   The Finite Element Method 1
       0.1.1   Generalities 1
       0.1.2   Historical Evolution of the Method 2
       0.1.3   State of the Art 2

   0.2   Object and Organization of the Book 3
       0.2.1   Teaching of the Finite Element Method 3
       0.2.2   Objectives of the Book 3
       0.2.3   Organization of the Book 4
   References 5

1. Approximations with Finite Elements 9

   1.0   Introduction 9
   1.1   Generalities 9
       1.1.1   Nodal Approximation 9
       1.1.2   Approximations with Finite Elements 14

   1.2   Geometrical Definition of the Elements 18
       1.2.1   Geometrical nodes 18
       1.2.2   Rules for the Partition of a Domain into Elements 18
       1.2.3   Shapes of Some Classical Elements 20
       1.2.4   Reference Elements 21
       1.2.5   Classical Elements 25
       1.2.6   Nodal Coordinates and Element Connectivities 27

   1.3   Approximation Based on a Reference Element 30
       1.3.1   Algebraic Form of Approximate Function $u(x)$ 30
       1.3.2   Properties of Approximate Function $u(\mathbf{x})$ 32

   1.4   Construction of Functions $N(\boldsymbol{\xi})$ and $\bar{N}(\boldsymbol{\xi})$ 36
       1.4.1   General Method of Construction 36
       1.4.2   Algebraic Properties of Functions N and $\bar{N}$ 41

1.5	Transformation of Differential Operators	42
	1.5.1 Generalities	42
	1.5.2 First Derivatives	43
	1.5.3 Second Derivatives	46
	1.5.4 Ill-conditioned Jacobian Matrix	48
1.6	Computations of Functions N, their Derivatives and the Jacobian Matrix	49
	1.6.1 Generalities	49
	1.6.2 Explicit Formulas for N	50
	1.6.3 Automatic Construction of Interpolation Function N	50
	1.6.4 Computation of the Jacobian Matrix and its Derivatives	53
1.7	Truncation Errors in an Element	64
	1.7.1 Introduction	64
	1.7.2 Evaluation of the Error	68
	1.7.3 Reduction of the Truncation Error	70
1.8	Numerical Example (Rainfall Estimation)	71
	Notation	74
	References	75

2. Various Types of Elements ... 77

2.0	Introduction	77
2.1	List of Elements Described in this Chapter	78
2.2	Line Elements	80
	2.2.1 Linear Element (two nodes, C^0)	80
	2.2.2 Lagrangian Elements (continuity C^0)	81
	2.2.2.1 Quadratic Elements with Equidistant nodes (three nodes, C^0)	82
	2.2.2.2 Cubic Elements with Equidistant nodes (four nodes, C^0)	83
	2.2.2.3 Element with n Equidistant nodes (n nodes, C^0)	84
	2.2.3 Elements with a Higher Degree of Inter-element Continuity	84
	2.2.3.1 Cubic Elements (two nodes, C^1)	85
	2.2.3.2 Fifth Order Element (two nodes, C^2)	86
	2.2.4 General Elements	87
	2.2.4.1 Lagrange–Hermite Element of fourth Order (three nodes, C^1)	88
	2.2.4.2 Hermite Element with One Non-nodal Degree of Freedom (two nodes, C^1)	88
2.3	Triangular Elements	89
	2.3.1 Coordinate Systems	89

2.3.2	Linear Element (triangular, three nodes, C^0)	91
2.3.3	Lagrangian Type (continuity C^0)	92
2.3.3.1	Quadratic Element (triangle, six nodes, C^0)	93
2.3.3.2	Element Using a Complete Polynomial of Order r (triangle, n nodes, C^0)	94
2.3.3.3	Complete Cubic Element (triangle, ten nodes, C^0)	95
2.3.3.4	Incomplete Cubic Element (triangle, nine nodes, C^0)	96
2.3.3.5	Curvilinear Elements	96
2.3.3.6	Non-compatible Element (triangle, three nodes, semi-C^0)	97
2.3.4	Hermite Elements	98
2.3.4.1	Complete Cubic Element (triangle, four nodes, semi-C^0)	98
2.3.4.2	Incomplete Cubic Element (triangle, three nodes, semi-C^0)	99
2.3.4.3	Fifth Order Element (triangle, three nodes, C^1)	100
2.4	Quadrilateral Elements	101
2.4.1	Coordinate Systems	101
2.4.2	Bilinear Element (quadrilateral, four nodes, C^0)	102
2.4.3	Higher Order Lagrangian Elements (C^0)	103
2.4.3.1	Complete Quadratic Element (quadrilateral, nine nodes, C^0)	103
2.4.3.2	Incomplete Quadratic (quadrilateral, eight nodes, C^0)	104
2.4.3.3	Incomplete Cubic Element (quadrilateral, 16 nodes, C^0)	105
2.4.3.4	Incomplete Cubic (quadrilateral, 12 nodes, C^0)	106
2.4.3.5	Curvilinear Elements	106
2.4.4	Higher Order Hermite Elements	107
2.4.4.1	Cubic Element (quadrilateral, four nodes, semi-C^1)	107
2.4.4.2	Continuous Rectangular Element (rectangle, four nodes, C^1)	108
2.5	Tetrahedrons (three-dimensional)	110
2.5.1	Coordinate System	110
2.5.2	Linear Element (tetrahedron, four nodes, C^0)	111
2.5.3	Higher Order Lagrangian Elements (C^0)	112
2.5.3.1	Complete Quadratic (tetrahedron, ten nodes, C^0)	112
2.5.3.2	Complete Cubic Element (tetrahedron, 20 nodes, C^0)	112
2.5.3.3	Curvilinear Elements	113
2.5.4	Hermite Elements	113

2.6 Hexahedrons (three-dimensional) 114
   2.6.1 Trilinear Element (hexahedron, eight nodes, C^0) 114
   2.6.2 Higher Order Lagrangian Elements (C^0) 114
      2.6.2.1 Complete Quadratic Element (hexahedron, 27 nodes, C^0) 115
      2.6.2.2 Incomplete Quadratic Element (hexahedron, 20 nodes, C^0) 115
      2.6.2.3 Incomplete Cubic (hexahedron, 32 nodes, C^0) 117
      2.6.2.4 Curvilinear Elements............................. 119
   2.6.3 Hermite Elements....................................... 120

2.7 Prisms (three-dimensional)................................. 120
   2.7.1 Six-node Prism (prism, six nodes, C^0)................... 120
   2.7.2 Element with 15 Nodes (prism, 15 nodes, C^0)............. 121

2.8 Other Elements 121
   2.8.1 Heterogeneous Elements................................. 121
   2.8.2 Modification of Elements 123
   2.8.3 Hierarchical Elements 124
   2.8.4 Superparametric Elements 126
   2.8.5 Infinite Elements....................................... 127
References 128

3. Variational Formulation of Engineering Problems 129

3.0 Introduction.. 129
3.1 Classification of Physical Problems 131
   3.1.1 Discrete and Continuous Systems......................... 131
   3.1.2 Equilibrium, Eigenvalue, and Propagation Problems 132

3.2 Weighted Residual Method.................................. 138
   3.2.1 Residuals 138
   3.2.2 Integral Forms 139

3.3 Integral Transformations 140
   3.3.1 Integration by Parts.................................... 140
   3.3.2 Weak Integral Forms.................................... 141
   3.3.3 Construction of Additional Integral Forms................ 143

3.4 Functionals... 145
   3.4.1 First Variation 145
   3.4.2 Functional Associated with an Integral Formulation 146
   3.4.3 Stationarity Principle................................... 148
   3.4.4 Lagrange Multipliers and Additional Functionals 149

3.5 Discretization of Integral Forms............................ 155
   3.5.1 Discretization of Function W........................... 155
   3.5.2 Approximation of Functions **u**......................... 156

		3.5.3	Choice of the Weighting Functions ψ	158

- 3.5.3 Choice of the Weighting Functions ψ 158
  - 3.5.3.1 Collocation by Points........................... 158
  - 3.5.3.2 Collocation by Sub-domains..................... 159
  - 3.5.3.3 Galerkin Method............................... 160
  - 3.5.3.4 Least Squares Method 162
- 3.5.4 Discretization of a Functional (Ritz Method).............. 163
- 3.5.5 Properties of the System of Equations.................... 165
- Notation.. 167
- References 168

4. Matrix Formulation of the Finite Element Method 169

- 4.0 Introduction... 169
- 4.1 The Finite Element Method 169
  - 4.1.1 Definition.. 169
  - 4.1.2 Conditions to be met for the Convergence to an Exact Result 173
  - 4.1.3 Patch Tests 174
- 4.2 Discretized Integral Forms W^e........................... 177
  - 4.2.1 Matrix Expression for W^e........................... 177
  - 4.2.2 Case of a Non-linear Differential Operator................ 179
  - 4.2.3 Integral Form W^e in the Space of the Element of Reference... 181
    - 4.2.3.1 Transformation of Derivatives with Respect to x 181
    - 4.2.3.2 Transformation of Nodal Variables................ 181
    - 4.2.3.3 Transformation of the Domain of Integration....... 182
    - 4.2.3.4 Transformation of the Contour Integrations 182
    - 4.2.3.5 Expression of $[k]$ and $\{f\}$ in the Space of the Element of Reference......................... 185
  - 4.2.4 Some Classical Forms of W^e and Element Matrices 185
- 4.3. Computational Techniques for Element Matrices 185
  - 4.3.1 Explicit Computations for a Triangular Element and Poisson's Equation................................. 185
  - 4.3.2 Matrix Organization for the Numerical Computations of Element Matrices 191
  - 4.3.3 Control Subroutines for the Computation of Element Matrices.. 192
  - 4.3.4 Subroutine ELEM01 (for quasi-harmonic problems)........ 193
  - 4.3.5 Subroutine ELEM02 (for plane elasticity) 199
- 4.4 Assembly of the Global Discretized Form W................. 207
  - 4.4.1 Assembly by Expansion of Element Matrices.............. 207
  - 4.4.2 Assembly for Elements of Structural Mechanics............ 212
- 4.5 Assembly Techniques 214

- 4.5.1 Phases of the Assembly Process... 214
- 4.5.2 Rules for the Assembly Process Used in this Book ... 214
- 4.5.3 Examples of Assembly Subroutine ... 216
- 4.5.4 Construction of Location Table LOCE... 217

4.6 Properties of Global Matrices... 219
- 4.6.1 Band Structure of Matrix $[K]$... 219
- 4.6.2 Symmetry... 222
- 4.6.3 Storage Strategies... 222

4.7 Global System of Equations... 227
- 4.7.1 Formulation of the System of Equations... 227
- 4.7.2 Introduction of Boundary Conditions ... 227
- 4.7.3 Reactions ... 230
- 4.7.4 Variable Transformations... 230
- 4.7.5 Linear Constraints Between Variables ... 232

4.8 Example: Poisson's Equation ... 235
Notation ... 238
References ... 239

5. Numerical Procedures... 240

5.0 Introduction... 240
5.1 Numerical Integration ... 240
- 5.1.1 Introduction ... 240
- 5.1.2 One-dimensional Numerical Integration ... 243
  - 5.1.2.1 Gauss Method ... 243
  - 5.1.2.2 Newton–Cotes Method ... 248
- 5.1.3 Numerical Integration in Two Dimensions... 252
  - 5.1.3.1 Quadrilateral Elements... 252
  - 5.1.3.2 Triangular Element... 255
- 5.1.4 Numerical Integration in Three Dimensions... 256
  - 5.1.4.1 Cubic (or brick) Elements ... 256
  - 5.1.4.2 Tetrahedronal Elements... 257
- 5.1.5 Approximate Integration... 258
- 5.1.6 Choice of the Number of Integration Points... 264
- 5.1.7 Numerical Integration Code... 265

5.2 Solution of Systems of Linear Equations ... 269
- 5.2.1 Introduction ... 269
- 5.2.2 Gaussian Elimination ... 269
  - 5.2.2.1 Triangularization Process... 270
  - 5.2.2.2 Solution of an Upper Triangular System ... 274
  - 5.2.2.3 Gaussian Elimination Computer Code ... 274
- 5.2.3 Matrix Factorization... 275
  - 5.2.3.1 Introduction ... 275

5.2.3.2	Matrix Formulation of Gaussian Elimination	276
5.2.3.3	Properties of Triangular Matrices $[l^s]$	277
5.2.3.4	Representation of the Factorized Matrix $[K]$ in Different Forms	278
5.2.3.5	Solution of a System of Equations after Decomposition of the Matrix of Coefficient	279
5.2.3.6	Decomposition Algorithms	280

5.2.4 Algorithm for Skyline Matrices in Compact Storage... 282
    5.2.4.1 Case of a Matrix $[K]$ Stored in Core 282
    5.2.4.2 Skyline Matrix Stored Out of Core by Blocks....... 287

5.3 Solution of Non-Linear Systems........................... 291
  5.3.1 Introduction 291
  5.3.2 Substitution Method 292
  5.3.3 Newton–Raphson Method 297
  5.3.4 Incremental Method (step by step) 300
  5.3.5 Change of Independent Variables 301
  5.3.6 Solution Strategy 303

5.4 Solution of Time-dependent Problems 305
  5.4.1 Introduction 305
  5.4.2 Direct Integration Methods for First Order Systems........ 307
    5.4.2.1 Explicit Euler's Method 307
    5.4.2.2 Implicit Euler's Method 311
    5.4.2.3 Semi-implicit Euler's Method 314
    5.4.2.4 Predictor–Corrector Formulas 316
    5.4.2.5 Explicit Methods of Runge–Kutta Type........... 319
  5.4.3 Modal Superposition for First Order System 321
  5.4.4 Direct Integration of Second Order Systems.............. 324
    5.4.4.1 Central Finite Difference Method................. 324
    5.4.4.2 Houbolt's Method............................. 325
    5.4.4.3 Methods of Newmark and Wilson................ 327
  5.4.5 Modal Superposition for Second Order Systems 330

5.5 Solution of the Matrix Eigenvalue Problem................. 333
  5.5.1 Introduction 333
  5.5.2 Fundamental Properties of Some Matrix Eigenvalue Problems 334
    5.5.2.1 Simplified Formulation......................... 334
    5.5.2.2 Eigenvalues................................... 334
    5.5.2.3 Eigenvectors.................................. 335
    5.5.2.4 Spectral Decomposition 336
    5.5.2.5 Transformation of $[K]$ and $[M]$ 336
    5.5.2.6 Rayleigh Quotient.............................. 338
    5.5.2.7 Separation of Eigenvalues 338
    5.5.2.8 Shifting of Eigenvalues......................... 339

5.5.3	Computations of Eigenvalues	340
	5.5.3.1 Inverse Iteration Method	340
	5.5.3.2 Jacobi Rotation Method	342
	5.5.3.3 Ritz Method	348
	5.5.3.4 Subspace Iteration	349
Notation		352
References		355

6. Coding Techniques ... 356

- 6.0 Introduction ... 356
- 6.1 Common Features of Finite Element Programs ... 356
- 6.2 Beginner's Program BBMEF ... 358
- 6.3 Multipurpose Programs ... 363
  - 6.3.1 Capabilities of General Codes ... 363
    - 6.3.1.1 Problem Types ... 363
    - 6.3.1.2 Problem Size ... 364
  - 6.3.2 Modularity ... 365
- 6.4 Program MEF ... 366
  - 6.4.1 Introduction ... 366
  - 6.4.2 Overall Organization ... 366
    - 6.4.2.1 Flow Chart of Functional Blocks ... 366
    - 6.4.2.2 Pseudo-dynamic Memory Management ... 367
    - 6.4.2.3 Programming Norms (programmer's reference guide) ... 367
  - 6.4.3 Organization of the Problem Data Base ... 370
    - 6.4.3.1 Entry and Execution Functional Blocks ... 370
    - 6.4.3.2 Core and Out of Core Storage ... 370
    - 6.4.3.3 Description of the Most Important Arrays and Variables in COMMON ... 376
- 6.5 Description of Functional Blocks ... 376
  - 6.5.1 Principal Program ... 376
  - 6.5.2 Functional Blocks Reading Data ... 382
    - 6.5.2.1 Block 'IMAG' ... 382
    - 6.5.2.2 Block 'COMT' ... 382
    - 6.5.2.3 Block 'COOR' ... 383
    - 6.5.2.4 Block 'DLPN' ... 384
    - 6.5.2.5 Block 'COND' ... 388
    - 6.5.2.6 Block 'PRND' ... 388
    - 6.5.2.7 Block 'PREL' ... 391
    - 6.5.2.8 Block 'ELEM' ... 394
    - 6.5.2.9 Block 'SOLC' ... 400
  - 6.5.3 Execution Functional Blocks ... 401
    - 6.5.3.1 Organization of Execution Blocks ... 402

6.5.3.2	Block 'SOLR'	402
6.5.3.3	Block 'LINM'	406
6.5.3.4	Block 'LIND'	418
6.5.3.5	Block 'NLIN'	418
6.5.3.6	Block 'TEMP'	425
6.5.3.7	Block 'VALP'	430

6.6 Description of Input Values for Program MEF 440
  6.6.1 Conventions 440
  6.6.2 Input Corresponding to Each Block 440

6.7 Applications of MEF 447
  6.7.1 Heat Conduction 447
  6.7.2 Plane Stress Problem 481

Index 505

INTRODUCTION

0.1 The Finite Element Method

0.1.1 *Generalities*

Modern engineering projects have become extremely complex, costly and subject to severe reliability and safety constraints. Space vehicles, aircraft and nuclear reactors are examples of projects where reliability and safety are of crucial importance. Other preoccupations of modern technology with the protection of the environment, the control of thermal, acoustical and chemical pollution, the management of natural resources like aquifers and weather predictions, have forced engineers to reexamine all methods of analysis. For a proper understanding, analysts need mathematical models that can be used to simulate the behaviour of complex physical systems. These models are then used during the design of the projects.

Engineering sciences (mechanics of solids, fluids, thermodynamics, etc.) allow physical systems to be described in the form of partial differential equations. Today, the finite element method has become the most popular method for solving such equations. The method coupled with developments in computer technology has successfully been applied to the solution of steady and transient problems in linear and non-linear regions for one-, two-, and three-dimensional domains. It can easily handle discontinuous geometrical shapes as well as material discontinuities.

The finite element discretization process, like the finite difference process, transforms partial differential equations into algebraic equations. It draws heavily on the following three disciplines:

— continuum mechanics, for the correct formulation of mathematical models;
— numerical analysis, for the elaboration of algorithmic solutions of the discretized equations;
— computer programming, to produce parametrized codes applicable to large classes of problems.

0.1.2 *Historical Evolution of the Method*

For the past hundred years structural mechanics has been used to analyse frames and trusses [1]. Stiffness matrices of bars and beams could be constructed using elementary strength of materials. The direct stiffness method then allowed the assembly of elementary matrices into global matrices of coefficients for the system of algebraic equations relating forces and displacements. Simple modifications of the global system of equations allowed boundary conditions to be satisfied. The solution of the system of linear equations gave the displacements at the nodes and the reactions at the supports. In the early nineteen-fifties, computers made it possible to solve structural problems very effectively, but the method was slowly accepted by industry. Turner, Clough, Martin, and Topp [2] introduced the finite element concepts in 1956. Almost simultaneously, Argyris and Kelsey [3] developed similar concepts in a series of publications on energy theorems. Courant [4], Hrennikoff [5] and McHenry [6] are also early precursors of the finite element methods.

During the nineteen-sixties, the finite element method became widely accepted, research was pursued simultaneously in various parts of the world in several directions:

— the method was reformulated as a special case of the weighted residual method [7–10];

— a wide variety of elements were developed including bending elements, curved elements and the isoparametric concept was introduced [11–15];

— the method was reorganized as a general method of solution for partial differential equations; its applicability to the solution of non-linear and dynamic problems of structures was amply demonstrated; extension in other domains, soil mechanics, fluid mechanics, thermodynamics, produced solutions to engineering problems hitherto intractable [14–26];

— a mathematical basis was established using concepts of functional analysis [27–28].

Starting in 1967, many books have been written on the finite element method [29–56]. The three editions of the book authored by Professor Zienkiewicz [30] received worldwide diffusion. Until now the books available in French were the translations of the books authored by Zienkiewicz [53] (second edition), Gallagher [54], Rockey *et al.* [55], as well as the books written by Absi [56] and Imbert [56a]. During the same period a number of journals devoted most of their pages to the finite element method [57–62].

0.1.3 *State of the Art*

The finite element method (FEM) has now been widely accepted for all kinds of structural engineering application in aerospace, aeronautics, naval architecture and nuclear-powered electrical generating stations. Fluid mechanics applications are currently being developed for studying tidal motions, thermal and chemical transport and diffusion problems as well as fluid structure interactions. A number of general purpose finite element computer codes have been successfully developed for industrial users in the field of solid and fluid mechanics. Without any pretence at being comprehensive we may mention: NASTRAN, ASKA, SAP, MARC, ANSYS, TITUS, ADINA [2, 65, 66, 67]. These codes were conceived to run on large computers. Today, many codes have been modified so they can run on mini- and microcomputers (Rammant in [25]).

A major defect of most codes has been the preparation of input information and the interpretation of voluminous outputs. The absence of control during executions of the various phases of a finite element analysis has always been felt to be a major problem as well as the ability to visualize what is being accomplished in the computer. Today the trend is to design software with facilities for model generation, program interaction and graphical display capabilities.

0.2 Object and Organization of the Book

0.2.1 *Teaching of the Finite Element Method*

Although the finite element method has gained wide acceptance and usage in industry, its teaching is not yet widespread. The method draws from many disciplines thereby complicating coherent teaching. Amongst the various skills to be mastered we may cite:

— deep understanding of the physical principles involved in the problem at hand;
— approximations by sub-domains and interpolation theory;
— mathematical modelling of principles involved leading to variational formulations; these are obtained using either energy principles or the method of weighted residuals;
— matrix algebra for the discretized formulation;
— numerical methods for integration, solution of systems of linear and non-linear algebraic equations;
— computer programming involving massive data files.

It is hard to conceive a balanced formation in all these diverse disciplines. Moreover, the necessary software adapted to finite element teaching, keeping at the same time essential features of a general purpose program, is still under active development. The teaching of the method requires so many prerequisites that it is almost invariably relegated to senior and graduate level courses.

0.2.2 *Objectives of the Book*

This text is an attempt to simplify the teaching of the finite element method to engineers. The mathematical preparation required is that commonly dispensed in an engineering undergraduate curriculum. Tables of formulas and constants are presented in sufficient detail to allow the reader to construct his own code.

After reading this text, the reader should be better equipped to use most of the large commercial finite element codes.

0.2.3 *Organization of the Book*

The book is divided into six chapters that are nearly independent of one another. The necessary numerical and programming techniques are contained in the body of the text.

Chapter 1

Approximation of continuous functions over sub-domains in terms of nodal values, introduction of interpolation concepts, reference elements and error estimations.

Chapter 2

Interpolation functions for classical elements in one, two, and three dimensions.

Chapter 3

Application of the weighted residual method for the construction of integral or variational forms from partial differential equations.

Chapter 4

Discretization of integral forms using matrix algebra. Fundamental element vectors and matrices and assemby techniques.

Chapter 5

Numerical methods needed to construct and solve linear and non-linear system of algebraic equations. Numerical methods of integration for propagation problems in the time domain. Matrix eigenvalue and eigenvector problem.

Chapter 6

FORTRAN programming techniques illustrated with the two programs (BBMEF) and (MEF).

Figure 0.1 shows the relationship between the various chapters. Note that Chapters 1, 3, and 4 are devoted to the fundamental concepts underlying the finite element method and that Chapters 2 and 5 are mostly devoted to prerequisite material. Examples of short FORTRAN programs are used throughout the book to illustrate the practical implementation of the method.

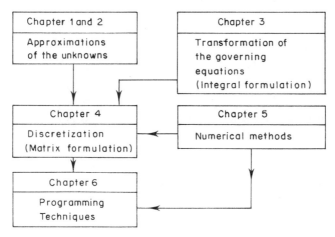

Figure 0.1 Logical flow of the chapters.

The list of references does not pretend to be exhaustive; it lists only the references mentioned in the text. More complete bibliographies can be found in Zienkiewicz [30], Gallagher [54], Norrie and de Vries [63], and Whiteman [64].

References

[1] N. J. Hoff, *Analysis of Structures*, Wiley, New York, 1956.
[2] M. J. Turner, R. W. Clough, H. C. Martin, and L. J. Topp, Stiffness and deflection analysis of complex structures, *Journal of Aeronautical Science*, **23**, 805–823, 1956.
[3] J. H. Argyris and S. Kelsey, *Energy Theorems and Structural Analysis*, Butterworth, London, 1960.
[4] R. Courant, Variational methods for the solution of problems of equilibrium and vibrations, *Bull. Am. Math. Soc.*, **49**, 1–23, 1943.
[5] A. Hrennikoff, Solution of problems in elasticity by the framework method, *J. appl. Mech.*, **8**, A169–A175, 1941.
[6] D. McHenry, A lattice analogy of the solution of plane stress problems, *J. Inst. Civil Eng.*, **21**, 59–82, 1943.
[7] O. C. Zienkiewicz and G. S. Holister, *Stress Analysis*, Wiley, New York, 1965.
[8] R. E. Greene, R. E. Jones, R. W. McLay, and D. R. Strome, Generalized variational principles in the finite-element method, *AIAA J.*, **7**, 7, 1254–1260, July, 1969.
[9] B. A. Finlayson, Weighted residual methods and their relation to finite element methods in flow problems, *Finite Elements in Fluids*, Vol. 2, pp. 1–31, Wiley, 1975.
[10] E. R. de Arantes e Oliveira, Theoretical foundations of the finite element method, *International Journal of Solids and Structures*, **4**, 929, 1968.
[11] C. A. Felippa, Refined finite element analysis of linear and non-linear two-dimensional structures, *Report UC SESM 66–22*, Department of Civil Engineering, University of California, Berkeley, October 1966.
[12] J. G. Ergatoudis, B. M. Irons, and O. C. Zienkiewicz, Three-dimensional analysis of arch dams and their foundations, *Symposium on Arch Dams*, Institute of Civil Engineering, London, March 1968.
[13] B. M. Irons and O. C. Zienkiewicz, The isoparametric finite element system—a new concept in finite element analysis, *Proceedings, Conference on Recent Advances in Stress Analysis*, Royal Aeronautical Society, London, 1968.

Proceedings of Conferences

[14] Proceedings of the 1st, 2nd, and 3rd Conferences on Matrix Methods in Structural Mechanics, Wright–Patterson A. F. B., Ohio, 1965, 1968, 1971.
[15] I. Holland and K. Bell (eds.), *Finite Element Methods in Stress Analysis*, Tapir, Trondheim, Norway, 1969.
[16] Proceedings of the 1st, 2nd, 3rd and 4th Conferences on Structural Mechanics in Reactor Technology, 1971, 1973, 1975, 1977.
[17] Symposium on Applied Finite Element Methods in Civil Engineering, Vanderbilt University, Nashville, ASCE, 1969.
[18] R. H. Gallagher, Y. Yamada, and J. T. Oden (eds.), *Recent Advances in Matrix Methods of Structural Analysis and Design*, University of Alabama Press, Huntsville, 1971.
[19] B. F. de Veubeke (ed.), *High Speed Computing of Elastic Structures*, University of Liège, 1971.
[20] C. A. Brebbia and H. Tottenham (eds.), *Variational Methods in Engineering*, Southampton University, 1973.
[21] S. J. Fenves, N. Perrone, J. Robinson, and W. C. Schnobrich (eds.), *Numerical and Computational Methods in Structural Mechanics*, Academic Press, New York, 1973.
[22] R. H. Gallagher, J. T. Oden, C. Taylor, and O. C. Zienkiewicz (eds.), *International Symposium on Finite Element Methods in Flow Problems*, Wiley, 1974.
[23] K. J. Bathe, J. T. Oden, and W. Wunderlich (eds.), *Formulation and Computational Algorithms in Finite Element Analysis (U.S.–Germany Symposium)*, MIT Press, 1977.
[24] W. G. Gray, G. F. Pinder, and C. A. Brebbia (eds.), *Finite Elements in Water Resources*, Pentech Press, London, 1977.
[25] J. Robinson (ed.), *Finite Element Methods in Commercial Environment*, Robinson and Associates, Dorset, England, 1978.
[26] R. Glowinski, E. Y. Rodin, and O. C. Zienkiewicz (eds.), *Energy Methods in Finite Element Analysis*, Wiley, Chichester, 1979.
[27] A. K. Aziz (ed.), *The Mathematical Foundations of the Finite Element Method with Applications to Partial Differential Equations*, Academic Press, New York, 1972.
[28] J. R. Whiteman (ed.), *The Mathematics of Finite Elements and Applications*, Academic Press, London, 1973.

Books

[29] J. S. Przemieniecki, *Theory of Matrix Structural Analysis*, McGraw-Hill, New York, 1968.
[30] O. C. Zienkiewicz, *The Finite Element Method in Engineering Science*, McGraw-Hill, New York, 1st edn, 1967, 3rd edn, 1977.
[31] C. S. Desai and J. F. Abel, *Introduction to the Finite Element Method*, Van Nostrand Reinhold, New York, 1972.
[32] J. T. Oden, *Finite Elements of Non-linear Continua*, McGraw-Hill, New York, 1972.
[33] H. C. Martin and G. F. Carey, *Introduction to Finite Element Analysis*, McGraw-Hill, New York, 1973.
[34] D. H. Norrie and G. de Vries, *The Finite Element Method*, Academic Press, New York, 1973.
[35] J. Robinson, *Integrated Theory of Finite Element Methods*, Wiley, London, 1973.
[36] G. Strang and O. J. Fix, *Analysis of the Finite Element Methods*, Prentice-Hall, New Jersey, 1973.
[37] O. Ural, *Finite Element Method, Basic Concepts and Applications*, Intext Educational Publishers, 1973.
[38] R. D. Cook, *Concepts and Applications of Finite Element Analysis*, Wiley, 1974.

[39] R. H. Gallagher, *Finite Element Analysis Fundamentals*, Prentice-Hall, New Jersey, 1975.
[40] K. H. Huebner, *The Finite Element Method for Engineers*, Wiley, 1975.
[41] K. Washizu, *Variational Methods in Elasticity and Plasticity*, Pergamon Press, 2nd edn, 1975.
[42] K. J. Bathe and E. L. Wilson, *Numerical Methods in Finite Element Analysis*, Prentice-Hall, New Jersey, 1976.
[43] Y. K. Cheung, *Finite Strip Method in Structural Analysis*, Pergamon Press, 1976.
[44] J. J. Connor and C. A. Brebbia, *Finite Element Technique for Fluid Flow*, Butterworth, 1976.
[45] L. J. Segerlind, *Applied Finite Element Analysis*, Wiley, 1976.
[46] A. R. Mitchell and R. Wait, *The Finite Element Method in Partial Differential Equations*, Wiley, London, 1977.
[47] G. F. Pinder and W. G. Gray, *Finite Element Simulation in Surface and Sub-surface Hydrology*, Academic Press, 1977.
[48] P. Tong and J. Rossetos, *Finite Element Method; Basic Techniques and Implementation*, MIT Press, 1977.
[49] T. J. Chung, *Finite Element Analysis in Fluid Dynamics*, McGraw-Hill, 1978.
[50] P. G. Ciarlet, *The Finite Element Method for Elliptic Problems*, North-Holland, Amsterdam, 1978.
[51] B. M. Irons and S. Ahmad, *Techniques of Finite Elements*, Ellis Horwood, Chichester, England, 1978.
[52] C. S. Desai, *Elementary Finite Element Method*, Prentice-Hall, New Jersey, 1979.
[53] O. C. Zienkiewicz, *La Methods des elements finis* (translated from the English), Pluralis, France, 1976.
[54] R. H. Gallagher, *Introduction aux elements finis* (translated from the English by J. L. Claudon), Pluralis, France, 1976.
[55] K. C. Rockey, H. R. Evans, D. W. Griffiths and D. A. Nethercot *Eléments finis* (translated from the English by C. Gomez), Eyrolles, 1979.
[56] E. Absi, *Méthode de calcul numérique en élasticité*, Eyrolles, France, 1978.
[56a] J.-F. Imbert, *Analyse des structures par éléments finis*, Cepadues Ed., France, 1979.

Journals and Periodicals

[57] *International Journal for Numerical Methods in Engineering* (eds. O. C. Zienkiewicz and R. H. Gallagher), Wiley.
[58] *International Journal of Computers and Structures* (ed. H. Liebowitz), Pergamon Press.
[59] *Computer Methods in Applied Mechanics and Engineering* (ed. J. H. Argyris), North Holland.
[60] *International Journal of Computers and Fluids* (ed. C. Taylor), Pergamon Press.
[61] *International Journal of Numerical Methods in Geotechnics* (ed. C. S. Desai), Wiley.
[62] *Finite Element News*, Robinson and Associates, Dorset, England.
[63] D. H. Norrie and G. de Vries, *Finite Element Bibliography*, IFI/Plenum, University of Calgary, 1976.
[64] J. R. Whiteman, *A Bibliography for Finite Elements*, Academic Press, 1975.

Comparisons of Programs

[65] W. Pilkey, K. Sczalski, and H. Schaeffer (eds.), *Structural Mechanics Computer Programs*, University Press of Virginia, Charlotteville, 1974.

[66] Frederiksson, Mackerle (eds.)
- *Structural Mechanics Finite Elements Computer Programs*
- *Structural Mechanics Pre and Post Processor Programs*
- *Finite Element Review*
- *Stress Analysis Programs for Fracture Mechanics*

Advanced Engineering Corp., Linkoping, Sweden, 1978.

[67] Grands codes de calcul de structures, Présentation et critère de choix, CTICM, Puteaux, 1978.

1

APPROXIMATIONS WITH FINITE ELEMENTS

1.0 Introduction

This chapter is devoted to the study of approximation techniques allowing the replacement of a continuous system by an equivalent discrete system. We first describe nodal approximations over a domain V then follow with approximations over sub-domains called finite elements. We present as well the technique of subdividing a given geometrical domain V into elements.

Reference element and their geometrical mapping into real sub-domains facilitate the construction of interpolation functions. The geometrical transformation (mapping) from the reference element to the real element is characterized by the Jacobian transformation matrix. A few pages are devoted to the estimation of the errors of approximation. The chapter ends with an application where the total rainfall over a given area is estimated from a few discrete measurements of rainfall.

1.1 Generalities

1.1.1 *Nodal Approximation*

A mathematical model of a physical system normally involves a number of variables and functions $u_{ex}(x)$ representing temperatures, velocities, thicknesses, etc. Where x represents the coordinates of a point of the domain. Let $u(x)$ be an approximation of $u_{ex}(x)$, then the error function is:

$$e(x) = u(x) - u_{ex}(x) \qquad (1.1)$$

and $e(x)$ is a measure of the quality of the approximation. To construct an approximation $u(x)$ it suffices to:

— write an expression containing n parameters a_i:

$$u(x, a_1, a_2, \ldots, a_n)$$

— determine these parameters using (1.1) and a plausible criterion of validity. For example, the error function $e(x)$ may be forced to be zero at 'n' points in the domain.

The approximate function should be so chosen that it is simple and smooth enough for numerical evaluation, derivation and integration within the domain. The approximation $u(x)$ may be employed, among others, to obtain:

— a simpler representation of a function difficult to manipulate or known only at certain data points;
— an approximate solution of a partial differential equation.

The two cases are illustrated below:

Example 1.1 *Approximation of a physical quantity $u(x)$*

Assume that a temperature distribution along a line x can be measured only at three points, as shown below, but that it is known that the distribution is continuous without jumps.

x	$u_{ex}(x)$
0.0	20 °C
0.5	25 °C
1.0	22 °C

An expression is desired giving the temperatures at all points between 0 and 1.0. Let us pass a quadratic polynomial through the three data points.

$$u_{ex}(x) \approx u(x, a_1, a_2, a_3) = a_1 + a_2 x + a_3 x^2$$

such that

$$u_{ex}(x=0) = u(x=0) = a_1 = 20$$
$$u_{ex}(x=0.5) = u(x=0.5) = a_1 + 0.5 a_2 + 0.25 a_3 = 25$$
$$u_{ex}(x=1.0) = u(x=1.0) = a_1 + a_2 + a_3 = 22$$

Solving the three simultaneous equations for a_1, a_2 and a_3 we get:

$$a_1 = 20, \quad a_2 = 18, \quad a_3 = -16$$

then $u(x) = 20 + 18x - 16x^2$ is an approximation of $u_{ex}(x)$. For example, the temperature at $x = 0.7$ is:

$$u(x = 0.7) = 20 + 20.6 - 7.84 = 24.76$$

Example 1.2 *Approximation solution of a differential equation*

We look for a function u_{ex} satisfying

— differential equation

$$\frac{d^2 u_{ex}(x)}{dx^2} = f(x) \quad \text{where} \quad 0 \le x \le 1$$

— boundary conditions
$$u_{ex}(x) = 0 \quad \text{for} \quad x = 0 \quad \text{and} \quad x = 1$$
$f(x)$ is a known function such that
$$f(x = 0.25) = 1$$
$$f(x) = 0.75) = 0.25$$
$$f(x) = 0 \text{ elsewhere}$$

Without saying why, let us try:
$$u_{ex}(x) \approx u(x) = a_1 \sin(\pi x) + a_2 \sin(2\pi x)$$
as an approximation for $u_{ex}(x)$.

Observe that the approximation satisfies the boundary conditions at $x = 0$ and $x = 1$. Let us also assume that the approximation satisfies the differential equation at points $x_1 = 0.25$ and $x_2 = 0.75$, then:

$$\left.\frac{d^2u}{dx^2}\right|_{x_1} = -a_1\pi^2 \sin(0.25\pi) - 4a_2\pi^2 \sin(0.5\pi) = f(x_1) = 1$$

$$\left.\frac{d^2u}{dx^2}\right|_{x_2} = -a_1\pi^2 \sin(0.75\pi) - 4a_2\pi^2 \sin(1.5\pi) = f(x_2) = 0.25$$

from which we find

$$a_1 = -\frac{5}{4\sqrt{2}}\frac{1}{\pi^2} \quad a_2 = -\frac{3}{32}\frac{1}{\pi^2}$$

$$u_{ex}(X) \approx u(x) = -\frac{5}{4\sqrt{2}}\frac{1}{\pi^2}\sin(\pi x) - \frac{3}{32}\frac{1}{\pi^2}\sin(2\pi x)$$

For $x = 0.25$ we find:

$$u(x = 0.25) = -\frac{23}{32}\frac{1}{\pi^2} = -0.0728$$

In this example the approximation is employed to discretize the differential equation. That is, it is replaced by a set of two algebraic equations with unknown a_1 and a_2.

Most frequently an approximate function is chosen to be a linear function of parameters a_i. It may thus be written as

$$u(x) = P_1(x)a_1 + P_2(x)a_2 + \ldots + P_n(x)a_n \tag{1.2}$$

$$u(x) = \langle P_1(x) \; P_2(x) \ldots P_n(x) \rangle \begin{Bmatrix} a_1 \\ a_2 \\ \vdots \\ a_n \end{Bmatrix} = \langle P \rangle \{a\} \tag{1.3}$$

where: $P_1, P_2, \ldots, P_n$ are linearly independent known functions chosen from a function set that is mathematically complete. Polynomials, Fourier series and others could be employed but most of the finite elements used so far have been constructed with polynomials. Parameters $a_1, a_2, \ldots, a_n$ are the generalized parameters of approximation.

Note that the parameters a_i have, in general, no direct physical meaning. Thus, for many applications, these are more conveniently replaced by nodal values of the function $u_{ex}(x)$ at n points with coordinates $x_1, x_2, \ldots, x_n$. The nodal approximation is furthermore supposed to satisfy the following relations:

$$u(x_1) = u_{ex}(x_1) = u_1$$
$$u(x_2) = u_{ex}(x_2) = u_2$$
$$\cdots\cdots\cdots\cdots\cdots$$
$$u(x_n) = u_{ex}(x_n) = u_n \qquad (1.4)$$

Function (1.2) can now be written as:

$$u(x) = N_1(x)u_1 + N_2(x)u_2 + \ldots + N_n(x)u_n$$

$$u(x) = \langle N_1(x) \; N_2(x) \ldots N_n(x) \rangle \begin{Bmatrix} u_1 \\ u_2 \\ \vdots \\ u_n \end{Bmatrix} = \langle N \rangle \{u_n\} \qquad (1.5)$$

Definitions

— Parameters a_i are the generalized parameters of the approximation.
— Parameters u_i are the nodal parameters.
— Equation (1.3) defines a non-nodal approximation.
— Equation (1.5) defines a nodal approximation (see Example 1.3).
— Functions $P(x)$ are called basis functions of the approximation.
— Functions $N(x)$ are called interpolation functions.
— The interpolation functions possess the following properties which derive from (1.4) and (1.5):

(a) as $u(x_i) = u_i$, the function N_i satisfies

$$N_j(x_i) = \begin{cases} 0 & \text{if } i \neq j \\ 1 & \text{if } i = j \end{cases} \qquad (1.6)$$

(b) The approximation error

$$e(x) = u(x) - u_{ex}(x) \qquad (1.7)$$

vanishes at all the node points, x_i:

$$e(x_i) = 0 \qquad (1.8)$$

Example 1.3 *Four-noded approximation*

Consider an arbitrary function $u_{ex}(x)$ known only at four nodes. Let us

approximate it by:
$$u(x) = N_1(x)u_1 + N_2(x)u_2 + N_3(x)u_3 + N_4(x)u_4$$
where N_i are Lagrange polynomials of third degree:
$$N_i(x) = \prod_{\substack{j=1 \\ j \neq i}}^{4} \frac{(x - x_j)}{(x_i - x_j)}$$

These polynomials satisfy (1.6). Take for example $N_i(x)$:
$$N_1(x) = \frac{(x - x_2)(x - x_3)(x - x_4)}{(x_1 - x_2)(x_1 - x_2)(x_1 - x_4)}$$
If $x_1 = 1.0$, $x_2 = 2.0$, $x_3 = 5.0$, $x_4 = 7.0$
$$N_1 = -\tfrac{1}{24}(x-2)(x-5)(x-7)$$

x	1	1.5	2	3	4	5	6	7
N_1	1	$\frac{77}{192}$	0	$-\frac{1}{3}$	$-\frac{1}{4}$	0	$\frac{1}{6}$	0

and the function plots as shown below:

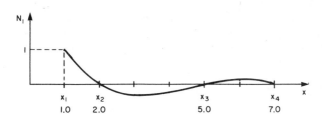

Functions u_{ex}, u and e look like:

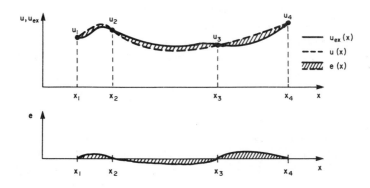

Nodal approximations of a function of one variable are easily extended to multivariable functions; for example,

$$u_{ex}(x, y, z) = u_{ex}(\mathbf{x})$$

where $\mathbf{x} = \langle x, y, z \rangle$ belong to domain V, then,

$$u(x, y, z) = u(\mathbf{x}) = \langle N_1(\mathbf{x}) \quad N_2(\mathbf{x}) \ldots N_n(\mathbf{x}) \rangle \begin{Bmatrix} u_1 \\ u_2 \\ \vdots \\ u_n \end{Bmatrix} = \langle N \rangle \{u_n\} \quad (1.9)$$

such that $u(\mathbf{x}_i) = u_{ex}(x_i) = u_i$
and $\mathbf{x}_i = \langle x_i, y_i, z_i \rangle$; $i = 1, 2, \ldots, n$ are the coordinates of the nodes.

1.1.2 Approximations with Finite Elements [1-3]

The construction of approximating functions $u(\mathbf{x})$ gets very difficult when the number of nodes increases. More complexity is introduced when the shape of the domain of definition V is irregular and some boundary conditions must be met as in Example 1.2. On the other hand, nodal approximations by sub-domain simplifies the construction of $u(\mathbf{x})$ and is very easily implemented in a computer. It consists of essentially two steps:

— subdivision of a given domain V, into sub-domains V^e
— choice of a different nodal approximation $u^e(\mathbf{x})$ on each sub-domain. In general it may as well depend on nodal values belonging to neighbouring sub-domains. Such is the case for supline-type approximation [3]. The finite element approximation is a special case of nodal approximation by sub-domain. Its main features are:

— the approximation over a sub-domain V^e depends only on the nodal values of that sub-domain or element;
— the approximation $u^e(x)$ is required to guarantee a certain minimum degree of continuity over each element and its interelement boundaries.

Definitions

— The points of the sub-domains where the function is evaluated are called interpolation nodes or simply nodes.
— The geometrical coordinates of such points are called nodal coordinates.
— The values of the function $u_i = u^e(\mathbf{x}_i) = u_{ex}(\mathbf{x}_i)$ at the nodes are called nodal variables.

Finite element approximations may be characterized by following distinct steps:

— the geometry of all elements must be defined analytically;
— appropriate interpolation functions $N_i(\mathbf{x})$ must be constructed for each element.

Example 1.4 One-dimensional finite element approximation

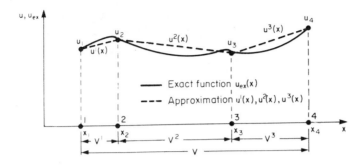

Geometrical definition of elements

Nodes 1, 2, 3, 4
Nodal coordinates x_1, x_2, x_3, x_4
Total domain $V: x_1 \leq x \leq x_4$
Elements $V^1: x_1 \leq x \leq x_2$
$V^2: x_2 \leq x \leq x_3$
$V^3: x_3 \leq x \leq x_4$

Construction of approximating functions $u^e(x)$

Nodal variables u_1, u_2, u_3, u_4
Linear functions $u^e(x)$ for each element
Element 1 (domain V^1):

$$u^1(x) = N_1 u_1 + N_2 u_2$$

where N_1 and N_2 are linear functions satisfying (1.6)

$$N_1 = \frac{x - x_2}{x_1 - x_2} \quad N_1(x_1) = 1 \quad N_1(x_2) = 0$$

$$N_2 = \frac{x - x_1}{x_2 - x_1} \quad N_2(x_1) = 0 \quad N_2(x_2) = 1$$

$$x_1 \leq x \leq x_2$$

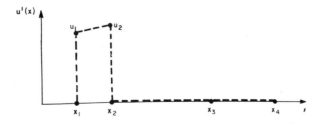

Element 2 (domain V^2):

$$u^2(x) = N_1 u_2 + N_2 u_3$$

$$N_1 = \frac{x - x_3}{x_2 - x_3}; \quad N_2 = \frac{x - x_2}{x_3 - x_2}$$

$$x_2 \leq x \leq x_3$$

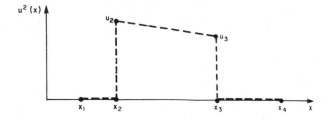

Element 3 (domain V^3):

$$u^3(x) = N_1 u_3 + N_2 u_4$$

$$N_1 = \frac{x - x_4}{x_3 - x_4}; \quad N_2 = \frac{x - x_3}{x_4 - x_3}$$

$$x_3 \leq x \leq x_4$$

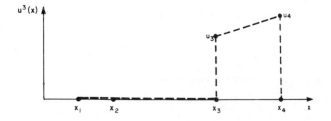

Functions u^e and $N_i(x)$ are different for each element V^e; these functions are null outside the element. The union of all the functions $u^1(x)$, $u^2(x)$ and $u^3(x)$ form the approximation $u(x)$ over the total domain V:

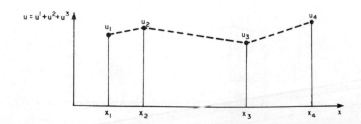

Example 1.5 Linear approximation over a two-dimensional domain

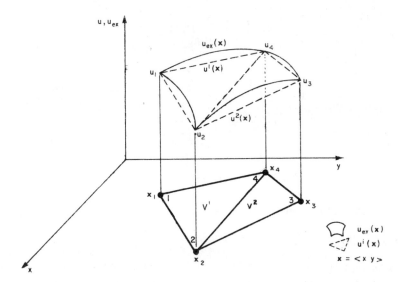

Geometrical definition of the elements

Nodes 1, 2, 3, 4
Nodal coordinates $(X_1 Y_1), (X_2 Y_2), (X_3 Y_3), (X_4 X_4)$
Overall domain V: quadrilateral 1–2–3–4
Elements V^1: triangle 1–2–4
          V^2: triangle 2–3–4

Construction of approximating functions $u^e(x)$

Nodal variables: u_1, u_2, u_3, u_4
Linear approximations over each element u^e
Element 1 (domain V^1):

$$u^1(\mathbf{x}) = N_1(\mathbf{x})u_1 + N_2(\mathbf{x})u_2 + N_3(\mathbf{x})u_4$$

where $u^1(\mathbf{x})$ is a linear function in x and y taking a value equal to u_1, u_2, u_4 at points 1, 2, and 4, the function is null outside the triangle; $N_1(\mathbf{x})$ is a function linear in x and y taking values of 1 at point 1 and 0 at points 2 and 4, the function is null outside the triangular domain.
Element 2 (domain V^2)

$$u^2(\mathbf{x}) = N_1(\mathbf{x})u_2 + N_2(\mathbf{x})u_3 + N_3(\mathbf{x})u_4$$

Figure 1.1 summarizes the various approximations mentioned so far. The next two paragraphs describe a systematic method to construct approximate functions needed for elements of complex shapes.

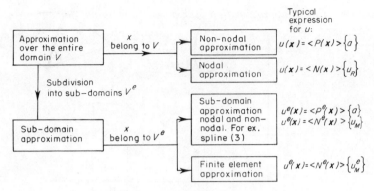

Figure 1.1 Approximation methods.

1.2 Geometrical Definition of the Elements

1.2.1 *Geometrical Nodes*

A set of n points is selected in the domain V to define the geometry of the elements. These points called geometrical nodes may sometimes coincide with the interpolation nodes. The domain V is then subdivided into a set of elements V^e of simple shapes. Each element is analytically and uniquely defined in terms of geometrical nodes belonging to that element and its boundary.

Example 1.6 *One-dimensional domain*

In Example 1.4, nodes 1, 2, 3, 4 are geometrical nodes chosen in domain V. Each element V^e is defined from the coordinates of two geometrical nodes located at its extremities:

$$x_2 \leq x \leq x_3$$

Example 1.7 *Two-dimensional triangular domain*

In example 1.5, nodes 1, 2, 3, 4 are geometrical nodes. Each element is defined from the coordinates of the three nodes situated at its vertices. For example, element 1 is defined by the coordinates of the three nodes 1, 2 and 4. This geometrical definition will be given explicitly in Example 1.9 using mapping concepts.

1.2.2 *Rules for the Partition of a Domain into Elements*

The subdivision of a domain V into finite element domain V^e should satisfy the following two requirements.

(a) two distinct elements can have common points only on their common boundaries if such boundaries exist; no overlapping is allowed. Common boundaries can be points, lines, or surfaces.

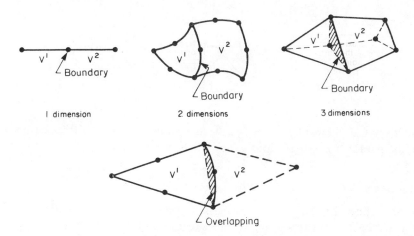

(b) the assembled elements should leave no holes within the domain and approximate the geometry of the real domain as closely as it is possible to do.

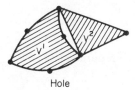

When the boundary of a domain V cannot be exactly represented by the elements selected, an error cannot be avoided. Such an error is called the geometrical discretization error and it can be decreased by reducing the size of the elements or by using elements allowing boundaries to become curved.

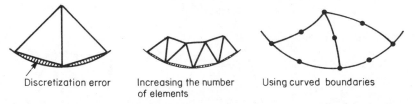

The two previous requirements are satisfied if the elements are constructed as follows.

— Each element is uniquely defined by geometrical nodes situated in that element. Most often these nodes are in the boundary of the element and could be common to other adjacent elements.

— The boundary of a two- or three-dimensional element is composed of curves and surfaces. Each portion of a boundary must be uniquely defined from the coordinates of the geometrical nodes belonging to that portion of boundary. Thus, a portion of boundary common to two different elements is defined identically for one or the other element sharing that common boundary.

Example 1.8 Boundary between two elements

Boundary 1–2–3 must be uniquely defined by nodes 1, 2, and 3. It is possible to use the parabola passing through these three nodes.

1.2.3 *Shapes of some Classical Elements*

We now show the shapes of some classical one, two-, and three-dimensional elements. Each element is given a name suggestive of its shape from the type of curves or surfaces bounding that element. Furthermore, the number of geometrical nodes needed to define a particular type of element is also suggestive of its shape.

(a) One-dimensional elements

(b) Two-dimensional elements
Triangular elements

Quadrilateral elements

(c) Three-dimensional elements
Tetrahedronal elements

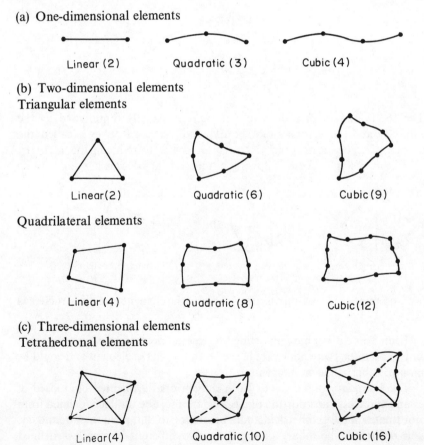

Hexahedronal elements

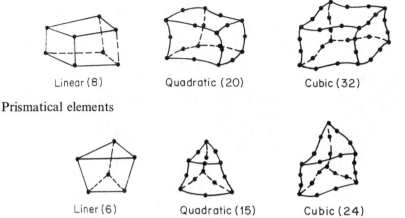

Linear (8)   Quadratic (20)   Cubic (32)

Prismatical elements

Liner (6)   Quadratic (15)   Cubic (24)

1.2.4 Reference Elements

To simplify the analytical expressions for elements of complex shapes an element of reference is introduced. Such an element V^r is defined in an abstract non-dimensional space with a very simple geometrical shape. The geometry of the reference element is then mapped into the geometry of the real element using geometrical transformation expressions. For example, in the case of a triangular region:

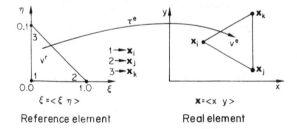

Reference element   Real element

Transformation τ^e defines the coordinates of each point of the real element in terms of the abstract coordinates ξ of the corresponding point of the element of reference.

$$\tau^e : \xi \to x^e = x^e(\xi) \qquad (1.10)$$

Transformation τ^e depends on the form and location of the real element. Thus, there is a different transformation τ^e for each real element:

$$\tau^e : \xi \to x^e = x^e(\xi, x_i, x_j, x_k, \ldots) \qquad (1.11)$$

Transformations τ^e must satisfy the rules established in paragraph 1.2.2. To achieve such a purpose each transformation τ^e is chosen so as to have the following properties:

— it must be a one to one mapping, i.e. for every point V^r of the reference element, there is one and only one point of the real element V^e.
— the geometrical nodes of the element of reference correspond to the geometrical nodes of the real element.
— every portion of the boundary of an element of reference, defined by the geometrical nodes of that boundary, corresponds to the portion of the boundary of the real element defined by the corresponding nodes.

Note that one element of reference of a particular type V^r (for example, a three-noded triangle) maps into all the real elements of the same type V^e using different transformations τ^e. To simplify the notation, the superscript e of a real element will not be shown from now on. We shall use a linear transformation τ with respect to the geometrical coordinates of a real element V^e.

$$\tau : \boldsymbol{\xi} \to \mathbf{x}(\boldsymbol{\xi}) = [\bar{N}(\boldsymbol{\xi})]\{\mathbf{x}_n\} \qquad (1.12)$$

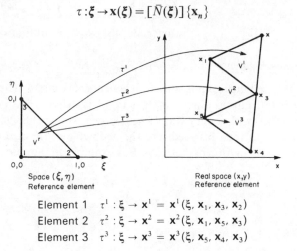

Space (ξ, η) Real space (x,y)
Reference element   Reference element

Element 1 $\tau^1 : \xi \to \mathbf{x}^1 = \mathbf{x}^1(\xi, \mathbf{x}_1, \mathbf{x}_3, \mathbf{x}_2)$
Element 2 $\tau^2 : \xi \to \mathbf{x}^2 = \mathbf{x}^2(\xi, \mathbf{x}_1, \mathbf{x}_5, \mathbf{x}_3)$
Element 3 $\tau^3 : \xi \to \mathbf{x}^3 = \mathbf{x}^3(\xi, \mathbf{x}_5, \mathbf{x}_4, \mathbf{x}_3)$

Furthermore, identical transformation functions will be used for all three coordinates.

$$\mathbf{x}(\boldsymbol{\xi}) = \langle \bar{N}(\boldsymbol{\xi}) \rangle \{\mathbf{x}_n\}$$
$$\mathbf{y}(\boldsymbol{\xi}) = \langle \bar{N}(\boldsymbol{\xi}) \rangle \{\mathbf{y}_n\}$$
$$\mathbf{z}(\boldsymbol{\xi}) = \langle \bar{N}(\boldsymbol{\xi}) \rangle \{\mathbf{z}_n\}$$

For example, in the case of a three-noded triangle (x_i, x_j, x_k),

$$x(\xi,\eta) = \bar{N}_1(\xi,\eta)x_i + \bar{N}_2(\xi,\eta)x_j + \bar{N}_3(\xi,\eta)x_k = \langle \bar{N} \rangle \begin{Bmatrix} x_i \\ x_j \\ x_k \end{Bmatrix}$$

$$y(\xi,\eta) = \bar{N}_1(\xi,\eta)y_i + \bar{N}_2(\xi,\eta)y_j + \bar{N}_3(\xi,\eta)y_k = \langle \bar{N} \rangle \begin{Bmatrix} y_i \\ y_j \\ y_k \end{Bmatrix}$$

where (ξ,η) belongs to V^r.

The functions N_i, normally chosen as polynomials in ξ are called geometrical transformation functions. We may now consider (1.12) as a nodal approximation by sub-domain for functions $x(\xi)$ and $y(\xi)$. Functions N_i must be so chosen as to satisfy the properties mentioned in paragraph 1.2.4. They can be constructed with the same techniques described in paragraphs 1.3 and 1.4.

With the geometrical transformation τ it is now possible to replace the analytical definition of each real element V^e in terms of real coordinates $\mathbf{x}$ by the easier analytical definition of its element of reference V^r in terms of non-dimensional coordinates ξ. From now on it will be more advantageous to use the non-dimensional space of the element of reference. We shall use functions $\mathbf{u}(\xi)$ instead of functions $\mathbf{u}(\mathbf{x})$ the relation between ξ and $\mathbf{x}$ being defined by (1.12). Functions $\mathbf{u}(\xi)$ and $\mathbf{u}(\mathbf{x})$ are different but take the same value at corresponding points. We have:

$\mathbf{u}(\mathbf{x}) = \mathbf{u}(\mathbf{x}(\xi))$ which, for simplicity, will be noted:
$\mathbf{u}(\mathbf{x}) = \mathbf{u}(\xi)$.

Example 1.9 Analytical definition of three-noded triangular element

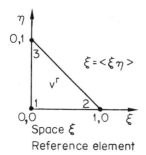

Space ξ
Reference element

Space x
Real element

The reference element is defined as:

$$\xi + \eta \leq 1$$
$$\xi \geq 0$$
$$\eta \geq 0$$

Consider the linear transformation τ.

τ:
$$x(\xi,\eta) = \langle 1-\xi-\eta \quad \xi \quad \eta \rangle \begin{Bmatrix} x_i \\ x_j \\ x_k \end{Bmatrix}$$

$$y(\xi,\eta) = \langle 1-\xi-\eta \quad \xi \quad \eta \rangle \begin{Bmatrix} y_i \\ y_j \\ y_k \end{Bmatrix}$$

It satisfies the following properties.

- The geometrical nodes of V^r with coordinates $\langle 0,0 \rangle$, $\langle 1,0 \rangle$, $\langle 0,1 \rangle$ transform into the geometrical nodes $\mathbf{x}_i, \mathbf{x}_j, \mathbf{x}_k$. For example,

$$\mathbf{x}(\xi=0, \eta=0) = \langle 1 \quad 0 \quad 0 \rangle \begin{Bmatrix} \mathbf{x}_i \\ \mathbf{x}_j \\ \mathbf{x}_k \end{Bmatrix} = \mathbf{x}_i$$

- Each boundary of V^r transforms into the corresponding boundary of V^e. For example, the boundary passing through nodes $\langle 1,0 \rangle$ and $\langle 0,1 \rangle$ which is described by equation

$$1 - \xi - \eta = 0$$

transforms into the boundary of V^e passing through $\mathbf{x}_i$ and $\mathbf{x}_j$ for which the parametric equations are:

$$x = \langle 0 \quad \xi \quad 1-\xi \rangle \begin{Bmatrix} x_i \\ x_j \\ x_k \end{Bmatrix} = \xi \quad x_j + (1-\xi) \quad x_k$$

$$y = \langle 0 \quad \xi \quad 1-\xi \rangle \begin{Bmatrix} y_i \\ y_j \\ y_k \end{Bmatrix} = \xi \quad y_j + (1-\xi) \quad y_k$$

We note that the equation is linear in ξ and η and depends only on the coordinates x_j and $\mathbf{x}_k$ of the nodes belonging to that boundary. The transformation is a one to one transformation if the Jacobian matrix of the transformation is non-singular.

$$[J] = \begin{Bmatrix} \frac{\partial x}{\partial \xi} & \frac{\partial y}{\partial \xi} \\ \frac{\partial x}{\partial \eta} & \frac{\partial y}{\partial \eta} \end{Bmatrix} = \begin{bmatrix} x_j - x_i & y_j - y_i \\ x_k - x_i & y_k - y_i \end{bmatrix}$$

$$\det(J) = (x_j - x_i)(y_k - y_i) - (x_k - x_i)(y_j - y_i)$$

The determinant of the Jacobian matrix is equal to twice the area of the triangle. It can vanish only when the three vertices are on the same straight line.

Remarks

- In this text, reference elements are also called parent elements.
- The geometrical transformation τ can be viewed as a simple change of variables $x \to \xi$.
- ξ and x can also be viewed as a system of local parametric lines attached to each element.

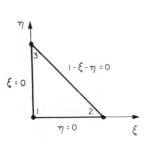
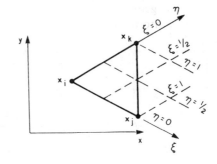

1.2.5 Classical Elements

We now illustrate the reference elements corresponding to the classical elements mentioned in paragraph 1.2.3.

(a) One-dimensional reference elements

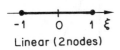
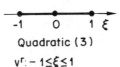
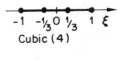

(b) Two-dimensional reference elements
Triangular

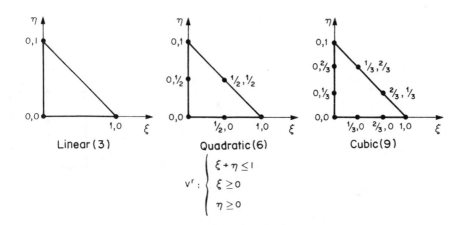

Square

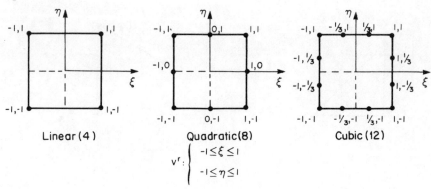

Linear (4)   Quadratic (8)   Cubic (12)

$$v^r : \begin{cases} -1 \leq \xi \leq 1 \\ -1 \leq \eta \leq 1 \end{cases}$$

(c) Three-dimensional reference elements

Tetrahedronal

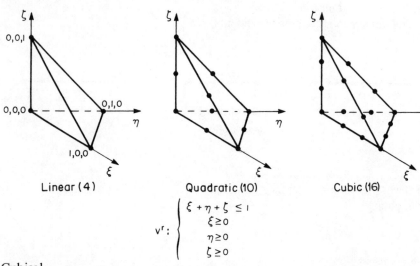

Linear (4)   Quadratic (10)   Cubic (16)

$$v^r : \begin{cases} \xi + \eta + \zeta \leq 1 \\ \xi \geq 0 \\ \eta \geq 0 \\ \zeta \geq 0 \end{cases}$$

Cubical

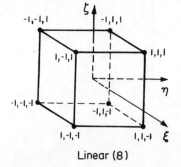 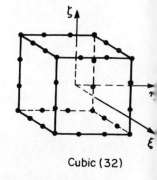

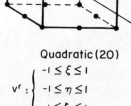

Linear (8)   Quadratic (20)   Cubic (32)

$$v^r : \begin{cases} -1 \leq \xi \leq 1 \\ -1 \leq \eta \leq 1 \\ -1 \leq \zeta \leq 1 \end{cases}$$

Prismatic

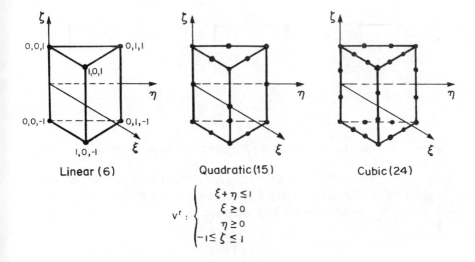

Linear (6)   Quadratic (15)   Cubic (24)

$$V^r : \begin{cases} \xi + \eta \leq 1 \\ \xi \geq 0 \\ \eta \geq 0 \\ -1 \leq \zeta \leq 1 \end{cases}$$

Remarks

— For quadratic elements mid-nodes are situated at the centre of the edges. For cubic elements, mid-nodes divide each edge in three equal parts. Explicit formulas for the geometrical transformations $\bar{N}(\xi)$ have not been mentioned; instead, a general numerical method of construction will be explained in paragraph 1.4.

— A standard organization of coordinates and connectivities is proposed in the next paragraph.

1.2.6 *Nodal Coordinates and Element Connectivities*

With nodes numbered sequentially from 1 to n and all the coordinates measured in a single global system of reference, the information can be grouped in an array CORG as shown below for a two-dimensional problem.

	Nodes				
	1	2	3	...	$\bar{n}$
x	x_1	x_2	x_3	...	$x_{\bar{n}}$
y	y_1	y_2	y_3	...	$y_{\bar{n}}$

Array CORG.

For elements numbered sequentially from 1 to n and the list of all the node numbers belonging to that element, the following array, called a connectivity matrix, can be constructed:

	Elements					
	1	2	...	e	...	n_{el}
1				j_1		
2				j_2		
Nodes 3			...	j_3	...	
⋮				⋮		
$\bar{n}^e$				$j_{\bar{n}}^e$		

Array CONEC.

where $\bar{n}^e$ is the maximum number of geometrical nodes per element, $i_1, i_2, \ldots, i_{\bar{n}^e}$ are the nodes of element e.

The two arrays CORG and CONEC are sufficient to completely define the transformation τ for all elements, i.e. to construct all the functions $\bar{N}(\xi)$ and the vector of nodal coordinates for each element.

$$\{x_n\} = \begin{Bmatrix} \mathbf{x}_1 \\ \mathbf{x}_2 \\ \vdots \\ \mathbf{x}_{n^e} \end{Bmatrix}$$

Example 1.10 *Arrays CORG and CONEC for a two-dimensional problem*

Consider a domain subdivided into two triangular and one square elements:

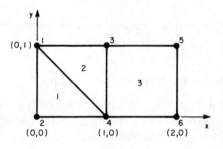

The nodes are numbered 1 to 6. The elements are numbered 1 to 3.

		Nodes					
		1	2	3	4	5	6
Array CORG :	x	0	0	1	1	2	2
	y	1	0	1	0	1	0

	Elements		
	1	2	3
1	1	3	3
2	2	1	4
3	4	4	6
4	—	—	5

Array CONEC :   Nodes

```
      SUBROUTINE GRILLE(NDIM,NNEL,MR,MP,NNT,NELT,VCORG,KCONEC)     GRIL  1
C================================================================GRIL  2
C                                                                 GRIL  3
C     INPUT NODAL COORDINATES AND ELEMENT CONNECTIVITY            GRIL  4
C     (TABLE CORG AND CONEC)                                      GRIL  5
C                                                                 GRIL  6
C     INPUT                                                       GRIL  6
C       NDIM    NUMBER OF DIMENSIONS OF THE PROBLEM (1,2 OR 3)    GRIL  8
C       NNEL    NUMBER OF NODES PER ELEMENT                       GRIL  9
C       MR      LOGICAL UNIT NUMBER FOR READING INPUT DATA        GRIL 10
C       MP      LOGICAL UNIT NUMBER FOR PRINTING                  GRIL 11
C                                                                 GRIL 12
C     OUTPUT                                                      GRIL 13
C       NNT     TOTAL NUMBER OF NODES                             GRIL 14
C       NELT    TOTAL NUMBER OF ELEMENTS                          GRIL 15
C       VCORG   NODAL COORDINATES TABLE                           GRIL 16
C       KCONEC  ELEMENT CONNECTIVITY TABLE                        GRIL 17
C                                                                 GRIL 18
C================================================================GRIL 19
      IMPLICIT REAL*8(A-H,O-Z)                                    GRIL 20
      DIMENSION VCORG(NDIM,1),KCONEC(NNEL,1)                      GRIL 21
C------- READ NODE AND ELEMENT NUMBERS                            GRIL 22
      READ(MR,1000) NNT,NELT                                      GRIL 23
 1000 FORMAT(16I5)                                                GRIL 24
      WRITE(MP,2000) NNT,NELT                                     GRIL 25
 2000 FORMAT(/'NUMBER OF NODES =',I5,' NUMBER OF ELEMENTS=',I5/)  GRIL 26
C------- READ COORDINATES                                         GRIL 27
      WRITE(MP,2010)                                              GRIL 28
 2010 FORMAT(' NODES          COORDINATES'/)                      GRIL 29
      DO 10 IN=1,NNT                                              GRIL 30
      READ(MR,1010)(VCORG(I,IN),I=1,NDIM)                         GRIL 31
 1010 FORMAT(8F10.0)                                              GRIL 32
   10 WRITE(MP,2020) IN,(VCORG(I,IN),I=1,NDIM)                    GRIL 33
 2020 FORMAT(1X,I5,8F10.5)                                        GRIL 34
C------- READ CONNECTIVITY                                        GRIL 35
      WRITE(MP,2030)                                              GRIL 36
 2030 FORMAT(/' ELEMENT           CONNECTIVITY'/)                 GRIL 37
      DO 20 IE=1,NELT                                             GRIL 38
      READ(MR,1000)(KCONEC(I,IE),I=1,NNEL)                        GRIL 39
   20 WRITE(MP,2040) IE,(KCONEC(I,IE),I=1,NNEL)                   GRIL 40
 2040 FORMAT(1X,I5,5X,14I5)                                       GRIL 41
      RETURN                                                      GRIL 42
      END                                                         GRIL 43
```

Figure 1.2 Subroutine GRILLE for reading the arrays of program BBMEF given in Chapter 6.

Remark.

It is important to recognize that, surfaces being oriented, the connectivities must follow a common order; for example, the positive sense of rotation about an outward unit normal vector.

The subroutine listing of Figure 1.2 is an example of a very simple code to read coordinate and connectivity arrays (CORG, CONEC). The subroutine is part of program BBMEF presented in Chapter 6. More elaborate subroutines are listed in Figures 6.13 and 6.18 for use in program MEF.

1.3 Approximation Based on a Reference Element

1.3.1 *Algebraic Form of Approximate Function u(x)*

Let us choose a set of n interpolation nodes of coordinates $\mathbf{x}_i$ in the domain V. These nodes do not need to coincide with the geometrical nodes. For each element V^e let us use a nodal approximation of type (1.5) for the exact function $u_{\text{ex}}(\mathbf{x})$:

$$u_{\text{ex}}(\mathbf{x}) \approx (\mathbf{x}) = \langle N_1(\mathbf{x}) \quad N_2(\mathbf{x}) \ldots N_{n^e}(\mathbf{x}) \rangle \begin{Bmatrix} u_1 \\ u_2 \\ \vdots \\ u_{n^e} \end{Bmatrix} = \langle N(x) \rangle \{u_n\} \quad (1.13)$$

where $\mathbf{x}$ belongs to V^e. $u_1, u_2, \ldots, u_{n^e}$ are the values of u_{ex} at the interpolation nodes n^e. $N(\mathbf{x})$ are the interpolation functions on the real element. Now, replace the approximation on the real element by the corresponding approximation on the reference element

$$u_{\text{ex}}(\boldsymbol{\xi}) \approx u(\boldsymbol{\xi}) = \langle N(\boldsymbol{\xi}) \rangle \{u_n\} \quad (1.14)$$

with (1.12):

$$\tau : \boldsymbol{\xi} \to \mathbf{x}(\boldsymbol{\xi}) = [\bar{N}(\boldsymbol{\xi})] \{\mathbf{x}_n\}$$

where $\{u_n\}$ are the nodal variables of the element. $\langle N(\boldsymbol{\xi}) \rangle$ are the interpolation functions on the reference element.

Remarks

— In general, functions $N(\mathbf{x})$ are used only for the most simple elements. Functions $N(\boldsymbol{\xi})$ where $\mathbf{x}$ and $\boldsymbol{\xi}$ are related by the transformation τ is defined by (1.12).

— In expression (1.13), functions $N(\mathbf{x})$ depend on the nodal coordinates of each element and thus are different for each element. On the other hand, in expression (1.14), functions $N(\boldsymbol{\xi})$ are independent of the geometry of the real element V^e. A unique set of functions $N(\boldsymbol{\xi})$ can thus be used for all elements having

the same reference or parent element. The parent element is characterized by:
- its shape,
- its geometrical nodes,
- its interpolation nodes.

Example 1.11   *Interpolation functions for a three-noded triangle*

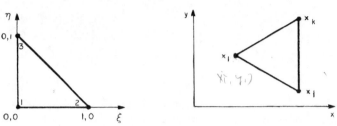

In this case, the three interpolation nodes are also geometrical nodes. The nodal variables are:

$$\{u_n\} = \begin{Bmatrix} u_i \\ u_j \\ u_k \end{Bmatrix}$$

The linear interpolation on the real element of the form (1.5) is:

$$u(x,y) = \langle N_1(x,y) \quad N_2(x,y) \quad N_3(x,y) \rangle \begin{Bmatrix} u_i \\ u_j \\ u_k \end{Bmatrix}$$

where

$$N_1(x,y) = \frac{1}{2A}[(y_k - y_j)(x_j - x) - (x_k - x_j)(y_j - y)]$$

$$N_2(x,y) = \frac{1}{2A}[(y_i - y_k)(x_k - x) - (x_i - x_k)(y_k - y)]$$

$$N_3(x,y) = \frac{1}{2A}[(y_j - y_i)(x_i - x) - (x_j - x_i)(y_i - y)]$$

$$2A = (x_k - x_j)(y_i - y_j) - (x_i - x_j)(y_k - y_j)$$

We note that functions $N_i(x,y)$ depend on the coordinates of the nodes. The interpolation on the element of reference is:

$$u(\xi,\eta) = \langle N_1(\xi,\eta) \quad N_2(\xi,\eta) \quad N_3(\xi,\eta) \rangle \begin{Bmatrix} u_i \\ u_j \\ u_k \end{Bmatrix}$$

$$N_1(\xi,\eta) = 1 - \xi - \eta$$
$$N_2(\xi,\eta) = \xi$$
$$N_3(\xi,\eta) = \eta$$

Expression $u(\xi,\eta)$ obtained by interpolation on the element of reference is identical to expression $u(x,y)$ obtained by interpolation on the real element provided points (ξ,η) and (x,y) are related by transformation τ:

$$\tau: \quad x(\xi,\eta) = \langle \bar{N}_1 \; \bar{N}_2 \; \bar{N}_3 \rangle \begin{Bmatrix} x_i \\ x_j \\ x_k \end{Bmatrix}$$

$$y(\xi,\eta) = \langle \bar{N}_1 \; \bar{N}_2 \; \bar{N}_3 \rangle \begin{Bmatrix} y_i \\ y_j \\ y_k \end{Bmatrix}$$

where $\bar{N}_1 \equiv N_1 \quad \bar{N}_2 \equiv N_2 \quad \bar{N}_3 \equiv N_3$.

Let us show that $u(\xi_0,\eta_0) \equiv u(x_0,y_0)$ if point (ξ_0,η_0) corresponds to point (x_0,y_0) in transformation τ. For example, let

$$\xi_0 = \tfrac{1}{4}, \quad \eta_0 = \tfrac{1}{2}$$

hence

$$u(\xi_0,\eta_0) = \langle \tfrac{1}{4} \; \tfrac{1}{4} \; \tfrac{1}{2} \rangle \begin{Bmatrix} u_i \\ u_j \\ u_k \end{Bmatrix} = \tfrac{1}{4}(u_i + u_j + 2u_k)$$

and

$$x_0(\xi_0,\eta_0) = \langle \tfrac{1}{4} \; \tfrac{1}{4} \; \tfrac{1}{2} \rangle \begin{Bmatrix} x_i \\ x_j \\ x_k \end{Bmatrix} = \tfrac{1}{4}(x_i + x_j + 2x_k)$$

$$y_0(\xi_0,\eta_0) = \tfrac{1}{4}(y_i + y_j + 2y_k)$$

then substituting into the expressions of $N_1(x,y), \; N_2(x,y), \; N_3(x,y)$:

$$N_1(x_0,y_0) = \tfrac{1}{4}$$
$$N_2(x_0,y_0) = \tfrac{1}{4}$$
$$N_3(x_0,y_0) = \tfrac{1}{2}$$

Therefore
$$u(x_0,y_0) = \langle \tfrac{1}{4} \; \tfrac{1}{4} \; \tfrac{1}{2} \rangle \begin{Bmatrix} u_i \\ u_j \\ u_k \end{Bmatrix} \equiv u(\xi_0,\eta_0).$$

1.3.2 Properties of Approximate Function $u(x)$

(a) *Fundamental property of the nodal approximation*

We may observe that the finite element approximation satisfies the properties of the nodal approximation of paragraph 1.1.1. Values of the approximate function

$u(x)$ coincide with values of the exact function $u_{ex}(x)$ at all the interpolation nodes.

$$u_{ex}(\mathbf{x}_i) = u(\mathbf{x}_i) = u_i = \langle N_1(\mathbf{x}_i) \quad N_2(\mathbf{x}_i) \ldots \rangle \begin{Bmatrix} u_1 \\ u_2 \\ \vdots \\ u_{ne} \end{Bmatrix}$$

From which $\quad N_j(\mathbf{x}_i) = \begin{cases} 0 & \text{if } i \neq j \\ 1 & \text{if } i = j \end{cases}$ (1.15)

Similarly, using the approximation on the element of reference:

$$u_{ex}(\boldsymbol{\xi}_i) = u(\boldsymbol{\xi}_i) = u_i = \langle N_1(\boldsymbol{\xi}_i) \quad N_2(\boldsymbol{\xi}_i) \ldots \rangle \begin{Bmatrix} u_1 \\ u_2 \\ \vdots \\ u_{ne} \end{Bmatrix}$$

From which $\quad N_j(\boldsymbol{\xi}_i) = \begin{cases} 0 & \text{if } i \neq j \\ 1 & \text{if } i = j \end{cases}$ (1.16)

Example 1.12 Fundamental property of the interpolation functions of a three-noded triangle

In Example 1.11 we verify that for $\mathbf{x} = \mathbf{x}_k = \langle x_k \quad y_k \rangle$

$$N_1(\mathbf{x} = \mathbf{x}_k) = 0; \quad N_2(\mathbf{x} = \mathbf{x}_k) = 0; \quad N_3(\mathbf{x} = \mathbf{x}_k) = 1$$

and for $\boldsymbol{\xi} = \boldsymbol{\xi}_3 = \langle 0 \quad 1 \rangle$

$$N_1(\boldsymbol{\xi} = \boldsymbol{\xi}_3) = 0; \quad N_2(\boldsymbol{\xi} = \boldsymbol{\xi}_3) = 0; \quad N_3(\boldsymbol{\xi} = \boldsymbol{\xi}_3) = 1$$

(b) Continuity within the element

If the approximate function $u(\mathbf{x})$ is required to be continuous together with all its derivatives up to order s, interpolation functions $N_i(\mathbf{x})$ of the same quality must be used.

(c) Inter-element continuity

If the approximate function $u(\mathbf{x})$ and its derivatives up to order s are required to be continuous on a common boundary with another element, then $u(\mathbf{x})$ and its derivatives up to order s can only depend upon the nodal variables on the common boundary. Consider first the continuity of a function across a common boundary with an adjacent element.

$$u(\mathbf{x}) = \langle N_1(\mathbf{x}) \quad N_2(\mathbf{x}) \ldots \rangle \begin{Bmatrix} u_1 \\ u_2 \\ \vdots \\ u_{ne} \end{Bmatrix}$$

Products $N_i(\mathbf{x})u_i$ must be null if u_i does not belong to the common boundary. Hence,

$$N_i(\mathbf{x}) = 0 \tag{1.17a}$$

when $\mathbf{x}$ is on a boundary and u_i does not belong to that boundary. Similarly, on the parent element $N_i(\boldsymbol{\xi}) = 0$ when $\boldsymbol{\xi}$ is on a boundary and u_i is not on that boundary. The continuity of the derivatives across a common boundary is similarly written as:

$$\frac{\partial u(\mathbf{x})}{\partial x} = \left\langle \frac{\partial N_1(\mathbf{x})}{\partial x} \quad \frac{\partial N_2(\mathbf{x})}{\partial x} \right\rangle \cdots \begin{Bmatrix} u_1 \\ u_2 \\ \vdots \\ u_{n^e} \end{Bmatrix}$$

where

$$\frac{\partial N_i(\mathbf{x})}{\partial x} = 0 \tag{1.17b}$$

when $\mathbf{x}$ is located on the common boundary and u_i is not. The previous condition for a two-dimensional parent element is:

$$\frac{\partial N_i(\boldsymbol{\xi})}{\partial \xi}\frac{\partial \xi}{\partial x} + \frac{\partial N_i(\boldsymbol{\xi})}{\partial \eta}\frac{\partial \eta}{\partial x} = 0$$

Continuity requirements between adjacent elements have played a major role in the development of the method. The degree of continuity to be maintained between adjacent elements is problem dependent and will be discussed in Chapters 3 and 4.

Example 1.13 *Continuous function for a three-noded triangle*

Consider edge $x_j - x_k$ of the element shown in Example 1.11

$$x - x_j = (y - y_j)\frac{(x_k - x_j)}{(y_k - y_j)}$$

We verify that the above expression nullifies function $N_i(x, y)$ corresponding to node i. The corresponding side of the element of reference is:

$$1 - \xi - \eta = 0$$

Such a relation nullifies $N_i(\xi, \eta)$. Thus, function $u^e(x)$ is continuous on edge $x_j - x_k$ since it depends linearly only on u_j and u_k on that edge:

$$u = \xi u_j + (1 - \xi)u_k$$

(d) *Complete polynomial interpolation functions*

Truncation error (1.7) can be reduced by decreasing the size of the elements. For many problems it is also necessary to decrease the errors on the derivatives of the approximate functions. To ensure that error $u - u_{\text{ex}}$ tends toward zero with a decrease in size of the element it is essential that the approximation u contains a non-zero constant term (see paragraph 1.5). Approximation u is then capable of representing exactly the constant function u_{ex} within an element. To ensure that error $\partial u / \partial x - \partial u_{\text{ex}} / \partial x$ tends toward zero with a decrease in the size of the element it is also essential that u contains a term in x. Thus, if $\partial u_{\text{ex}} / \partial x$ is constant, $\partial u / \partial x$ will be capable of representing that constant exactly. In general, if the errors on u_{ex} and its derivatives up to order s are to decrease with the size of an element, expression (1.13) must contain a complete polynomial of order s. Furthermore if function u and its derivatives, up to degree $s - 1$, are continuous across common boundaries with adjacent elements, then the truncation errors for u_{ex} and its derivatives up to order s will tend toward zero everywhere in domain V, including its boundaries. When these inter-element continuity conditions are not satisfied, convergence can still be obtained in some cases. Patch tests [3, 5, 6] described in paragraph 4.1.3 can be used in such cases to predict whether convergence will be obtained or not.

For a linear transformation τ the conclusions with respect to approximation $u(\mathbf{x})$ on a real element hold as well for the approximation $u(\boldsymbol{\xi})$ on the parent element. The expression $u(\boldsymbol{\xi})$ must include a complete polynomial of order s in ξ, η, ζ for a convergence up to s th order of u_{ex}. When transformation τ is not linear, complete polynomials in x, y, z correspond to complete polynomials in ξ, η, ζ where $\langle N \rangle = \langle \bar{N} \rangle$ and $S \leq 1$ [7].

Example 1.14 Complete polynomial for a three-noded triangle

We verify that expressions $u(x, y)$ and $u(\xi, \eta)$ of Example 1.11 contain complete polynomial of order 1 in x, y and ξ, η. Thus, the errors on u_{ex} and its first derivatives tend toward zero with a reduction of the size of each element. To verify that approximations $u(x, y)$ and $u(\xi, \eta)$ are able to represent the exact function $u_{\text{ex}}(x, y) = u_0$ (constant), let $u_i = u_j = u_k = u_0$ in the expressions for $u(x, y)$ and $u(\xi, \eta)$. We get:

$$u(x, y) = (N_1(x, y) + N_2(x, y) + N_3(x, y))u_0 = u_0$$
$$u(\xi, \eta) = (N_1(\xi, \eta) + N_2(\xi, \eta) + N_3(\xi, \eta))u_0 = u_0$$

as the functions N verify:

$$N_1(x, y) + N_2(x, y) + N_3(x, y) = 1$$
$$N_1(\xi, \eta) + N_2(\xi, \eta) + N_3(\xi, \eta) = 1$$

Definitions

— If only the values of the functions are continuous across boundaries, the function is said to be of class C^0. When the function and its first derivatives are continuous, it is of class C^1. In general, if the function and all its derivatives up to order α are continuous, it is said to be of class C^α.

— An element is isoparametric if the geometrical transformation functions

$\bar{N}(\xi)$ are identical to the interpolation functions $N(\xi)$. For each element, geometrical and interpolation nodes are identical [2].

— An element is pseudo-parametric if functions $\bar{N}(\xi)$ and $N(\xi)$ are different polynomials built with the same monomials.

— An element is sub-parametric when the geometrical polynomials $\bar{N}(\xi)$ are of a lower order than its interpolation polynomials $N(\xi)$. It is super-parametric in the opposite case. Super-parametric elements do not possess property (d) mentioned above (see paragraph 2.8.4).

— The number of nodal variables associated with the total number of interpolation nodes of an element is called the number of degrees of freedom (n_d). All nodes (geometrical and interpolation) are stored in the same array **CORG** described in paragraph 1.2.6. Similarly, the list of the two kinds of nodes is stored in array **CONEC** described in paragraph 1.2.6. The distinction between geometrical and interpolation nodes will be made, if necessary, in the subroutines calculating the functions $\bar{N}(\xi)$ and $N(\xi)$.

1.4 Construction of Functions $N(\xi)$ and $\bar{N}(\xi)$

Geometrical transformations $\bar{N}(\xi)$ and interpolation function $N(\xi)$ have identical properties. They can sometimes be constructed with polynomials having properties described in paragraph 1.2.4 and 1.3.2. Such polynomials are often Lagrange or Hermite polynomials; however, no systematic method of construction has been found for all cases. A number of well-known formulas have been found for classical elements. In the following paragraphs we describe a systematic numerical method of construction for all elements.

1.4.1 *General Method of Construction*

(*a*) *Choice of a polynomial basis*

Let us write $u(\xi)$ for the reference element in the form of a linear combination of known independent functions $P_1(\xi), P_2(\xi)\ldots$ which are most frequently independent monomials. The choice of functions $P_i(\xi)$ is one of the most important operations in the finite element method.

$$u(\xi) = \langle P_1(\xi) \quad P_2(\xi) \ldots \rangle \begin{Bmatrix} a_1 \\ a_2 \\ \vdots \\ a_{n_d} \end{Bmatrix} = \langle P(\xi) \rangle \{a\} \qquad (1.18)$$

The set of functions $P(\xi)$ constitute the polynomial basis of the approximation. The number of terms in the basis must be equal to the number of degrees of liberty n_d of the element. A complete polynomial basis is always preferred but this is possible only for a few integer values of n_d. The following table lists typical cases giving the number of monomials necessary for a complete polynomial of degree r.

Degree of the polynomial r	1 dimension n_d	2 dimensions n_d	3 dimensions n_d
1	2	3	4
2	3	6	10
3	4	10	20
4	5	15	35
5	6	21	56

Example 1.15 *Complete and incomplete polynomial basis*

Number of dimensions	Degree of the polynomial	Polynomial basis $\langle P \rangle$		n_d
Complete basis				
1	1	$\langle 1 \; \xi \rangle$	(linear)	2
1	2	$\langle 1 \; \xi \; \xi^2 \rangle$	(quadratic)	3
2	1	$\langle 1 \; \xi \; \eta \rangle$	(linear)	3
2	2	$\langle 1 \; \xi \; \eta \; \xi^2 \; \xi\eta \; \eta^2 \rangle$	(quadratic)	6
3	1	$\langle 1 \; \xi \; \eta \; \zeta \rangle$	(linear)	4
3	2	$\langle 1 \; \xi \; \eta \; \zeta \; \xi^2 \; \xi\eta \; \eta^2 \; \eta\zeta \; \zeta^2 \; \xi\zeta \rangle$	(quadratic)	10
Incomplete basis				
2	2	$\langle 1 \; \xi \; \eta \; \zeta\eta \rangle$	(bi-linear)	4
3	3	$\langle 1 \; \xi \; \eta \; \zeta \; \xi\eta \; \eta\zeta \; \xi\zeta \; \xi\eta\zeta \rangle$	(tri-linear)	8

To construct geometrical transformation functions $\bar{N}$, we select expressions of the same form for x, y and z.

$$x(\xi) = \langle \bar{P}(\xi) \rangle \{a_x\}$$
$$y(\xi) = \langle \bar{P}(\xi) \rangle \{a_y\} \quad (1.19)$$
$$z(\xi) = \langle \bar{P}(\xi) \rangle \{a_z\}$$

The number of functions $\bar{P}(\xi)$ and coefficients $\{a_x\}, \{a_y\}$ and $\{a_z\}$ is equal to the number of geometrical nodes of the element.

Definitions

— Coefficients $\{a\}$ are called generalized variables of the element to distinguish them from nodal variables $\{u_n\}$.

— Expression $u(\xi) = \langle P(\xi) \rangle \{a\}$ defines a generalized approximation to be distinguished from the nodal approximation $u(\xi) = \langle N(\xi) \rangle \{u_n\}$.

— Coefficients $\{a_x\}, \{a_y\}, \{a_z\}$ are sometimes called generalized coordinates of the element to distinguish them from the nodal coordinates $\{x_n\}, \{y_n\}, \{z_n\}$.

(b) Relation between generalized and nodal variables

At each interpolation node of coordinates $\{\xi_i\}$, function $u(\xi)$ takes its nodal value $u_i = u_{ex}(\xi_i)$:

$$\begin{Bmatrix} u_1 \\ u_2 \\ \vdots \\ u_{n_d} \end{Bmatrix} = \{u_n\} = \begin{bmatrix} \langle P_1(\xi_1) & P_2(\xi_1) & \cdots & P_{n_d}(\xi_1) \rangle \\ \langle P_1(\xi_2) & P_2(\xi_2) & \cdots & P_{n_d}(\xi_2) \rangle \\ \hline \langle P_1(\xi_{n_d}) & P_2(\xi_{n_d}) & \cdots & P_{n_d}(\xi_{n_d}) \rangle \end{bmatrix} \{a\}$$

$$\{u_n\} = [P_n]\{a\} \tag{1.20}$$

thus inverting nodal matrix $[P_n]$ of order n_d

$$\{a\} = [P_n]^{-1}\{u_n\} \tag{1.21}$$

In order to define $\{a\}$ uniquely in terms of $\{u_n\}$ the matrix $[P_n]$ in (1.21) should not be singular. This depends on the choice of polynomial basis and coordinates (ξ_i) of the nodes of the reference element. Since $[P_n]$ is independent of the geometry of the real element, once it has been determined that $[P_n]$ is non-singular it will be so for all real elements having the same reference element. In a similar manner, we write equation (1.19) at the coordinates of the geometrical nodes to obtain

$$\begin{aligned} \{x_n\} &= [\bar{P}_n]\{a_x\} \\ \{y_n\} &= [\bar{P}_n]\{a_y\} \\ \{z_n\} &= [\bar{P}_n]\{a_z\} \end{aligned} \tag{1.22}$$

thus, after inversion of $[\bar{P}_n]$

$$\begin{aligned} \{a_x\} &= [\bar{P}_n]^{-1}\{x_n\} \\ \{a_y\} &= [\bar{P}_n]^{-1}\{y_n\} \\ \{a_z\} &= [\bar{P}_n]^{-1}\{z_n\} \end{aligned} \tag{1.23}$$

(c) Analytical expressions for N and $\bar{N}$

Substituting (1.12) into (1.18)

$$u(\xi) = \langle P(\xi) \rangle [P_n]^{-1}\{u_n\}$$

or

$$u(\xi) = \langle N(\xi) \rangle \{u_n\} \tag{1.24}$$

where

$$\langle N(\xi) \rangle = \langle P(\xi) \rangle [P_n]^{-1}$$

In a similar manner, we get:
$$x(\xi) = \langle \bar{N}(\xi) \rangle \{x_n\}$$
$$y(\xi) = \langle \bar{N}(\xi) \rangle \{y_n\} \quad (1.25)$$
$$z(\xi) = \langle \bar{N}(\xi) \rangle \{z_n\}$$

where
$$\langle \bar{N}(\xi) \rangle = \langle \bar{P}(\xi) \rangle [\bar{P}_n]^{-1}$$

(d) Differentiation of function $u(\xi)$

Differentiating (1.24) we obtain

$$\begin{Bmatrix} u_{,\xi} \\ u_{,\eta} \\ u_{,\zeta} \end{Bmatrix} = \begin{bmatrix} \langle P_{,\xi} \rangle \\ \langle P_{,\eta} \rangle \\ \langle P_{,\zeta} \rangle \end{bmatrix} [P_n]^{-1} \{u_n\} = \begin{bmatrix} \langle N_{,\xi} \rangle \\ \langle N_{,\eta} \rangle \\ \langle N_{,\zeta} \rangle \end{bmatrix} \{u_n\} = [B_\xi]\{u_n\} \quad (1.26)$$

Summary of operations required to construct $\langle N \rangle$

— Choice of polynomial basis $\langle P(\xi) \rangle$
— Evaluation of nodal matrix $[P_n(i,j)] = [P_j(\xi_i)]_{ji, j=1,2,\ldots,n_d}$
— Inversion of $[P_n]$
— Computation of $\langle N \rangle$

$$\langle N \rangle = \langle P(\xi) \rangle [P_n]^{-1}$$

These operations need to be performed only once for each different element of reference.

Example 1.16 Construction of shape functions $N(\xi)$ for an isoparametric four-noded quadrilateral

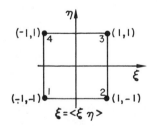
Reference element

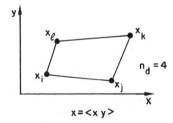
Real element

(a) Choice of polynomial basis

Since $n_d = 4$ we cannot use a complete polynomial. The best choice is a bilinear polynomial since it would respect both the symmetry and inter-element continuity

$$\langle P \rangle = \langle 1, \quad \xi, \quad \eta, \quad \xi\eta \rangle$$

Note that $u(\xi) = \langle P \rangle \{a\}$ becomes linear on each side of the element $\xi = \pm 1$ and $\eta = \pm 1$.

(b) *Evaluation of* $[P_m]$

$$[P_n] = \begin{bmatrix} 1 & -1 & -1 & 1 \\ 1 & 1 & -1 & -1 \\ 1 & 1 & 1 & 1 \\ 1 & -1 & 1 & -1 \end{bmatrix} \quad \{\xi_n\} = \begin{Bmatrix} -1 \\ 1 \\ 1 \\ -1 \end{Bmatrix}$$

$$\{\eta_n\} = \begin{Bmatrix} -1 \\ -1 \\ 1 \\ 1 \end{Bmatrix}$$

(c) *Inversion of* $[P_n]$

In this case, matrix $[P_n]$ is orthogonal since the scalar products of its different columns are null. Each column vector has a norm of value 4. Therefore:

$$[P_n]^{-1} = \frac{1}{4}[P_n]^T = \frac{1}{4}\begin{bmatrix} 1 & 1 & 1 & 1 \\ -1 & 1 & 1 & -1 \\ -1 & -1 & 1 & 1 \\ 1 & -1 & 1 & -1 \end{bmatrix}$$

(d) *Expression of* $\langle N \rangle$:

$$\langle N \rangle = \langle N_1 \; N_2 \; N_3 \; N_4 \rangle = \langle P \rangle [P_n]^{-1}$$

$$\langle N \rangle = \left\langle \frac{1-\xi-\eta+\xi\eta}{4}; \frac{1+\xi-\eta-\xi\eta}{4}; \frac{1+\xi+\eta+\xi\eta}{4}; \frac{1-\xi+\eta-\xi\eta}{4} \right\rangle$$

$$\langle N \rangle = \tfrac{1}{4}\langle (1-\xi)(1-\eta); \; (1+\xi)(1-\eta); \; (1+\xi)(1+\eta); \; (1-\xi)(1+\eta) \rangle$$

The element is isoparametric:

$$\langle \bar{N} \rangle \equiv \langle N \rangle$$

$$x(\xi,\eta) = \langle N_1 \; N_2 \; N_3 \; N_4 \rangle \begin{Bmatrix} x_1 \\ x_2 \\ x_3 \\ x_4 \end{Bmatrix}$$

$$y(\xi,\eta) = \langle N_1 \; N_2 \; N_3 \; N_4 \rangle \begin{Bmatrix} y_1 \\ y_2 \\ y_3 \\ y_4 \end{Bmatrix}$$

1.4.2 Algebraic Properties of Functions N and $\bar{N}$

(a) Each function of interpolation $N_i(\xi)$ is formed as th[e]
polynomial basis $\langle P(\xi) \rangle$ and the column i of matrix $[P_n$

$$N_i(\xi) = \langle P(\xi) \rangle \{C_i\}$$

where $\quad [P_n]^{-1} = [\{C_1\} \ \{C_2\} \ \ldots \ \{C_i\} \ \ldots$

Function $N_i(\xi)$ is then a linear combination of the function$\ldots$ in the polynomial basis $\langle P(\xi) \rangle$, the coefficients being the terms of the column i. The matrix $[P_n]^{-1}$ can be considered as a convenient method of storage for the coefficients of all the shape functions N_i.

(b) Post-multiplying the last of equation of (1.24)

$$\langle N(\xi) \rangle = \langle P(\xi) \rangle [P_n]^{-1}$$

by $[P_n]$, we get:

$$\langle N(\xi) \rangle [P_n] = \langle P(\xi) \rangle \tag{1.29}$$

Upon using the definition of $[P_n]$ in equation (1.20), we get:

$$\sum_{i=1}^{n_d} N_i(\xi) P_j(\xi_i) \equiv P_j(\xi) \quad j = 1, 2, \ldots, n_d \tag{1.30}$$

This last expression is a characteristic of the algebraic structure of shape functions N_i. It shows that the terms $P_j(\xi)$ belong to the polynomial basis used to construct N_i. This relation may be employed to verify if any given polynomial $p(\xi)$ is included independently in the basis of the interpolation functions $\langle N \rangle$. The following identity must result:

$$\sum_{i=1}^{n_d} N_i(\xi) p(\xi_i) \equiv p(\xi) \tag{1.31}$$

In some convergence studies, it is sometimes required if certain particular monomials are independently included in the approximation of u. The relation (1.31) may thus be usefully employed. For example if monomials $1, \xi, \eta$ are contained in $\langle N \rangle$, we must verify that:

$$\sum_{i=1}^{n_d} N_i(\xi) = 1$$

$$\sum_{i=1}^{n_d} N_i(\xi) \xi_i = \xi$$

$$\sum_{i=1}^{n_d} N_i(\xi) \eta_i = \eta$$

Example 1.17 Properties of the shape functions of a linear quadrilateral

We may employ (1.13) to verify if the functions N_i of Example 1.16 contain the nials 1, $\xi, \eta, \xi\eta$ as a basis.

$$\sum_{i=1}^{4} N_i(\xi,\eta) = N_1 + N_2 + N_3 + N_4 = 1$$

$$\sum_{i=1}^{4} N_i(\xi,\eta)\xi_i = -N_1 + N_2 + N_3 - N_4 = \xi$$

$$\sum_{i=1}^{4} N_i(\xi,\eta)\eta_i = -N_1 - N_2 + N_3 + N_4 = \eta$$

$$\sum_{i=1}^{4} N_i(\xi,\eta)\xi_i\eta_i = N_1 - N_2 + N_3 - N_4 = \xi\eta$$

(c) Differentiation of equation (1.30) gives:

$$\sum_{i=1}^{n_d} \frac{\partial N_i(\xi)}{\partial \xi} P_j(\xi_i) \equiv \frac{\partial P_j(\xi)}{\partial \xi} \tag{1.32}$$

Such expressions together with (1.15), (1.16), (1.17), and (1.30) are very useful to verify explicit forms of the interpolation functions N_i and their derivatives.

1.5 Transformation of Differential Operators [2]

1.5.1 *Generalities*

The governing equations of a physical problem are written in the real domain and involve unknown functions u_{ex} and their derivatives: $\dfrac{\partial u_{ex}}{\partial x}$, $\dfrac{\partial u_{ex}}{\partial y}$, etc. Since approximation (1.13) in the space of the real element is often very complicated, it is more convenient to work in the space of the element of reference (1.14)

$$u_{ex} \approx u(\xi) = \langle N(\xi) \rangle \{u_n\} \tag{1.33}$$

together with transformation (1.12).

$$\tau : \xi \to \mathbf{x} = \mathbf{x}(\xi) = [\bar{N}(\xi)]\{\mathbf{x}_n\}$$
$$\mathbf{x} = \langle x \quad y \quad z \rangle \tag{1.34}$$
$$\xi = \langle \xi \quad \eta \quad \zeta \rangle$$

The transformation being one to one, we thus have

$$\tau^{-1} : \mathbf{x} \to \xi = \xi(\mathbf{x}) \tag{1.35}$$

Since the inverse transformation τ^{-1} is generally very difficult to construct, except for the most simple elements, it is more convenient to perform all operations in the space of the reference element. For expressions containing derivatives with respect to the real space (x,y,z) it is necessary to obtain equivalent expressions in the space of the reference element (ξ,η,ζ). Such expressions depend on the Jacobian matrix $[J]$ of the transformation.

1.5.2 First Derivatives

Using the chain rule of calculus we get the following expression:

$$\left\{\begin{array}{c}\dfrac{\partial}{\partial \xi}\\[4pt] \dfrac{\partial}{\partial \eta}\\[4pt] \dfrac{\partial}{\partial \zeta}\end{array}\right\} = \begin{bmatrix}\dfrac{\partial x}{\partial \xi} & \dfrac{\partial y}{\partial \xi} & \dfrac{\partial z}{\partial \xi}\\[4pt] \dfrac{\partial x}{\partial \eta} & \dfrac{\partial y}{\partial \eta} & \dfrac{\partial z}{\partial \eta}\\[4pt] \dfrac{\partial x}{\partial \zeta} & \dfrac{\partial y}{\partial \zeta} & \dfrac{\partial z}{\partial \zeta}\end{bmatrix} \left\{\begin{array}{c}\dfrac{\partial}{\partial x}\\[4pt] \dfrac{\partial}{\partial y}\\[4pt] \dfrac{\partial}{\partial z}\end{array}\right\} \quad (1.36a)$$

or

$$\{\partial_\xi\} = [J]\{\partial_x\} \quad (1.36b)$$

where $[J]$ is the Jacobian matrix of the geometrical transformation. A similar expression exists which replaces derivatives of the real space by those of the reference space.

$$\left\{\begin{array}{c}\dfrac{\partial}{\partial x}\\[4pt] \dfrac{\partial}{\partial y}\\[4pt] \dfrac{\partial}{\partial z}\end{array}\right\} = \begin{bmatrix}\dfrac{\partial \xi}{\partial x} & \dfrac{\partial \eta}{\partial x} & \dfrac{\partial \zeta}{\partial x}\\[4pt] \dfrac{\partial \xi}{\partial y} & \dfrac{\partial \eta}{\partial y} & \dfrac{\partial \zeta}{\partial y}\\[4pt] \dfrac{\partial \xi}{\partial z} & \dfrac{\partial \eta}{\partial z} & \dfrac{\partial \zeta}{\partial z}\end{bmatrix} \left\{\begin{array}{c}\dfrac{\partial}{\partial \xi}\\[4pt] \dfrac{\partial}{\partial \eta}\\[4pt] \dfrac{\partial}{\partial \zeta}\end{array}\right\} \quad (1.37a)$$

thus:
$$\{\partial_x\} = [j]\{\partial_\xi\} \quad (1.37b)$$

substituting (1.37b) into (1.36b):

$$[j] = [J]^{-1} \quad (1.38)$$

One may note that, in practice it is the matrix $[J]$ which is explicitly defined, the matrix $[j]$ being obtained numerically from $[J]^{-1}$ which is supposed to exist for a one-to-one transformation τ

Determination of $[j] = [J]^{-1}$

We show explicit forms for $[j]$ in the case of one-, two-, and three-dimensional spaces.

* One dimension

$$[J] = J_{11}; \quad [j] = [J]^{-1} = \dfrac{1}{J_{11}} \quad (1.39)$$

* Two dimensions

$$[J] = \begin{bmatrix} J_{11} & J_{12}\\ J_{21} & J_{22}\end{bmatrix}; \quad [J]^{-1} = \dfrac{1}{\det(J)}\begin{bmatrix} J_{22} & -J_{12}\\ -J_{21} & J_{11}\end{bmatrix} \quad (1.40)$$

*Three dimensions

$$[J] = \begin{bmatrix} J_{11} & J_{12} & J_{13} \\ J_{21} & J_{22} & J_{23} \\ J_{31} & J_{32} & J_{33} \end{bmatrix}$$

$$[J]^{-1} = \frac{1}{\det(J)} \begin{bmatrix} J_{22}J_{33} - J_{32}J_{23}; & J_{13}J_{32} - J_{12}J_{33}; & J_{12}J_{23} - J_{13}J_{22} \\ J_{31}J_{23} - J_{21}J_{33}; & J_{11}J_{33} - J_{13}J_{31}; & J_{21}J_{13} - J_{23}J_{11} \\ J_{21}J_{32} - J_{31}J_{22}; & J_{12}J_{31} - J_{32}J_{11}; & J_{11}J_{22} - J_{12}J_{21} \end{bmatrix}$$
(1.41)

$$\det(J) = H_{11}(J_{22}J_{23} - J_{32}J_{23}) + J_{12}(J_{31}J_{23} - J_{21}J_{33}) + J_{13}(J_{21}J_{32} - J_{31}J_{22})$$

Computations of the terms of $[J]$

The various terms of $[J]$ are obtained directly by differentiation of expression (1.12) rewritten as follows:

$$\langle x \quad y \quad z \rangle = \langle \bar{N}(\boldsymbol{\xi}) \rangle [\{x_n\} \quad \{y_n\} \quad \{z_n\}] \quad (1.42)$$

$\{x_n\}, \{y_n\}, \{z_n\}$ being the geometrical node coordinates. The Jacobian matrix is:

$$[J] = \begin{Bmatrix} \dfrac{\partial}{\partial \xi} \\ \dfrac{\partial}{\partial \eta} \\ \dfrac{\partial}{\partial \zeta} \end{Bmatrix} \langle x \quad y \quad z \rangle = \begin{bmatrix} \langle \bar{N}_{,\xi} \rangle \\ \langle \bar{N}_{,\eta} \rangle \\ \langle \bar{N}_{,\zeta} \rangle \end{bmatrix} [\{x_n\} \quad \{y_n\} \quad \{z_n\}] \quad (1.43)$$
$$(3 \times \bar{n}^e) \qquad (\bar{n}^e \times 3)$$

It is obtained as the product of two matrices, one containing the derivatives of the geometrical transformation functions with respect to the space of the reference element, and the other containing the real coordinates of the geometrical nodes of the element.

Transformation of an integral

The change of variables (1.34) allows us to change the integration of a function f in the real geometrical domain V^e into a simpler integration in the space of the reference element V^r

$$\int_{V^e} f(\mathbf{x}) \, dx \, dy \, dz = \int_{V^r} f(\mathbf{x}(\boldsymbol{\xi})) \det(J) \, d\xi \, d\eta \, d\zeta \quad (1.44)$$

$\det(J)$ being the determinant of the Jacobian matrix $[J]$. An element of volume is defined by the mixed product

$$dV = (d\bar{x} \times d\bar{y}) \cdot d\bar{z}$$

In a cartesian system we may write
$$d\vec{x} = dx.\vec{i}; \quad d\vec{y} = dy.\vec{j}; \quad d\vec{z} = dz.\vec{k}$$
where $\vec{i}, \vec{j}, \vec{k}$ are the unit vectors
This leads to
$$dV = dx \ dy \ dz.$$
In the space of the reference element
$$dV = (d\vec{\xi} \times d\vec{\eta}) \cdot d\vec{\zeta}$$
where
$$d\vec{\xi} = (J_{11}\vec{i} + J_{21}\vec{j} + J_{31}\vec{k}) d\xi$$
$$d\vec{\eta} = (J_{12}\vec{i} + J_{22}\vec{j} + J_{32}\vec{k}) d\eta$$
$$d\vec{\zeta} = (J_{13}\vec{i} + J_{23}\vec{j} + J_{33}\vec{k}) d\zeta$$
therefore
$$dV = \det(J) d\xi \ d\eta \ d\zeta$$

Example 1.18 *Jacobian matrix of a four-noded quadrilateral element*

Using (1.43) the functions N are obtained

$$[J] = \frac{1}{4} \begin{bmatrix} -(1-\eta) & (1-\eta) & (1+\eta) & -(1+\eta) \\ -(1-\xi) & -(1+\xi) & (1+\xi) & (1-\xi) \end{bmatrix} \begin{bmatrix} x_1 & y_1 \\ x_2 & y_2 \\ x_3 & y_3 \\ x_4 & y_4 \end{bmatrix}$$

$$[J] = \frac{1}{4} \begin{bmatrix} -x_1 + x_2 + x_3 - x_4 & -y_1 + y_2 + y_3 - y_4 \\ +\eta(x_1 - x_2 + x_3 - x_4) & +\eta(y_1 - y_2 + y_3 - y_4) \\ \hline -x_1 - x_2 + x_3 + x_4 & -y_1 - y_2 + y_3 + y_4 \\ +\xi(x_1 - x_2 + x_3 - x_4) & +\xi(y_1 - y_2 + y_3 - y_4) \end{bmatrix}$$

$$\det(J) = A_0 + A_1 \xi + A_2 \eta$$

$$A_0 = \tfrac{1}{8}[(y_4 - y_2)(x_3 - x_1) - (y_3 - y_1)(x_4 - x_2)]$$
$$A_1 = \tfrac{1}{8}[(y_3 - y_4)(x_2 - x_1) - (y_2 - y_1)(x_3 - x_4)]$$
$$A_2 = \tfrac{1}{8}[(y_4 - y_1)(x_3 - x_2) - (y_3 - y_2)(x_4 - x_1)]$$

If the quadrilateral is rectangular of sides equal to:
$x_2 - x_1 = 2a$ and $y_2 - y_4 = 2b$:

$$x_1 = x_4 \qquad x_2 = x_3$$
$$y_1 = y_2 \qquad y_3 = y_4$$

$$[J] = \begin{bmatrix} a & 0 \\ 0 & b \end{bmatrix}, \quad \det(J) = ab$$

and

$$\int_{V^e} f(\mathbf{x}) \, dx \, dy = \int_{-1}^{1} \int_{-1}^{1} f(\mathbf{x}(\boldsymbol{\xi})) a \, b \, d\xi \, d\eta$$

1.5.3 Second Derivatives

We now look for expressions of second derivatives with respect to x in terms of derivatives with respect to ξ. The result will be:

$$\begin{Bmatrix} \dfrac{\partial^2}{\partial x^2} \\ \dfrac{\partial^2}{\partial y^2} \\ \dfrac{\partial^2}{\partial z^2} \\ \dfrac{\partial^2}{\partial x \partial y} \\ \dfrac{\partial^2}{\partial y \partial z} \\ \dfrac{\partial^2}{\partial x \partial z} \end{Bmatrix} = [T_1] \begin{Bmatrix} \dfrac{\partial}{\partial \xi} \\ \dfrac{\partial}{\partial \eta} \\ \dfrac{\partial}{\partial \zeta} \end{Bmatrix} + [T_2] \begin{Bmatrix} \dfrac{\partial^2}{\partial \xi^2} \\ \dfrac{\partial^2}{\partial \eta^2} \\ \dfrac{\partial^2}{\partial \zeta^2} \\ \dfrac{\partial^2}{\partial \xi \partial \eta} \\ \dfrac{\partial^2}{\partial \eta \partial \zeta} \\ \dfrac{\partial^2}{\partial \xi \partial \zeta} \end{Bmatrix} \quad (1.45a)$$

which we note:

$$\{\partial_x^2\} = [T_1]\{\partial_\xi\} + [T_2]\{\partial_\xi^2\} \tag{1.45b}$$

(a) Computation of $[T_1]$

Matrix $[T_1]$ depends on the terms of $[j]$ which are not explicitly known in general. The following computations resolve the difficulty.

From (1.36) we get

$$\{\partial_\xi^2\} = [C_1]\{\partial_x\} + [C_2]\{\partial_x^2\} \tag{1.46a}$$

then, using (1.37b)

$$\{\partial_\xi^2\} = [C_1][j]\{\partial_\xi\} + [C_2]\{\partial_x^2\} \tag{1.46b}$$

Substituting (1.46b) into (1.45b):

$$\{\partial_x^2\} = ([T_1] + [T_2][C_1][j])\{\partial_\xi\} + [T_2][C_2]\{\partial_x^2\}$$

From which we get

$$[T_2][C_2] = [I] \text{ therefore } [T_2] = [C_2]^{-1} \tag{1.47a}$$

$$[T_1] + [T_2][C_1][j] = 0 \text{ therefore } [T_1] = -[T_2][C_1][j] \tag{1.47b}$$

Matrices $[T_2]$ and $[C_1]$ are given explicitly in (1.48) and (1.49); matrix $[I]$ is the identity matrix.

(b) Computation of $[T_2]$ and $[C_1]$

Matrix $[T_2]$ defined by (1.45b) is written explicitly below where j_{ij} are terms of matrix $[j]$ given in (1.38) and (1.43).

$$[T_2] = \begin{bmatrix} j_{11}^2 & j_{12}^2 & j_{13}^2 & 2j_{11}j_{12} & 2j_{12}j_{13} & 2j_{13}j_{11} \\ j_{21}^2 & j_{22}^2 & j_{23}^2 & 2j_{21}j_{22} & 2j_{22}j_{23} & 2j_{23}j_{21} \\ j_{31}^2 & j_{32}^2 & j_{33}^2 & 2j_{31}j_{32} & 2j_{32}j_{33} & 2j_{33}j_{31} \\ j_{11}j_{21} & j_{12}j_{22} & j_{13}j_{23} & j_{11}j_{22}+j_{12}j_{21} & j_{12}j_{23}+j_{13}j_{22} & j_{11}j_{23}+j_{13}j_{21} \\ j_{21}j_{31} & j_{22}j_{32} & j_{23}j_{33} & j_{21}j_{32}+j_{22}j_{31} & j_{22}j_{33}+j_{23}j_{32} & j_{21}j_{33}+j_{23}j_{31} \\ j_{31}j_{11} & j_{32}j_{12} & j_{33}j_{13} & j_{31}j_{12}+j_{32}j_{11} & j_{32}j_{13}+j_{33}j_{12} & j_{31}j_{13}+j_{33}j_{11} \end{bmatrix} = [C_2]^{-1}$$

(1.48)

Matrix $[C_1]$ defined by (1.46a) is:

$$[C_1] = \begin{bmatrix} \frac{\partial}{\partial \xi}\langle J_{11} \quad J_{12} \quad J_{13} \rangle \\ \frac{\partial}{\partial \eta}\langle J_{21} \quad J_{22} \quad J_{23} \rangle \\ \frac{\partial}{\partial \zeta}\langle J_{31} \quad J_{32} \quad J_{33} \rangle \\ \frac{1}{2}\left(\frac{\partial}{\partial \eta}\langle J_{11} \quad J_{12} \quad J_{13} \rangle + \frac{\partial}{\partial \xi}\langle J_{21} \quad J_{22} \quad J_{23} \rangle\right) \\ \frac{1}{2}\left(\frac{\partial}{\partial \zeta}\langle J_{21} \quad J_{22} \quad J_{23} \rangle + \frac{\partial}{\partial \eta}\langle J_{31} \quad J_{32} \quad J_{33} \rangle\right) \\ \frac{1}{2}\left(\frac{\partial}{\partial \xi}\langle J_{31} \quad J_{32} \quad J_{33} \rangle + \frac{\partial}{\partial \xi}\langle J_{11} \quad J_{12} \quad J_{13} \rangle\right) \end{bmatrix}$$

(1.49)

Matrix $[C_2]$ is obtained from $[T_2]$ where the terms of $[j]$ are replaced by the corresponding terms of $[J]$.

In two dimensions we find:

$$[T_2] = \begin{bmatrix} j_{11}^2 & j_{12}^2 & 2j_{11}j_{12} \\ j_{21}^2 & j_{22}^2 & 2j_{21}j_{22} \\ j_{11}j_{21} & j_{12}j_{22} & j_{11}j_{22}+j_{12}j_{21} \end{bmatrix}$$

(1.50a)

$$[C_1] = \begin{bmatrix} \frac{\partial}{\partial \xi}\langle J_{11} \quad J_{12} \rangle \\ \frac{\partial}{\partial \eta}\langle J_{21} \quad J_{22} \rangle \\ \frac{1}{2}\left(\frac{\partial}{\partial \eta}\langle J_{11} \quad J_{12} \rangle + \frac{\partial}{\partial \xi}\langle J_{21} \quad J_{22} \rangle\right) \end{bmatrix}$$

(1.50b)

1.5.4 Ill-Conditioned Jacobian Matrix

For severely distorted elements, the determinant of $[J]$ may become zero or negative at certain points of the element. The sign of $[J]$ should always be checked and flagged if it changes at some points.

Example 1.19 *Ill-conditioning for a quadratic isoparametric line element*

Consider the following three-noded line element:

$$\langle N \rangle = \langle \bar{N} \rangle = \langle \tfrac{1}{2}\xi(\xi-1) \ -(\xi-1)(\xi+1) \ \tfrac{1}{2}\xi(\xi+1) \rangle$$

$$[J] = \left\langle \frac{2\xi-1}{2} \ -2\xi \ \frac{2\xi+1}{2} \right\rangle \begin{Bmatrix} x_1 \\ x_2 \\ x_3 \end{Bmatrix}$$

Let $x_1 = 0$ and $x_3 = L$, then

$$\det(J) = (\xi + 0.5)l - 2\xi x_2$$

For any choice of x_2 this determinant is null if $\xi_0 = 0.5l/2x_2 - l$ and since for an interior point $-1 \le \xi_0 \le 1$ we get the inadmissible interior regions. One may solve this relation for different values of x_2/l which lead $k - 1 \le \xi_0 \le 1$, thus identifying the range of inadmissible positioning of the interior node x_2 as shown below ($\det J \le 0$):

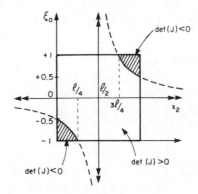

Admissible values of x_2 for which $[J]$ is not singular are:

$$\frac{l}{4} < x_2 < \frac{3l}{4}$$

To avoid such difficulties in eight-noded elements (paragraph 2.4.3.2) Zienkiewicz [2, page 186] proposes the following conditions:

- the four interior angles α are less than 180°;
- middle nodes must lie in the middle third portion of each side.

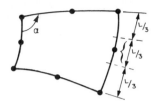

1.6 Computations of Functions N, their Derivatives and the Jacobian Matrix

1.6.1 *Generalities*

In the subsequent chapters we need approximations of

$$u(\mathbf{x}), \frac{\partial u(\mathbf{x})}{\partial x}, \frac{\partial u(\mathbf{x})}{\partial y} \text{etc.}$$

These approximations are used to evaluate integrals over the volume of an element (see Example 1.8 and Chapter 4):

$$k = \int_{V^r} f\left(u(\mathbf{x}), \frac{\partial u(\mathbf{x})}{\partial x} \ldots \right) dV^e. \tag{1.51a}$$

Such integrals are replaced by integrals in the space of the reference element with the help of (1.37) and (1.44).

$$k = \int_{V^r} f\left(u(\xi), \frac{\partial u(\xi)}{\partial \xi}, \ldots, [j]\right) \det(J(\xi)) dV^r. \tag{1.51b}$$

Furthermore, these integrals are computed by numerical techniques to be examined in paragraph 5.1

$$k \approx \sum_r W_r f\left(u(\xi_r), \frac{\partial u(\xi_r)}{\partial \xi}, \ldots, [j(\xi_r)]\right) \det(J(\xi_r)) \tag{1.51c}$$

where ξ_r are the coordinates of a set of integration points (for example Gauss points); W_r are weighting factors of the numerical integration formula

$$u(\xi_r) = \langle N(\xi_r) \rangle \{u_n\}$$

$$\frac{\partial u(\xi_r)}{\partial \xi} = \left\langle \frac{\partial N(\xi_r)}{\partial \xi} \right\rangle \{u_n\}$$

$[j(\xi_r)]$ and $\det(J(\xi_r))$ are the inverse of the Jacobian matrix and its determinant evaluated at point ξ_r.

Note that the expressions for $\langle N(\xi_r) \rangle$ and $\langle \partial N(\xi_r)/\partial \xi \rangle$ are independent of the real geometry of the element. All computations are in the space of the reference element. It is then expeditious to evaluate $N(\xi_r)$ and its derivatives only once for

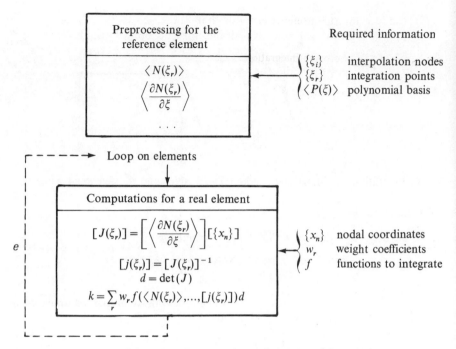

Figure 1.3 Computations of N, $[j]$, and k.

each type of element. The Jacobian matrix and its determinant, however, depend on the geometrical coordinates of each element and must therefore be calculated for each element. If the element is not isoparametric one must remember that the geometrical interpolation functions $\bar{N}$ are not the same as the interpolation functions N used for u. Figure 1.3 describes the sequence of computations.

1.6.2 Explicit Formulas for N

Closed form formulas of linear quadratic and cubic elements for lines, surfaces and volumes are well known. With such formulas, the computations of N, $\partial N/\partial \xi$, $[j]$ and det $[J]$ can all be written explicitly. Figure 1.4 shows a FORTRAN program for isoparametric quadrilateral of four or eight nodes. The preparatory algebraic task is massive and prone to errors; for a 32 nodal point brick one must construct 128 cubic and quadratic functions.

In the following paragraph we describe an alternative numerical method applicable to any element.

1.6.3 Automatic Construction of Interpolation Function N

Four subroutines using expression (1.24) are proposed for the computations of N and its derivatives. A testing program is also included in Figures 1.5 to 1.8. The FORTRAN variables are described in Chapter 6 (Figures 6.6 to 6.9).

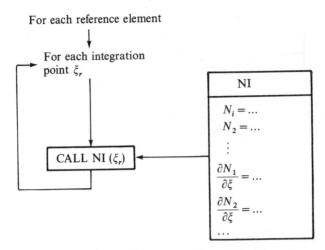

Figure 1.4(a) Explicit computations of functions N.

```
      SUBROUTINE NIQ(VKPG,IPG,VNI)                                   NIQ   1
C================================================================NIQ   2
C     TO EVALUATE INTERPOLATION FUNCTIONS N AND THEIR DERIVATIVES    NIQ   3
C     D(N)/D(KSI) D(N)/D(ETA) BY EXPLICIT FORMULATION                NIQ   4
C     (QUADRILATERAL ELEMENTS WITH 4 OR 8 NODES)                     NIQ   5
C     INPUT                                                          NIQ   6
C        VKPG      COORDINATES OF INTEGRATION POINTS                 NIQ   7
C        IPG       NUMBER OF INTEGRATION POINTS                      NIQ   8
C     OUTPUT                                                         NIQ   9
C        VNI       FUNCTIONS N AND THEIR DERIVATIVES                 NIQ  10
C================================================================NIQ  11
      IMPLICIT REAL*8(A-H,O-Z)                                       NIQ  12
      COMMON/COOR/NDIM                                               NIQ  13
      COMMON/RGDT/IEL,ITPE,ITPE1,IGRE,IDLE,ICE,IPRNE,IPREE,INEL,IDEG NIQ  14
      COMMON/ES/M,MR,MP                                              NIQ  15
      DIMENSION VKPG(1),VNI(1)                                       NIQ  16
      DATA P25/.25D0/,P5/.5D0/,UN/1.D0/,DE/2.D0/                     NIQ  17
C------- LOOP OVER INTEGRATION POINTS                                NIQ  18
      II=0                                                           NIQ  19
      I1=0                                                           NIQ  20
      DO 10 IG=1,IPG                                                 NIQ  21
      XG=VKPG(I1+1)                                                  NIQ  22
      YG=VKPG(I1+2)                                                  NIQ  23
C-------  4 NODES ELEMENTS        NIQ   24
      IF(INEL.NE.4) GO TO 20                                         NIQ  25
C-------  FUNCTIONS N     NIQ   26
      VNI(II+1)=P25*(UN-XG)*(UN-YG)                                  NIQ  27
      VNI(II+2)=P25*(UN+XG)*(UN-YG)                                  NIQ  28
      VNI(II+3)=P25*(UN+XG)*(UN+YG)                                  NIQ  29
      VNI(II+4)=P25*(UN-XG)*(UN+YG)                                  NIQ  30
C-------  KSI DERIVATIVES       NIQ   31
      VNI(II+5)=-P25*(UN-YG)                                         NIQ  32
      VNI(II+6)=P25*(UN-YG)                                          NIQ  33
      VNI(II+7)=P25*(UN+YG)                                          NIQ  34
      VNI(II+8)=-P25*(UN+YG)                                         NIQ  35
C-------  ETA DERIVATIVES       NIQ   36
      VNI(II+9)=-P25*(UN-XG)                                         NIQ  37
      VNI(II+10)=-P25*(UN+XG)                                        NIQ  38
      VNI(II+11)=P25*(UN+XG)                                         NIQ  39
```

Figure 1.4(b) (*Contd.*)

```
      VNI(II+12)=P25*(UN-XG)                                    NIQ  40
      II=II+12                                                  NIQ  41
      I1=I1+2                                                   NIQ  42
      GO TO 10                                                  NIQ  43
C------- 8 NODES ELEMENTS        NIQ   44
20    IF(INEL.NE.8) GO TO 100                                   NIQ  45
C------- FUNCTIONS N                                            NIQ  46
      VNI(II+1)=P25*(UN-XG)*(UN-YG)*(-XG-YG-UN)                 NIQ  47
      VNI(II+2)=P5*(UN-(XG*XG))*(UN-YG)                         NIQ  48
      VNI(II+3)=P25*(UN+XG)*(UN-YG)*(XG-YG-UN)                  NIQ  49
      VNI(II+4)=P5*(UN+XG)*(UN-(YG*YG))                         NIQ  50
      VNI(II+5)=P25*(UN+XG)*(UN+YG)*(XG+YG-UN)                  NIQ  51
      VNI(II+6)=P5*(UN-(XG*XG))*(UN+YG)                         NIQ  52
      VNI(II+7)=P25*(UN-XG)*(UN+YG)*(YG-XG-UN)                  NIQ  53
      VNI(II+8)=P5*(UN-XG)*(UN-(YG*YG))                         NIQ  54
C------- KSI DERIVATIVES                                        NIQ  55
      II=II+8                                                   NIQ  56
      VNI(II+1)=P25*((DE*XG)+YG)*(UN-YG)                        NIQ  57
      VNI(II+2)=-XG*(UN-YG)                                     NIQ  58
      VNI(II+3)=P25*((DE*XG)-YG)*(UN-YG)                        NIQ  59
      VNI(II+4)=P5*(UN-(YG*YG))                                 NIQ  60
      VNI(II+5)=P25*((DE*XG)+YG)*(UN+YG)                        NIQ  61
      VNI(II+6)=-XG*(UN+YG)                                     NIQ  62
      VNI(II+7)=-P25*(YG-(DE*XG))*(UN+YG)                       NIQ  63
      VNI(II+8)=-P5*(UN-(YG*YG))                                NIQ  64
C------- ETA DERIVATIVES         NIQ   65
      II=II+8                                                   NIQ  66
      VNI(II+1)=P25*(UN-XG)*(XG+(DE*YG))                        NIQ  67
      VNI(II+2)=-P5*(UN-(XG*XG))                                NIQ  68
      VNI(II+3)=-P25*(UN+XG)*(XG-(DE*YG))                       NIQ  69
      VNI(II+4)=-YG*(UN+XG)                                     NIQ  70
      VNI(II+5)=P25*(UN+XG)*(XG+(DE*YG))                        NIQ  71
      VNI(II+6)=P5*(UN-(XG*XG))                                 NIQ  72
      VNI(II+7)=P25*(UN-XG)*((DE*YG)-XG)                        NIQ  73
      VNI(II+8)=-YG*(UN-XG)                                     NIQ  74
      II=II+8                                                   NIQ  75
      I1=I1+2                                                   NIQ  76
      GO TO 10                                                  NIQ  77
100   WRITE(MP,2000)                                            NIQ  78
2000  FORMAT(' ***, ERROR, FUNCTION N NOT DEFINED FOR ELEMENT TYPE')  NIQ  79
10    CONTINUE                                                  NIQ  80
      RETURN                                                    NIQ  81
      END                                                       NIQ  82
```

Figure 1.4(b) Subroutine NIQ for explicit computation of functions N and their derivatives for a quadrilateral element with four to eight nodes.

Remarks

— If we use explicit formulas for $\langle N \rangle$ in BASEP, matrix $[P_n]$ becomes a unit matrix since $\langle P \rangle \equiv \langle N \rangle$

— For commercial codes it would be more efficient to replace all computations of $\langle N \rangle$ and its derivatives by an interpolation function data base. A simple reading from the data base would replace all shape function calculations. The data base could be enlarged as needed when new elements become necessary.

— If the element is not isoparametric the computations must be performed independently for $\bar{N}$ and N.

Figures	Description
1.5	Flow of subroutines
1.6(a)	Subroutine PNINV for the computation of P_n^{-1}
1.6(b)	*Subroutine NI* for the computation of functions N and their derivatives at a point
1.6(c)	Subroutine BASEP for the computation of the polynomial basis P and its derivatives at a point
1.6(d)	Subroutine INVERS for the inversion of a non-symmetrical matrix
1.7(a)	Demonstration program for the eight-noded quadrilateral element described in paragraph 2.4.3.2
1.7(b)	Result of the demonstration program
1.8(a)	Modifications of the demonstration program and of PNINV for the four-noded tetrahedronal element of Hermite type described in paragraph 2.5.4
1.8(b)	Results of the previous demonstration program

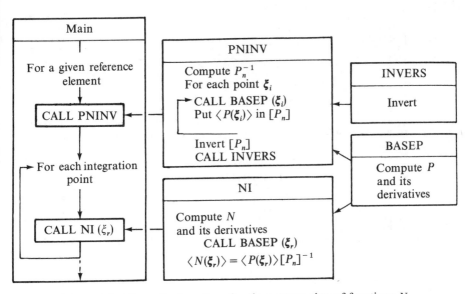

Figure 1.5 Flow of subroutines for the computation of functions N.

1.6.4 *Computation of the Jacobian Matrix and its Derivatives*

In Figure 1.9 a subroutine named JACOB is listed. The subroutine computes the Jacobian matrix, its inverse and its determinant for one-, two-, and three-dimensional elements.

Subroutine DNIDX listed in Figure 1.10 computes the derivatives of N using formula (1.37).

```
      SUBROUTINE PNINV(VKSI,KEXP,VP,K1,VPN)                     PNIN   1
C=====================================================================PNIN   2
C     EVALUATE   THE PN-INVERSE MATRIX WHICH                    PNIN   3
C     CONTAINS THE COEFFICIENTS OF FUNCTIONS N                  PNIN   4
C        INPUT       VKSI,KEXP,INEL,IDLE,ITPE,M,MP              PNIN   5
C        WORKSPACE   VP,K1                                      PNIN   6
C        OUTPUT      VPN                                        PNIN   7
C=====================================================================PNIN   8
      IMPLICIT REAL*8(A-H,O-Z)                                  PNIN   9
      COMMON/COOR/NDIM                                          PNIN  10
      COMMON/RGDT/IEL,ITPE,ITPE1,IGRE,IDLE,ICE,IPRNE,IPREE,INEL,IDEG,IPGPNIN 11
      COMMON/ES/M,MR,MP                                         PNIN  12
      DIMENSION VKSI(1),KEXP(1),VP(1),K1(1),VPN(1),KDER(3)      PNIN  13
      DATA ZERO/0.D0/,KDER/3*0/                                 PNIN  14
C--------------------------------------------------------------------PNIN  15
C                                                               PNIN  16
C....... FORM PN MATRIX (FOR ANY LAGRANGE TYPE ELEMENT)         PNIN  17
C                                                               PNIN  18
C                                                               PNIN  19
      IO=1                                                      PNIN  20
      I1=1                                                      PNIN  21
      DO 20 IN=1,INEL                                           PNIN  22
      CALL BASEP(VKSI(I1),KEXP,KDER,VP)                         PNIN  23
      I2=IO                                                     PNIN  24
      DO 10 IJ=1,INEL                                           PNIN  25
      VPN(I2)=VP(IJ)                                            PNIN  26
   10 I2=I2+INEL                                                PNIN  27
      IO=IO+1                                                   PNIN  28
   20 I1=I1+NDIM                                                PNIN  29
C                                                               PNIN  30
C....... END OF PN FORMATION                                    PNIN  31
C                                                               PNIN  32
C------- PRINT THE PN MATRIX                                    PNIN  33
      IF(M.LT.4) GO TO 40                                       PNIN  34
      WRITE(MP,2000)                                            PNIN  35
 2000 FORMAT(/' PN MATRIX'/)                                    PNIN  36
      ID=(INEL-1)*INEL                                          PNIN  37
      DO 30 IO=1,INEL                                           PNIN  38
      I1=IO+ID                                                  PNIN  39
   30 WRITE(MP,2010) (VPN(IJ),IJ=IO,I1,INEL)                    PNIN  40
 2010 FORMAT(1X,10E13.5/(14X,9E13.5))                           PNIN  41
C------- INVERSE THE PN MATRIX                                  PNIN  42
   40 CALL INVERS(VPN,INEL,INEL,K1,DET)                         PNIN  43
      IF(DET.NE.ZERO) GO TO 50                                  PNIN  44
      WRITE(MP,2020) ITPE                                       PNIN  45
 2020 FORMAT(' *** ERROR, PN SINGULAR, ELEMENT TYPE:',I3)       PNIN  46
      STOP                                                      PNIN  47
C------- PRINT THE PN-INVERSE MATRIX                            PNIN  48
   50 IF(M.LT.4) GO TO 70                                       PNIN  49
      WRITE(MP,2030)                                            PNIN  50
 2030 FORMAT(/' PN-INVERSE MATRIX'/)                            PNIN  51
      DO 60 IO=1,INEL                                           PNIN  52
      I1=IO+ID                                                  PNIN  53
   60 WRITE(MP,2010) (VPN(IJ),IJ=IO,I1,INEL)                    PNIN  54
   70 RETURN                                                    PNIN  55
      END                                                       PNIN  56
```

Figure 1.6(a) Subroutine PNINV for the computation of $[P_n]^{-1}$, used in program MEF, Chapter 6.

```
      SUBROUTINE NI(VKSI,KEXP,KDER,VP,VPN,VNI)                    NI    1
C======================================================================NI    2
C     TO EVALUATE FUNCTIONS N OR THEIR DERIVATIVES                NI    3
C     AT POINT VKSI ON THE REFERENCE ELEMENT                      NI    4
C        INPUT    VKSI,KEXP,KDER,VP,VPN,IDLE,M,MP                 NI    5
C        OUTPUT   VNI                                             NI    6
C======================================================================NI    7
      IMPLICIT REAL*8(A-H,O-Z)                                    NI    8
      COMMON/COOR/NDIM                                            NI    9
      COMMON/RGDT/IEL,ITPE,ITPE1,IGRE,IDLE,ICE,IPRNE,IPREE,INEL,IDEG,IPG NI  10
      COMMON/ES/M,MR,MP                                           NI   11
      DIMENSION VKSI(1),KEXP(1),KDER(1),VP(1),VPN(1),VNI(1)       NI   12
      DATA ZERO/0.D0/                                             NI   13
C----------------------------------------------------------------------NI   14
C------- COMPUTE THE POLYNOMIAL BASIS AT POINT VKSI               NI   15
      CALL BASEP(VKSI,KEXP,KDER,VP)                               NI   16
C-------   P*(PN-INVERSE) PRODUCT                                 NI   17
      I0=1                                                        NI   18
      DO 20 IJ=1,INEL                                             NI   19
      I1=I0                                                       NI   20
      C=ZERO                                                      NI   21
      DO 10 II=1,INEL                                             NI   22
      C=C+VP(II)*VPN(I1)                                          NI   23
   10 I1=I1+1                                                     NI   24
      VNI(IJ)=C                                                   NI   25
   20 I0=I0+INEL                                                  NI   26
C-------   PRINT FUNCTIONS N                                      NI   27
      IF(M.LT.3) GO TO 30                                         NI   28
      WRITE(MP,2000) (KDER(I),I=1,NDIM)                           NI   29
 2000 FORMAT(/' DERIVATIVE OF N WITH ORDER ',3I2)                 NI   30
      WRITE(MP,2010) (VKSI(I),I=1,NDIM)                           NI   31
 2010 FORMAT(14X,'AT POINT ',3E13.5)                              NI   32
      WRITE(MP,2020) (VNI(I),I=1,INEL)                            NI   33
 2020 FORMAT(/(1X,10E13.5))                                       NI   34
   30 RETURN                                                      NI   35
      END                                                         NI   36
```

Figure 1.6(b) Subroutine NI for the computation of functions N and their derivatives, used in program MEF, Chapter 6.

```
      SUBROUTINE BASEP(VKSI,KEXP,KDER,VP)                              BASE   1
C=====================================================================BASE   2
C       TO EVALUATE THE POLYNOMIAL BASIS AND ITS DERIVATIVES AT POINT VKSIBASE 3
C       INPUT    VKSI,KEXP,KDER,IDLE,IDEG,NDIM,M,MP                    BASE   4
C       OUTPUT   VP                                                    BASE   5
C=====================================================================BASE   6
      IMPLICIT REAL*8(A-H,O-Z)                                         BASE   7
      COMMON/COOR/NDIM                                                 BASE   8
      COMMON/RGDT/IEL,ITPE,ITPE1,IGRE,IDLE,ICE,IPRNE,IPREE,INEL,IDEG,IPGBASE 9
      COMMON/ES/M,MR,MP                                                BASE  10
      DIMENSION VKSI(1),KEXP(1),KDER(1),VP(1)                          BASE  11
      DIMENSION PUISS(3,10)                                            BASE  12
      DATA ZERO/0.D0/,UN/1.D0/                                         BASE  13
C---------------------------------------------------------------------BASE  14
C------ FORM SUCCESSIVE POWERS OF KSI,ETA,DZETA                        BASE  15
      DO 10 I=1,NDIM                                                   BASE  16
      PUISS(I,1)=UN                                                    BASE  17
      DO 10 ID=1,IDEG                                                  BASE  18
10    PUISS(I,ID+1)=PUISS(I,ID)*VKSI(I)                                BASE  19
C------ DERIVATIVES OF ORDER KDER WITH RESPECT TO KSI,ETA,DZETA        BASE  20
      DO 50 IDL=1,INEL                                                 BASE  21
      C1=UN                                                            BASE  22
      IO=(IDL-1)*NDIM                                                  BASE  23
      DO 30 I=1,NDIM                                                   BASE  24
      IDR=KDER(I)                                                      BASE  25
      IO=IO+1                                                          BASE  26
      IXP=KEXP(IO)+1                                                   BASE  27
      J=IXP-IDR                                                        BASE  28
      IF(J.LE.0) GO TO 40                                              BASE  29
      IF(IDR.LE.0) GO TO 30                                            BASE  30
      DO 20 ID=1,IDR                                                   BASE  31
20    C1=C1*(IXP-ID)                                                   BASE  32
30    C1=C1*PUISS(I,J)                                                 BASE  33
      GO TO 50                                                         BASE  34
40    C1=ZERO                                                          BASE  35
50    VP(IDL)=C1                                                       BASE  36
C------ PRINT POLYNOMIAL BASIS                                         BASE  37
      IF(M.LT.4) GO TO 60                                              BASE  38
      WRITE(MP,2000) (KDER(I),I=1,NDIM)                                BASE  39
2000  FORMAT(/' POLYNOMIAL BASIS, DERIVATIVE OF ORDER ',3I2)           BASE  40
      WRITE(MP,2010) (VKSI(I),I=1,NDIM)                                BASE  41
2010  FORMAT(19X,'AT POINT ',3E13.5)                                   BASE  42
      WRITE(MP,2020) (VP(I),I=1,INEL)                                  BASE  43
2020  FORMAT(/(1X,10E12.5))                                            BASE  44
60    RETURN                                                           BASE  45
      END                                                              BASE  46
```

Figure 1.6(c) Subroutine BASEP for the computation of the polynomial basis, used in program MEF, Chapter 6

```fortran
      SUBROUTINE INVERS(VP,N,IVP,K,DET)                          INVE  1
C================================================================INVE  2
C     TO INVERT A NON-SYMMETRIC MATRIX WITH SEARCH OF A          INVE  3
C     NON-ZERO PIVOT IN A COLUMN                                 INVE  4
C     INPUT                                                      INVE  5
C        VP        MATRIX TO BE INVERTED                         INVE  6
C        N         ORDER OF THE MATRIX                           INVE  7
C        IVP       DIMENSION OF THE MATRIX IN THE CALLING PROGRAM INVE 8
C        K         INTEGER WORKING ARRAY WITH LENGTH N           INVE  9
C     OUTPUT                                                     INVE 10
C        VP        INVERSE MATRIX                                INVE 11
C        DET       DETERMINANT                                   INVE 12
C================================================================INVE 13
      IMPLICIT REAL*8(A-H,O-Z)                                   INVE 14
      DIMENSION VP(IVP,IVP),K(N)                                 INVE 15
      DATA ZERO/0.D0/,UN/1.D0/,EPS/1.D-13/                       INVE 16
      ABS(X)=DABS(X)                                             INVE 17
C----------------------------------------------------------------INVE 18
      DET=UN                                                     INVE 19
      DO 5 I=1,N                                                 INVE 20
5     K(I)=I                                                     INVE 21
C------- START INVERSION                                         INVE 22
      DO 80 II=1,N                                               INVE 23
C------- SEARCH FOR NON-ZERO PIVOT IN COLUMN II                  INVE 24
      DO 10 I=II,N                                               INVE 25
      PIV=VP(I,II)                                               INVE 26
      IF(ABS(PIV).GT.EPS) GO TO 20                               INVE 27
10    CONTINUE                                                   INVE 28
      DET=ZERO                                                   INVE 29
      RETURN                                                     INVE 30
C------- EXCHANGE LINES II AND I                                 INVE 31
20    DET=DET*PIV                                                INVE 32
      IF(I.EQ.II) GO TO 40                                       INVE 33
      I1=K(II)                                                   INVE 34
      K(II)=K(I)                                                 INVE 35
      K(I)=I1                                                    INVE 36
      DO 30 J=1,N                                                INVE 37
      C=VP(I,J)                                                  INVE 38
      VP(I,J)=VP(II,J)                                           INVE 39
30    VP(II,J)=C                                                 INVE 40
      DET=-DET                                                   INVE 41
C------- NORMALIZE PIVOT LINE                                    INVE 42
40    C=UN/PIV                                                   INVE 43
      VP(II,II)=UN                                               INVE 44
      DO 50 J=1,N                                                INVE 45
50    VP(II,J)=VP(II,J)*C                                        INVE 46
```

```
C------- ELIMINATION                              INVE  47
        DO 70 I=1,N                               INVE  48
        IF(I.EQ.II) GO TO 70                      INVE  49
        C=VP(I,II)                                INVE  50
        VP(I,II)=ZERO                             INVE  51
        DO 60 J=1,N                               INVE  52
   60   VP(I,J)=VP(I,J)-C*VP(II,J)                INVE  53
   70   CONTINUE                                  INVE  54
   80   CONTINUE                                  INVE  55
C------ REORDER THE COLUMNS OF INVERSE MATRIX     INVE  56
        DO 120 J=1,N                              INVE  57
C------- FIND J1 SUCH THAT K(J1)=J                INVE  58
        DO 90 J1=J,N                              INVE  59
        JJ=K(J1)                                  INVE  60
        IF(JJ.EQ.J) GO TO 100                     INVE  61
   90   CONTINUE                                  INVE  62
  100   IF(J.EQ.J1) GO TO 120                     INVE  63
C------ EXCHANGE COLUMNS J AND J1                 INVE  64
        K(J1)=K(J)                                INVE  65
        DO 110 I=1,N                              INVE  66
        C=VP(I,J)                                 INVE  67
        VP(I,J)=VP(I,J1)                          INVE  68
  110   VP(I,J1)=C                                INVE  69
  120   CONTINUE                                  INVE  70
        RETURN                                    INVE  71
        END                                       INVE  72
```

Figure 1.6(d) Subroutine INVERS for the inversion of a non-symmetrical matrix, used in program MEF, Chapter 6.

```
C=====================================================================TENI   1
C       TEST PROGRAM :                                               TENI   2
C       TO COMPUTE FUNCTIONS N AND THEIR DERIVATIVES                 TENI   3
C=====================================================================TENI   4
        IMPLICIT REAL*8(A-H,O-Z)                                     TENI   5
        COMMON/COOR/NDIM                                             TENI   6
        COMMON/ES/M,MR,MP                                            TENI   7
        COMMON/RGDT/IEL,ITPE,ITPE1,IGRE,IDLE,ICE,IPRNE,IPREE,INEL,IDEG,IPGTENI 8
C                                                                    TENI   9
C......  DEFINE THE 8 NODES QUADRILATERAL REFERENCE ELEMENT          TENI  10
C                                                                    TENI  11
C       DIMENSIONS OF ARRAYS REQUIRED FOR COMPUTATION OF N           TENI  12
C               VKSI(NDIM*INEL), KEXP(NDIM*INEL), VP(INEL),          TENI  13
        DIMENSION VKSI(      16), KEXP(      16), VP(    8),         TENI  14
C               VKPG(NDIM*IPG), KDER(NDIM), VNI(INEL*IPG*INI),       TENI  15
     1          VKPG(     8), KDER(   2), VNI(         64),          TENI  16
C               VPN(IDLE*IDLE), K1(IDLE)                             TENI  17
     2          VPN(    64), K1(    8)                               TENI  18
C       COORDINATES OF INTERPOLATION NODES ON REFERENCE ELEMENT      TENI  19
        DATA VKSI/-1.D0,-1.D0, +0.D0,-1.D0, +1.D0,-1.D0, +1.D0,+0.D0,TENI  20
     1            +1.D0,+1.D0, +0.D0,+1.D0, -1.D0,+1.D0, -1.D0,+0.D0/TENI  21
C       EXPONENTS OF EACH TERM OF THE POLYNOMIAL BASIS               TENI  22
        DATA KEXP/0,0, 1,0, 0,1, 2,0, 1,1, 0,2, 2,1, 1,2/            TENI  23
C       COORDINATES OF INTEGRATION POINTS                            TENI  24
        DATA VKPG/-0.577350269189626D0,-0.577350269189626D0,         TENI  25
     1            +0.577350269189626D0,-0.577350269189626D0,         TENI  26
     2            +0.577350269189626D0,+0.577350269189626D0,         TENI  27
     3            -0.577350269189626D0,+0.577350269189626D0/         TENI  28
C       ELEMENT PARAMETERS                                           TENI  29
        INEL=8                                                       TENI  30
        IDLE=8                                                       TENI  31
        IDEG=2                                                       TENI  32
        IPG=4                                                        TENI  33
        NDIM=2                                                       TENI  34
C                                                                    TENI  35
C......  END OF ELEMENT DEFINITION                                   TENI  36
C                                                                    TENI  37
C-------  INITIALIZE COMMON VARIABLES                                TENI  38
        M=0                                                          TENI  39
        MR=5                                                         TENI  40
        MP=6                                                         TENI  41
C-------  COMPUTE PN INVERSE                                         TENI  42
        CALL PNINV(VKSI,KEXP,VP,K1,VPN)                              TENI  43
C-------  COMPUTE FUNCTION N AT EACH INTEGRATION POINT               TENI  44
        I0=1                                                         TENI  45
        I1=1                                                         TENI  46
        DO 10 I=1,3                                                  TENI  47
 10     KDER(I)=0                                                    TENI  48
        DO 20 IP=1,IPG                                               TENI  49
        CALL NI(VKPG(I0),KEXP,KDER,VP,VPN,VNI(I1))                   TENI  50
        I0=I0+NDIM                                                   TENI  51
 20     I1=I1+IDLE                                                   TENI  52
C-------  COMPUTE D(N)/D(KSI) DERIVATIVE AT EACH INTEGRATION POINT   TENI  53
        I0=1                                                         TENI  54
        KDER(1)=1                                                    TENI  55
        DO 30 IP=1,IPG                                               TENI  56
        CALL NI(VKPG(I0),KEXP,KDER,VP,VPN,VNI(I1))                   TENI  57
        I0=I0+NDIM                                                   TENI  58
 30     I1=I1+IDLE                                                   TENI  59
C-------  PRINT RESULTS                                              TENI  60
        I0=1                                                         TENI  61
        I1=IDLE                                                      TENI  62
        I2=1                                                         TENI  63
        I3=NDIM                                                      TENI  64
        WRITE(MP,2000)                                               TENI  65
```

```
2000      FORMAT(/' FUNCTIONS N AT INTEGRATION POINTS'/)              TENI  66
          DO 40 IP=1,IPG                                              TENI  67
          WRITE(MP,2010) (VKPG(I),I=I2,I3)                            TENI  68
2010      FORMAT(3X,'COORDINATES ',3F10.5)                            TENI  69
          WRITE(MP,2015) (VNI(I),I=I0,I1)                             TENI  70
2015      FORMAT(13X,'N ',5F10.5/(15X,5F10.5))                        TENI  71
          I2=I2+NDIM                                                  TENI  72
          I3=I3+NDIM                                                  TENI  73
          I0=I0+IDLE                                                  TENI  74
40        I1=I1+IDLE                                                  TENI  75
          I2=1                                                        TENI  76
          I3=NDIM                                                     TENI  77
          WRITE(MP,2020)                                              TENI  78
2020      FORMAT(/' FUNCTIONS D(N)/D(KSI) AT INTEGRATION POINTS'/)    TENI  79
          DO 50 IP=1,IPG                                              TENI  80
          WRITE(MP,2025) (VKPG(I),I=I2,I3)                            TENI  81
2025      FORMAT(3X,'COORDINATES ',3F10.5)                            TENI  82
          WRITE(MP,2030) (VNI(I),I=I0,I1)                             TENI  83
2030      FORMAT(3X,'D(N)/D(KSI) ',5F10.5/(15X,5F10.5))               TENI  84
          I2=I2+NDIM                                                  TENI  85
          I3=I3+NDIM                                                  TENI  86
          I0=I0+IDLE                                                  TENI  87
50        I1=I1+IDLE                                                  TENI  88
          STOP                                                        TENI  89
          END                                                         TENI  90
```

Figure 1.7(a) Demonstration program corresponding to the eight-node quadrilateral element of paragraph 2.4.3.2.

```
FUNCTIONS N AT INTEGRATION POINTS

   COORDINATES   -0.57735   -0.57735
             N    0.09623    0.52578   -0.16667    0.14088   -0.09623
                  0.14088   -0.16667    0.52578
   COORDINATES    0.57735   -0.57735
             N   -0.16667    0.52578    0.09623    0.52578   -0.16667
                  0.14088   -0.09623    0.14088
   COORDINATES    0.57735    0.57735
             N   -0.09623    0.14088   -0.16667    0.52578    0.09623
                  0.52578   -0.16667    0.14088
   COORDINATES   -0.57735    0.57735
             N   -0.16667    0.14088   -0.09623    0.14088   -0.16667
                  0.52578    0.09623    0.52578

FUNCTIONS D(N)/D(KSI) AT INTEGRATION POINTS

   COORDINATES   -0.57735   -0.57735
    D(N)/D(KSI)  -0.68301    0.91068   -0.22767    0.33333   -0.18301
                  0.24402   -0.06100   -0.33333
   COORDINATES    0.57735   -0.57735
    D(N)/D(KSI)   0.22767   -0.91068    0.68301    0.33333    0.06100
                 -0.24402    0.18301   -0.33333
   COORDINATES    0.57735    0.57735
    D(N)/D(KSI)   0.18301   -0.24402    0.06100    0.33333    0.68301
                 -0.91068    0.22767   -0.33333
   COORDINATES   -0.57735    0.57735
    D(N)/D(KSI)  -0.06100    0.24402   -0.18301    0.33333   -0.22767
                  0.91068   -0.68301   -0.33333
```

Figure 1.7(b) Results of the program of Figure 1.7(a).

Modification of the main Program

```
C....... DEFINE TETRAHEDRIC REFERENCE ELEMENT WITH 16 D.O.F.        TENI 10
C                                                                    TENI ..
C        DIMENSIONS OF ARRAYS REQUIRED TO COMPUTE N                  TENI ..
         DIMENSION VKSI( 12),    KEXP( 48),      VP( 16),            TENI ..
        1           VKPG( 15),   KDER(  3),     VNI( 160),           TENI ..
        2           VPN( 256),     K1( 16)                           TENI ..
C        COORDINATES OF INTERPOLATION NODES ON THE REFERENCE ELEMENT TENI ..
         DATA VKSI/0.D0,0.D0,0.D0,  1.D0,0.D0,0.D0,  0.D0,1.D0,0.D0, TENI ..
        1          0.D0,0.D0,1.D0/                                   TENI ..
C        EXPONENTS OF EACH TERM IN THE POLYNOMIAL BASIS              TENI ..
         DATA KEXP/0,0,0,  1,0,0,  0,1,0,  0,0,1,  2,0,0,  1,1,0,  0,2,0, TENI ..
        1         0,1,1,  0,0,2,  1,0,1,  3,0,0,  1,2,0,  0,3,0,  0,1,2, TENI ..
        2         0,0,3,  2,0,1/                                     TENI ..
C        COORDINATES OF INTEGRATION POINTS                           TENI ..
         DATA VKPG/                                                  TENI ..
        1 0.25            D0,0.25            D0,0.25            D0, TENI ..
        2 0.166666666666666D0,0.166666666666666D0,0.166666666666666D0, TENI ..
        3 0.333333333333333D0,0.166666666666666D0,0.166666666666666D0, TENI ..
        4 0.166666666666666D0,0.333333333333333D0,0.166666666666666D0, TENI ..
        5 0.166666666666666D0,0.166666666666666D0,0.333333333333333D0/ TENI ..
C        ELEMENT PARAMETERS                                          TENI ..
         INEL=16                                                     TENI ..
         IDLE=16                                                     TENI ..
         IDEG=3                                                      TENI ..
         IPG=5                                                       TENI ..
         NDIM=3                                                      TENI ..
C                                                                    TENI ..
C....... END OF ELEMENT DEFINITION                                   TENI ..
C                                                                    TENI 37
```

Modification of PNINV

```
C....... COMPUTE MATRIX PN (HERMITE ELEMENT :                        PNIN 17
C                WITH 16 D.O.F. TRETRAHEDRA)                         PNIN ..
C                                                                    PNIN ..
         I0=1                                                        PNIN ..
         I1=1                                                        PNIN ..
         DO 20 IN=1,4                                                PNIN ..
         DO 5 I=1,3                                                  PNIN ..
       5 KDER(I)=0                                                   PNIN ..
         DO 15 ID=1,4                                                PNIN ..
         IF(ID.GT.1) KDER(ID-1)=1                                    PNIN ..
         IF(ID.GT.2) KDER(ID-2)=0                                    PNIN ..
         CALL BASEP(VKSI(I1),KEXP,KDER,VP)                           PNIN ..
         I2=I0                                                       PNIN ..
         DO 10 IJ=1,IDLE                                             PNIN ..
         VPN(I2)=VP(IJ)                                              PNIN ..
      10 I2=I2+IDLE                                                  PNIN ..
      15 I0=I0+1                                                     PNIN ..
      20 I1=I1+NDIM                                                  PNIN ..
C                                                                    PNIN ..
C....... END OF PN COMPUTATION                                       PNIN ..
C                                                                    PNIN 32
```

Figure 1.8(a) Modifications of the demonstration program of Figure 1.7(a) and of the sub-routine PNINV of Figure 1.6(a) for the four-node tetrahedronal element of paragraph 2.5.4.

FUNCTIONS N AT INTEGRATION POINTS

COORDINATES	0.25000	0.25000	0.25000		
N	0.53125	0.07813	0.07813	0.07813	0.15625
	-0.04688	0.04688	0.01563	0.15625	0.01563
	-0.04688	0.04688	0.15625	0.04688	0.01563
	-0.04688				
COORDINATES	0.16667	0.16667	0.16667		
N	0.77778	0.08796	0.08796	0.08796	0.07407
	-0.02315	0.02315	0.00463	0.07407	0.00463
	-0.02315	0.02315	0.07407	0.02315	0.00463
	-0.02315				
COORDINATES	0.33333	0.16667	0.16667		
N	0.59259	0.10185	0.06481	0.07407	0.25926
	-0.07407	0.04630	0.01852	0.07407	0.00926
	-0.02315	0.02315	0.07407	0.03704	0.00463
	-0.02315				
COORDINATES	0.16667	0.33333	0.16667		
N	0.59259	0.07407	0.10185	0.06481	0.07407
	-0.02315	0.03704	0.00463	0.25926	0.01852
	-0.07407	0.04630	0.07407	0.02315	0.00926
	-0.02315				
COORDINATES	0.16667	0.16667	0.33333		
N	0.59259	0.06481	0.07407	0.10185	0.07407
	-0.02315	0.02315	0.00926	0.07407	0.00463
	-0.02315	0.03704	0.25926	0.04630	0.01852
	-0.07407				

FUNCTIONS D(N)/D(KSI) AT INTEGRATION POINTS

COORDINATES	0.25000	0.25000	0.25000		
D(N)/D(KSI)	-1.12500	0.00000	-0.18750	-0.12500	1.12500
	-0.31250	0.18750	0.12500	0.00000	0.06250
	0.00000	0.00000	0.00000	0.12500	0.00000
	0.00000				
COORDINATES	0.16667	0.16667	0.16667		
D(N)/D(KSI)	-0.83333	0.27778	-0.13889	-0.05556	0.83333
	-0.25000	0.13889	0.05556	0.00000	0.02778
	0.00000	0.00000	0.00000	0.11111	0.00000
	0.00000				
COORDINATES	0.33333	0.16667	0.16667		
D(N)/D(KSI)	-1.33333	-0.08333	-0.13889	-0.11111	1.33333
	-0.33333	0.13889	0.11111	0.00000	0.02778
	0.00000	0.00000	0.00000	0.05556	0.00000
	0.00000				
COORDINATES	0.16667	0.33333	0.16667		
D(N)/D(KSI)	-0.83333	0.19444	-0.22222	-0.05556	0.83333
	-0.25000	0.22222	0.05556	0.00000	0.11111
	0.00000	0.00000	0.00000	0.11111	0.00000
	0.00000				
COORDINATES	0.16667	0.16667	0.33333		
D(N)/D(KSI)	-0.83333	0.16667	-0.13889	-0.11111	0.83333
	-0.25000	0.13889	0.11111	0.00000	0.02778
	0.00000	0.00000	0.00000	0.22222	0.00000
	0.00000				

Figure 1.8(b) Results of the program of Figure 1.7(a) modified according to Figure 1.8(a).

```
      SUBROUTINE JACOB(VNI,VCORE,NDIM,INEL,VJ,VJ1,DETJ)            JACB   1
C========================================================================JACB   2
C        TO EVALUATE THE JACOBIAN MATRIX, ITS DETERMINANT AND       JACB   3
C        ITS INVERSE (1,2,3 DIMENSIONS)                             JACB   4
C        INPUT                                                      JACB   5
C           VNI      DERIVATIVES OF INTERPOLATION FUNCTION W.R.T.   JACB   6
C                    KSI,ETA,DZETA                                  JACB   7
C           VCORE    ELEMENT NODAL COORDINATES                      JACB   8
C           NDIM     NUMBER OF DIMENSIONS                           JACB   9
C           INEL     NUMBER OF NODES PER ELEMENT                    JACB  10
C        OUTPUT                                                     JACB  11
C           VJ       JACOBIAN MATRIX                                JACB  12
C           VJ1      INVERSE OF JACOBIAN MATRIX                     JACB  13
C           DETJ     DETERMINANT OF JACOBIAN MATRIX                 JACB  14
C========================================================================JACB  15
      IMPLICIT REAL*8(A-H,O-Z)                                      JACB  16
      DIMENSION VNI(INEL,1),VCORE(NDIM,1),VJ(1),VJ1(1)              JACB  17
      DATA ZERO/0.D0/,UN/1.D0/                                      JACB  18
C------------------------------------------------------------------ JACB  19
C------- FORM THE JACOBIAN MATRIX                                   JACB  20
      J=1                                                           JACB  21
      DO 20 JJ=1,NDIM                                               JACB  22
      DO 20 II=1,NDIM                                               JACB  23
      C=ZERO                                                        JACB  24
      DO 10 IJ=1,INEL                                               JACB  25
   10 C=C+VNI(IJ,II)*VCORE(JJ,IJ)                                   JACB  26
      VJ(J)=C                                                       JACB  27
   20 J=J+1                                                         JACB  28
C------- 1, 2, OR 3 DIMENSIONAL INVERSION                           JACB  29
      GO TO (40,50,60),NDIM                                         JACB  30
   40 DETJ=VJ(1)                                                    JACB  31
      IF(DETJ.EQ.ZERO) RETURN                                       JACB  32
      VJ1(1)=UN/DETJ                                                JACB  33
      RETURN                                                        JACB  34
   50 DETJ=VJ(1)*VJ(4)-VJ(2)*VJ(3)                                  JACB  35
      IF(DETJ.EQ.ZERO) RETURN                                       JACB  36
      VJ1(1)=VJ(4)/DETJ                                             JACB  37
      VJ1(2)=-VJ(2)/DETJ                                            JACB  38
      VJ1(3)=-VJ(3)/DETJ                                            JACB  39
      VJ1(4)=VJ(1)/DETJ                                             JACB  40
      RETURN                                                        JACB  41
   60 DETJ=VJ(1)*(VJ(5)*VJ(9)-VJ(8)*VJ(6))                          JACB  42
     1    +VJ(4)*(VJ(8)*VJ(3)-VJ(2)*VJ(9))                          JACB  43
     2    +VJ(7)*(VJ(2)*VJ(6)-VJ(5)*VJ(3))                          JACB  44
      IF(DETJ.EQ.ZERO) RETURN                                       JACB  45
      VJ1(1)=(VJ(5)*VJ(9)-VJ(6)*VJ(8))/DETJ                         JACB  46
      VJ1(2)=(VJ(3)*VJ(8)-VJ(2)*VJ(9))/DETJ                         JACB  47
      VJ1(3)=(VJ(2)*VJ(6)-VJ(3)*VJ(5))/DETJ                         JACB  48
      VJ1(4)=(VJ(7)*VJ(6)-VJ(4)*VJ(9))/DETJ                         JACB  49
      VJ1(5)=(VJ(1)*VJ(9)-VJ(7)*VJ(3))/DETJ                         JACB  50
      VJ1(6)=(VJ(4)*VJ(3)-VJ(6)*VJ(1))/DETJ                         JACB  51
      VJ1(7)=(VJ(4)*VJ(8)-VJ(7)*VJ(5))/DETJ                         JACB  52
      VJ1(8)=(VJ(2)*VJ(7)-VJ(8)*VJ(1))/DETJ                         JACB  53
      VJ1(9)=(VJ(1)*VJ(5)-VJ(4)*VJ(2))/DETJ                         JACB  54
      RETURN                                                        JACB  55
      END                                                           JACB  56
```

Figure 1.9 Subroutine JACOB for computing the Jacobian matrix, its inverse and its determinant, used in program MEF, Chapter 6.

```
      SUBROUTINE DNIDX(VNI,VJ1,NDIM,INEL,VNIX)                  DNID   1
C==============================================================DNID   2
C     COMPUTE THE DERIVATIVES OF INTERPOLATION FUNCTIONS WITH   DNID   3
C     RESPECT TO X,Y,Z                                          DNID   4
C     (1,2 OR 3 DIMENSIONS)                                     DNID   5
C         INPUT                                                 DNID   6
C             VNI    DERIVATIVES OF INTERPOLATION FUNCTIONS WITH RESPECT DNID 7
C                    TO  KSI,ETA,DZETA                          DNID   8
C             VJ1    INVERSE OF THE JACOBIAN                    DNID   9
C             NDIM   NUMBER OF DIMENSIONS (1,2 OR 3)            DNID  10
C             INEL   NUMBER OF INTERPOLATION FUNCTIONS (OR NODES) DNID 11
C         OUTPUT                                                DNID  12
C             VNIX   X,Y,Z DERIVATIVES OF INTERPOLATION FUNCTIONS DNID 13
C==============================================================DNID  14
      IMPLICIT REAL*8(A-H,O-Z)                                  DNID  15
      DIMENSION VNI(INEL,1),VJ1(NDIM,1),VNIX(INEL,1)            DNID  16
      DATA ZERO/0.D0/                                           DNID  17
C--------------------------------------------------------------DNID  18
      DO 20 I=1,NDIM                                            DNID  19
      DO 20 J=1,INEL                                            DNID  20
      C=ZERO                                                    DNID  21
      DO 10 IJ=1,NDIM                                           DNID  22
   10 C=C+VJ1(I,IJ)*VNI(J,IJ)                                   DNID  23
   20 VNIX(J,I)=C                                               DNID  24
      RETURN                                                    DNID  25
      END                                                       DNID  26
```

Figure 1.10 Subroutine DNIDX for computing the derivatives of functions N with respect to **x**, used in program MEF, Chapter 6.

These two routines are used in program MEF described in Chapter 6. The FORTRAN variables are all described in Figures 6.6 to 6.9.

1.7 Truncation Errors in an Element

1.7.1 *Introduction*

In this paragraph we introduce the fundamental principles required to estimate the truncation error for an element. We find the conditions necessary to force the error to tend towards zero as the size of the element is decreased. We also develop an estimate of the error in the form of $e \leq cl^\alpha$ where c and α are constants depending on the type of approximation used. The truncation error at any point in the element is defined by (1.7):

$$e(\mathbf{x}) = u(\mathbf{x}) - u_{\text{ex}}(\mathbf{x}) \qquad (1.52)$$

For the reference element we have:

$$e(\boldsymbol{\xi}) = u(\boldsymbol{\xi}) - u_{\text{ex}}(\boldsymbol{\xi}) \qquad (1.53)$$

For two corresponding points, errors $e(\mathbf{x})$ and $e(\boldsymbol{\xi})$ take the same value. To characterize the maximum error in an element we use a norm of the maximum of function $e(\mathbf{x})$

$$|e| = \text{Maximum on } V^e \text{ of } |e(\mathbf{x})| \qquad (1.54)$$

Define the error for each derivatives of order s by:

$$e_s(\mathbf{x}) = D^s(e(\mathbf{x})) = \frac{\partial^s e(\mathbf{x})}{\partial^a x \partial^b y \partial^c z}; \quad \begin{array}{l} a+b+c = s \\ a,b,c,s \geq 0 \end{array}$$

The corresponding norm is:

$$|e|_s = \text{Maximum on } V^e \text{ of } |D^s(e(\mathbf{x}))| \qquad (1.55)$$

for all a, b, c such that $a + b + c = s$. The root mean square value for the error is:

$$\|e\|_s^2 = \sum_{a+b+c=s} \int_{V^e} (D^s(e(\mathbf{x})))^2 \, dV^e \qquad (1.56)$$

According to Strang, reference [3], pages 142–144, the error norms $|e|_s$ and $\|e\|_s^2$ may be presented in the following standard forms:

$$|e|_s \leq c l^{n-s} |u_{\text{ex}}(\mathbf{x})|_n \qquad (1.57a)$$

$$\|e\|_s^2 \leq C l^{2(n-s)} |u_{\text{ex}}(\mathbf{x})|_n^2 \qquad (1.57b)$$

where:

— c and C are constants depending upon the type of element and approximation used;
— the polynomial basis of the approximation is complete up to order $n-1$;
— l is the maximum characteristic length in the element;
— all derivatives of the approximation $u(\mathbf{x})$ up to order s are bounded;
— $|u_{\text{ex}}(\mathbf{x})|_n$ is the norm (1.55) of $u_{\text{ex}}(\mathbf{x})$ with $s = n$.

We give below certain aspects of error estimation which may be useful, from an engineer's point of view, to evaluate the quality of an approximation. To estimate error $e(\xi)$, we use a Taylor series development of function u_{ex} in the neighbourhood of point ξ. For one dimension we have:

$$u_{\text{ex}}(\xi + h) = u_{\text{ex}}(\xi) + h \frac{\partial u_{\text{ex}}}{\partial \xi}\bigg|_\xi + \cdots + \frac{h^{n-1}}{(n-1)!} \frac{\partial^{n-1} u_{\text{ex}}}{\partial \xi^{n-1}}\bigg|_\xi + \frac{h^n}{n!} R_n \qquad (1.58)$$

$$R_n = \frac{\partial^n u_{\text{ex}}}{\partial \xi^n}\bigg|_{\bar{\xi} \text{ on } [\xi, \xi+h]}$$

Assume that the element has n interpolation nodes with coordinates $\xi_1, \xi_2, \ldots, \xi_n$. For a value of $h = \xi_i - \xi$ expression (1.58) becomes:

$$u_{\text{ex}}(\xi_i) = u_i = u_{\text{ex}}(\xi) + (\xi_i - \xi) \frac{\partial u_{\text{ex}}}{\partial \xi}\bigg|_\xi +$$

$$+ \cdots + \frac{(\xi_i - \xi)^{n-1}}{(n-1)!} \frac{\partial^{n-1} u_{\text{ex}}}{\partial \xi^{n-1}}\bigg|_\xi + \frac{(\xi_i - \xi)^n}{n!} R_i \qquad (1.59)$$

$$R_i = \frac{\partial^n u_{\text{ex}}}{\partial \xi^n}\bigg|_{\bar{\xi}_i \text{ on } [\xi_i, \xi]}$$

Using (1.59) to evaluate nodal values u_n and substituting in the approximate function $u(\xi) = \langle N(\xi) \rangle \{u_n\}$ we get:

$$u(\xi) = \left(\sum_i N_i\right) u_{\text{ex}}(\xi) + \left(\sum_i N_i \cdot (\xi_i - \xi)\right) \frac{\partial u_{\text{ex}}}{\partial \xi}\bigg|_\xi +$$

$$+ \cdots + \frac{1}{(n-1)!} \left(\sum_i N_i \cdot (\xi_i - \xi)^{n-1}\right) \frac{\partial^{n-1} u_{\text{ex}}}{\partial \xi^{n-1}}\bigg|_\xi + \frac{1}{n!} \sum_i N_i \cdot (\xi_i - \xi)^n \cdot R_i \quad (1.60)$$

From (1.31) we have:

$$\sum_i N_i = 1$$

$$\sum_i N_i \cdot (\xi_i - \xi) = \sum_i N_i \xi_i - \sum_i N_i \xi = 0 \quad (1.61)$$

$$\cdots\cdots\cdots\cdots\cdots\cdots\cdots\cdots\cdots\cdots\cdots$$

$$\sum_i N_i \cdot (\xi_i - \xi)^{n-1} = 0$$

$$e(\xi) = u(\xi) - u_{\text{ex}}(\xi) = \frac{1}{n!} \sum_i N_i \cdot (\xi_i - \xi)^n R_i \quad (1.62a)$$

This last expression contains derivatives of order n with respect to ξ in the terms R_i. Using the chain rule of differentiation we have

$$\frac{\partial}{\partial \xi} = \frac{\partial x}{\partial \xi} \cdot \frac{\partial}{\partial x}$$

For a linear transformation τ

$$\frac{\partial x}{\partial \xi} = \text{constant} = \frac{l}{2}$$

l being the length of the element. Then

$$\frac{\partial^n}{\partial \xi^n} = \left(\frac{l}{2}\right)^n \frac{\partial^n}{\partial x^n}$$

$$R_i = \left(\frac{l}{2}\right)^n \frac{\partial^n u_{\text{ex}}}{\partial x^n}\bigg|_{x(\bar{\xi}_i)}$$

Expression (1.62a) becomes

$$e(x) = \frac{1}{n!} \left(\frac{l}{2}\right)^n \sum_i N_i \cdot (\xi_i - \xi)^n \frac{\partial^n u_{\text{ex}}}{\partial x^n}\bigg|_{x(\bar{\xi}_i)} \quad (1.62b)$$

This last expression tends toward zero with a decreasing value of l provided $n \geq 1$. Similar computations can be made for errors of the derivatives of u_{ex}. The norm of (1.62b) can be written under its generalized form (1.57a) for $s = 0$

$$|e|_0 \leq c l^n |u_{\text{ex}}(x)|_n$$

Example 1.20   Error for a linear line element

$$u_{ex}(\xi=-1) = u_1 = u_{ex}(\bar\xi) + h_1 \left.\frac{\partial u_{ex}}{\partial \xi}\right|_{\bar\xi} + \frac{h_1^2}{2}R_1$$

where:
$$h_1 = -1-\bar\xi \quad R_1 = \left.\frac{\partial^2 u_{ex}}{\partial \xi^2}\right|_{-1<\bar{\bar\xi}<\bar\xi}$$

$$u_{ex}(\xi=1) = u_2 = u_{ex}(\bar\xi) + h_2 \left.\frac{\partial u_{ex}}{\partial \xi}\right|_{\bar\xi} + \frac{h_2^2}{2}R_2$$

where:
$$h_2 = 1-\bar\xi \quad R_2 = \left.\frac{\partial^2 u_{ex}}{\partial \xi^2}\right|_{\bar\xi \leq \bar{\bar\xi} \leq 1}.$$

The approximate function over the element V^r is:

$$u(\xi) = \langle N_1(\xi) \; N_2(\xi) \rangle \begin{Bmatrix} u_1 \\ u_2 \end{Bmatrix} = \frac{1-\xi}{2}u_1 + \frac{1+\xi}{2}u_2$$

or using the preceding expressions of u_1 and u_2

$$u(\xi) = u_{ex}(\bar\xi) + \tfrac{1}{4}(1-\xi^2)((1+\xi)R_1 + (1-\xi)R_2).$$

The error is:

$$e(\xi) = u(\xi) - u_{ex}(\xi) = \tfrac{1}{4}(1-\xi^2)((R_1+R_2) + \xi(R_1-R_2))$$

the norm of the maximum verifies:

$$|e|_0 \leq \tfrac{1}{2}\text{Max}\left.\frac{\partial^2 u_{ex}}{\partial \xi^2}\right|_{V_r}$$

Since $(1-\xi^2) \leq 1$ for all ξ on V^r:

$$R_1 + R_2 + \xi(R_1 - R_2) \leq 2\,\text{Max}\left.\frac{\partial^2 u_{ex}}{\partial \xi^2}\right|_{V_r}$$

To obtain the above inequality, one makes use of the fact that R_1 and R_2 are defined only over disjoint parts of an element. Thus one would consider the maximum of either $(R_1 + \xi R_1)$ or $(R_2 - \xi R_2)$. To write the error as a function of the derivatives with respect to x we use the following geometrical transformation

$$x = \left\langle \frac{1-\xi}{2} \; \frac{1+\xi}{2} \right\rangle \begin{Bmatrix} x_1 \\ x_2 \end{Bmatrix}$$

Hence

$$\frac{\partial u}{\partial \xi} = \frac{\partial x}{\partial \xi}\frac{\partial u}{\partial x} = \frac{x_2 - x_1}{2}\frac{\partial u}{\partial x} = \frac{l}{2}\frac{\partial u}{\partial x}$$

$$\frac{\partial^2 u}{\partial \xi^2} = \frac{l^2}{4}\frac{\partial^2 u}{\partial x^2}$$

Therefore:
$$|e|_0 \le \frac{l^2}{8} \text{Max} \left| \frac{\partial^2 u_{ex}}{\partial x^2} \right|_{V_e}$$

This last expression is of the form (1.57a) with:
$$c = \tfrac{1}{8}, \quad n = 2, \quad s = 0$$

$$e_1 = \frac{\partial u}{\partial x} - \frac{\partial u_{ex}}{\partial x}$$

$$= \frac{2}{l}\left(\frac{\partial u}{\partial \xi} - \frac{\partial u_{ex}}{\partial \xi} \right)$$

$$|e|_1 \le \frac{l}{2} \text{Max} \left| \frac{\partial^2 u_{ex}}{\partial x^2} \right|_{V_e}$$

This expression is of the form (1.57a) with:
$$c = \tfrac{1}{2}, \quad n = 2, \quad s = 1$$

1.7.2 Evaluation of the Error

We describe here a general method for estimating the truncation error of line elements. The method can be generalized for surface and volume elements as will be shown in paragraph 2.3.2

Using a Taylor series development of $u_{ex}(\xi)$ in the neighbourhood of $\xi = 0$ we write:

$$u_{ex}(\xi) = u_{ex}(0) + \xi \frac{\partial u_{ex}}{\partial \xi}\bigg|_0 + \cdots + \frac{\xi^{n-1}}{(n-1)!} \frac{\partial^{n-1} u_{ex}}{\partial \xi^{n-1}}\bigg|_0 + \frac{\xi^n}{n!} R \quad (1.63)$$

$$R = \frac{\partial^n u_{ex}}{\partial \xi^n}\bigg|_{\bar\xi \text{ on } [0,\xi]}$$

$$u_{ex}(\xi) = \langle 1 \; \xi \; \xi^2 \ldots \xi^{n-1} \rangle \begin{Bmatrix} u_{ex} \\ \dfrac{\partial u_{ex}}{\partial \xi} \\ \vdots \\ \dfrac{1}{(n-1)!} \dfrac{\partial^{n-1} u_{ex}}{\partial \xi^{n-1}} \end{Bmatrix}_{\xi=0} + \xi^n \frac{1}{n!} R$$

(1.64a)

using matrix notation:

$$u_{ex}(\xi) = \langle P \rangle \{\partial u_{ex}\} + \xi^n \frac{1}{n!} R \quad (1.64b)$$

The finite element approximate function is written with the help of (1.18):

$$u(\xi) = \langle P \rangle \{a\} \quad (1.65)$$

The value of u and u_{ex} are the same at the nodes.

$$\{u_n\} = [P_n]\{a\} = [P_n]\{\partial u_{ex}\} + \frac{1}{N!}\begin{Bmatrix} \xi_1^n & R_1 \\ \xi_2^n & R_2 \\ \cdots & \cdots \\ \xi_n^n & R_n \end{Bmatrix} \quad (1.66)$$

$$R_i = \frac{\partial^n u_{ex}}{\partial \xi^n}\bigg|_{\bar\xi \text{on}[0,\xi_i]}$$

Hence:

$$\{\partial u_{ex}\} = \{a\} - \frac{1}{n!}[P_n]^{-1}\begin{Bmatrix} \xi_1^n & R_1 \\ \xi_2^n & R_2 \\ \cdots & \cdots \\ \xi_n^n & R_n \end{Bmatrix} \quad (1.67)$$

Substituting (1.67) into (1.64b) and using (1.65) and (1.24):

$$e(\xi) = u(\xi) - u_{ex}(\xi) = \frac{1}{n!}\langle N \rangle \begin{Bmatrix} \xi_1^n & R_1 \\ \xi_2^n & R_2 \\ \cdots & \cdots \\ \xi_n^n & R_n \end{Bmatrix} - \frac{1}{n!}\xi^n R \quad (1.68)$$

The norm of the error $e(\xi) = u(\xi) - u_{ex}(\xi)$ is written after replacing $R_1, R_2, \ldots, R_n$ by $R = \text{Max}|\partial^n u_{ex}/\partial \xi^n|_{V^r}$:

$$|e|_0 \leq \frac{1}{n!} \text{Max}\left|\langle N \rangle \begin{Bmatrix} \xi_1^n \\ \xi_2^n \\ \vdots \\ \xi_n^n \end{Bmatrix} - \xi^n\right|_{V^r} \cdot \text{Max}\left|\frac{\partial^n u_{ex}}{\partial \xi^n}\right|_{V^r} \quad (1.69)$$

The geometry of the real element is taken into account through the values of derivatives with respect to ξ using formulas of paragraph 1.5. The errors on derivatives of the approximations are obtained by differentiating equation (1.68). For example, in the case of the first derivative:

$$e_1(\xi) = \frac{\partial u}{\partial x} - \frac{\partial u_{ex}}{\partial x} = \left(\frac{\partial u}{\partial \xi} - \frac{\partial u_{ex}}{\partial \xi}\right)\frac{\partial \xi}{\partial x} \quad (1.70)$$

$$e_1(\xi) = \left(\frac{1}{n!}\left\langle\frac{\partial N}{\partial \xi}\right\rangle \begin{Bmatrix} \xi_1^n & R_1 \\ \xi_2^n & R_2 \\ \cdots & \cdots \\ \xi_n^n & R_n \end{Bmatrix} - \frac{1}{(n-1)!}\xi^{n-1}R\right)\frac{\partial \xi}{\partial x} \quad (1.71)$$

Example 1.21 Error for a linear line element

$$\langle N \rangle = \left\langle \frac{1-\xi}{2} \; \frac{1+\xi}{2} \right\rangle \quad n=2$$

Let us use (1.69):

$$|e|_0 \le \tfrac{1}{2} \operatorname{Max} \left| \left\langle \frac{1-\xi}{2} \; \frac{1+\xi}{2} \right\rangle \begin{Bmatrix} 1 \\ 1 \end{Bmatrix} - \xi^2 \right| \cdot \operatorname{Max} \left| \frac{\partial^2 u_{ex}}{\partial \xi^2} \right|$$

$$\le \tfrac{1}{2} \operatorname{Max} |\xi^2 - 1| \cdot \operatorname{Max} \left| \frac{\partial^2 u_{ex}}{\partial \xi^2} \right|$$

$$\le \frac{l^2}{8} \operatorname{Max} \left| \frac{\partial^2 u_{ex}}{\partial x^2} \right|$$

$$|e|_1 \le \tfrac{1}{2} \operatorname{Max} \left| \left\langle -\frac{1}{2} \; \frac{1}{2} \right\rangle \begin{Bmatrix} 1 \\ 1 \end{Bmatrix} - 2\xi \right| \cdot \operatorname{Max} \left| \frac{\partial^2 u_{ex}}{\partial \xi^2} \right| \cdot \operatorname{Max} \left| \frac{\partial \xi}{\partial x} \right|$$

$$\le \frac{l}{2} \operatorname{Max} \left| \frac{\partial^2 u_{ex}}{\partial x^2} \right|$$

This last expression is of the form (1.57a) with:

$$c = \tfrac{1}{2}, \quad n = 2, \quad s = 1$$

1.7.3 Reduction of the Truncation Error

To reduce the errors defined in (1.57) one must:

— decrease l, i.e. use a greater number of smaller elements;
— replace the polynomial basis by one of higher degree n.

To preserve the geometrical size of an element we have two choices:

— increase the number of nodes per element while preserving the same degrees of freedom per node (for example, see paragraph 2.2.2);
— increase the number of degrees of freedom per node (using Hermite interpolation formulas) while preserving the same number of nodes per element (for example, see paragraph 2.2.3).

Combinations of the previous two techniques are also possible but result in very complex coding. Bubble functions taking zero value on all boundaries and nodes of an element can also be added for the interior region of that element.

$$u(\xi) = \langle N(\xi) \rangle \{u_n\} + \langle P_I(\xi) \rangle \{a_I\} \qquad (1.72)$$

Where $P_I(\xi_i) = 0$ if ξ_i is the point of nodal approximation
  $P_I(\xi) = 0$ if ξ is on the boundary.

This expression is in effect a combination of nodal and non-nodal approximation for each element (paragraph 2.2.4.2).

1.8 Numerical Example (Rainfall Estimation)

While finite element approximations are most often used for the discretization of partial differential equations, they can also be used to approximate any continuous function that is known only at a few discrete points of its domain. We propose to compute the total volume of rain having fallen over a certain region of area A using the information provided by a network of rain gauge readings u_i.

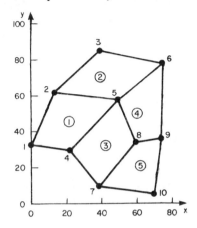

The total rainfall Q is defined by

$$Q = \int_A u(x, y) \, dA \qquad (1.73)$$

where $u(x, y)$ is the continuous function giving the depth of precipitation over the area A. Rain gauge readings u_i from 10 gauges situated at coordinates x_i, y_i are available. The data is taken from reference [8]. We use a finite element approximation to evaluate (1.73).

(a) Choice of nodes and elements

We select stations $1, 2, \ldots, 10$ as geometrical and interpolation nodes.

Nœud	x_i(km)	y_i(km)
1	0.0	33.3
2	13.2	62.3
3	39.3	84.5
4	22.2	30.1
5	49.9	57.6
6	78.8	78.2
7	39.3	10.0
8	59.7	34.3
9	73.9	36.2
10	69.8	5.1

The region is divided into five quadrilateral elements with connectivities shown below.

Element	Nodes			
	i	j	k	l
1	1	4	5	2
2	2	5	6	3
3	4	7	8	5
4	5	8	9	6
5	7	10	9	8

The vector of nodal values in centimetres is:

$$\{U_n\}^T = \langle 4.62\ 3.81\ 4.76\ 5.45\ 4.90\ 10.35\ 4.96\ 4.26\ 18.36\ 15.69 \rangle$$

(b) Approximation of $u(x, y)$ over each element

Choose a bilinear quadrilateral element similar to the one described in Example (1.16). For each element, the approximation $u(\xi, \eta)$ is:

$$u(\xi, \eta) = \langle P \rangle [P_n]^{-1} \{u_n\} \qquad (1.74)$$

the geometrical transformation is:

$$x(\xi, \eta) = \langle P \rangle [P_n]^{-1} \{x_n\}$$
$$y(\xi, \eta) = \langle P \rangle [P_n]^{-1} \{y_n\}$$

where:

$$\{u_n\}^T = \langle u_i\ u_j\ u_k\ u_l \rangle$$
$$\{x_n\}^T = \langle x_i\ x_j\ x_k\ x_l \rangle$$
$$\{y_n\}^T = \langle y_i\ y_j\ y_k\ y_l \rangle$$

i, j, k, l being the four nodes given in array CONEC. The determinant of the Jacobian matrix was found in Example (1.18.)

$$\det(J) = A_0 + A_1 \xi + A_2 \eta \qquad (1.75)$$

(c) Evaluation of Q

The total volume of rain is obtained by summing the contributions over each element.

$$Q = \sum_{e=1}^{5} Q^e \qquad (1.76)$$

$$Q^e = \int_{A^e} u(x, y)\, dA$$

$$= \int_{-1}^{1} \int_{-1}^{1} u(\xi, \eta) \det(J)\, d\xi\, d\eta$$

which, after replacing u by approximation (1.74), gives:

$$Q^e = \int_{-1}^{1} \int_{-1}^{1} \langle P \rangle [P_n]^{-1} \{u_n\} \det(J) d\xi \, d\eta$$

$$Q^e = \int_{-1}^{1} \int_{-1}^{1} (A_0 + A_1\xi + A_2\eta) \langle 1 \quad \xi \quad \eta \quad \xi\eta \rangle d\xi \, d\eta \cdot [P_n]^{-1} \{u_n\} \quad (1.77)$$

After an explicit integration we get:

$$Q^e = \left\langle A_0 \frac{A_1}{3} \frac{A_2}{3} \right\rangle \begin{Bmatrix} u_i + u_j + u_k + u_l \\ -u_i + u_j + u_k - u_l \\ -u_i - u_j + u_k + u_l \end{Bmatrix} \quad (1.78)$$

where coefficients A_0, A_1 and A_2 are functions of the coordinates of the nodes given in Example (1.18) and u_i, u_j, u_k, u_l are the values of rainfall at the four nodes of the element extracted from $\{u_n\}$. The final results are given in the table shown below.

Element	A_0(km²)	A_1	A_2	Q^e(cm km²)
1	228.18	1.64	55.04	4261.41
2	241.65	− 5.70	12.99	5771.07
3	217.56	− 25.18	− 14.01	4272.97
4	182.79	− 65.72	29.70	6954.45
5	159.37	15.94	− 66.85	6983.87

Total volume of rainfall $Q = \sum Q^e = 28243.78$ cm km².
Total area $A = 4 \sum A_0 = 4118.21$ km².
Average rainfall height $u_m = Q/A = 6.86$ cm.

Important Results

Nodal approximation of a function:

$$u(x) = \langle N \rangle \{u_n\} \quad (1.9)$$

Transformation of a reference element into a real element:

$$\tau : \boldsymbol{\xi} \to \mathbf{x}(\boldsymbol{\xi}) = (\bar{N}(\boldsymbol{\xi}))]\{\mathbf{x}_n\} \quad (1.12)$$

Approximation of u on the element of reference:

$$u(\boldsymbol{\xi}) = \langle N(\boldsymbol{\xi}) \rangle \{u_n\} \quad (1.14)$$

Properties of interpolation functions:

$$N_j(\boldsymbol{\xi}_i) = \begin{cases} 0 & \text{if } i \neq j \\ 1 & \text{if } i = j \end{cases} \quad (1.16)$$

$$\sum_{i=1}^{n_d} N_i(\boldsymbol{\xi}) p(\boldsymbol{\xi}_i) = p(\boldsymbol{\xi}) \quad (1.31)$$

Construction of interpolation functions:

$$u(\pmb{\xi}) = \langle P(\pmb{\xi}) \rangle \{a\} \qquad (1.18)$$

$$\{u_n\} = [P_n]\{a\} \qquad (1.20)$$

$$\langle N(\pmb{\xi}) \rangle = \langle P(\pmb{\xi}) \rangle [P_n]^{-1} \qquad (1.24)$$

Transformation of first derivatives:

$$\{\partial_x\} = [j]\{\partial_\xi\} \qquad (1.37b)$$

$$[j] = [J]^{-1} \qquad (1.38)$$

$$[J] = \begin{bmatrix} \langle \bar{N}_{,\xi} \rangle \\ \langle \bar{N}_{,\eta} \rangle \\ \langle \bar{N}_{,\zeta} \rangle \end{bmatrix} [\{x_n\} \ \{y_n\} \ \{z_n\}] \qquad (1.43)$$

Transformation of an integral:

$$\int_{V^e} f(\mathbf{x})\,dx\,dy\,dz = \int_{V^r} f(\mathbf{x}(\pmb{\xi}))\det(J)\,d\xi\,d\eta\,d\zeta \qquad (1.44)$$

Approximation errors:

$$|e|_s \le c\, l^{n-s} |u_{\text{ex}}(\mathbf{x})|_n \qquad (1.57a)$$

Notation

$\langle a \rangle = \langle a_1 \ a_2 \ldots a_n \rangle$	generalized parameters of an approximation
$\langle a_x \rangle, \langle a_y \rangle, \langle a_z \rangle$	generalized coordinates of the element
$e(x) = u(x) - u_{\text{ex}}(x)$	approximation error
$[J], [j], \det(J)$	Jacobian matrix, its inverse and determinant
n	number of interpolation nodes
n_d	number of degrees of freedom of an element
n^e	number of interpolation nodes of an element
n_{el}	number of elements
$\bar{n}$	number of geometrical nodes
$\bar{n}^e$	number of geometrical nodes of an element
$\langle N(x) \rangle = \langle N_1(x) \ N_2(x)\ldots \rangle$	nodal interpolation functions over the real element
$\langle N(\xi) \rangle = \langle N_1(\xi) \ N_2(\xi)\ldots \rangle$	interpolation functions over the element of reference
$\langle \bar{N}(\xi) \rangle = \langle \bar{N}_1(\xi) \ \bar{N}_2(\xi)\ldots \rangle$	geometrical transformation functions
$[P_n], [\bar{P}_n]$	nodal interpolation matrices
$\langle P(x) \rangle = \langle P_1(x) \ P_2(x)\ldots \rangle$	polynomial basis over the real element
$\langle P(\xi) \rangle = \langle P_1(\xi) \ P_2(\xi)\ldots \rangle$	polynomial basis over the reference element
$\langle \bar{P}(\xi) \rangle$	polynomial basis of the geometrical transformation

$[T_1], [T_2], [C_1], [C_2]$	transformation matrices for second derivatives
$u(x)$	approximate functions
$u_{ex}(x)$	exact functions
$u^e(x)$ or sometimes $u(x)$	approximate function over an element
$\langle u_n \rangle = \langle u_1 \; u_2 \ldots \rangle$	nodal variables
V	domain under study
V^e	element domain
V^r	reference element domain
$\mathbf{x} = \langle x \; y \; z \rangle$	Cartesian coordinates of a point
$\mathbf{x}_i = \langle x_i \; y_i \; z_i \rangle$	coordinates of node i
$\langle \mathbf{x}_n \rangle$	coordinates of the nodes of an element
$\langle \bar{\mathbf{x}}_n \rangle$	coordinates of the geometrical nodes
$\langle \partial_x \rangle, \langle \partial_\xi \rangle$	differential operators:

$$\left\langle \frac{\partial}{\partial x} \frac{\partial}{\partial y} \frac{\partial}{\partial z} \right\rangle \text{ and } \left\langle \frac{\partial}{\partial \xi} \frac{\partial}{\partial \eta} \frac{\partial}{\partial \zeta} \right\rangle$$

$\langle \partial_x^2 \rangle, \langle \partial_\xi^2 \rangle$	second order differential operator from (1.45a) and (1.45b)
$\boldsymbol{\xi} = \langle \xi \; \eta \; \zeta \rangle$	coordinates of a point of the element of reference
$\langle \boldsymbol{\xi}_n \rangle$	coordinates of the nodes of an element of reference
$\langle \bar{\boldsymbol{\xi}}_n \rangle$	coordinates of the geometrical nodes of an element of reference
τ^e or τ	geometrical transformation for element e
$\| \; \|_s, \; \| \; \|_s^2$	maximum and root mean square norm for derivatives of order s of a function
CONEC	connectivity array
CORG	nodal coordinates array

Remarks

A column matrix (also called a vector) can be noted in three different ways:

$$\mathbf{a}, \quad \{a\}, \quad \langle a \rangle^T$$

A matrix T, its transpose, and its inverse are noted:

$$[T], \quad [T]^T, \quad [T]^{-1}$$

References

[1] J. T. Oden, *Finite Elements of Non-Linear Continua*, McGraw-Hill, New York, 1972.
[2] O. C. Zienkiewicz, *The Finite Element Method in Engineering Science*, McGraw-Hill, New York, 1st edn, 1967, 3rd edn, 1977.
[3] G. Strang and O. J. Fix, *Analysis of the Finite Element Method*, Prentice-Hall, New Jersey, 1973.

[4] W. J. Gordon and C. A. Hall, Construction of curvilinear coordinate systems and application to mesh generation, *Int. J. Num. Meth. Eng.*, 7, pp. 461–477, 1973.
[5] B. M. Irons and A. Razzaque, Experience with the patch test, in *Mathematical foundations of the F.E.M.*, pp. 557–587, Academic Press, 1972.
[6] B. F. de Veubeke, Variational principles and the patch test, *Int. J. Num. Meth. Eng.*, 8, pp. 783–801, 1974.
[7] P. G. Ciarlet, *The Finite Element Method for Elliptic Problems*, North-Holland, Amsterdam, 1978.
[8] J. E. Akin, Calculation of mean areal depth of precipitation, *Journal of Hydrology*, 12, pp. 363–376, 1971.

2

VARIOUS TYPES Of ELEMENTS

2.0 Introduction

In the first chapter we studied finite element approximation techniques. We have particularly discussed the notions of elements of reference and interpolation functions. In this chapter we develop the interpolation functions of the elements used most often in practice. A particular type of element is defined by:

— its shape; for example, triangular,
— the coordinates $\{\xi_n\}$ of its n interpolation nodes,
— the number of degrees of freedom n_d,
— the definition of its nodal variables $\{u_n\}$,
— the polynomial basis of the approximation $\langle P \rangle$,
— the degree of inter-element continuity that must be satisfied: C^0, C^1, C^2
(see definition in paragraph 1.3.2).

From the information contained in the previous chapter, it is possible to construct the interpolation functions $\langle N(\xi) \rangle$ and their derivatives with respect to ξ, η, ζ:

$$u = \langle N \rangle \{u_n\} \quad \text{(relation (1.9))}$$
$$\langle N \rangle = \langle P \rangle [P_n]^{-1} \quad \text{(relation (1.24))}$$

where $[P_n]$ is defined in (1.20). For non-isoparametric elements, the same techniques are used to construct the geometrical transformation functions $\langle \bar{N} \rangle$; the Jacobian matrix can be constructed with the help of (1.43). For isoparametric elements $\langle N \rangle \equiv \langle \bar{N} \rangle$.

Interpolation functions for most of the elements described in this chapter can also be found in Connor and Brebbia [1], Mitchell and Wait [2] and Zienkiewicz [3].

2.1 List of Elements Described in this Chapter

One-dimensional elements

	Degree of polynomial basis	Continuity (see § 1.3.2)	Number of Nodes n	Number of DOF n_d	Paragraph (§)
Lagrange elements	1	C^0	2	2	2.2.1
	2	C^0	3	3	2.2.2.1
	3	C^0	4	4	2.2.2.2
	$n-1$	C^0	n	n	2.2.2.3
Hermite elements	3	C^1	2	4	2.2.3.1
	5	C^2	2	6	2.2.3.2
Lagrange–Hermite	4	C^1	3	5	2.2.4.1
Hermite with non-nodal DOF	4	C^1	2	5	2.2.4.2

Two-dimensional elements

	Degree of polynomial	Continuity (see § 1.3.2)	Number of Nodes n	Number of DOF n_d	Paragraph (§)
Triangles					
Lagrange	1	C^0	3	3	2.3.2
	2	C^0	6	6	2.3.3.1
	r	C^0	$\frac{(r+1)(r+2)}{2}$	$\frac{(r+1)(r+2)}{2}$	2.3.3.2
	3	C^0	10	10	2.3.3.2
	3 (incomplete)	C^0	9	9	2.3.3.4
	1	semi-C^0	3	3	2.3.3.6
Hermite	3	semi-C^1	4	10	2.3.4.1
	3 (incomplete)	semi-C^1	3	9	2.3.4.2
	5	C^1	3	18	2.3.4.3

Quadrilaterals

Lagrange	1	C^0	4	4	2.4.2
	2	C^0	9	9	2.4.3.1
	3	C^0	16	16	2.4.3.3
Lagrange (incomplete)	2	C^0	8	8	2.4.3.2
	3	C^0	12	12	2.4.3.4
Hermite	3	semi-C^1	4	12	2.4.4.1
Hermite (rectangular)	3	C^1	4	16	2.4.4.2

Three-dimensional elements

	Degree of polynomial basis	Continuity (see §1.3.2)	Number of Nodes n	Number of DOF n_d	Paragraph (§)
Tetrahedrons					
Lagrange	1	C^0	4	4	2.5.2
	2	C^0	10	10	2.5.3.1
	3	C^0	20	20	2.5.3.2
Hermite	3	semi-C^1	8	20	2.5.4
Hexahedrons					
Lagrange	1	C^0	8	8	2.6.1
	2	C^0	27	27	2.6.2.1
Lagrange (incomplete)	2	C^0	20	20	2.6.2.2
	3	C^0	32	32	2.6.2.3
Hermite	3	semi-C^1	8	32	2.6.3
Prisms					
Lagrange	1	C^0	6	6	2.7.1
	2	C^0	15	15	2.7.2

Remark.

Semi-C^0 implies that the approximation respects C^0 continuity on the boundary nodes but not along the boundary of an element. The same applies to semi-C^1.

2.2 Line Elements

2.2.1 *Linear Element (two nodes, C^0)*

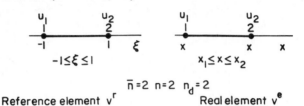

Reference element v^r Real element v^e

$\bar{n}=2$ $n=2$ $n_d=2$

Geometrical and interpolation nodes are identical; the element is isoparametric. By convention, nodes are numbered from left to right.

$$\langle P \rangle = \langle 1 \quad \xi \rangle \tag{2.1a}$$

$$\{P_n\} = \begin{bmatrix} 1 & -1 \\ 1 & 1 \end{bmatrix}; \quad [P_n]^{-1} = \frac{1}{2}\begin{bmatrix} 1 & 1 \\ -1 & 1 \end{bmatrix} \tag{2.1b}$$

	$\frac{1}{c}\{N\}$	$\frac{1}{c}\{\partial N/\partial \xi\}$	c
1	$1-\xi$	-1	
2	$1+\xi$	1	$1/2$

$$[J] = \frac{\partial x}{\partial \xi} = \frac{x_2 - x_1}{2} = \frac{l}{2}; \quad [j] = [J]^{-1} = \frac{2}{l} \tag{2.1c}$$

Remark.

To obtain N multiply functions in the table by c. For example,

$$N_1 = \tfrac{1}{2}(1-\xi) \text{ etc.}$$

Functions N have the following shape:

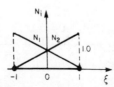

The truncation error computed in Example 1.20 verifies that:

$$e(\xi) \leq \tfrac{1}{2}(1-\xi^2) R \quad \text{where} \quad R = \text{Max} \left|\frac{\partial^2 u_{ex}}{\partial \xi^2}\right|_{V_r}$$

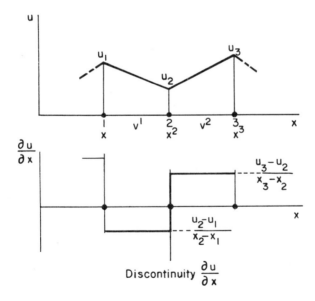

$$|e|_0 \leq \frac{l^2}{8} \text{Max} \left|\frac{\partial^2 u_{ex}}{\partial x^2}\right|_{V_e} \tag{2.1d}$$

where:
$$l = x_2 - x_1$$

The error for $\partial u/\partial x$ is:

$$|e|_1 \leq \frac{l}{2} \text{Max} \left|\frac{\partial^2 u_{ex}}{\partial x^2}\right|_{V_e} \tag{2.1e}$$

Function u and $\partial u/\partial x$ are continuous within the element but only u is continuous across the boundaries of the element.

2.2.2 Lagrangian Elements (continuity C^0)

In this family of elements the number of interpolation nodes is increased while keeping only one variable u_i per node. The geometrical nodes, functions $\bar{N}$ and the Jacobian matrix $[J]$ remain similar to those of paragraph 2.2.1. These elements are sub-parametric.

2.2.2.1 Quadratic elements with equidistant nodes (three nodes, C^0)

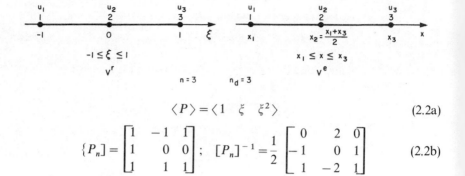

$$\langle P \rangle = \langle 1 \quad \xi \quad \xi^2 \rangle \tag{2.2a}$$

$$[P_n] = \begin{bmatrix} 1 & -1 & 1 \\ 1 & 0 & 0 \\ 1 & 1 & 1 \end{bmatrix}; \quad [P_n]^{-1} = \frac{1}{2}\begin{bmatrix} 0 & 2 & 0 \\ -1 & 0 & 1 \\ 1 & -2 & 1 \end{bmatrix} \tag{2.2b}$$

$\frac{1}{c}\{N\}$	$\frac{1}{c}\{\partial N/\partial \xi\}$	c
1 $-\xi(1-\xi)$	$-1+2\xi$	
2 $2(1-\xi^2)$	-4ξ	1/2
3 $\xi(1+\xi)$	$1+2\xi$	

The functions N have the following form:

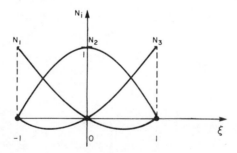

The approximation error from (1.68) verifies:

$$e(\xi) \leq \tfrac{1}{6}\xi(1-\xi^2)R \quad \text{where} \quad R = \text{Max}\left|\frac{\partial^3 u_{ex}}{\partial \xi^3}\right|_{V^r}$$

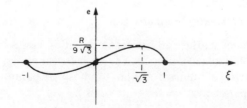

$$|e|_0 \leq \frac{l^3}{72\sqrt{3}} \text{Max} \left|\frac{\partial^3 u_{ex}}{\partial x^3}\right|_{Ve} \tag{2.2c}$$

$$|e|_1 \leq \frac{l^2}{24} \text{Max} \left|\frac{\partial^3 u_{ex}}{\partial x^3}\right|_{Ve} \tag{2.2d}$$

As for linear elements, u and $\partial u/\partial x$ are continuous within the element and only u is continuous across boundaries.

2.2.2.2 Cubic elements with equidistant nodes (four nodes, C^0)

$$\langle P \rangle = \langle 1 \quad \xi \quad \xi^2 \quad \xi^3 \rangle \tag{2.3a}$$

$$\{P_n\}^{-1} = \frac{1}{16} \begin{bmatrix} -1 & 9 & 9 & -1 \\ 1 & -27 & 27 & -1 \\ 9 & -9 & -9 & 9 \\ -9 & 27 & -27 & 9 \end{bmatrix} \tag{2.3b}$$

	$\frac{1}{c}\{N\}$	$\frac{1}{c}\{\partial N/\partial \xi\}$	c
1	$-(1-\xi)(1-9\xi^2)$	$1 + 18\xi - 27\xi^2$	
2	$9(1-\xi^2)(1-3\xi)$	$-27 - 18\xi + 81\xi^2$	1/16
3	$9(1-\xi^2)(1+3\xi)$	$27 - 18\xi - 81\xi^2$	
4	$-(1+\xi)(1-9\xi^2)$	$-1 + 18\xi + 27\xi^2$	

$$e(\xi) \leq \tfrac{1}{9}(-1 + 10\xi^2 - 9\xi^4)\tfrac{1}{24} \text{Max} \left|\frac{\partial^4 u_{ex}}{\partial \xi^4}\right|_{Vr}$$

$$|e|_0 \leq \frac{l^4}{1944} \text{Max} \left|\frac{\partial^4 u_{ex}}{\partial x^4}\right|_{Ve} \tag{2.3c}$$

$$|e|_1 \leq \frac{l^3}{108} \text{Max} \left|\frac{\partial^4 u_{ex}}{\partial x^4}\right|_{Ve} \tag{2.3d}$$

2.2.2.3 Element with n equidistant nodes (n nodes, C^0)

Functions N are Lagrange polynomials of degree $n-1$

$$N_i(\xi) = \prod_{\substack{j=1 \\ j \neq i}}^{n} \frac{(\xi_j - \xi)}{(\xi_j - \xi_i)} \tag{2.4a}$$

Because nodes are equidistant we get:

$$\xi_i = -1 + 2\frac{j-1}{n-1}$$

$$N_i(\xi) = \prod_{\substack{j=1 \\ j \neq i}}^{n} \frac{(2j - n - 1) - (n-1)\xi}{2(j-i)} \tag{2.4b}$$

These functions can also be obtained by the general method described in paragraph (1.5) with:

$$\langle P \rangle = \langle 1 \quad \xi \quad \xi^2 \quad \ldots \quad \xi^{n-1} \rangle \tag{2.4c}$$

$$\langle \xi_n \rangle = \langle -1; \; -1 + \Delta; \; -1 + 2\Delta; \ldots \; -1 + (n-1)\Delta \rangle$$

where:

$$\Delta = \frac{2}{n-1}$$

The errors are of the form (1.57a):

$$|e|_0 = C_0 l^n \operatorname{Max} \left| \frac{\partial^n u_{ex}}{\partial x^n} \right|_{V^e} \tag{2.4d}$$

$$|e|_1 = C_1 l^{n-1} \operatorname{Max} \left| \frac{\partial^n u_{ex}}{\partial x^n} \right|_{V^e} \tag{2.4e}$$

2.2.3 Elements with a Higher Degree of Inter Element Continuity

These elements are obtained by using Hermite interpolation polynomial and nodal variables that include derivatives as well.

$$\left. \frac{\partial u_{ex}}{\partial x} \right|_{\text{node } i}, \quad \left. \frac{\partial^2 u_{ex}}{\partial x^2} \right|_{\text{node } i} \quad \text{etc.}$$

We shall note:

$$\partial_x u_i = \left. \frac{\partial u}{\partial x} \right|_{x=x_i} \quad \partial_x^2 u_i = \left. \frac{\partial^2 u}{\partial x^2} \right|_{x=x_i}$$

$$\partial_\xi u_i = \left. \frac{\partial u}{\partial \xi} \right|_{\xi=\xi_i} \quad \partial_\xi^2 u_i = \left. \frac{\partial^2 u}{\partial \xi^2} \right|_{\xi=\xi_i}$$

The geometrical nodes, functions $\bar{N}$ and the Jacobian matrix $[J]$ remain identical to those of the linear element in paragraph 2.2.1.

2.2.3.1 Cubic Elements (two nodes, C^1)

Number of variables per node: 2

$$\langle P \rangle = \langle 1 \quad \xi \quad \xi^2 \quad \xi^3 \rangle \tag{2.5a}$$

$$[P_n] = \begin{bmatrix} \langle P(\xi_1) \rangle \\ \langle \frac{\partial P}{\partial \xi}(\xi_1) \rangle \\ \hdashline \langle P(\xi_2) \rangle \\ \langle \frac{\partial P}{\partial \xi}(\xi_2) \rangle \end{bmatrix} = \begin{bmatrix} 1 & -1 & 1 & -1 \\ 0 & 1 & -2 & 3 \\ 1 & 1 & 1 & 1 \\ 0 & 1 & 2 & 3 \end{bmatrix}$$

$$[P_n]^{-1} = \frac{1}{4} \begin{bmatrix} 2 & 1 & 2 & -1 \\ -3 & -1 & 3 & -1 \\ 0 & -1 & 0 & 1 \\ 1 & 1 & -1 & 1 \end{bmatrix} \tag{2.5b}$$

$$u(\xi) = \langle N \rangle \{u_n\}_\xi \quad \text{or} \quad u(\xi) = \langle N \rangle \{u_n\} \tag{2.5c}$$

where functions $\langle N \rangle$ differ for $\{u_n\}_\xi$ and $\{u_n\}$ by simple multiplication factors (see 2.5d)

	$\frac{1}{c}\{N\}$	$\frac{1}{c}\{\partial N/\partial \xi\}$	c for $\{u_n\}_\xi$	c for $\{u_n\}$
1	$(1-\xi)^2(2+\xi)$	$-3(1-\xi^2)$		1/4
2	$(1-\xi^2)(1-\xi)$	$(-1+\xi)(1+3\xi)$	1/4	1/8
3	$(1+\xi)^2(2-\xi)$	$3(1-\xi^2)$		1/4
4	$(-1+\xi^2)(1+\xi)$	$(-1-\xi)(1-3\xi)$		1/8

These last functions are Hermite polynomials that do not satisfy equation (1.30) because of the derivatives $\langle \partial P/\partial \xi \rangle$ in $[P_n]$. However, it is possible to obtain equations of the same type, for instance:

$$N_1 + N_3 = 1 \quad \text{and} \quad -N_1 + N_2 + N_3 + N_4 = \xi$$

Nodal variables on the reference element are different from those on the real element because of the differentiations with respect to ξ and x:

$$\{u_n\}_\xi = \begin{Bmatrix} u_1 \\ \partial_\xi u_1 \\ u_2 \\ \partial_\xi u_2 \end{Bmatrix}; \quad \{u_n\} = \begin{Bmatrix} \mathbf{u}_1 \\ \mathbf{u}_2 \end{Bmatrix} = \begin{Bmatrix} u_1 \\ \partial_x u_1 \\ u_2 \\ \partial_x u_2 \end{Bmatrix}$$

$$\{u_n\}_\xi = \begin{bmatrix} 1 & 0 & 0 & 0 \\ 0 & \dfrac{l}{2} & 0 & 0 \\ 0 & 0 & 1 & 0 \\ 0 & 0 & 0 & \dfrac{l}{2} \end{bmatrix} \{u_n\}; \quad l = x_2 - x_1 \tag{2.5d}$$

The derivatives with respect to x are kept as final nodal variables for the element. The four functions N_i are illustrated below by their plot in reference space:

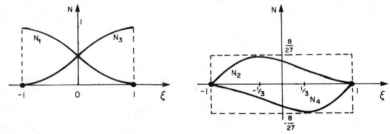

Using the method of paragraph 1.7.2, we get:

$$e(\xi) \le \tfrac{1}{24}(1-\xi^2)^2 \operatorname{Max} \left| \frac{\partial^4 u_{ex}}{\partial \xi^4} \right|_{V^r}$$

$$|e|_0 \le \frac{l^4}{384} \operatorname{Max} \left| \frac{\partial^4 u_{ex}}{\partial x^4} \right|_{V^e} \tag{2.5e}$$

$$|e|_1 \le \frac{l^3}{72\sqrt{3}} \operatorname{Max} \left| \frac{\partial^4 u_{ex}}{\partial x^4} \right|_{V^e} \tag{2.5f}$$

We note that the order of magnitude of the errors of this element is the same as that of the four-noded Lagrangian element, with differences in the numerical multiplication factors. The continuity is of class C^1, u and $\partial u/\partial x$ are continuous across boundaries.

2.2.3.2 Fifth order Element (two nodes, C^2)

We now have three variables per node: $u_i, \partial_x u_i, \partial_x^2 u_i$

$$\langle P \rangle = \langle 1 \quad \xi \quad \xi^2 \quad \xi^3 \quad \xi^4 \quad \xi^5 \rangle \tag{2.6a}$$

$$[P_n] = \begin{bmatrix} \langle P(\xi_1) \rangle \\ \langle \frac{\partial P}{\partial \xi}(\xi_1) \rangle \\ \langle \frac{\partial^2 P}{\partial \xi^2}(\xi_1) \rangle \\ \text{---------} \\ \langle P(\xi_2) \rangle \\ \langle \frac{\partial P}{\partial \xi}(\xi_2) \rangle \\ \langle \frac{\partial^2 P}{\partial \xi^2}(\xi_2) \rangle \end{bmatrix}; \quad [P_n]^{-1} = \frac{1}{16} \begin{bmatrix} 8 & 5 & 1 & 8 & -5 & 1 \\ -15 & -7 & -1 & 15 & -7 & 1 \\ 0 & -6 & -2 & 0 & 6 & -2 \\ 10 & 10 & 2 & -10 & 10 & -2 \\ 0 & 1 & 1 & 0 & -1 & 1 \\ -3 & -3 & -1 & 3 & -3 & 1 \end{bmatrix}$$

(2.6b)

$$u(\xi) = \langle N \rangle \{u_n\}_\xi$$

	$\frac{1}{c}\{N\}$	$\frac{1}{c}\{\partial N/\partial \xi\}$	c for $\{u_n\}_\xi$	c for $\{u_n\}$
1	$(1-\xi)^3(8+9\xi+3\xi^2)$	$-15(1-\xi^2)^2$		1/16
2	$(1-\xi)^3(1+\xi)(5+3\xi)$	$-(1-\xi)^2(1+3\xi)(7+5\xi)$		$l/32$
3	$(1-\xi)^3(1+\xi)^2$	$-(1-\xi)^2(1+\xi)(1+5\xi)$		1/64
4	$(1+\xi)^3(8-9\xi+3\xi^2)$	$15(1-\xi^2)^2$	1/16	1/16
5	$(1+\xi)^3(-1+\xi)(5-3\xi)$	$-(1+\xi)^2(1-3\xi)(7-5\xi)$		$l/32$
6	$(1+\xi)^3(1-\xi)^2$	$(1+\xi)^2(1-\xi)(1-5\xi)$		$l^2/64$

$$e(\xi) \leq \tfrac{1}{720}(1-\xi^2)^3 \, \text{Max} \left| \frac{\partial^6 u_{ex}}{\partial \xi^6} \right|_{V_r}$$

$$|e|_0 \leq \frac{1}{720} \frac{l^6}{64} \, \text{Max} \left| \frac{\partial^6 u_{ex}}{\partial x^6} \right|_{V_e} \tag{2.6c}$$

$$|e|_1 \leq \frac{1}{720} \frac{l^5}{8\sqrt{3}} \, \text{Max} \left| \frac{\partial^6 u_{ex}}{\partial x^6} \right|_{V_e} \tag{2.6d}$$

2.2.4 General Elements

Elements of any complexity can be constructed at will, using the following techniques:

— increase the number of geometrical nodes as well as the number of variables per node; the element can thus be a cross between a Lagrange and a Hermite element,

— add non-nodal approximations; the added degrees of freedom are not attached to any node.

The geometrical approximation $\langle \bar{N} \rangle$ and its Jacobian matrix $[J]$ are again identical to those of paragraph 2.2.1.

2.2.4.1 Lagrange–Hermite Element of fourth Order (three nodes, C^1)

$$\langle P \rangle = \langle 1 \quad \xi \quad \xi^2 \quad \xi^3 \quad \xi^4 \rangle \tag{2.7a}$$

$$[P_n] = \begin{bmatrix} \langle P(\xi_1) \rangle \\ \langle \frac{\partial P}{\partial \xi}(\xi_1) \rangle \\ \hline \langle P(\xi_2) \rangle \\ \hline \langle P(\xi_3) \rangle \\ \langle \frac{\partial P}{\partial \xi}(\xi_3) \rangle \end{bmatrix} ; [P_n]^{-1} = \frac{1}{4} \begin{bmatrix} 0 & 0 & 4 & 0 & 0 \\ -3 & -1 & 0 & 3 & -1 \\ 4 & 1 & -8 & 4 & -1 \\ 1 & 1 & 0 & -1 & 1 \\ -2 & -1 & 4 & -2 & 1 \end{bmatrix} \tag{2.7b}$$

	$\frac{1}{c}\{N\}$	$\frac{1}{c}\{\partial N/\partial \xi\}$	c for $\{u_n\}_\xi$	c for $\{u_n\}$
1	$-\xi(1-\xi)^2(3+2\xi)$	$(1-\xi^2)(-3+8\xi)$		1/4
2	$-\xi(1-\xi)(1-\xi^2)$	$(1-\xi)(-1+\xi+4\xi^2)$		1/8
3	$4(1-\xi^2)^2$	$-16\xi(1-\xi^2)$	1/4	1/4
4	$\xi(1+\xi)^2(3-2\xi)$	$(1-\xi^2)(3+8\xi)$		1/4
5	$-\xi(1+\xi)(1-\xi^2)$	$(1+\xi)(-1-\xi+4\xi^2)$		1/8

2.2.4.2 Hermite Element with One Non-nodal Degree of Freedom (two nodes, C^1)

To the nodal variables of the element described in 2.2.3.1 we add a generalized variable a_1:

$$\langle P \rangle = \langle 1 \quad \xi \quad \xi^2 \quad \xi^3 \quad \xi^4 \rangle \tag{2.8a}$$

$$[P_n] = \begin{bmatrix} \langle P(\xi_1) \rangle \\ \langle \frac{\partial P}{\partial \xi}(\xi_1) \rangle \\ \hline \langle P(\xi_2) \rangle \\ \langle \frac{\partial P}{\partial \xi}(\xi_2) \rangle \\ \hline \langle 0\ 0\ 0\ 0\ 1 \rangle \end{bmatrix}; \quad [P_n]^{-1} = \frac{1}{4}\begin{bmatrix} 2 & 1 & 2 & -1 & | & 4 \\ -3 & -1 & 3 & -1 & | & 0 \\ 0 & -1 & 0 & 1 & | & -8 \\ 1 & 1 & -1 & 1 & | & 0 \\ \hline 0 & 0 & 0 & 0 & | & 4 \end{bmatrix}$$

(2.8b)

$$u(\xi) = \langle N_1 \quad N_2 \quad N_3 \quad P_1 \rangle \{u_n\}_\xi$$

Functions N_1 to N_4 correspond to functions N_1 to N_4 of 2.2.3.1. Furthermore:

$$P_1 = (1 - \xi^2)^2$$

P_1 and $\partial P_1/\partial \xi$ vanish at the two nodes. P_1 is identical to function N_3 of the preceding paragraph.

$$\{u_n\}_\xi = \begin{bmatrix} u_1 \\ u_2 \\ a_1 \end{bmatrix}_\xi ; \quad \{u_n\} = \begin{Bmatrix} u_1 \\ u_2 \\ a_1 \end{Bmatrix}; \quad \{u_n\}_\xi = \begin{bmatrix} 1 & 0 & 0 & 0 & 0 \\ 0 & \frac{l}{2} & 0 & 0 & 0 \\ 0 & 0 & 1 & 0 & 0 \\ 0 & 0 & 0 & \frac{l}{2} & 0 \\ 0 & 0 & 0 & 0 & 1 \end{bmatrix} \{u_n\} \quad (2.8c)$$

2.3 Triangular Elements

2.3.1 *Coordinate Systems*

For all triangular elements the following reference element will be used:

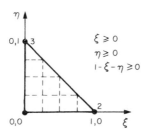

Coordinates (ξ, η) can also be interpreted as curvilinear coordinates on the real element:

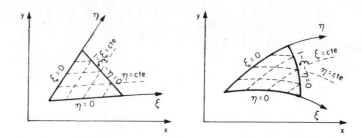

Area coordinates, also named natural coordinates and barycentric by others, have been extensively used in finite element literature. They are defined below:

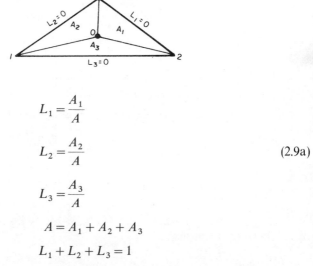

$$L_1 = \frac{A_1}{A}$$

$$L_2 = \frac{A_2}{A} \qquad (2.9a)$$

$$L_3 = \frac{A_3}{A}$$

$$A = A_1 + A_2 + A_3$$

$$L_1 + L_2 + L_3 = 1$$

A_1, A_2, A_3 are the areas of subtriangles 0–2–3, 0–3–1 and 0–1–2.
A is the total area of triangles 1–2–3.

Coordinates L_1, L_2 and L_3 are related to coordinates ξ, η by:

$$\begin{aligned} L_1 &\equiv 1 - \xi - \eta \\ L_2 &\equiv \xi \\ L_3 &\equiv \eta \end{aligned} \qquad (2.9b)$$

The reference element shown below illustrates the relationship between area and reference coordinates. By convention, the nodes of the element of reference are numerated in the positive sense of rotation on the oriented surface, starting with a corner node.

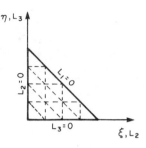

2.3.2 Linear Element (triangle, three nodes, C^0)

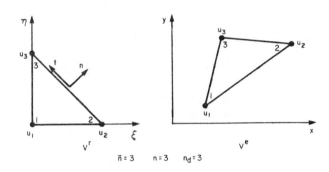

$\bar{n} = 3 \qquad n = 3 \qquad n_d = 3$

Geometrical and interpolation nodes are identical.

$$\langle P \rangle = \langle 1 \quad \xi \quad \eta \rangle \qquad (2.10\text{a})$$

$$[P_n] = \begin{bmatrix} \langle P(\xi_1) \rangle \\ \langle P(\xi_2) \rangle \\ \langle P(\xi_3) \rangle \end{bmatrix}; \quad [P_n]^{-1} = \begin{bmatrix} 1 & 0 & 0 \\ -1 & 1 & 0 \\ -1 & 0 & 1 \end{bmatrix} \qquad (2.10\text{b})$$

	$\{N\}$	$\{\partial N/\partial \xi\}$	$\{\partial N/\partial \eta\}$
1	$1 - \xi - \eta$	-1	-1
2	ξ	1	0
3	η	0	1

$$[J] = \begin{bmatrix} x_2 - x_1 & y_2 - y_1 \\ x_3 - x_1 & y_3 - y_1 \end{bmatrix}; \det(J) = 2A = (x_2 - x_1)(y_3 - y_1) -$$

$$- (x_3 - x_1)(y_2 - y_1) \qquad (2.10\text{c})$$

The graphs of functions N are:

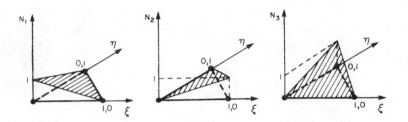

The truncation error can be obtained from equation (1.68) extended to two-dimensional space.

$$e(\xi, \eta) = \tfrac{1}{2} \langle \xi(1-\xi); -\xi\eta; \eta(1-\eta) \rangle \left\{ \begin{array}{c} \dfrac{\partial^2 u_{ex}}{\partial \xi^2} \\ 2\dfrac{\partial^2 u_{ex}}{\partial \xi\, \partial \eta} \\ \dfrac{\partial^2 u_{ex}}{\partial \eta^2} \end{array} \right\}_{\bar\xi \text{on} Vr}$$

$$|e|_0 \leq C_0 l^2 \, \text{Max} \, |D_x^2 u_{ex}|_{V^e} \qquad (2.10\text{d})$$

where
$$\text{Max} \, |D_x^2 u_{ex}|_{V^e} = \underset{V^e}{\text{Max}} \left(\left|\dfrac{\partial^2 u_{ex}}{\partial x^2}\right|, \left|2\dfrac{\partial^2 u_{ex}}{\partial x\, \partial y}\right|, \left|\dfrac{\partial^2 u_{ex}}{\partial y^2}\right| \right)$$

From reference [4 and 5] we can show that:

$$|e|_1 \leq C_1 \dfrac{l}{\sin\theta} \text{Max} \, |D_x^2 u_{ex}|_{V^e} \qquad (2.10\text{e})$$

where l is the largest characteristic length of the element, and θ is the largest interior angle of the triangle.

Term $1/\sin\theta$ comes from the transformation of $\partial u/\partial \xi$ into $\partial u/\partial x$. Function $u(x, y)$ and its first derivative are continuous within the element. However, only $u(x)$ is continuous across the boundary. Note then that tangential derivative $\partial u/\partial t$ will be continuous; only normal derivative $\partial u/\partial n$ could be discontinuous (see paragraph 2.3.3.1).

2.3.3 *Lagrangian Type (continuity C^0)*

These elements are obtained by increasing the number of interpolation nodes on the boundary and the interior of the element. The same geometrical nodes, geometrical functions $\bar N$ and Jacobian matrix $[j]$ as those of the previous paragraph are preserved. These elements are therefore sub-parametric. Curvilinear elements will be introduced in paragraph 2.3.3.5.

2.3.3.1 Quadratic Element (triangle, six nodes, C^0)

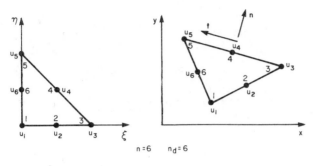

$n = 6 \quad n_d = 6$

$$\langle P \rangle = \langle 1 \quad \xi \quad \eta \quad \xi^2 \quad \xi\eta \quad \eta^2 \rangle \tag{2.11a}$$

$$\langle \xi_i \rangle = \langle 0\ 0;\ \tfrac{1}{2}\ 0;\ 1\ 0;\ \tfrac{1}{2}\ \tfrac{1}{2};\ 0\ 1;\ 0\ \tfrac{1}{2} \rangle$$

$$\langle \mathbf{x}_i \rangle = \langle x_1\ y_1;\ x_2\ y_2;\ x_3\ y_3;\ x_4\ y_4;\ x_5\ y_5;\ x_6\ y_6 \rangle$$

where

$$\mathbf{x}_2 = \tfrac{1}{2}(\mathbf{x}_1 + \mathbf{x}_3)$$
$$\mathbf{x}_4 = \tfrac{1}{2}(\mathbf{x}_3 + \mathbf{x}_5)$$
$$\mathbf{x}_6 = \tfrac{1}{2}(\mathbf{x}_5 + \mathbf{x}_1)$$

$$[P_n]^{-1} = \begin{bmatrix} 1 & 0 & 0 & 0 & 0 & 0 \\ -3 & 4 & -1 & 0 & 0 & 0 \\ -3 & 0 & 0 & 0 & -1 & 4 \\ 2 & -4 & 2 & 0 & 0 & 0 \\ 4 & -4 & 0 & 4 & 0 & -4 \\ 2 & 0 & 0 & 0 & 2 & -4 \end{bmatrix} \tag{2.11b}$$

	$\{N\}$	$\{\partial N/\partial \xi\}$	$\{\partial N/\partial \eta\}$
1	$-\lambda(1-2\lambda)$	$1-4\lambda$	$1-4\lambda$
2	$4\xi\lambda$	$4(\lambda-\xi)$	-4ξ
3	$-\xi(1-2\xi)$	$-1+4\xi$	0
4	$4\xi\eta$	4η	4ξ
5	$-\eta(1-2\eta)$	0	$-1+4\eta$
6	$4\eta\lambda$	-4η	$4(\lambda-\eta)$

$$\lambda = 1 - \xi - \eta$$

$$|e|_0 \leq C_0 l^3 \operatorname{Max}|D_x^3 u_{\text{ex}}| \tag{2.11c}$$

$$|e|_1 \leq C_1 \frac{l^2}{\sin\theta} \operatorname{Max}|D_x^3 u_{\text{ex}}| \quad (\text{see [5] page 134}) \tag{2.11d}$$

Inter-element continuity of u and $\partial u/\partial t$ is preserved but $\partial u/\partial n$ may be discontinuous. For example, on side 3–5, $\eta = 1 - \xi$

$$u_{3-5} = \langle 0; 0; -\xi(1-2\xi); 4\xi(1-\xi); (-1+\xi)(-1+2\xi); 0\rangle\{u_n\}$$

This last expression depends only on variables u_3, u_4, u_5 belonging to side 3–5. Function u is therefore continuous across that side. Since ξ and t are related by:

$$t = \frac{l}{2} - l\xi \quad 0 \le \xi \le 1 \quad -\frac{l}{2} \le t \le \frac{l}{2}$$

where l is the length of side 3–5. Differentiating u_{3-5} with respect to t gives:

$$\frac{\partial u_{3-5}}{\partial t} = \frac{\partial u}{\partial \xi}\frac{\partial \xi}{\partial t} = -l\langle 0; 0; -1+4\xi; 4-8\xi; -3+4\xi; 0\rangle\{u_n\}$$

This last expression depends only on u_3, u_4, u_5, therefore, $\partial u/\partial t$ is going to be continuous across 3–5. For $\partial u/\partial n$, however, we obtain a linear combination of $[\partial u/\partial \xi]_{3-5}$ and $[\partial u/\partial \eta]_{3-5}$ depending on all the nodal variables u_i showing that continuity is not going to be maintained in general.

2.3.3.2 *Element Using a Complete Polynomial of order r (triangle, n nodes, C^0)*

A complete polynomial of order r contains $n = (r+1)(r+2)/2$ terms and therefore, needs n nodes having one degree of freedom. A number of $3r$ nodes are distributed evenly on the boundary of the triangle and the remaining nodes distributed in the interior. For example, with $r = 4$:

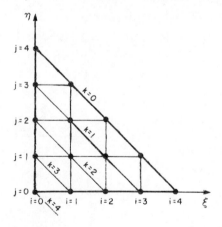

A node can be identified with three integers i, j, k related to the nodal coordinates (ξ_i, η_j) by the following expressions:

$$\xi_i = \frac{i}{r} \quad \eta_j = \frac{j}{r} \quad 0 \le i+j \le r$$

$$k = r - i - j$$

Interpolation functions can then be written explicitly for each node (i, j, k) using a product of the straight lines passing through all the nodes except (i, j):

$$N(i,j,k) = \prod_{l=0}^{i-1} \frac{l - r\xi}{l - i} \prod_{m=0}^{j-1} \frac{m - r\eta}{m - j} \prod_{n=0}^{k-1} \frac{n - r(1 - \xi - \eta)}{n - k} \qquad (2.12a)$$

These functions could also be obtained using the method of paragraph 1.4.1. Truncation errors are similar to equation (2.10d).

$$|e|_0 \leq C_0 l^{r+1} \, \text{Max} \, |D_x^{r+1} u_{\text{ex}}|_{V^e} \qquad (2.12b)$$

$$|e|_1 \leq C_1 \frac{l^r}{\sin \theta} \, \text{Max} \, |D_x^{r+1} u_{\text{ex}}|_{V^e} \qquad (2.12c)$$

2.3.3.3 Complete Cubic Element (triangle, ten nodes, C^0)

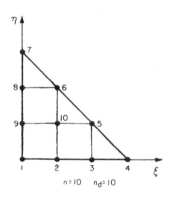

$n = 10 \quad n_d = 10$

Functions N for such an element are constructed directly from the formula (2.12a)

$$\langle P \rangle = \langle 1 \quad \xi \quad \eta \quad \xi^2 \quad \xi\eta \quad \eta^2 \quad \xi^3 \quad \xi^2\eta \quad \xi\eta^2 \quad \eta^3 \rangle$$

$$[P_n]^{-1} = \begin{bmatrix} 1 & 0 & 0 & 0 & 0 & 0 & 0 & 0 & 0 & 0 \\ -5.5 & 9 & -4.5 & 1 & 0 & 0 & 0 & 0 & 0 & 0 \\ -5.5 & 0 & 0 & 0 & 0 & 0 & 1 & -4.5 & 9 & 0 \\ 9 & -22.5 & 18 & -4.5 & 0 & 0 & 0 & 0 & 0 & 0 \\ 18 & -22.5 & 4.5 & 0 & -4.5 & -4.5 & 0 & 4.5 & -22.5 & 27 \\ 9 & 0 & 0 & 0 & 0 & 0 & -4.5 & 18 & -22.5 & 0 \\ -4.5 & 13.5 & -13.5 & 4.5 & 0 & 0 & 0 & 0 & 0 & 0 \\ -13.5 & 27 & -13.5 & 0 & 13.5 & 0 & 0 & 0 & 13.5 & -27 \\ -13.5 & 13.5 & 0 & 0 & 0 & 13.5 & 0 & -13.5 & 27 & -27 \\ -4.5 & 0 & 0 & 0 & 0 & 0 & 4.5 & -13.5 & 13.5 & 0 \end{bmatrix}$$

$$(2.13b)$$

	$\frac{1}{c}(N)$	$\frac{1}{c}\{\partial N/\partial \xi\}$	$\frac{1}{c}\{\partial N/\partial \eta\}$	c
1	$\lambda(-1+3\lambda)(-2+3\lambda)$	$-2+18\lambda-27\lambda^2$	$-2+18\lambda-27\lambda^2$	
2	$9\lambda\xi(-1+3\lambda)$	$9\lambda(-1+3\lambda-6\xi)+9\xi$	$-9\xi(-1+6\lambda)$	
3	$9\lambda\xi(-1+3\xi)$	$9\xi(1+6\lambda-3\xi)-9\lambda$	$-9\xi(-1+3\xi)$	
4	$\xi(-1+3\xi)(-2+3\xi)$	$2-18\xi+27\xi^2$	0	
5	$9\xi\eta(-1+3\xi)$	$9\eta(-1+6\xi)$	$9\xi(-1+3\xi)$	1/2
6	$9\xi\eta(-1+3\eta)$	$9\eta(-1+3\eta)$	$9\xi(-1+6\eta)$	
7	$\eta(-1+3\eta)(-2+3\eta)$	0	$2-18\eta+27\eta^2$	
8	$9\lambda\eta(-1+3\eta)$	$-9\eta(-1+3\eta)$	$9\eta(1+6\lambda-3\eta)-9\lambda$	
9	$9\lambda\eta(-1+3\lambda)$	$-9\eta(-1+6\lambda)$	$9\lambda(-1+3\lambda-6\eta)+9\eta$	
10	$54\xi\eta\lambda$	$54\eta(\lambda-\xi)$	$54\xi(\lambda-\eta)$	

where $\lambda = 1 - \xi - \eta$.

For area coordinates $L_1 = \lambda$, $L_2 = \xi$, $L_3 = \eta$.

2.3.3.4 Incomplete Cubic Element (triangle, nine nodes, C^0)

Interior nodes, like node 10 of the previous element, are often very troublesome; they can be eliminated but at the expense of the functional basis completeness. We propose the following strategy taken from reference [2].

Variable u_{10} is replaced by a linear combination of the other nine variables:

$$u_{10} = \tfrac{1}{4}(u_2 + u_3 + u_5 + u_6 + u_8 + u_9) - \tfrac{1}{6}(u_1 + u_4 + u_7) \quad (2.14a)$$

Coefficients $\tfrac{1}{4}$ and $-\tfrac{1}{6}$ are chosen so that the modified functions $\langle N_1, N_2, \ldots, N_9 \rangle$ contain complete quadratic polynomial as basis (verified by equation (1.31)) and preserve symmetry in the element of reference.

$$\{N\} = \{N\}^{(1)} + \{a\} N_{10}; \quad \left\{\frac{\partial N}{\partial \xi}\right\} = \left\{\frac{\partial N}{\partial \xi}\right\}^{(1)} + \{a\} \frac{\partial N_{10}}{\partial \xi};$$

$$\left\{\frac{\partial N}{\partial \eta}\right\} = \left\{\frac{\partial N}{\partial \eta}\right\}^{(1)} + \{a\} \frac{\partial N_{10}}{\partial \xi} \quad (2.14b)$$

where $\{N\}^{(1)}$, $\{\partial N/\partial \xi\}^{(1)}$ and $\{\partial N/\partial \eta\}^{(1)}$ are the first nine functions of the previous table; and

$$a^T = \langle -\tfrac{1}{6} \quad \tfrac{1}{4} \quad \tfrac{1}{4} \quad -\tfrac{1}{6} \quad \tfrac{1}{4} \quad \tfrac{1}{4} \quad -\tfrac{1}{6} \quad \tfrac{1}{4} \quad \tfrac{1}{4} \rangle \quad (2.14c)$$

2.3.3.5 Curvilinear Elements

When the region to be modelled by finite elements has curved boundaries, the modelling is more effectively accomplished by using elements with curved boundaries. This can be accomplished by increasing the number of geometrical nodes of the element.

(a) Element with parabolic boundaries

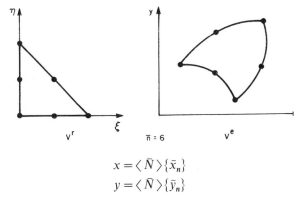

$$x = \langle \bar{N} \rangle \{\bar{x}_n\}$$
$$y = \langle \bar{N} \rangle \{\bar{y}_n\}$$

where functions $\langle \bar{N} \rangle$ are identical to functions $\langle N \rangle$ of paragraph 2.3.3.1, and $\{\bar{x}_n\}$ and $\{\bar{y}_n\}$ are the coordinates of the six geometrical nodes of the real element.

This element is isoparametric if we use the approximation of u defined in paragraph 2.3.3.1.

(b) Element with cubic boundaries

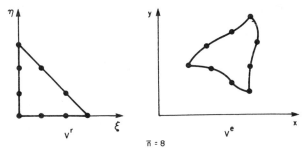

The geometrical interpolation functions $\langle \bar{N} \rangle$ are identical to the interpolation functions $\langle N \rangle$ of paragraph 2.3.3.4. Note that the geometrical distortion should not be so severe as to render $\det(J) \leq 0$.

2.3.3.6 *Non-compatible Element* (*triangle, three nodes, semi-C^0*)

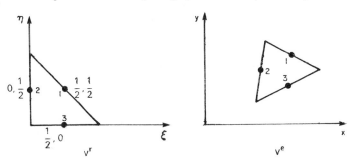

$$\langle P \rangle = \langle 1 \quad \xi \quad \eta \rangle \qquad (2.15a)$$

$$[P_n] = \begin{bmatrix} P(\xi_1) \\ P(\xi_2) \\ P(\xi_3) \end{bmatrix} ; \quad [P_n]^{-1} = \begin{bmatrix} -1 & 1 & 1 \\ 2 & -2 & 0 \\ 2 & 0 & -2 \end{bmatrix} \qquad (2.15b)$$

	$\{N\}$	$\{\partial N/\partial \xi\}$	$\{\partial N/\partial \eta\}$
1	$-1 + 2\xi + 2\eta$	2	2
2	$1 - 2\xi$	-2	0
3	$1 - 2\eta$	0	-2

This element does not preserve inter-element continuity of the function u except at the nodes.

2.3.4 Hermite Elements

The geometry of the element can be linear (paragraph 2.2.1), quadratic (2.3.3.5(a)) or cubic (2.3.3.5(b)). In the last two cases, the number of geometrical nodes is greater than the number of functional interpolation nodes.

2.3.4.1 Complete Cubic Element *(triangle, four nodes, semi-C^1)*

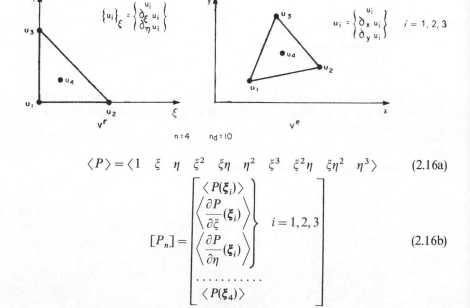

$$\langle P \rangle = \langle 1 \quad \xi \quad \eta \quad \xi^2 \quad \xi\eta \quad \eta^2 \quad \xi^3 \quad \xi^2\eta \quad \xi\eta^2 \quad \eta^3 \rangle \qquad (2.16a)$$

$$[P_n] = \begin{bmatrix} \langle P(\xi_i) \rangle \\ \left\langle \dfrac{\partial P}{\partial \xi}(\xi_i) \right\rangle \\ \left\langle \dfrac{\partial P}{\partial \eta}(\xi_i) \right\rangle \\ \cdots\cdots\cdots \\ \langle P(\xi_4) \rangle \end{bmatrix} \quad i = 1, 2, 3 \qquad (2.16b)$$

$$u(\xi) = \langle N \rangle \{u_n\}_\xi$$

		$\{N\}$	$\{\partial N/\partial \xi\}$	$\{\partial N/\partial \eta\}$
Node 1	1	$\lambda^2(3-2\lambda) - 7a$	$6\lambda(-1+\lambda) - 7b$	$6\lambda(-1+\lambda) - 7c$
	2	$\xi\lambda^2 - a$	$\lambda(\lambda - 2\xi) - b$	$-2\xi\lambda - c$
	3	$\eta\lambda^2 - a$	$-2\lambda\eta - b$	$\lambda(\lambda - 2\eta) - c$
Node 2	4	$\xi^2(3-2\xi) - 7a$	$6\xi(1-\xi) - 7b$	$-7c$
	5	$\xi^2(-1+\xi) + 2a$	$\xi(-2+3\xi) + 2b$	$2c$
	6	$\xi^2\eta - a$	$2\xi\eta - b$	$\xi^2 - c$
Node 3	7	$\eta^2(3-2\eta) - 7a$	$-7b$	$6\eta(1-\eta) - 7c$
	8	$\xi\eta^2 - a$	$\eta^2 - b$	$2\xi\eta - c$
	9	$\eta^2(-1+\eta) + 2a$	$2b$	$\eta(-2+3\eta) + 2c$
Node 4	10	$27a$	$27b$	$27c$

with $\lambda = 1 - \xi - \eta$; $a = \xi\eta\lambda$; $b = \eta(\lambda - \xi)$; $c = \xi(\lambda - \eta)$

$$\langle u_n \rangle_\xi = \langle \langle u_1 \rangle_\xi \langle u_2 \rangle_\xi \langle u_3 \rangle_\xi u_4 \rangle$$

$$\langle u_n \rangle = \langle \mathbf{u}_1 \ \mathbf{u}_2 \ \mathbf{u}_3 \ \mathbf{u}_4 \rangle$$

$$\{u_n\}_\xi = [T]\{u_n\}$$

$$[T] = \begin{bmatrix} [T_1] & & & \\ & [T_2] & & \\ & & [T_3] & \\ & & & 1 \end{bmatrix} \quad \text{where} \quad \{T_i\} = \begin{bmatrix} 1 & 0 & 0 \\ 0 & & \\ 0 & [J(\xi_i)] & \end{bmatrix} \quad i = 1, 2, 3$$

(2.16c)

2.3.4.2 Incomplete Cubic Element (triangle, three nodes, semi-C^0)

The central node of the previous element can be eliminated by using the following value for nodal variable u_4 [2]:

$$u_4 = \tfrac{1}{3}(u_1 + u_2 + u_3) + \tfrac{1}{18}(\partial_\xi u_1 - 2\partial_\xi u_2 + \partial_\xi u_3) + \tfrac{1}{18}(\partial_\eta u_1 + \partial_\eta u_2 - 2\partial_\eta u_3)$$

the numerical factors 1/3, 1/18 and -1/9 are obtained from the following conditions: modified function $\langle N_1 \ldots N_9 \rangle$ contain complete quadratic polynomial as basis (verified by equation 1.31)) and preserve symmetry in the element of reference.

Functions N and their derivatives are obtained using a technique similar to the one used in paragraph 2.3.3.4. Functions u and its first derivatives are continuous at the three nodes; u and $\partial u/\partial t$ are continuous across boundaries but $\partial u/\partial n$ is in general discontinuous.

2.3.4.3 Fifth Order Element (triangle, three nodes, C^1)

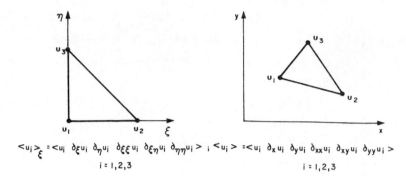

$\langle u_i \rangle_\xi = \langle u_i \; \partial_\xi u_i \; \partial_\eta u_i \; \partial_{\xi\xi} u_i \; \partial_{\xi\eta} u_i \; \partial_{\eta\eta} u_i \rangle$; $\langle u_i \rangle = \langle u_i \; \partial_x u_i \; \partial_y u_i \; \partial_{xx} u_i \; \partial_{xy} u_i \; \partial_{yy} u_i \rangle$

$i = 1,2,3$; $i = 1,2,3$

The polynomial basis $\langle P \rangle$ is complete up to terms of fourth order and contains three terms of fifth order. The last three terms are chosen to force the normal derivative on each side to be cubic in ξ and η:

$$\langle P \rangle = \langle 1 \quad \xi \quad \eta; \quad \xi^2 \quad \xi\eta \quad \eta^2; \quad \xi^3 \quad \xi^2\eta \quad \xi\eta^2 \quad \eta^3;$$
$$\xi^4 \quad \xi^3\eta \quad \xi^2\eta^2 \quad \xi\eta^3 \quad \eta^4; \quad \xi^5 - 5\xi^3\eta^2;$$
$$\xi^2\eta^3 - \xi^3\eta^2; \quad \eta^5 - 5\xi^3\eta^2 \rangle \tag{2.17a}$$

Matrix $[P_n]$ of dimension (18×18) is then non-singular and can be inverted to obtain the coefficients of the 18 functions $\langle N \rangle$. With the degree of freedom ordered as:

$$u, \frac{\partial u}{\partial \xi}, \frac{\partial u}{\partial \eta}, \frac{\partial^2 u}{\partial \xi^2}, \frac{\partial^2 u}{\partial \xi \partial \eta}, \frac{\partial^2 u}{\partial \eta^2}$$

for each node the interpolation functions are [2]:

Node 1
$$\begin{cases} N_1 = \lambda^2(10\lambda - 15\lambda^2 + 6\lambda^3 + 30\xi\eta(\xi + \eta)) \\ N_2 = \xi\lambda^2(3 - 2\lambda - 3\xi^2 + 6\xi\eta) \\ N_3 = \eta\lambda^2(3 - 2\lambda - 3\eta^2 + 6\xi\eta) \\ N_4 = \tfrac{1}{2}\xi^2\lambda^2(1 - \xi + 2\eta) \\ N_5 = \xi\eta\lambda^2 \\ N_6 = \tfrac{1}{2}\eta^2\lambda^2(1 + 2\xi - \eta) \end{cases}$$

Node 2
$$\begin{cases} N_7 = \xi^2(10\xi - 15\xi^2 + 6\xi^3 + 15\eta^2\lambda) \\ N_8 = \dfrac{\xi^2}{2}(-8\xi + 14\xi^2 - 6\xi^3 - 15\eta^2\lambda) \\ N_9 = \dfrac{\xi^2\eta}{2}(6 - 4\xi - 3\eta - 3\eta^2 + 3\xi\eta) \\ N_{10} = \dfrac{\xi^2}{4}(2\xi(1-\xi)^2 + 5\eta^2\lambda) \end{cases}$$

$$\left| \begin{array}{l} N_{11} = \dfrac{\xi^2 \eta}{2}(-2 + 2\xi + \eta + \eta^2 - \xi\eta) \\[6pt] N_{12} = \dfrac{\xi^2 \eta^2 \lambda}{4} + \dfrac{\xi^3 \eta^2}{2} \end{array} \right.$$

Node 3
$$\left\{ \begin{array}{l} N_{13} = \eta^2(10\eta - 15\eta^2 + 6\eta^3 + 15\xi^2\lambda) \\[4pt] N_{14} = \dfrac{\xi\eta^2}{2}(6 - 3\xi - 4\eta - 3\xi^2 + 3\xi\eta) \\[4pt] N_{15} = \dfrac{\eta^2}{2}(-8\eta + 14\eta^2 - 6\eta^3 - 15\xi^2\lambda) \\[4pt] N_{16} = \dfrac{\xi^2 \eta^2 \lambda}{4} + \dfrac{\xi^2 \eta^2}{2} \\[4pt] N_{17} = \dfrac{\xi\eta^2}{2}(-2 + \xi + 2\eta + \xi^2 - \xi\eta) \\[4pt] N_{18} = \dfrac{\eta^2}{4}(2\eta(1-\eta)^2 + 5\xi^2\lambda) \end{array} \right.$$

$$\lambda = 1 - \xi - \eta$$

To transform the nodal variables of the element of reference into nodal variables of the real element, the Jacobian matrix must be used for the derivative degrees of freedom.

$$\{u_i\}_\xi = [T_i]\{u_i\} \qquad (2.17b)$$

For an element with straight sides, the transformation matrix is:

$$[T_i] = \begin{bmatrix} 1 & 0 & 0 & 0 & 0 & 0 \\ 0 & J_{11} & J_{12} & 0 & 0 & 0 \\ 0 & J_{21} & J_{22} & 0 & 0 & 0 \\ 0 & 0 & 0 & J_{11}^2 & 2J_{11}J_{12} & J_{12}^2 \\ 0 & 0 & 0 & J_{11}J_{21} & J_{12}J_{21} + J_{11}J_{22} & J_{12}J_{22} \\ 0 & 0 & 0 & J_{21}^2 & 2J_{21}J_{22} & J_{22}^2 \end{bmatrix} \qquad (2.17c)$$

2.4 Quadrilateral Elements

2.4.1 Coordinate Systems

For all reference elements, the following system of reference is used:

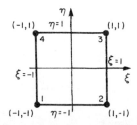

Coordinates (ξ, η) can also be viewed as curvilinear coordinates on the real element.

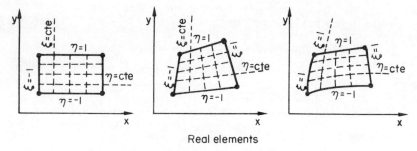

Real elements

By convention, nodes are numbered in the positive sense of rotation on the oriented surface of the element, starting with a corner node.

2.4.2 Bilinear Element (*quadrilateral, four nodes, C^0*)

The element is described in Examples 1.16 and 1.18.

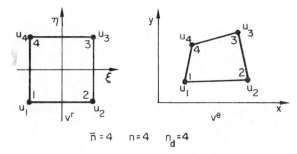

$\bar{n} = 4 \quad n = 4 \quad n_d = 4$

The element is isoparametric:

$$\langle P \rangle = \langle 1 \quad \xi \quad \eta \quad \xi\eta \rangle \qquad (2.18a)$$

$$[P_n] = \begin{bmatrix} \langle P(\xi_1) \rangle \\ \langle P(\xi_2) \rangle \\ \langle P(\xi_3) \rangle \\ \langle P(\xi_4) \rangle \end{bmatrix} \qquad [P_n]^{-1} = \frac{1}{4} \begin{bmatrix} 1 & 1 & 1 & 1 \\ -1 & 1 & 1 & -1 \\ -1 & -1 & 1 & 1 \\ 1 & -1 & 1 & -1 \end{bmatrix} \qquad (2.18b)$$

	$\frac{1}{c}\{N\}$	$\frac{1}{c}\{\partial N/\partial \xi\}$	$\frac{1}{c}\{\partial N/\partial \eta\}$	c
1	$(1-\xi)(1-\eta)$	$-1+\eta$	$-1+\xi$	
2	$(1+\xi)(1-\eta)$	$1-\eta$	$-1-\xi$	$\frac{1}{4}$
3	$(1+\xi)(1+\eta)$	$1+\eta$	$1+\xi$	
4	$(1-\xi)(1+\eta)$	$-1-\eta$	$1-\xi$	

An explicit expression for the Jacobian matrix $[J]$ is found in Example 1.18.

Truncation error can be obtained by a generalization of expression (1.68) in two-dimensional space.

2.4.3 Higher Order Lagrangian Elements (C^0)

For the elements of this paragraph we use the geometrical interpolation and Jacobian matrix of paragraph 2.4.2. Curvilinear elements will be described in paragraph 2.4.3.5.

2.4.3.1 Complete Quadratic Element (quadrilateral, nine nodes, C^0)

For this element, the complete quadratic basis of (2.19) is used. Such elements have often been used for fluid mechanics problems.

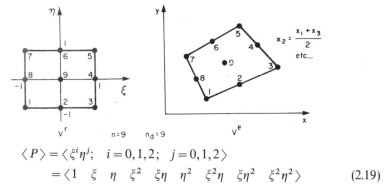

$$\langle P \rangle = \langle \xi^i \eta^j;\ i=0,1,2;\ j=0,1,2 \rangle$$
$$= \langle 1\ \ \xi\ \ \eta\ \ \xi^2\ \ \xi\eta\ \ \eta^2\ \ \xi^2\eta\ \ \xi\eta^2\ \ \xi^2\eta^2 \rangle \qquad (2.19)$$

Functions $N(\xi, \eta)$ and their derivatives are shown in the following table.

	$\{N\}$	$\{\partial N/\partial \xi\}$	$\{\partial N/\partial \eta\}$
1	$\dfrac{(1-\xi)(1-\eta)\xi\eta}{4}$	$\dfrac{(1-2\xi)(1-\eta)\eta}{4}$	$\dfrac{(1-\xi)(1-2\eta)\xi}{4}$
2	$\dfrac{-(1-\xi^2)(1-\eta)\eta}{2}$	$(1-\eta)\xi\eta$	$\dfrac{-(1-\xi^2)(1-2\eta)}{2}$
3	$\dfrac{-(1+\xi)(1-\eta)\xi\eta}{4}$	$\dfrac{-(1+2\xi)(1-\eta)\eta}{4}$	$\dfrac{-(1+\xi)(1-2\eta)\xi}{4}$
4	$\dfrac{(1+\xi)(1-\eta^2)\xi}{2}$	$\dfrac{(1+2\xi)(1-\eta^2)}{2}$	$-(1+\xi)\xi\eta$
5	$\dfrac{(1+\xi)(1+\eta)\xi\eta}{4}$	$\dfrac{(1+2\xi)(1+\eta)\eta}{4}$	$\dfrac{(1+\xi)(1+2\eta)\xi}{4}$
6	$\dfrac{(1-\xi^2)(1+\eta)\eta}{2}$	$-(1+\eta)\xi\eta$	$\dfrac{(1-\xi^2)(1+2\eta)}{2}$
7	$\dfrac{-(1-\xi)(1+\eta)\xi\eta}{4}$	$\dfrac{-(1-2\xi)(1+\eta)\eta}{4}$	$\dfrac{-(1-\xi)(1+2\eta)\xi}{4}$
8	$\dfrac{-(1-\xi)(1-\eta^2)\xi}{2}$	$\dfrac{-(1-2\xi)(1-\eta^2)}{2}$	$(1-\xi)\xi\eta$
9	$(1-\xi^2)(1-\eta^2)$	$-2(1-\eta^2)\xi$	$-2(1-\xi^2)\eta$

2.4.3.2 Incomplete Quadratic (quadrilateral, eight nodes, C^0)

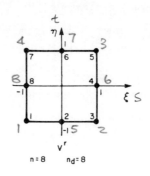

$$\langle P \rangle = \langle 1 \quad \xi \quad \eta \quad \xi^2 \quad \xi\eta \quad \eta^2 \quad \xi^2\eta \quad \xi\eta^2 \rangle \tag{2.20}$$

This element has continuity of class C^0

	$\{N\}$	$\{\partial N/\partial \xi\}$	$\{\partial N/\partial \eta\}$
1	$-\dfrac{(1-\xi)(1-\eta)(1+\xi+\eta)}{4}$	$\dfrac{(1-\eta)(2\xi+\eta)}{4}$	$\dfrac{(1-\xi)(\xi+2\eta)}{4}$
2	$\dfrac{(1-\xi^2)(1-\eta)}{2}$	$-(1-\eta)\xi$	$\dfrac{-(1-\xi^2)}{2}$
3	$-\dfrac{(1-\xi)(1-\eta)(1+\xi+\eta)}{4}$	$\dfrac{(1-\eta)(2\xi-\eta)}{4}$	$\dfrac{-(1+\xi)(\xi-2\eta)}{4}$
4	$\dfrac{(1+\xi)(1-\eta^2)}{2}$	$\dfrac{(1-\eta^2)}{2}$	$-(1+\xi)\eta$
5	$-\dfrac{(1+\xi)(1+\eta)(1-\xi-\eta)}{4}$	$\dfrac{(1+\eta)(2\xi+\eta)}{4}$	$\dfrac{(1+\xi)(\xi+2\eta)}{4}$
6	$\dfrac{(1-\xi^2)(1+\eta)}{2}$	$-(1+\eta)\xi$	$\dfrac{(1-\xi^2)}{2}$
7	$-\dfrac{(1-\xi)(1+\eta)(1+\xi-\eta)}{4}$	$\dfrac{(1+\eta)(2\xi-\eta)}{4}$	$\dfrac{-(1-\xi)(\xi-2\eta)}{4}$
8	$\dfrac{(1-\xi)(1-\eta^2)}{2}$	$\dfrac{-(1-\eta^2)}{2}$	$-(1-\xi)\eta$

2.4.3.3 Incomplete Cubic Element (quadrilateral, 16 nodes, C^0)

n = 16 n_d = 16

$$\langle P \rangle = \langle \xi^i \eta^j; \ i=0,1,2,3; \ j=0,1,2,3 \rangle \qquad (2.21)$$

	$\{N(\xi,\eta)\}$	$\{\partial N(\xi,\eta)/\partial \xi\}$	$\{\partial N(\xi,\eta)/\partial \eta\}$
1	$N_1(\xi).N_1(\eta)$	$B_1(\xi).N_1(\eta)$	$N_1(\xi).B_1(\eta)$
2	$N_2(\xi).N_1(\eta)$	$B_2(\xi).N_1(\eta)$	$N_2(\xi).B_1(\eta)$
3	$N_3(\xi).N_1(\eta)$	$B_3(\xi).N_1(\eta)$	$N_3(\xi).B_1(\eta)$
4	$N_4(\xi).N_1(\eta)$	$B_4(\xi).N_1(\eta)$	$N_4(\xi).B_1(\eta)$
5	$N_4(\xi).N_2(\eta)$	$B_4(\xi).N_2(\eta)$	$N_4(\xi).B_2(\eta)$
6	$N_4(\xi).N_3(\eta)$	$B_4(\xi).N_3(\eta)$	$N_4(\xi).B_3(\eta)$
7	$N_4(\xi).N_4(\eta)$	$B_4(\xi).N_4(\eta)$	$N_4(\xi).B_4(\eta)$
8	$N_3(\xi).N_4(\eta)$	$B_3(\xi).N_4(\eta)$	$N_3(\xi).B_4(\eta)$
9	$N_2(\xi).N_4(\eta)$	$B_2(\xi).N_4(\eta)$	$N_2(\xi).B_4(\eta)$
10	$N_1(\xi).N_4(\eta)$	$B_1(\xi).N_4(\eta)$	$N_1(\xi).B_4(\eta)$
11	$N_1(\xi).N_3(\eta)$	$B_1(\xi).N_3(\eta)$	$N_1(\xi).B_3(\eta)$
12	$N_1(\xi).N_2(\eta)$	$B_1(\xi).N_2(\eta)$	$N_1(\xi).B_2(\eta)$
13	$N_2(\xi).N_2(\eta)$	$B_2(\xi).N_2(\eta)$	$N_2(\xi).B_2(\eta)$
14	$N_3(\xi).N_2(\eta)$	$B_3(\xi).N_2(\eta)$	$N_3(\xi).B_2(\eta)$
15	$N_3(\xi).N_3(\eta)$	$B_3(\xi).N_3(\eta)$	$N_3(\xi).B_3(\eta)$
16	$N_2(\xi).N_3(\eta)$	$B_2(\xi).N_3(\eta)$	$N_2(\xi).B_3(\eta)$

where

$$\langle N_i(\xi) \rangle = \tfrac{1}{16} \langle -(1-\xi)(1-9\xi^2); \ 9(1-\xi^2)(1-3\xi); $$
$$9(1-\xi^2)(1+3\xi); \ -(1+\xi)(1-9\xi^2) \rangle$$
$$= \langle N_1(\xi) \ \ N_2(\xi) \ \ N_3(\xi) \ \ N_4(\xi) \rangle$$

$$\langle B_i(\xi) \rangle = \tfrac{1}{16} \langle 1+18\xi-27\xi^2; \ -27-18\xi+81\xi^2;$$
$$27-18\xi-81\xi^2; \ -1+18\xi+27\xi^2 \rangle$$
$$= \langle B_1(\xi) \ \ B_2(\xi) \ \ B_3(\xi) \ \ B_4(\xi) \rangle$$

Lagrangian elements having $n \times n$ nodes could easily be obtained by simple multiplication of Lagrange polynomials of order $(n - 1)$ in the two directions ξ and η.

2.4.3.4 Incomplete Cubic (quadrilateral, 12 nodes, C^0)

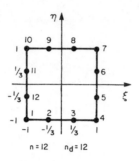

$$\langle P \rangle = \langle 1 \quad \xi \quad \eta \quad \xi^2 \quad \xi\eta \quad \eta^2 \quad \xi^3 \quad \xi^2\eta \quad \xi\eta^2 \quad \eta^3 \quad \xi^3\eta \quad \xi\eta^3 \rangle$$

	$\frac{1}{c}\{N\}$	$\frac{1}{c}\{\partial N/\partial\xi\}$	$\frac{1}{c}\{\partial N/\partial\eta\}$	c
1	$(1-\xi)(1-\eta)\lambda$	$(1-\eta)(\frac{10}{9}+2\xi-3\xi^2-\eta^2)$	$(1-\xi)(\frac{10}{9}+2\eta-\xi^2-3\eta^2)$	
2	$(1-3\xi)(1-\xi^2)(1-\eta)$	$(1-\eta)(-3-2\xi+9\xi^2)$	$(-1+\xi^2)(1-3\xi)$	
3	$(1+3\xi)(1-\xi^2)(1-\eta)$	$(1-\eta)(3-2\xi-9\xi^2)$	$(-1+\xi^2)(1+3\xi)$	
4	$(1+\xi)(1-\eta)\lambda$	$(1-\eta)(-\frac{10}{9}+2\xi+3\xi^2+\eta^2)$	$(1+\xi)(\frac{10}{9}+2\eta-\xi^2-3\eta^2)$	
5	$(1+\xi)(1-3\eta)(1-\eta^2)$	$(1-\eta^2)(1-3\eta)$	$(1+\xi)(-3-2\eta+9\eta^2)$	
6	$(1+\xi)(1+3\eta)(1-\eta^2)$	$(1-\eta^2)(1+3\eta)$	$(1+\xi)(3-2\eta-9\eta^2)$	9/32
7	$(1+\xi)(1+\eta)\lambda$	$(1+\eta)(-\frac{10}{9}+2\xi+3\xi^2+\eta^2)$	$(1+\xi)(-\frac{10}{9}+2\eta+\xi^2+3\eta^2)$	
8	$(1+3\xi)(1-\xi^2)(1+\eta)$	$(1+\eta)(3-2\xi-9\xi^2)$	$(1-\xi^2)(1+3\xi)$	
9	$(1-3\xi)(1-\xi^2)(1+\eta)$	$(1+\eta)(-3-2\xi+9\xi^2)$	$(1-\xi^2)(1-3\xi)$	
10	$(1-\xi)(1+\eta)\lambda$	$(1+\eta)(\frac{10}{9}+2\xi-3\xi^2-\eta^2)$	$(1-\xi)(-\frac{10}{9}+2\eta+\xi^2+3\eta^2)$	
11	$(1-\xi)(1+3\eta)(1-\eta^2)$	$(-1+\eta^2)(1+3\eta)$	$(1-\xi)(3-2\eta-9\eta^2)$	
12	$(1+\xi)(1-3\eta)(1-\eta^2)$	$(-1+\eta^2)(1-3\eta)$	$(1-\xi)(-3-2\eta+9\eta^2)$	

$$\lambda = (-\tfrac{10}{9} + \xi^2 + \eta^2) \tag{2.22}$$

2.4.3.5 Curvilinear Elements

Curvilinear quadrilaterals are obtained by increasing the number of geometrical nodes on the boundaries. Isoparametric elements have been the most popular such elements.

(a) Quadratic elements

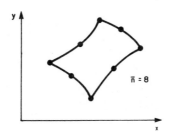

Functions $\bar{N}$ are identical to functions N of paragraph 2.4.3.2.

(b) Cubic elements

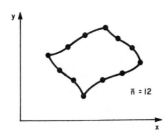

Functions $\bar{N}$ are identical to functions N of paragraph 2.4.3.4

2.4.4 Higher Order Hermite Elements

The geometrical interpolation can again be linear, quadratic or cubic as in the case of Lagrangian elements.

2.4.4.1 *Cubic Element (quadrilateral, four nodes, Semi-C^1)*

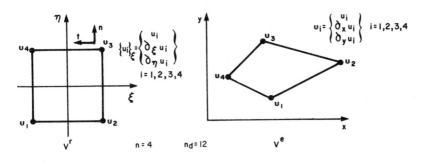

$$\langle P \rangle = \langle 1 \quad \xi \quad \eta \quad \xi^2 \quad \xi\eta \quad \eta^2 \quad \xi^3 \quad \xi^2\eta \quad \xi\eta^2 \quad \eta^3 \quad \xi^3\eta \quad \xi\eta^3 \rangle \quad (2.23a)$$

	$\dfrac{1}{e}\{N\}$	$\dfrac{1}{e}\{\partial N/\partial \xi\}$	$\dfrac{1}{e}\{\partial N/\partial \eta\}$	e
Node 1 $\begin{cases}1\\2\\3\end{cases}$	$a(\alpha-\xi-\eta)$ $a(1-\xi^2)$ $a(1-\eta^2)$	$(1-\eta)(-3+3\xi^2+\eta^2+\eta)$ $-a(1+3\xi)$ $(-1+\eta)(1-\eta^2)$	$(1-\xi)(-3+\xi^2+3\eta^2+\xi)$ $(-1+\xi)(1-\xi^2)$ $-a(1+3\eta)$	
Node 2 $\begin{cases}4\\5\\6\end{cases}$	$b(\alpha+\xi-\eta)$ $-b(1-\xi^2)$ $b(1-\eta^2)$	$(1-\eta)(3-3\xi^2-\eta^2-\eta)$ $-b(1-3\xi)$ $(1-\eta)(1-\eta^2)$	$(1+\xi)(-3-\xi+\xi^2+3\eta^2)$ $(1+\xi)(1-\xi^2)$ $-b(1+3\eta)$	$1/8$
Node 3 $\begin{cases}7\\8\\9\end{cases}$	$c(\alpha+\xi+\eta)$ $-c(1-\xi^2)$ $-c(1-\eta^2)$	$(1+\eta)(3-3\xi^2-\eta^2+\eta)$ $-c(1-3\xi)$ $(-1-\eta)(1-\eta^2)$	$(1+\xi)(3+\xi-\xi^2-3\eta^2)$ $(-1-\xi)(1-\xi^2)$ $-c(1-3\eta)$	
Node 4 $\begin{cases}10\\11\\12\end{cases}$	$d(\alpha-\xi+\eta)$ $d(1-\xi^2)$ $-d(1-\eta^2)$	$(1+\eta)(-3+3\xi^2+\eta^2-\eta)$ $-d(1+3\xi)$ $(1+\eta)(1-\eta^2)$	$(1-\xi)(3-\xi-\xi^2-3\eta^2)$ $(1-\xi)(1-\xi^2)$ $-d(1-3\eta)$	

with $a=(1-\xi)(1-\eta);\quad b=(1+\xi)(1-\eta);\quad c=(1+\xi)(1+\eta);$
$d=(1-\xi)(1+\eta);\quad \alpha=2-\xi^2-\eta^2.$

Function $\langle N \rangle$ are given for nodal variables evaluated in the reference geometry $\{u_i\}_\xi$. The transformation between the two sets of nodal variables is given by:

$$\{u_n\}_\xi = \begin{Bmatrix} \mathbf{u}_1 \\ \mathbf{u}_2 \\ \mathbf{u}_3 \\ \mathbf{u}_4 \end{Bmatrix}_\xi \quad \{u_n\} = \begin{Bmatrix} \mathbf{u}_1 \\ \mathbf{u}_2 \\ \mathbf{u}_3 \\ \mathbf{u}_4 \end{Bmatrix}_x$$

$$\{u_n\}_\xi = \begin{bmatrix} [T_1] & & & \\ & [T_2] & & \\ & & [T_3] & \\ & & & [T_4] \end{bmatrix} \{u_n\} \quad \text{where} \quad [T_i] = \begin{bmatrix} 1 & 0 & 0 \\ 0 & & \\ 0 & & [J(\xi_i)] \end{bmatrix}$$
(2.23b)

Across boundaries, u and $\partial u/\partial t$ are continuous but $\partial u/\partial n$ is not. At the nodes, however, continuity of u, $\partial u/\partial x$ and $\partial u/\partial y$ is preserved.

2.4.4.2 Continuous Rectangular Element (rectangle, four nodes, C^1)

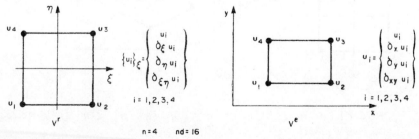

For this element continuity of class C^1 is rigorously preserved but the geometry must be rectangular with sides parallel to the global system of reference (x, y).

$$\langle P \rangle = \langle \xi^i \eta^j ; \quad i = 0, 1, 2, 3; \quad j = 0, 1, 2, 3 \rangle \tag{2.24}$$

Interpolation functions N are given below.

		$\{N(\xi,\eta)\}$	$\{\partial N(\xi,\eta)/\partial \xi\}$	$\{\partial N(\xi,\eta)/\partial \eta\}$
Node 1	1	$N_1(\xi).N_1(\eta)$	$B_1(\xi).N_1(\eta)$	$N_1(\xi).B_1(\eta)$
	2	$N_2(\xi).N_1(\eta)$	$B_2(\xi).N_1(\eta)$	$N_2(\xi).B_1(\eta)$
	3	$N_1(\xi).N_2(\eta)$	$B_1(\xi).N_2(\eta)$	$N_1(\xi).B_2(\eta)$
	4	$N_2(\xi).N_2(\eta)$	$B_2(\xi).N_2(\eta)$	$N_2(\xi).B_2(\eta)$
Node 2	5	$N_3(\xi).N_1(\eta)$	$B_3(\xi).N_1(\eta)$	$N_3(\xi).B_1(\eta)$
	6	$N_4(\xi)/N_1(\eta)$	$B_4(\xi).N_1(\eta)$	$N_4(\xi).B_1(\eta)$
	7	$N_3(\xi).N_2(\eta)$	$B_3(\xi).N_2(\eta)$	$N_3(\xi).B_2(\eta)$
	8	$N_4(\xi).N_2(\eta)$	$B_4(\xi).N_2(\eta)$	$N_4(\xi).B_2(\eta)$
Node 3	9	$N_3(\xi).N_3(\eta)$	$B_3(\xi).N_3(\eta)$	$N_3(\xi).B_3(\eta)$
	10	$N_4(\xi).N_3(\eta)$	$B_4(\xi).N_3(\eta)$	$N_4(\xi).B_3(\eta)$
	11	$N_3(\xi).N_4(\eta)$	$B_3(\xi).N_4(\eta)$	$N_3(\xi).B_4(\eta)$
	12	$N_4(\xi).N_4(\eta)$	$B_4(\xi).N_4(\eta)$	$N_4(\xi).B_4(\eta)$
Node 4	13	$N_1(\xi).N_3(\eta)$	$B_1(\xi).N_3(\eta)$	$N_1(\xi).B_3(\eta)$
	14	$N_2(\xi).N_3(\eta)$	$B_2(\xi).N_3(\eta)$	$N_2(\xi).B_3(\eta)$
	15	$N_1(\xi).N_4(\eta)$	$B_1(\xi).N_4(\eta)$	$N_1(\xi).B_4(\eta)$
	16	$N_2(\xi).N_4(\eta)$	$B_2(\xi).N_4(\eta)$	$N_2(\xi).B_4(\eta)$

$$\langle N_i(\xi) \rangle = \tfrac{1}{4} \langle (1-\xi)^2(2+\xi); \quad (1-\xi^2)(1-\xi); \quad (1+\xi^2)(2-\xi); \\ (-1+\xi^2)(1+\xi) \rangle$$
$$= \langle N_1(\xi) \quad N_2(\xi) \quad N_3(\xi) \quad N_4(\xi) \rangle$$
$$\langle B_i(\xi) \rangle = \tfrac{1}{4} \langle -3(1-\xi^2); \quad (-1+\xi)(1+3\xi); \quad 3(1-\xi^2); \\ (-1-\xi)(1-3\xi) \rangle$$
$$= \langle B_1(\xi) \quad B_2(\xi) \quad B_3(\xi) \quad B_4(\xi) \rangle$$

The nodal transformation is similar to (2.23b)

$$[T_i] = \begin{bmatrix} 1 & 0 & 0 & 0 \\ 0 & \dfrac{a}{2} & 0 & 0 \\ 0 & 0 & \dfrac{b}{2} & 0 \\ 0 & 0 & 0 & \dfrac{ab}{4} \end{bmatrix}$$

where $a = x_2 - x_1$, $b = y_4 - y_1$.

2.5 Tetrahedrons (Three-Dimensional)

2.5.1 *Coordinate System*

For all elements using such a shape, the following reference element is used:

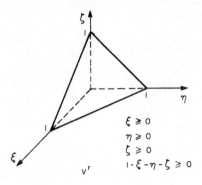

$\xi \geq 0$
$\eta \geq 0$
$\zeta \geq 0$
$1 - \xi - \eta - \zeta \geq 0$

As in other cases, coordinates (ξ, η, ζ) can be interpreted as curvilinear coordinates on the real element. Surfaces $\xi = $ constant as well as $\eta = $ constant and $\zeta = $ constant are surfaces parallel to the face of the real element. Barycentric coordinates L_1, L_2, L_3 and L_4 are also used to characterize the position of a point 0 in the element.

$$L_1 = \frac{V_1}{V}$$

$$L_2 = \frac{V_2}{V}$$

$$L_3 = \frac{V_3}{V} \quad (2.25a)$$

$$L_4 = \frac{V_4}{V}$$

$$V = V_1 + V_2 + V_3 + V_4$$

$$L_1 + L_2 + L_3 + L_4 = 1$$

V_i is the volume of tetrahedron i–j–k–l ($i, j, k, l = 1, 2, 3, 4$). For example, V_3 is the volume of tetrahedron 0–1–2–3–4. Transformation between coordinates ξ, η, ζ and L_1, L_2, L_3, L_4 are given by:

$$L_1 \equiv 1 - \xi - \eta - \zeta$$
$$L_2 \equiv \xi$$
$$L_3 \equiv \eta \quad (2.25b)$$
$$L_4 \equiv \zeta$$

Note that the nodes must be ordered in a coherent manner. Here, the first three nodes are taken in the positive sense of rotation about a unit normally oriented towards the interior of the element.

2.5.2 Linear Element (tetrahedron, four nodes, C^0)

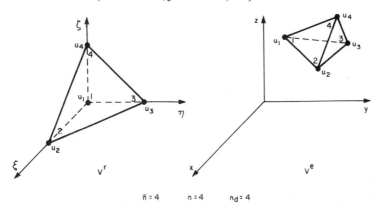

$\bar{n} = 4 \qquad n = 4 \qquad n_d = 4$

The four geometrical nodes are identical to the interpolation nodes (i.e. the element is isoparametric):

$$\langle P \rangle = \langle 1 \quad \xi \quad \eta \quad \zeta \rangle \qquad (2.26a)$$

$$[P_n] = \begin{bmatrix} \langle P(\xi_1) \rangle \\ \langle P(\xi_2) \rangle \\ \langle P(\xi_3) \rangle \\ \langle P(\xi_4) \rangle \end{bmatrix} ; \quad [P_n]^{-1} = \begin{bmatrix} 1 & 0 & 0 & 0 \\ -1 & 1 & 0 & 0 \\ -1 & 0 & 1 & 0 \\ -1 & 0 & 0 & 1 \end{bmatrix} \qquad (2.26b)$$

	$\{N\}$	$\{\partial N/\partial \xi\}$	$\{\partial N/\partial \eta\}$	$\{\partial N/\partial \zeta\}$
1	$1 - \xi - \eta - \zeta$	-1	-1	-1
2	ξ	1	0	0
3	η	0	1	0
4	ζ	0	0	1

The Jacobian matrix is:

$$[J] = \begin{bmatrix} x_2 - x_1 & y_2 - y_1 & z_2 - z_1 \\ x_3 - x_1 & y_3 - y_1 & z_3 - z_1 \\ x_4 - x_1 & y_4 - y_1 & z_4 - z_1 \end{bmatrix} \qquad (2.26c)$$

$$\det(J) = 6V$$

where V is the volume of the real element.

2.5.3 *Higher Order Lagrangian Elements* (C^0)

2.5.3.1 *Complete Quadratic (tetrahedron, ten nodes, C^0)*

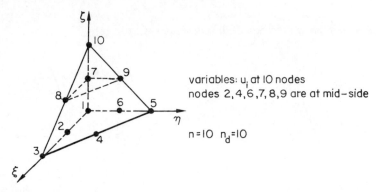

variables: u_i at 10 nodes
nodes 2,4,6,7,8,9 are at mid-side

$n = 10 \quad n_d = 10$

The linear geometrical representation of the previous element is preserved. The complete polynomial basis is:

$$\langle P \rangle = \langle 1 \quad \xi \quad \eta \quad \zeta \quad \xi^2 \quad \xi\eta \quad \eta^2 \quad \eta\zeta \quad \zeta^2 \quad \xi\zeta \rangle \tag{2.27}$$

Functions N are easily derived from those of the quadratic triangle (paragraph 2.3.3.1):

	$\{N\}$	$\{\partial N/\partial \xi\}$	$\{\partial N/\partial \eta\}$	$\{\partial N/\partial \zeta\}$
1	$-\lambda(1-2\lambda)$	$1-4\lambda$	$1-4\lambda$	$1-4\lambda$
2	$4\xi\lambda$	$4(\lambda-\xi)$	-4ξ	-4ξ
3	$-\xi(1-2\xi)$	$-1+4\xi$	0	0
4	$4\xi\eta$	4η	4ξ	0
5	$-\eta(1-2\eta)$	0	$-1+4\eta$	0
6	$4\eta\lambda$	-4η	$4(\lambda-\eta)$	-4η
7	$4\zeta\lambda$	-4ζ	-4ζ	$4(\lambda-\zeta)$
8	$4\xi\zeta$	4ζ	0	4ξ
9	$4\eta\zeta$	0	4ζ	4η
10	$-\zeta(1-2\zeta)$	0	0	$-1+4\zeta$

with $\quad \lambda = 1 - \xi - \eta - \zeta.$

2.5.3.2 *Complete Cubic Element (tetrahedron, 20 nodes, C^0)*

The interpolation functions N are obtained from the functions of paragraph 2.3.3.3 in which λ is replaced by $(1-\xi-\eta-\zeta)$: functions N_1 to N_{10} are then identical. Functions $N_{11}, N_{12}, N_{13}, N_{17}, N_{18}$ and N_{20} are identical to functions $N_9, N_{10}, N_5, N_8, N_6$ and N_7 of paragraph 2.3.3.3 in which η is replaced by ζ. Functions N_{15}, N_{16} and N_{19} are obtained from functions N_5, N_{10} and N_6 of paragraph 2.3.3.3 by replacing ξ by ζ. Functions N_{14} is $27\xi\eta\zeta$.

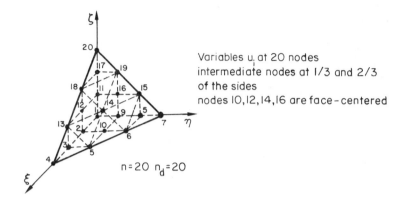

Variables u_i at 20 nodes
intermediate nodes at 1/3 and 2/3 of the sides
nodes 10,12,14,16 are face-centered

n=20 n_d=20

$$\langle P \rangle = \langle 1 \quad \xi \quad \eta \quad \zeta \quad \xi^2 \quad \xi\eta \quad \eta^2 \quad \eta\zeta \quad \zeta^2 \quad \xi\zeta \quad \xi^3 \quad \xi^2\eta \quad \xi\eta^2$$
$$\eta^3 \quad \eta^2\zeta \quad \eta\zeta^2 \quad \zeta^3 \quad \zeta^2\xi \quad \zeta\xi^2 \quad \xi\eta\zeta \rangle \tag{2.28}$$

$\langle P \rangle$ is a complete cubic polynomial basis in ξ, η, ζ.

Remark

Nodes 10, 12, 14, 16 in the middle of the faces could be eliminated as in paragraph 2.3.3.4 to reduce the number of nodes to 16.

2.5.3.3. *Curvilinear Elements*

Elements with curved surfaces can be constructed by using geometrical mapping functions $\bar{N}$ from paragraphs 2.5.3.1 and 2.5.3.2.

2.5.4 **Hermite Elements**

As in the case of the triangle, an element of class semi-C^1 would use the following nodal variables:

— $u_1, \partial_\xi u_i, \partial_\eta u_i, \partial_\zeta u_i$ at the four vertices,
— u_i at the centre of each face.

Such an element would then have eight nodes and 20 degrees of freedom, and would use a complete cubic polynomial basis. It would also be possible to eliminate the nodes in the middle of the faces by using an incomplete polynomial basis in which terms $\xi^2\eta, \eta^2\zeta, \zeta^2\xi, \xi\eta\zeta$ would be suppressed.

2.6 Hexahedrons (three-dimensional)

2.6.1 Trilinear Element (hexahedron, eight nodes, C^0)

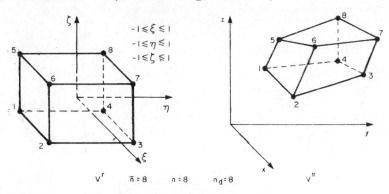

$-1 \leq \xi \leq 1$
$-1 \leq \eta \leq 1$
$-1 \leq \zeta \leq 1$

$V^r \quad \bar{n} = 8 \quad n = 8 \quad n_d = 8 \quad V^u$

This element is isoparametric; its polynomial basis is:

$$\langle P \rangle = \langle 1 \quad \xi \quad \eta \quad \zeta \quad \xi\eta \quad \eta\zeta \quad \zeta\xi \quad \xi\eta\zeta \rangle \tag{2.29}$$

Functions N are obtained by multiplications of the appropriate functions of the corresponding linear one-dimensional element.

	$\frac{1}{c}\{N\}$	$\frac{1}{c}\{\partial N/\partial \xi\}$	$\frac{1}{c}\{\partial N/\partial \eta\}$	$\frac{1}{c}\{\partial N/\partial \zeta\}$	c
1	$a_2 b_2 c_2$	$-b_2 c_2$	$-a_2 c_2$	$-a_2 b_2$	
2	$a_1 b_2 c_2$	$b_2 c_2$	$-a_1 c_2$	$-a_1 b_2$	
3	$a_1 b_1 c_2$	$b_1 c_2$	$a_1 c_2$	$-a_1 b_1$	
4	$a_2 b_1 c_2$	$-b_1 c_2$	$a_2 c_2$	$-a_2 b_1$	1/8
5	$a_2 b_2 c_1$	$-b_2 c_1$	$-a_2 c_1$	$a_2 b_2$	
6	$a_1 b_2 c_1$	$b_2 c_1$	$-a_1 c_1$	$a_1 b_2$	
7	$a_1 b_1 c_1$	$b_1 c_1$	$a_1 c_1$	$a_1 b_1$	
8	$a_2 b_1 c_1$	$-b_1 c_1$	$a_2 c_1$	$a_2 b_1$	

$$a_1 = 1 + \xi; \quad a_2 = 1 - \xi$$
$$b_1 = 1 + \eta; \quad b_2 = 1 - \eta$$
$$c_1 = 1 + \zeta; \quad c_2 = 1 - \zeta$$

2.6.2 Higher Order Lagrangian Elements (C^0)

In this family of elements the geometrical functions $\bar{N}$ are the same as in the previous element.

2.6.2.1 Complete Quadratic Element (hexahedron, 27 nodes, C^0)

The element uses a one-dimensional quadratic Lagrange polynomial in each of the three directions ξ, η, and ζ.

$$n = 27 \qquad n_d = 27$$
$$\langle P \rangle = \langle \xi^i \ \eta^j \ \zeta^k; \quad i = 0, 1, 2; \quad j = 0, 1, 2; \quad k = 0, 1, 2 \rangle \tag{2.30a}$$

Nodal coordinates are obtained from the 27 combinations of $-1, 0, 1$. The functions have the following form:

$$N(\xi, \eta, \zeta) = N(\xi).N(\eta).N(\zeta) \tag{2.30b}$$

where $N(\xi)$, $N(\eta)$ and $N(\zeta)$ are identical to function $N(\xi)$ given in paragraph 2.2.2.1.

2.6.2.2 Incomplete Quadratic Element (hexahedron, 20 nodes, C^0)

Such an element is very popular with isoparametric geometrical mapping:

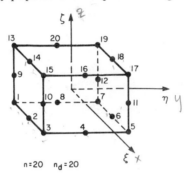

n=20 n_d=20

$$\langle P \rangle = \langle 1 \ \xi \ \eta \ \zeta; \ \xi^2 \ \xi\eta \ \eta^2 \ \eta\zeta \ \zeta^2 \ \xi\zeta;$$
$$\xi^2\eta \ \xi\eta^2 \ \eta^2\zeta \ \eta\zeta^2 \ \xi\zeta^2 \ \xi^2\zeta \ \xi^2\zeta \ \xi\eta\zeta; \ \xi^2\eta\zeta \ \xi\eta^2\zeta \ \xi\eta\zeta^2 \rangle \tag{2.31}$$

Functions N_i and their derivatives are given below:

— corner nodes

Node i	1	3	5	7	13	15	17	19
ξ_i	-1	1	1	-1	-1	1	1	-1
η_i	-1	-1	1	1	-1	-1	1	1
ζ_i	-1	-1	-1	-1	1	1	1	1

$$N_i = \tfrac{1}{8}(1 + \xi\xi_i)(1 + \eta\eta_i)(1 + \zeta\zeta_i)(-2 + \xi\xi_i + \eta\eta_i + \zeta\zeta_i)$$

$$\frac{\partial N_i}{\partial \xi} = \tfrac{1}{8}\xi_i(1 + \eta\eta_i)(1 + \zeta\zeta_i)(-1 + 2\xi\xi_i + \eta\eta_i + \zeta\zeta_i)$$

$$\frac{\partial N_i}{\partial \eta} = \tfrac{1}{8}\eta_i(1 + \xi\xi_i)(1 + \zeta\zeta_i)(-1 + \xi\xi_i + 2\eta\eta_i + \zeta\zeta_i)$$

$$\frac{\partial N_i}{\partial \zeta} = \tfrac{1}{8}\zeta_i(1 + \xi\xi_i)(1 + \eta\eta_i)(-1 + \xi\xi_i + \eta\eta_i + 2\zeta\zeta_i)$$

— nodes on the sides parallel to axis ξ

Node i	2	6	14	18
$\xi_i = 0$; η_i	-1	1	-1	1
ζ_i	-1	-1	1	1

$$N_i = \tfrac{1}{4}(1 - \xi^2)(1 + \eta\eta_i)(1 + \zeta\zeta_i)$$

$$\frac{\partial N_i}{\partial \xi} = -\tfrac{1}{2}\xi(1 + \eta\eta_i)(1 + \zeta\zeta_i)$$

$$\frac{\partial N_i}{\partial \eta} = \tfrac{1}{4}\eta_i(1 - \xi^2)(1 + \zeta\zeta_i)$$

$$\frac{\partial N_i}{\partial \zeta} = \tfrac{1}{4}\zeta_i(1 - \xi^2)(1 + \eta\eta_i)$$

— nodes on the sides parallel to axis η

Node i	4	8	16	20
$\eta_i = 0$; ξ_i	1	-1	1	-1
ζ_i	-1	-1	1	1

$$N_i = \tfrac{1}{4}(1 + \xi\xi_i)(1 - \eta^2)(1 + \zeta\zeta_i)$$

$$\frac{\partial N_i}{\partial \xi} = \tfrac{1}{4}\xi_i(1 - \eta^2)(1 + \zeta\zeta_i)$$

$$\frac{\partial N_i}{\partial \eta} = -\tfrac{1}{2}\eta(1 + \xi\xi_i)(1 + \zeta\zeta_i)$$

$$\frac{\partial N_i}{\partial \zeta} = \tfrac{1}{4}\zeta_i(1 + \xi\xi_i)(1 - \eta^2)$$

— nodes on the sides parallel to axis ζ

Nodes i	9	10	11	12
$\zeta_i = 0$; ξ_i	-1	1	1	-1
η_i	-1	-1	1	1

$$N_i = \tfrac{1}{4}(1 + \xi\xi_i)(1 + \eta\eta_i)(1 - \zeta^2)$$

$$\frac{\partial N_i}{\partial \xi} = \tfrac{1}{4}\xi_i(1 + \eta\eta_i)(1 - \zeta^2)$$

$$\frac{\partial N_i}{\partial \eta} = \tfrac{1}{4}\eta_i(1 + \xi\xi_i)(1 - \zeta^2)$$

$$\frac{\partial N_i}{\partial \zeta} = -\tfrac{1}{2}\zeta(1 + \xi\xi_i)(1 + \eta\eta_i)$$

2.6.2.3 Incomplete Cubic (hexahedron, 32 nodes, C^0)

The element has eight corner nodes and 24 mid-side nodes dividing each edge in three equal parts.

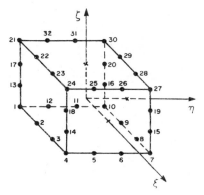

The polynomial basis is a complete cubic (20 terms) with the addition of the following terms:

$$\begin{array}{cccccc}\xi^3\eta & \xi\eta^3 & \eta^3\zeta & \eta\zeta^3 & \xi\zeta^3 & \xi^3\zeta \\ \xi^2\eta\zeta & \xi\eta^2\zeta & \xi\eta\zeta^2 & & & \\ \xi^3\eta\zeta & \xi\eta^3\zeta & \xi\eta\zeta^3. & & & \end{array} \qquad (2.32)$$

Functions N_i and their derivatives are:

— corner nodes

Node i	1	4	7	10	21	24	27	30
ξ_i	-1	1	1	-1	-1	1	1	-1
η_i	-1	-1	1	1	-1	-1	1	1
ζ_i	-1	-1	-1	-1	1	1	1	1

$$N_i = \tfrac{9}{64}(1 + \xi\xi_i)(1 + \eta\eta_i)(1 + \zeta\zeta_i)(-\tfrac{19}{9} + \xi^2 + \eta^2 + \zeta^2)$$

$$\frac{\partial N_i}{\partial \xi} = \tfrac{9}{64}(1 + \eta\eta_i)(1 + \zeta\zeta_i)(\xi_i(-\tfrac{19}{9} + 3\xi^2 + \eta^2 + \zeta^2) + 2\xi)$$

$$\frac{\partial N_i}{\partial \eta} = \tfrac{9}{64}(1+\xi\xi_i)(1+\zeta\zeta_i)(\eta_i(-\tfrac{19}{9}+\xi^2+3\eta^2+\zeta^2)+2\eta)$$

$$\frac{\partial N_i}{\partial \zeta} = \tfrac{9}{64}(1+\xi\xi_i)(1+\eta\eta_i)(\zeta_i(-\tfrac{19}{9}+\xi^2+\eta^2+3\zeta^2)+2\zeta)$$

— nodes on sides parallel to axis ξ

Node i	2	3	8	9	22	23	28	29
ξ_i	$-\tfrac{1}{3}$	$\tfrac{1}{3}$	$\tfrac{1}{3}$	$-\tfrac{1}{3}$	$-\tfrac{1}{3}$	$\tfrac{1}{3}$	$\tfrac{1}{3}$	$-\tfrac{1}{3}$
η_i	-1	-1	1	1	-1	-1	1	1
ζ_i	-1	-1	-1	-1	1	1	1	1

$$N_i = \tfrac{81}{64}(1-\xi^2)(\tfrac{1}{9}+\xi\xi_i)(1+\eta\eta_i)(1+\zeta\zeta_i)$$

$$\frac{\partial N_i}{\partial \xi} = \tfrac{81}{64}(1+\eta\eta_i)(1+\zeta\zeta_i)\left(\xi_i - \frac{2\xi}{9} - 3\xi^2\xi_i\right)$$

$$\frac{\partial N_i}{\partial \eta} = \tfrac{81}{64}\eta_i(1-\xi^2)(\tfrac{1}{9}+\xi\xi_i)(1+\zeta\zeta_i)$$

$$\frac{\partial N_i}{\partial \zeta} = \tfrac{81}{64}\zeta_i(1-\xi^2)(\tfrac{1}{9}+\xi\xi_i)(1+\eta\eta_i)$$

— nodes on sides parallel to axis η

Node i	5	6	11	12	25	26	31	32
ξ_i	1	1	-1	-1	1	1	-1	-1
η_i	$-\tfrac{1}{3}$	$\tfrac{1}{3}$	$\tfrac{1}{3}$	$-\tfrac{1}{3}$	$-\tfrac{1}{3}$	$\tfrac{1}{3}$	$\tfrac{1}{3}$	$-\tfrac{1}{3}$
ζ_i	-1	-1	-1	-1	1	1	1	1

$$N_i = \tfrac{81}{64}(1+\xi\xi_i)(1-\eta^2)(\tfrac{1}{9}+\eta\eta_i)(1+\zeta\zeta_i)$$

$$\frac{\partial N_i}{\partial \xi} = \tfrac{81}{64}\xi_i(1-\eta^2)(\tfrac{1}{9}+\eta\eta_i)(1+\zeta\zeta_i)$$

$$\frac{\partial N_i}{\partial \eta} = \tfrac{81}{64}(1+\xi\xi_i)(1+\zeta\zeta_i)\left(\eta_i - \frac{2\eta}{9} - 3\eta^2\eta_i\right)$$

$$\frac{\partial N_i}{\partial \zeta} = \tfrac{81}{64}\zeta_i(1+\xi\xi_i)(1-\eta^2)(\tfrac{1}{9}+\eta\eta_i)$$

— nodes on sides parallel to axis ζ

Node i	13	14	15	16	17	18	19	20
ξ_i	-1	1	1	-1	-1	1	1	-1
η_i	-1	-1	1	1	-1	-1	1	1
ζ_i	$-\frac{1}{3}$	$-\frac{1}{3}$	$-\frac{1}{3}$	$-\frac{1}{3}$	$\frac{1}{3}$	$\frac{1}{3}$	$\frac{1}{3}$	$\frac{1}{3}$

$$N_i = \tfrac{81}{64}(1 + \xi\xi_i)(1 + \eta\eta_i)(1 - \zeta^2)(\tfrac{1}{9} + \zeta\zeta_i)$$

$$\frac{\partial N_i}{\partial \xi} = \tfrac{81}{64}\xi_i(1 + \eta\eta_i)(1 - \zeta^2)(\tfrac{1}{9} + \zeta\zeta_i)$$

$$\frac{\partial N_i}{\partial \eta} = \tfrac{81}{64}\eta_i(1 + \xi\xi_i)(1 - \zeta^2)(\tfrac{1}{9} + \zeta\zeta_i)$$

$$\frac{\partial N_i}{\partial \zeta} = \tfrac{81}{64}(1 + \xi\xi_i)(1 + \eta\eta_i)\left(\zeta_i - \frac{2\zeta}{9} - 3\zeta^2\zeta_i\right)$$

2.6.2.4 Curvilinear Elements

(a) Element with quadratic surfaces

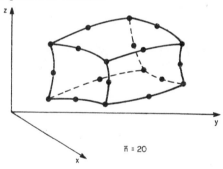

$\bar{n} = 20$

Functions $\bar{N}$ are identical to functions N of paragraph 2.6.2.2.

(b) Elements with cubic surfaces

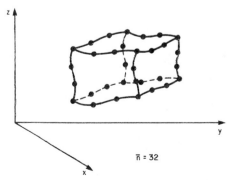

$\bar{n} = 32$

Functions $\bar{N}$ are identical to the functions N of paragraph 2.6.2.3.

2.6.3 Hermite Elements

As in other cases before, it is also possible here to construct a semi-C^1 element with eight nodes and the following nodal variables at each node:

$$u_i \quad \partial_\xi u_i \quad \partial_\eta u_i \quad \partial_\zeta u_i.$$

Elements of class C^1 are rarely used because of their very high number of degrees of freedom.

2.7 Prisms (three-dimensional)

2.7.1 Six-node Prism (*prism, six nodes, C^0*)

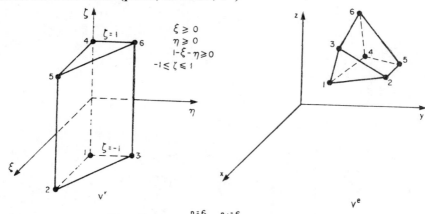

$$\langle P \rangle = \langle 1 \quad \xi \quad \eta \quad \zeta \quad \xi\zeta \quad \eta\zeta \rangle \tag{2.33}$$

	$\{N\}$	$\{\partial N/\partial \xi\}$	$\{\partial N/\partial \eta\}$	$\{\partial N/\partial \zeta\}$
1	λa	$-a$	$-a$	$-\dfrac{\lambda}{2}$
2	ξa	a	0	$-\dfrac{\xi}{2}$
3	ηa	0	a	$-\dfrac{\eta}{2}$
4	λb	$-b$	$-b$	$\dfrac{\lambda}{2}$
5	ξb	b	0	$\dfrac{\xi}{2}$
6	ηb	0	b	$\dfrac{\eta}{2}$

$$\lambda = 1 - \xi - \eta$$
$$a = \frac{1-\zeta}{2}$$
$$b = \frac{1+\zeta}{2}$$

2.7.2 Element with 15 Nodes (prism, 15 nodes, C^0)

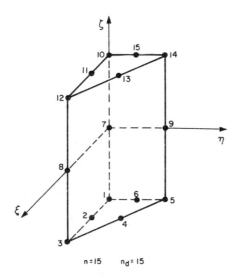

n = 15 n_d = 15

$$P = \langle 1 \;\; \xi \;\; \eta \;\; \zeta \;\; \xi\zeta \;\; \eta\zeta \;\; \xi^2\zeta \;\; \xi\eta\zeta \;\; \eta^2\zeta \;\; \zeta^2 \;\; \xi\zeta^2 \;\; \eta\zeta^2 \;\; \xi^2\zeta^2 \;\; \xi\eta\zeta^2 \;\; \eta^2\zeta^2 \rangle \quad (2.34)$$

Let us note that the polynomial basis reduces to $\langle 1, \xi, \eta \rangle$ when $\zeta = 0$. For nodes 1 to 6 ($\zeta = -1$), functions N are the same as in paragraph 2.3.3.1 multiplied by $-1/2\zeta(1-\zeta)$. For nodes 10 to 15 ($\zeta = 1$) again use the functions of paragraph 2.3.3.1 but this time, multiply by $1/2\zeta(1+\zeta)$. Finally, functions N_7, N_8 and N_9 corresponding to nodes ($\zeta = 0$) are:

$$(1 - \xi - \eta)(1 - \zeta^2); \; \xi(1 - \zeta^2); \; \eta(1 - \zeta^2)$$

2.8 Other Elements

2.8.1 Heterogeneous Elements

Elements can be constructed using groupings of functions of different nature:

$$\mathbf{u} = \begin{Bmatrix} u \\ v \\ p \end{Bmatrix}$$

where u and v could be velocities and p a pressure. Approximation for each variable may be different.

$$u = \langle N_u \rangle \{u_n\}$$
$$v = \langle N_v \rangle \{v_n\} \quad (2.35a)$$
$$p = \langle N_p \rangle \{p_n\}$$

which can be written:

$$\mathbf{u} = \begin{Bmatrix} u \\ v \\ p \end{Bmatrix} = \begin{bmatrix} \langle N_u \rangle & & 0 \\ & \langle N_v \rangle & \\ 0 & & \langle N_p \rangle \end{bmatrix} \begin{Bmatrix} \{u_n\} \\ \{v_n\} \\ \{p_n\} \end{Bmatrix} \quad (2.35b)$$

$$\mathbf{u} = [N]\{u_n\}$$

When some variables are of the same nature, identical interpolation functions are often preferred; for instance, in the example above: $\langle N_u \rangle \equiv \langle N_v \rangle$.

Example 2.1   Eight-node quadrilateral element for fluid flows

For a broad class of two-dimensional fluid flow problems we have to construct approximations for a two-dimensional velocity field as well as pressure field. The nature of the governing equation suggests a quadratic approximation for the velocity components but only a linear approximation for the pressure field.

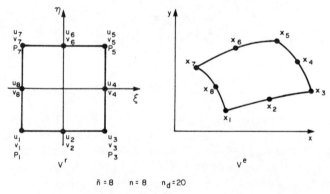

$\tilde{n} = 8 \quad n = 8 \quad n_d = 20$

We select the following approximations:

$$\begin{Bmatrix} u \\ v \\ p \end{Bmatrix} = \begin{bmatrix} [N_u\rangle \\ (1 \times 8) & 0 & 0 \\ 0 & \langle N_u \rangle \\ & (1 \times 8) & 0 \\ 0 & 0 & \langle N_p \rangle \\ & & (1 \times 4) \end{bmatrix} \begin{Bmatrix} u_n \\ (8 \times 1) \\ v_n \\ (8 \times 1) \\ P_n \\ (4 \times 1) \end{Bmatrix}$$

$$\begin{Bmatrix} x \\ y \end{Bmatrix} = \begin{bmatrix} \langle \bar{N} \rangle & 0 \\ 0 & \langle \bar{N} \rangle \end{bmatrix} \begin{Bmatrix} \{x_n\} \\ \{y_n\} \end{Bmatrix}$$

where: $\langle N_u \rangle \equiv \langle \bar{N} \rangle$ is given in paragraph 2.4.3.2, $\langle N_p \rangle$ is given in paragraph 2.4.2.

$$\{u_n\}^T = \langle u_1 \ldots u_8 \rangle$$
$$\{v_n\}^T = \langle v_1 \ldots v_8 \rangle$$
$$\{p_n\}^T = \langle p_1 \; p_3 \; p_5 \; p_7 \rangle$$
$$\{x_n\}^T = \langle x_1 \ldots x_8 \rangle$$
$$\{y_n\}^T = \langle y_1 \ldots y_p \rangle$$

2.8.2 *Modifications of Elements*

Elements with a variable number of nodes are frequently needed to obtain a transition between high and low order elements or between elements of different shapes:

— transition between linear triangle and quadratic quadrilateral;

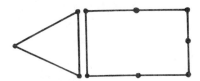

3-node triangle 7node quadrilateral

— linear interpolation in η direction and quadratic in ξ direction.

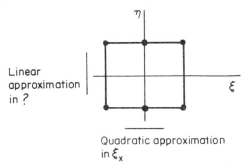

Example 2.2 Seven-node quadrilateral

We start with the full eight-node quadrilateral and write a linear constraint equation for node 8 in terms of nodes 1 and 7.

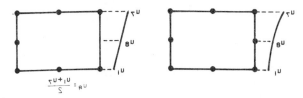

For the eight-node element:
$$\langle N \rangle = \langle N_1 N_2 \ldots N_8 \rangle$$
For a seven-node element (1 2 3 4 5 6 7):
$$\langle N \rangle = \left\langle \left(N_1 + \frac{N_8}{2}\right) N_2 N_3 N_4 N_5 N_6 \left(N_7 + \frac{N_8}{2}\right) \right\rangle$$
For a six-node element: (1 2 3 4 5 7):

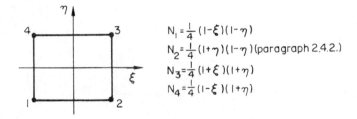

$$\langle N \rangle = \left\langle \left(N_1 + \frac{N_8}{2}\right) N_2 N_3 N_4 \left(N_5 + \frac{N_6}{2}\right) \left(N_7 + \frac{N_6}{2} + \frac{N_8}{2}\right) \right\rangle$$

2.8.3 Hierarchical Elements 7

Quadratic functions for one-, two-, and three-dimensional elements can be constructed in a hierarchy starting from the linear elements. For example, in the case of four-node elements:

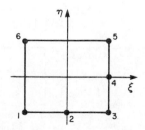

$N_1 = \frac{1}{4}(1-\xi)(1-\eta)$
$N_2 = \frac{1}{4}(1+\xi)(1-\eta)$ (paragraph 2.4.2.)
$N_3 = \frac{1}{4}(1+\xi)(1+\eta)$
$N_4 = \frac{1}{4}(1-\xi)(1+\eta)$

Adding mid-side nodes to sides $\eta = -1$ and $\xi = 1$ requires the following additive modifications:

— corner nodes:

$$N_1^* = N_1 - \frac{a}{2}$$

$$N_3^* = N_2 - \frac{a}{2} - \frac{b}{2}$$

$$N_5^* = N_3 \quad - \frac{b}{2}$$
(2.36)

$$N_6^* = N_4$$

— side nodes:

$$N_2^* = a$$
$$N_4^* = b$$

where

$$a = \tfrac{1}{2}(1 - \xi^2)(1 - \eta)$$
$$b = \tfrac{1}{2}(1 + \xi)(1 - \eta^2)$$

are the interpolation functions N_2 and N_4 of the corresponding mid-side nodes of the eight-node element of paragraph 2.4.3.2.

	Terms of functions N		
Nodes i (ξ_i)	Terms to add to all elem. with 2, 3, 4 nodes	Terms to add to elements with 3 to 4 nodes	Terms to add to elements having 4 nodes
-1	$\dfrac{1-\xi}{2}$	$-\dfrac{a}{2}$	$\tfrac{1}{16}(-1 + 9\xi + \xi^2 - 9\xi^3)$
1	$\dfrac{1+\xi}{2}$	$-\dfrac{a}{2}$	$\tfrac{1}{16}(-1 - 9\xi + \xi^2 + 9\xi^3)$
Elements (3 nodes) $\xi_i = 0$ Elements (4 nodes) $\xi_i = -\tfrac{1}{3}$	0	a	$\tfrac{1}{16}(9 - 27\xi - 9\xi^2 + 27\xi^3)$
Elements (4 nodes) $\xi_i = \tfrac{1}{3}$	0	0	$\tfrac{1}{16}(9 + 27\xi - 9\xi^2 - 27\xi^3)$

$a = 1 - \xi^2 = $ funct. N_2 of paragraph 2.2.2.1.

Figure 2.1 Functions N for one-dimensional elements having two to four nodes.

Nodes i (ξ_i, η_i)	Terms of functions N				
	Terms common to all elements (paragraph 2.4.2.)	Terms to add for each node added to one side			
		node $0,-1$	node $1,0$	node $0,1$	node $-1,0$
Corner					
$-1\quad -1$	$\frac{1}{4}(1-\xi)(1-\eta)$	$-\frac{a}{2}$	0	0	$-\frac{d}{2}$
$1\quad -1$	$\frac{1}{4}(1+\xi)(1-\eta)$	$-\frac{a}{2}$	$-\frac{b}{2}$	0	0
$1\quad 1$	$\frac{1}{4}(1+\xi)(1+\eta)$	0	$-\frac{b}{2}$	$-\frac{c}{2}$	0
$-1\quad 1$	$\frac{1}{4}(1-\xi)(1+\eta)$	0	0	$-\frac{c}{2}$	$-\frac{d}{2}$
Side					
$0\quad -1$		a	0	0	0
$1\quad 0$		0	b	0	0
$0\quad 1$		0	0	c	0
$-1\quad 0$		0	0	0	d

$\left.\begin{array}{l} a = \frac{1}{2}(1-\xi^2)(1-\eta) = \text{funct. } N_2 \\ b = \frac{1}{2}(1+\xi)(1-\eta^2) = \text{funct. } N_4 \\ c = \frac{1}{2}(1-\xi^2)(1+\eta) = \text{funct. } N_6 \\ d = \frac{1}{2}(1-\xi)(1-\eta^2) = \text{funct. } N_8 \end{array}\right\}$ paragraph 2.4.3.2.

Figure 2.2 Functions N for quadrilateral elements having four to eight nodes.

In general, the addition of a mid-side node modifies only the corresponding end nodes of that side. Term $-a/2$ must be added to each end node where a is the interpolation function of the added mid-side node. This is valid for one-, two-, and three-dimensional elements. (See Figures 2.1 and 2.2).

2.8.4 Superparametric Elements

So far, isoparametric and sub-parametric elements have been described. For superparametric elements, the degree of geometric interpolation functions $\bar{N}$ is higher than the degree of N. Using such elements poses some problems of convergence. To ensure convergence by mesh refinement, the values of u and its derivatives must have truncation errors tending towards zero as the element average size is decreased. For example, in two dimensions, this is the same as saying that the function approximation must contain a linear polynomial of the form:

$$u_0(x, y) = a_1 + a_2 x + a_3 y \tag{2.37a}$$

The geometrical transformation being

$$x = \langle \bar{N} \rangle \{x_n\}$$
$$y = \langle \bar{N} \rangle \{y_n\} \qquad (2.37b)$$

we get, for u_0, after substitution:

$$u_0(\xi, \eta) = a_1 + a_2 \langle \bar{N} \rangle \{x_n\} + a_3 \langle \bar{N} \rangle \{y_n\}$$

$$= \langle \bar{N} \rangle \begin{Bmatrix} b_1 \\ b_2 \\ \vdots \\ b_n \end{Bmatrix} \quad \text{where:} \quad b_i = a_1 + a_2 x_i + a_3 y_i$$

$$\left(\sum_i \bar{N}_i = 1 \right) \qquad (2.37c)$$

If the approximation

$$u(\xi, \eta) = \langle N \rangle \{u_n\}$$

is to contain $u_0(\xi,\eta)$ then, $\bar{N}_i$ must be a linear combination of functions N_i. This means that the polynomial basis $\langle \bar{P} \rangle$ should be included in $\langle P \rangle$ for a proper convergence. That means, for superparametric elements, the convergence criterion is violated.

2.8.5 Infinite Elements

For functions monotonically converging towards zero at infinity, it is possible to map the infinite region into a reference element using a singular mapping function.

Consider the following one-dimensional infinite element and its reference representation:

Element of reference   Real element

$$x = x_1 + \alpha \frac{1 + \xi}{1 - \zeta} \qquad (2.38a)$$

For a classical linear approximation on the element of reference with $u_2 = 0$ at infinity:

$$u(\xi) = \left\langle \frac{1-\xi}{2} \frac{1+\xi}{2} \right\rangle \begin{Bmatrix} u_1 \\ u_2 = 0 \end{Bmatrix} = \frac{1-\xi}{2} u_1 \qquad (2.38b)$$

Using transformation (2.38a)

$$\xi = \frac{\alpha(x - x_1) - 1}{\alpha(x - x_1) + 1}$$

we get

$$u(x) = \frac{1}{1 + \alpha(x - x_1)} u_1 \qquad (2.38c)$$

Such an approximation approaches zero as x tends to infinity in $1/\alpha x$. It is also possible to modify the form of $u(x)$ by multiplying interpolation functions $N_1(\xi)$ by a function $f_1(\xi)$ that becomes null for $\xi = 1$; for example

$$f(\xi) = \exp\left[-\frac{\beta(1+\xi)}{1-\xi}\right]$$

then:

$$u(\xi) = \exp\left[-\frac{\beta(1+\xi)}{1-\xi}\right] \cdot \frac{1-\xi}{2} \cdot u_1$$

$$u(x) = e^{-(\beta/\alpha)(x-x_1)} \frac{1}{1 + \alpha(x - x_1)} u_1$$

The same techniques can be extended to multidimensional regions.

Other choices of functions are contained in reference [8].

References

[1] J. J. Connor and C. A. Brebbia, *Finite Element Technique for Fluid Flow*, Butterworth, 1976.
[2] A. R. Mitchell and R. Wait, *The Finite Element Method in Partial Differential Equations*, Wiley, Chichester 1977.
[3] O. C. Zienkiewicz, *The Finite Element Method in Engineering Science*, McGraw-Hill, New York, 1st edn, 1967, 3rd edn, 1977.
[4] J. L. Synge, *The Hypercircle Method in Mathematical Physics*, Cambridge University Press, London, 1957.
[5] J. T. Oden, *Finite Elements of Non-Linear Continua*, McGraw-Hill, New York, 1972.
[6] M. Zlamal, Some recent advances in the mathematics of finite elements, *Mathematics of Finite Elements and Applications*, pp. 59–81, Academic Press, New York, 1973.
[7] K. J. Bathe and E. L. Wilson, *Numerical Methods in Finite Element Analysis*, Prentice-Hall, New Terey, 1976.
[8] P. Bettess, Infinite elements, *Int. J. Num. Meth. Eng.*, **11**, 53–64, 1977.

3

VARIATIONAL FORMULATION OF ENGINEERING PROBLEMS

3.0 Introduction

In the first two chapters, the finite element methodology is used to study the construction of continuous approximating functions that are known only at discrete points over their domain of definition. The most commonly used 'elements' are studied in detail. In the current chapter, we devote our attention to integral or variational formulations of physical problems leading to systems of simultaneous algebraic equations (linear and non-linear) after discretization. The finite element method is one of the discretization procedures that can be used to generate the systems of algebraic equations; the classical finite difference method is another one. (See Figure 3.1).

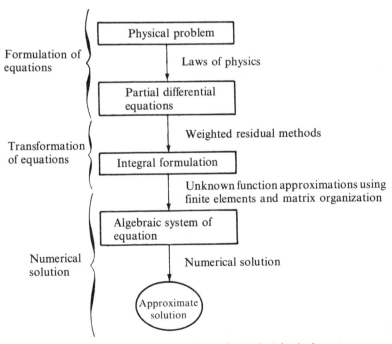

Figure 3.1 Transformation of equations of a physical system.

Physical systems may be broadly classified into two categories: discrete systems or continuous systems. Based on laws of physics, an engineering problem is thus represented

— either by a discrete system which is characterized by a set of algebraic equations involving a finite number of unknowns or degrees of freedom;

— or by a continuous system which is very often characterized by a set of partial differential equations with corresponding boundary conditions. For certain cases, one may obtain as well an equivalent formulation in terms of boundary integral equations.

The main purpose of the finite element method is to solve numerically complex continuous systems for which it is not possible to construct any analytic solution. In order to employ finite element discretization, a physical problem should be characterized by a variational formulation. We therefore present in this chapter, a general method called method of weighted residues for transforming the differential formulation into an equivalent variational formulation. By this method, any set of partial differential equations is transformed into an integral expression which depends on the choice of weighting functions employed. Some classical formulas of integral transformations are then often employed to construct modified integral formulations which would contain derivative terms of lowest order.

In continuum mechanics, one is familiar with the energy theorems and the principle of virtual work or power. A family of variational expressions are obtained by a direct application of these principles, for example, the principle of the minimum potential or complementary energy. The weighted residual method is a powerful and versatile tool for analysts to obtain a wider class of variational formulations, a sub-class of which may be simply identified with those derived through energy principles. The techniques of Lagrange multiplier and penalty parameter are employed to obtain mixed and complementary integral formulations.

For the weighted residual methods, the choice of weighting functions and the subdivision of the domain of definition have given a rich variety of procedures on the same theme for a validation criterion. The most important of these variations are:

— Galerkin or Ritz formulation;
— Point and sub-domain collocation formulation;
— Least squares formulation;
— Boundary integral formulations employing Galerkin or collocation weighting.

The method of undetermined parameters consists in replacing the unknown functions in any of the above-mentioned integral formulations by approximations of types (1.2) that depend on a finite number of parameters. This method is called the finite element method if one employs nodal approximations by subdomains presented in paragraph 1.1.2. The discretization of any of the integral

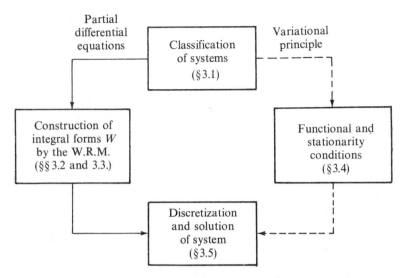

Figure 3.2 Organization of Chapter 3.

formulations thus leads to a set of algebraic equations, the solution of which is an approximate solution to the original problem.

The organization of this chapter is schematized in Figure 3.2.

3.1 Classification of Physical Problems [1, 2]

3.1.1 *Discrete and Continuous Systems*

Any physical system can be characterized by a set of variables that are functions of spatial coordinates $\mathbf{x} = (x, y, z)$ and time t. The system is said to be stationary if none of the variables are time dependent.

A number of variables $\mathbf{d}$ of the system are known *a priori*: material properties, geometrical dimensions, applied forces, boundary conditions, etc. Other variables $\mathbf{u}$ are unknown: displacements, velocities, temperatures, stresses, etc.

A mathematical model of this system results in relations between $\mathbf{u}$ and $\mathbf{d}$ after applying the law of physics to the system under study. These relations form a system of equations in u to be solved. The number of degrees of liberty of the system is equal to the number of parameters required to define u at a given instant of time.

The system is discrete if the number of degrees of liberty is finite; it is continuous if the number of degrees of liberty is infinite.

A discrete system is described by a set of algebraic equations. A continuous system is governed by differential equations, partial differential equations and/or integro-differential equations accompanied by appropriate spatial and temporal boundary conditions.

Algebraic systems of equations can be solved directly by numerical methods as described in Chapter 5. On the other hand, continuous systems must first be discretized, i.e. replaced by approximately equivalent system of algebraic equations.

3.1.2 *Equilibrium, Eigenvalue, and Propagation Problems*

Discrete and continuous systems can be subdivided into three broad classes that we briefly describe below.

(a) *Equilibrium or Boundary Value Problems*

For such problems one must find the unknown values **u** in a steady state system. For a discrete system, the governing equations can be written in matrix form as follows:

$$[K]\{U\} = \{F\} \tag{3.1a}$$

where $[K]$ is a matrix of coefficients characterizing the system, $\{U\}$ are the unknown variables, and $\{F\}$ are the applied generalized forces.

Continuous systems, on the other hand, are described by partial differential equations:

$$\begin{aligned} \mathcal{L}(\mathbf{u}) + \mathbf{f}_V &= 0 \quad \text{on a domain } V \\ \mathcal{C}(\mathbf{u}) &= \mathbf{f}_S \quad \text{on boundary } S \text{ of } V \end{aligned} \tag{3.1b}$$

where $\mathcal{L}$ and $\mathcal{C}$ are differential operators characterizing the system, **u** are the unknown functions, and $\mathbf{f}_V$ and $\mathbf{f}_S$ are given functions.

(b) *Eigenvalue Problems*

This is a sub-class of the general problem of equilibrium obtained when the formulation leads to homogeneous equations depending upon supplementary unknown parameters λ, called eigenvalues.

— For a discrete system:

$$[K]\{U\} = \lambda[M]\{U\} \tag{3.2a}$$

where $[M]$ is a mass matrix.

— For a continuous system, the formulation generally leads to:

$$\begin{aligned} \mathcal{L}_1(\mathbf{u}) &= \lambda \mathcal{L}_2(\mathbf{u}) \quad \text{on the domain } V \\ \mathcal{C}_1(\mathbf{u}) &= \lambda \mathcal{C}_2(\mathbf{u}) \quad \text{on boundary } S \end{aligned} \tag{3.2b}$$

where $\mathcal{L}_1, \mathcal{L}_2, \mathcal{C}_1, \mathcal{C}_2$ are differential operators.

(c) *Propagation or Initial Value Problems*

In that class of problems one must evaluate $\mathbf{u}(x, t)$ for $t > t_0$ when $u(\mathbf{x}, t_0)$ is known.

— For a discrete system we get equations of the following type:

$$[M]\frac{d^2}{dt^2}\{U\} + [C]\frac{d}{dt}\{U\} + [K]\{U\} = \{F(t)\} \quad \text{for} \quad t > t_0 \quad (3.3a)$$

with initial conditions:

$$\{U\} = \{U_0\} \quad \text{and} \quad \frac{d}{dt}[U] = \{\dot{U}_0\} \quad \text{for} \quad t = t_0$$

where $[C]$ is the damping matrix.
— For a continuous system we get:

$$m\frac{\partial^2 \mathbf{u}}{\partial t^2} + c\frac{\partial \mathbf{u}}{\partial t} + \mathscr{L}(\mathbf{u}) + \mathbf{f}_V = 0 \quad \text{on} \quad V$$

$$\mathscr{C}(\mathbf{u}) = \mathbf{f}_S \quad \text{on} \quad S \quad (3.3b)$$

with initial conditions:

$$\mathbf{u} = \mathbf{u}_0 \quad \text{and} \quad \frac{\partial \mathbf{u}}{\partial t} = \dot{\mathbf{u}}_0 \quad \text{for} \quad t = t_0$$

Figure 3.3 summarizes the classification system that will be used. We now introduce a few definitions to describe the equations of physical systems.

— A discrete system is linear if the terms of $[K]$, $[M]$, $[C]$ and $\{F\}$ are constants independent of $\mathbf{u}$.

— A continuous system is linear if the expressions $\mathscr{L}(\mathbf{u})$ and $\mathscr{C}(\mathbf{u})$ are linear in $\mathbf{u}$ and its derivatives. Furthermore, f_V, f_S, m, c are independent of $\mathbf{u}$ and its derivatives.

We may now write:

$$\mathscr{L}(\mathbf{u}) = [\mathscr{L}]\{u\}; \quad \mathscr{C}(\mathbf{u}) = [\mathscr{C}]\{u\}$$

where $[\mathscr{L}]$ and $[\mathscr{C}]$ are matrices of differential operators independent of $\mathbf{u}$.

For example, the Laplacian operator is:

$$\mathscr{L}(u) = \Delta u = \frac{\partial^2 u}{\partial x^2} + \frac{\partial^2 u}{\partial y^2} = \left[\frac{\partial}{\partial x^2} + \frac{\partial}{\partial y^2}\right] \cdot u$$

— A system of partial differential equations is of order m if it contains derivatives up to order m only.

— A differential operator is homogeneous if:

$$\mathscr{L}(u = 0) = 0$$

— A system of linear partial differential equations:

$$[\mathscr{L}]\{u\} + \{f_V\} = 0$$

is homogeneous if:

$$\{f_V\} = 0$$

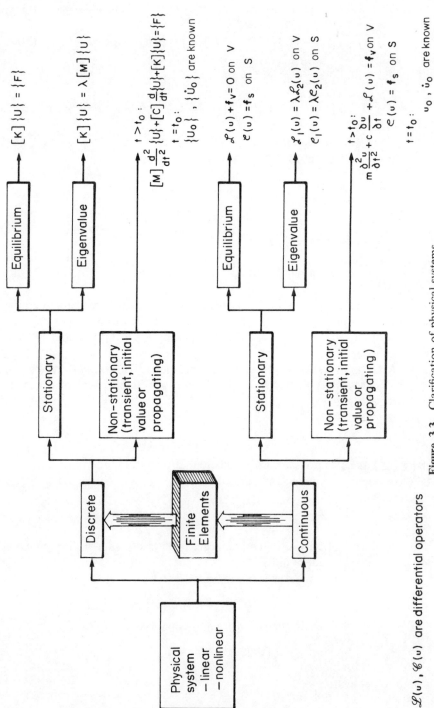

Figure 3.3 Clarification of physical systems.

and the boundary conditions

$$[\mathscr{C}]\{u\} = \{f_S\}$$

are homogeneous if:

$$\{f_S\} = 0$$

— A system of partial differential equations is self-adjoint or symmetrical if:

$$\int_V \langle u \rangle [\mathscr{L}]\{V\} \, dV = \int_V \langle v \rangle [\mathscr{L}]\{u\} \, dV \qquad (3.4a)$$

where u and v are sufficiently continuous function satisfying the homogeneous boundary conditions:

$$\mathscr{C}(\mathbf{u}) = \mathscr{C}(\mathbf{v}) = 0 \qquad (3.4b)$$

— A system of partial differential equations is positive if:

$$\int_V \langle u \rangle [\mathscr{L}]\{u\} \, dV \geq 0 \qquad (3.4c)$$

for any function $\mathbf{u}$ satisfying (3.4b).

If (3.4c) is null for $u = 0$ only, the system is positive definite.

Example 3.1 Two-dimensional continuous problems

Equilibrium problem:

Poisson's equation in two dimensions is:

$$\frac{\partial^2 u}{\partial x^2} + \frac{\partial^2 u}{\partial y^2} + f_V = 0 \quad \text{on} \quad V$$

That equation governs many physical problems; for example, the problem of steady state heat conduction in a two-dimensional homogeneous and isotropic region. The solution becomes unique after one of the following two types of boundary conditions are applied on every point of the boundary S of the domain V.

— Condition on u (Dirichlet):

$$u = u_S \quad \text{on} \quad S_u$$

where S_u represents the part of S where u is specified.

— Condition on $\partial u/\partial n$ or flux condition:

$$\frac{\partial u}{\partial n} + \alpha u = f_S \quad \text{on} \quad S_f$$

where S_f is the remaining part of boundary S where the flux is specified.

If $\alpha \neq 0$, the boundary condition is also known as a Cauchy condition; when $\alpha = 0$ it is frequently called a Neuman condition.

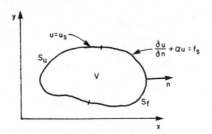

The governing equation and its boundary conditions can be written in matrix form as:

$$\left[\frac{\partial^2}{\partial x^2} + \frac{\partial^2}{\partial y^2}\right] \cdot u + f_V = 0 \quad \text{or} \quad [\mathscr{L}]\{u\} + \{f_V\} = 0 \quad \text{on} \quad V$$

$$\left[\frac{\partial}{\partial n} + \alpha\right] \cdot u = f_S \quad \text{or} \quad [\mathscr{C}_f]\{u\} = \{f_S\} \quad \text{on} \quad S_f$$

$$[1] \cdot u = u_S \quad \text{or} \quad [\mathscr{C}_u]\{u\} = \{f_u\} \quad \text{on} \quad S_u$$

Solving such a problem requires finding u satisfying all three equations.

The system is self-adjoint since we can show (using integration by parts) that:

$$\int_V u\left[\frac{\partial^2 v}{\partial x^2} + \frac{\partial^2 v}{\partial y^2}\right] dV = \int_V v\left[\frac{\partial^2 u}{\partial x^2} + \frac{\partial^2 u}{\partial y^2}\right] dV$$

when u and v satisfy homogeneous boundary conditions:

$$\left.\begin{array}{l}\frac{\partial u}{\partial n} + \alpha u = 0 \\ \frac{\partial v}{\partial n} + \alpha v = 0\end{array}\right\} \text{ on } S_f; \quad \left.\begin{array}{l}u = 0 \\ v = 0\end{array}\right\} \text{ on } S_u$$

We can also show in a similar fashion that the system is positive definite because:

$$\int_V u\left[\frac{\partial^2 u}{\partial x^2} + \frac{\partial^2 u}{\partial y^2}\right] dV > 0 \quad \text{if } \alpha \geq 0$$

for any u different from zero satisfying homogeneous boundary conditions.

Eigenvalue problem

The Helmholtz equation is:

$$\frac{\partial^2 u}{\partial x^2} + \frac{\partial^2 u}{\partial y^2} + \lambda u = 0 \quad \text{on} \quad V$$

It is associated with Neuman or Dirichlet boundary conditions. For a complete solution one needs to find both λ and **u**. This formulation is obtained for many physical problems in electro-magnetism, acoustics and vibrations of elastic membranes under tension.

Propagation problems (transient)

A typical transient problem can be described as:

$t > t_0$

$$-\frac{\partial u}{\partial t} + \frac{\partial^2 u}{\partial x^2} + \frac{\partial^2 u}{\partial y^2} + f(t) = 0 \quad \text{on} \quad V$$

with boundary conditions

$$\frac{\partial u}{\partial n} + \alpha u = f_s \quad \text{on} \quad S_f$$

$$u = u_s \quad \text{on} \quad S_u$$

and initial conditions:

$$u = u_0 \quad \text{on} \quad V \quad \text{for} \quad t = t_0$$

For instance, this equation describes the transient temperature distribution over a two-dimensional heat conduction problem.

Example 3.2 Navier–Stokes equations

A transient, two-dimensional, incompressible laminar viscous flow is governed by:

$$\frac{\partial u}{\partial t} + u\frac{\partial u}{\partial x} + v\frac{\partial u}{\partial y} + \frac{1}{\rho}\frac{\partial p}{\partial x} - \frac{\mu}{\rho}\left(\frac{\partial^2 u}{\partial x^2} + \frac{\partial^2 u}{\partial y^2}\right) + f_x = 0$$

$$\frac{\partial v}{\partial t} + u\frac{\partial v}{\partial x} + v\frac{\partial v}{\partial y} + \frac{1}{\rho}\frac{\partial p}{\partial y} - \frac{\mu}{\rho}\left(\frac{\partial^2 v}{\partial x^2} + \frac{\partial^2 v}{\partial y^2}\right) + f_y = 0$$

$$\frac{\partial u}{\partial x} + \frac{\partial v}{\partial y} = 0$$

where u and v are the components of the velocity vector field, p is the static pressure, f_x, f_y are components of a body vector field, μ is the dynamic viscosity factor of the fluid, and ρ is the mass density of the fluid.

The boundary conditions for $t > t_0$ are:

$$\left.\begin{array}{r}-p + 2\mu\dfrac{\partial u_n}{\partial n} = f_n \\[6pt] \mu\left(\dfrac{\partial u_s}{\partial n} + \dfrac{\partial u_n}{\partial s}\right) = f_s\end{array}\right\} \quad \text{on} \quad S_f$$

$$\left.\begin{array}{r}u_n = u_{ns} \\ u_s = u_{ss}\end{array}\right\} \quad \text{on} \quad S_u$$

where n and s are normal and tangent directions on the boundary.

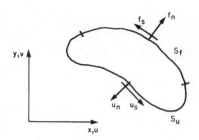

Initial conditions are:

$$\left.\begin{array}{l} u = u_0 \\ v = v_0 \\ p = p_0 \end{array}\right\} \text{ on } V$$

The non-linear equations are written as:

where:
$$\{\dot{u}\} + [\mathscr{L}]\{u\} + \{f_V\} = 0$$

$$\{\dot{u}\} = \left\{\begin{array}{c} \dfrac{\partial u}{\partial t} \\ \dfrac{\partial v}{\partial t} \\ 0 \end{array}\right\}; \quad \{u\} = \left\{\begin{array}{c} u \\ v \\ p \end{array}\right\}; \quad \{f_V\} = \left\{\begin{array}{c} f_x \\ f_y \\ 0 \end{array}\right\}$$

$$[\mathscr{L}] = \left[\begin{array}{c|c|c} u\dfrac{\partial}{\partial x} + v\dfrac{\partial}{\partial y} - \dfrac{\mu}{\rho}\left(\dfrac{\partial^2}{\partial x^2} + \dfrac{\partial^2}{\partial y^2}\right) & 0 & \dfrac{1}{\rho}\dfrac{\partial}{\partial x} \\ \hline 0 & u\dfrac{\partial}{\partial x} + v\dfrac{\partial}{\partial y} - \dfrac{\mu}{\rho}\left(\dfrac{\partial^2}{\partial x^2} + \dfrac{\partial^2}{\partial y^2}\right) & \dfrac{1}{\rho}\dfrac{\partial}{\partial y} \\ \hline \dfrac{\partial}{\partial x} & \dfrac{\partial}{\partial y} & 0 \end{array}\right]$$

For steady state problems, suppress $\{\dot{u}\}$.

3.2 Weighted Residual Method [3]

3.2.1 *Residuals*

Consider a steady state continuous physical system leading to the following system of partial differential equations of order m (linear or non-linear):

$$\mathscr{L}(\mathbf{u}) + \mathbf{f}_V = 0 \quad \text{on domain } V \qquad (3.5a)$$

The boundary conditions are:

$$\mathcal{C}(\mathbf{u}) = \mathbf{f}_S \quad \text{on boundary } S \tag{3.5b}$$

Variables **u** are functions of the geometrical coordinates **x**. Solutions are functions **u** satisfying both (3.5a) and (3.5b).

Define a residual function as follows:

$$\mathbf{R}(\mathbf{u}) = \mathcal{L}(\mathbf{u}) + \mathbf{f}_V \tag{3.6}$$

Obviously, the residual function vanishes when u is a solution of (3.5).

3.2.2 Integral Forms

The weighted residual method consists in finding functions **u** that satisfy the following integral equation:

$$W(\mathbf{u}) = \int_V \langle \psi \rangle \{\mathbf{R}(\mathbf{u})\} \, dV = \int_V \langle \psi \rangle \{\mathcal{L}(\mathbf{u}) + f_V\} \, dV = 0 \tag{3.7}$$

for any weighting function ψ belonging to a set of functions E_ψ. Solutions **u** belong to an admissible set E_u satisfying the boundary conditions (3.5b).

Any solution **u** satisfying (3.5a) and (3.5b) also satisfies (3.7) no matter what is the weighting function. However, a solution of (3.7) depends on the choice of the weighting function. For example, if E_ψ contains a complete set of Dirac functions $\delta(x)$ over V, then the solution u from equation (3.7) will also be the solution of (3.5a) since the residue R becomes identically zero at all points of V. If the set E_ψ is finite, a function **u** satisfying (3.7) is in general an approximate solution of (3.5a). That is, it does not satisfy (3.5a) at all points of V, exactly.

Example 3.3   Integral form of the Poisson's equation

The integral form of Poisson's equation is:

$$W = \int_V \psi(x,y) \left(\frac{\partial^2 u}{\partial x^2} + \frac{\partial^2 u}{\partial y^2} + f_V \right) dV = 0$$

where **u** must be twice differentiable and it should satisfy all the boundary conditions on S_u and S_f.

Example 3.4   Integral form of the Navier–Stokes equations

The integral form of the Navier–Stokes equations can be written in matrix form as:

$$W = \int_V \langle \psi_u(x,y) \psi_v(x,y) \psi_p(x,y) \rangle \left\{ [\mathcal{L}] \begin{Bmatrix} u \\ v \\ p \end{Bmatrix} + f_V \right\} dV = 0$$

where u, v possess second derivatives and p possesses first derivatives, u, v, p satisfy all boundary conditions on S_u and S_f, and $[\mathcal{L}]$ is defined in Example 3.2

3.3 Integral Transformations

3.3.1 *Integration by Parts*

Integration by parts is used in expressions like (3.7) to lower the order of the highest derivatives contained in the integrand. Let us first recall the basic equations of integration by parts.

(a) One dimension

$$\int_{x_1}^{x_2} \psi \frac{du}{dx} dx = -\int_{x_1}^{x_2} \frac{d\psi}{dx} u \, dx + (\psi u)\Big|_{x_1}^{x_2} \tag{3.8a}$$

$$\int_{x_1}^{x_2} \psi \frac{d^2 u}{dx^2} dx = -\int_{x_1}^{x_2} \frac{d\psi}{dx} \frac{du}{dx} dx + \left(\psi \frac{du}{dx}\right)\Big|_{x_1}^{x_2} \tag{3.8b}$$

(b) Two dimensions

$$\int_A \psi \frac{\partial u}{\partial x} dx\, dy = -\int_A \frac{\partial \psi}{\partial x} u \, dx\, dy + \oint_S \psi u \, dy$$

$$= -\int_A \frac{\partial \psi}{\partial x} u \, dx\, dy + \int_S \psi u l \, dS \tag{3.9a}$$

contour integration

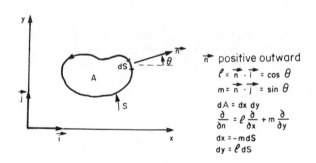

$\vec{n}$ positive outward
$\ell = \vec{n} \cdot \vec{i} = \cos\theta$
$m = \vec{n} \cdot \vec{j} = \sin\theta$
$dA = dx\, dy$
$\frac{\partial}{\partial n} = \ell \frac{\partial}{\partial x} + m \frac{\partial}{\partial y}$
$dx = -m\, dS$
$dy = \ell\, dS$

$$\int_A \psi \frac{\partial u}{\partial y} dx\, dy = -\int_A \frac{\partial \psi}{\partial y} u \, dx\, dy - \oint_S \psi u \, dx$$

$$= -\int_A \frac{\partial \psi}{\partial y} u \, dx\, dy + \oint_S \psi u m \, dS \tag{3.9b}$$

$$\int_A \psi \frac{\partial^2 u}{\partial x^2} dx\, dy = -\int_A \frac{\partial \psi}{\partial x} \frac{\partial u}{\partial x} dx\, dy + \oint_S \psi \frac{\partial u}{\partial x} l \, dS$$

$$\int_A (\psi \Delta u - u \Delta \psi) dx\, dy = \oint_S \left(\psi \frac{\partial u}{\partial n} - u \frac{\partial \psi}{\partial n}\right) dS \quad \text{where} \quad \Delta = \frac{\partial^2}{\partial x^2} + \frac{\partial^2}{\partial y^2}$$

$$\tag{3.9c}$$

(c) Three dimensions

$$\int_V \psi \frac{\partial u}{\partial x} dx\,dy\,dz = -\int_V \frac{\partial \psi}{\partial x} u\,dx\,dy\,dz + \int_S \psi\,u\,l\,dS \qquad (3.10)$$

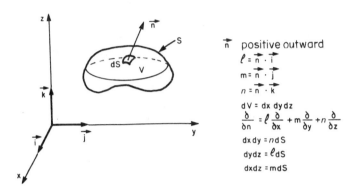

$\vec{n}$ positive outward
$\ell = \vec{n} \cdot \vec{i}$
$m = \vec{n} \cdot \vec{j}$
$n = \vec{n} \cdot \vec{k}$
$dV = dx\,dy\,dz$
$\frac{\partial}{\partial n} = \ell \frac{\partial}{\partial x} + m \frac{\partial}{\partial y} + n \frac{\partial}{\partial z}$
$dx\,dy = n\,dS$
$dy\,dz = \ell\,dS$
$dx\,dz = m\,dS$

3.3.2 Weak Integral Forms

A given integral form may be transformed to obtain a so-called weak form through integration by parts. If necessary, one may even perform successive integration by parts to obtain any desired integral population. Such expressions are sought for the following reasons:

— the order of the highest derivative of u is reduced; this then relaxes the continuity conditions required for convergence;
— some boundary conditions appearing in the weak integral do not have to be satisfied identically by **u**.

On the other hand, integration by parts introduces derivatives of the weighting function ψ. The continuity conditions on ψ are then more severe than before such an operation.

The net effect of integration by parts is a relaxing of continuity conditions for admissible functions **u** to be used in the solution of (3.5).

Example 3.5 Weak integral form for the Poisson equation

In the integral form of Example 3.3, u must:

— be differentiable twice;
— satisfy all boundary conditions on S_u and S_f; the choice of weighting functions ψ is unrestricted.

After an integration by parts:

$$W = -\int_V \left(\frac{\partial \psi}{\partial x}\frac{\partial u}{\partial x} + \frac{\partial \psi}{\partial y}\frac{\partial u}{\partial y} - \psi f_V\right) dV +$$

$$+ \int_{S_f} \psi \frac{\partial u}{\partial n} dS + \int_{S_u} \psi \frac{\partial u}{\partial n} dS = 0$$

Functions ψ and u must be differentiable once. The introduction of contour integrations on S_f and S_u allows us to use the boundary condition on S_f:

$$\frac{\partial u}{\partial n} = f_S - \alpha u \text{ on } S_f$$

to replace:

$$\int_{S_f} \psi \frac{\partial u}{\partial n} dS \text{ by } \int_{S_f} \psi(f_S - \alpha u) dS$$

Furthermore, the contour integral on S_u can be eliminated by using

$$\psi = 0 \text{ on } S_u$$

The weak integral form then reduces to:

$$W = -\int_V \left(\frac{\partial \psi}{\partial x}\frac{\partial u}{\partial x} + \frac{\partial \psi}{\partial y}\frac{\partial u}{\partial y} - \psi f_V\right) dV + \int_{S_f} \psi(f_S - \alpha u) dS = 0 \qquad (1)$$

where u and ψ must satisfy the following boundary conditions:

$$u = u_S \text{ on } S_u$$
$$\psi = 0 \text{ on } S_u$$

After two integrations by parts of the integral form of Example 3.3:

$$W = \int_V \left(\left(\frac{\partial^2 \psi}{\partial x^2} + \frac{\partial^2 \psi}{\partial y^2}\right) u + \psi f_V\right) dV + \oint_S \left(\psi \frac{\partial u}{\partial n} - \frac{\partial \psi}{\partial n} u\right) dS = 0 \qquad (2)$$

If we choose weighting functions satisfying $\partial^2 \psi/\partial x^2 + \partial^2 \psi/\partial y^2 = 0$ at all points of V and $f_V = 0$, then equation (2) no longer contains any volume integral.

$$W = \oint_S \left(\psi \frac{\partial u}{\partial n} - \frac{\partial \psi}{\partial n} u\right) dS = 0 \qquad (3)$$

This last expression is a contour integral on the boundary of the region. The above expression is in effect a basis of the 'boundary integral method'.

Summary

Formulation	Conditions on u			Conditions on ψ		
	Derivat. order	condition on S_f	condition on S_u	Derivat. order	condition on S_f	condition S_u
Partial differential equations	2	$\dfrac{\partial u}{\partial n} + \alpha u = f_S$	$u = u_S$	null	null	null
Internal form of Example 3.3	2	$\dfrac{\partial u}{\partial n} + \alpha u = f_S$	$u = u_S$	null	null	null
↓ Integral form (1)	1	null	$u = u_S$	1	null	$\psi = 0$
↓ If $f_V = 0$ integral form (3) (no volume integral)	null	$\dfrac{\partial u}{\partial n} + \alpha u = f_S$	$u = u_S$	ψ satisfy $\Delta\psi \equiv 0$ (no condition on S)		

For a problem described by a partial differential system of order m such as (3.5) and for its integral form (3.7), admissible functions must be differentiable m times and satisfy all boundary conditions. After integration by parts, the conditions of admissibility for u and ψ are:

— u must be $m-s$ differentiable;
— ψ must be s differentiable;
— u must satisfy only those boundary conditions containing derivatives up to order $m-s-1$;
— ψ must be null on the boundaries where u must satisfy the remaining boundary conditions; boundary conditions containing derivatives of order higher than $m-s$ are then automatically taken care of by the integral form, in an approximate manner.

3.3.3 *Construction of Additional Integral Forms*

In practice, the system $\mathscr{L}(\mathbf{u}) + f_V = 0$ is often obtained after elimination of a number of variables $\mathbf{q}$ between various field equations, for example:

$$\mathscr{L}_e(\mathbf{q}) - \mathbf{f}_V = 0: \text{equilibrium or conservation laws}$$
$$\mathscr{L}_c(\mathbf{q}, \mathbf{u}) = 0: \text{constitutive laws} \qquad (3.11)$$

The operator $\mathscr{L}$ is generally of a higher order than $\mathscr{L}_e$ and $\mathscr{L}_c$. If the integral forms are constructed directly from (3.11), supplementary variables **q** will appear in the formulation but the conditions of differentiability of **u** will be reduced.

Example 3.6 Construction of a supplementary form for Poisson's equation applied to heat flow

The two-dimensional heat flow equation is:

$$\mathscr{L}_e(\mathbf{q}) - f_V = \frac{\partial q_x}{\partial x} + \frac{\partial q_y}{\partial y} - f_V = 0$$

where **q** is the heat flow and f_V a heat source.

Heat flow q and temperature are related by:

$$\mathscr{L}_e(\mathbf{q}, \mathbf{u}) = 0 \begin{cases} q_x + \dfrac{\partial u}{\partial x} = 0 \\ q_y + \dfrac{\partial u}{\partial y} = 0 \end{cases}$$

where $u(x, y)$ is the temperature at point (x, y).

Poisson's equation is obtained after eliminating q_x and q_y between the three previous equations:

$$\frac{\partial^2 u}{\partial x^2} + \frac{\partial^2 u}{\partial y^2} + f_V = 0$$

Applying the weighted residual method directly to operators $\mathscr{L}_e$ and $\mathscr{L}_c$ we get:

$$W_r = \int_V \langle \psi_u \rangle \{\mathscr{L}_e(\mathbf{q}) - f_V\} dV + \int_V \langle \psi_q \rangle \{\mathscr{L}_c(\mathbf{q}, \mathbf{u})\} dV = 0 \qquad (3.12a)$$

Since ψ_u and ψ_q are independent

$$\int_V \langle \psi_u \rangle \{\mathscr{L}_e(\mathbf{q}) - f_V\} dV = 0$$

$$\int_V \langle \psi_q \rangle \{\mathscr{L}_c(\mathbf{q}, \mathbf{u})\} dV = 0 \qquad (3.12b)$$

where **u** and **q** must satisfy all the boundary conditions on S_u and S_f.

Example 3.7 Mixed integral form for Poisson's equation

Using the equations of Example 3.6 to construct W_r, we get:

$$W_r = \int_V \left(\psi_u \left(\frac{\partial q_x}{\partial x} + \frac{\partial q_y}{\partial y} - f_V \right) + \psi_{q_x} \left(q_x + \frac{\partial u}{\partial x} \right) \right.$$
$$\left. + \psi_{q_y} \left(q_y + \frac{\partial u}{\partial y} \right) \right) dV = 0$$

After choosing weighting functions similar to the functions u, q_x and q_y and the notation:

$$\psi_u = \delta u$$
$$\psi_{q_x} = \delta q_x$$
$$\psi_{q_y} = \delta q_y$$

we get:

$$W_r = \int_V \left(\delta u \left(\frac{\partial q_x}{\partial x} + \frac{\partial q_y}{\partial y} - f \right) + \delta q_x \left(q_x + \frac{\partial u}{\partial x} \right) + \right.$$
$$\left. + \delta q_y \left(q_y + \frac{\partial u}{\partial y} \right) \right) dV = 0$$

3.4 Functionals [4, 5]

We now show that the weighted residual method is equivalent to minimizing a functional in certain cases. For example, in mechanics of solids, the functional could be the total potential energy of the mechanical system. This gives an integral formulation directly from the stationarity property of the functional. The last method is very useful when the total potential energy is easier to write than the partial differential equations of equilibrium (3.5)

3.4.1 First Variation

A functional π is a function of an ensemble of functions and their derivatives:

$$\pi = \pi \left(\mathbf{u}, \frac{\partial \mathbf{u}}{\partial x}, \ldots \right) \tag{3.13a}$$

The first variation of π is defined as:

$$\delta \pi = \frac{\partial \pi}{\partial \mathbf{u}} \delta \mathbf{u} + \frac{\partial \pi}{\partial \left(\frac{\partial \mathbf{u}}{\partial x} \right)} \delta \left(\frac{\partial \mathbf{u}}{\partial x} \right) + \cdots \tag{3.13b}$$

where $\delta \mathbf{u}, \delta(\partial u/\partial x)$ are arbitrary variations of $\mathbf{u}$ and $\partial u/\partial x$.
The variation operator δ has the following properties:

$$\delta \left(\frac{\partial u}{\partial x} \right) = \frac{\partial (\delta u)}{\partial x}$$

$$\delta(\delta u) = 0$$

$$\delta \left(\int_V u \, dV \right) = \int_V \delta u \, dV \tag{3.14}$$

$$\delta(u + v) = \delta u + \delta v$$
$$\delta(uv) = u \delta v + v \delta u = \delta(vu)$$
$$\delta(cu) = c \delta u \, (c = \text{constant})$$

Example 3.8   One dimensional functional

Consider the following one-dimensional functional:

$$\pi\left(u, \frac{du}{dx}\right) = \int_{x_1}^{x_2} \left(\frac{1}{2}\left(\frac{du}{dx}\right)^2 - uf\right) dx, \quad (f \text{ is constant})$$

Its first variation from (3.13b) is:

$$\delta\pi = \delta \int_{x_1}^{x_2} \left(\frac{1}{2}\left(\frac{du}{dx}\right)^2 - uf\right) dx$$

and using the properties of δ:

$$\delta\pi = \int_{x_1}^{x_2} \left(\delta\left(\frac{du}{dx}\right)\frac{du}{dx} - \delta u f\right) dx = \int_{x_1}^{x_2} \left(\frac{d(\delta u)}{dx}\frac{du}{dx} - \delta u f\right) dx$$

The second variation of π can be obtained by repeating the previous steps for ($\delta\pi$):

$$\delta^2\pi = \delta(\delta\pi) = \int_{x_1}^{x_2} \left(\delta\left(\frac{du}{dx}\right)\right)^2 dx = \int_{x_1}^{x_2} \left(\frac{d(\delta u)}{dx}\right)^2 dx$$

because

$$\delta(\delta u) = 0$$

3.4.2 Functional Associated with an Integral Formulation

For certain problems defined by (3.5a) and (3.5b) it is possible to construct a functional $\pi(\mathbf{u}, \partial\mathbf{u}/\partial x, \ldots)$ such that:

$$\delta\pi \equiv W = 0 \tag{3.15a}$$

where W is a particular integral form (for example, of Galerkin type) obtained when ψ is chosen to be $\delta\mathbf{u}$ in equation (3.7). Integration by parts is used if necessary on the expression

$$W = \int_V \langle \delta\mathbf{u} \rangle \{\mathscr{L}(\mathbf{u}) + f_V\} dV = 0 \tag{3.15b}$$

to obtain the desired formulation. This is possible under the following circumstances:

— $\mathscr{L}$ and $\mathscr{C}$ are linear operators and all derivatives are of even order;
— $\mathbf{f}_S$ and $\mathbf{f}_V$ are independent of $\mathbf{u}$.

These conditions are sufficient for a functional to exist but they are not necessary.

Example 3.9   Functional for Poisson's equation

$$\mathscr{L}(u) + f_V = \frac{\partial^2 u}{\partial x^2} + \frac{\partial^2 u}{\partial y^2} + f_V = 0$$

The corresponding integral form is obtained in Example 3.5:

$$W = \int_V \left(\frac{\partial \psi}{\partial x} \frac{\partial u}{\partial x} + \frac{\partial \psi}{\partial y} \frac{\partial u}{\partial y} - \psi f_V \right) dV + \int_{S_f} \psi(\alpha u - f_S) dS = 0$$

Choosing $\psi = \delta u$ for weighting functions, we get:

$$W = \int_V \left(\frac{\partial(\delta u)}{\partial x} \frac{\partial u}{\partial x} + \frac{\partial(\delta u)}{\partial y} \frac{\partial u}{\partial y} - \delta u f_V \right) dV + \int_{S_f} \delta u(\alpha u - f_S) dS = 0$$

Defining a functional π by:

$$\pi\left(u, \frac{\partial u}{\partial x}, \frac{\partial u}{\partial y}\right) = \int_V \left(\frac{1}{2}\left(\frac{\partial u}{\partial x}\right)^2 + \frac{1}{2}\left(\frac{\partial u}{\partial y}\right)^2 - u f_V \right) dV + \int_{S_f} (\tfrac{1}{2}\alpha u^2 - u f_S) dS$$

we verify readily that

$$\delta \pi \equiv W = 0$$

Equation (3.15) can be interpreted as a stationarity condition for functional π. A solution $\mathbf{u}$ for $W=0$ also renders the functional stationary, i.e. $\delta \pi = 0$. The functional is either a minimum or a maximum depending upon whether the sign of the second variation $\delta^2 \pi$ is either positive or negative, respectively.

$$\delta^2 \pi\left(\mathbf{u}, \frac{\partial \mathbf{u}}{\partial x}, \ldots\right) = \frac{\partial^2 \pi}{\partial \mathbf{u}} \delta\mathbf{u}\delta\mathbf{u} + \frac{\partial^2 \pi}{\partial\left(\frac{\partial \mathbf{u}}{\partial x}\right)^2} \delta\left(\frac{\partial \mathbf{u}}{\partial x}\right) \delta\left(\frac{\partial \mathbf{u}}{\partial x}\right) + \ldots \quad (3.16)$$

Example 3.10 Second variation of the functional corresponding to Poisson's equation

From Example 3.9 we get:

$$\delta^2 \pi = \int_V \left(\left(\frac{\partial(\delta u)}{\partial x}\right)^2 + \left(\frac{\partial(\delta u)}{\partial y}\right)^2 \right) dV + \int_{S_f} \alpha(\delta u)^2 dS$$

The value of δ^2 is always positive for non-zero value of δu and α positive. The solution is then a minimum of π.

A functional $\pi(\mathbf{u}, \partial \mathbf{u}/\partial x, \ldots)$ is linear if it contains only linear terms of $\mathbf{u}$ and its derivative; for example:

$$\pi\left(\mathbf{u}, \frac{\partial \mathbf{u}}{\partial x}\right) = \int_V \left(a_1 u + a_2 \frac{\partial u}{\partial x}\right) dV \quad (3.17)$$

A functional is quadratic if all its terms are of second order; for example:

$$\pi = \int_V \left(a_1 \left(\frac{\partial u}{\partial x}\right)^2 + a_2 u^2\right) dV \quad (3.18)$$

Some quadratic functionals may contain linear terms. For a purely quadratic functional it is possible to use the following matrix representation:

$$\pi = \frac{1}{2}\int_V \left\langle \mathbf{u} \; \frac{\partial \mathbf{u}}{\partial x} \; \ldots \right\rangle [D] \begin{Bmatrix} \mathbf{u} \\ \frac{\partial \mathbf{u}}{\partial x} \\ \vdots \end{Bmatrix} dV \qquad (3.19)$$

where $[D]$ is a symmetrical matrix independent of $\mathbf{u}$. The first and second variations are then given by:

$$\delta\pi = \int_V \left\langle \delta\mathbf{u} \; \frac{\partial(\delta\mathbf{u})}{\partial x} \; \ldots \right\rangle [D] \begin{Bmatrix} \mathbf{u} \\ \frac{\partial \mathbf{u}}{\partial x} \\ \vdots \end{Bmatrix} dV \qquad (3.20a)$$

$$\delta^2\pi = \int_V \left\langle \delta\mathbf{u} \; \frac{\partial(\delta\mathbf{u})}{\partial x} \; \ldots \right\rangle [D] \begin{Bmatrix} \delta\mathbf{u} \\ \frac{\partial(\delta\mathbf{u})}{\partial x} \\ \vdots \end{Bmatrix} dV \qquad (3.20b)$$

The functional π is positive definite if matrix $[D]$ is positive definite, i.e. all the eigenvalues of matrix $[D]$ are positive. Then the second variation $\delta^2\pi$ is also positive and π is a minimum.

Example 3.11 *Matrix form of the functional formulation of Poisson's equation*

Functional π of Example 3.9 can be written in matrix form as follows:

$$\pi = \frac{1}{2}\int_V \left(\left\langle \frac{\partial u}{\partial x} \; \frac{\partial u}{\partial y} \right\rangle \begin{bmatrix} 1 & 0 \\ 0 & 1 \end{bmatrix} \begin{Bmatrix} \frac{\partial u}{\partial x} \\ \frac{\partial u}{\partial y} \end{Bmatrix} - 2uf_V \right) dV + \int_{S_f} \left(\frac{\alpha u^2}{2} - uf_S \right) dS$$

Its second variation is:

$$\delta W = \delta^2\pi = \int_V \left(\left\langle \frac{\partial(\delta u)}{\partial x} \; \frac{\partial(\delta u)}{\partial y} \right\rangle \begin{bmatrix} 1 & 0 \\ 0 & 1 \end{bmatrix} \begin{Bmatrix} \frac{\partial(\delta u)}{\partial x} \\ \frac{\partial(\delta u)}{\partial y} \end{Bmatrix} \right) dV + \int_{S_f} \alpha(\delta u)^2 dS$$

In this case, $[D]$ is a unit matrix and therefore positive definite.

3.4.3 Stationarity Principle

Rewriting partial differential equations (3.5) and separating boundary conditions of S_f and S_u:

$$\mathcal{L}(\mathbf{u}) + \mathbf{f}_V = 0 \quad \text{on} \quad V \qquad (3.21a)$$

$$\mathscr{C}_f(\mathbf{u}) = \mathbf{f}_S \quad \text{on} \quad S_f \tag{3.21b}$$

$$\mathscr{C}_u(\mathbf{u}) = \mathbf{f}_u \quad \text{on} \quad S_u \tag{3.21c}$$

The integral form obtained using the weighted residual method is:

$$W(\mathbf{u}) = \int_V \langle \psi \rangle \{\mathscr{L}(\mathbf{u}) + f_V\} \, dV = 0 \quad \text{for all } \psi \tag{3.22}$$

$$\mathscr{C}_f(\mathbf{u}) = \mathbf{f}_S \quad \text{on} \quad S_f$$
$$\mathscr{C}_u(\mathbf{u}) = \mathbf{f}_u \quad \text{on} \quad S_u$$

Choosing the weighting function as: $\psi = \delta\mathbf{u}$ and integrating by parts it is possible, in certain cases, to construct a functional π which is stationary for a solution $\mathbf{u}$:

$$\delta\pi(\mathbf{u}) \equiv W(\mathbf{u}) = 0 \tag{3.23}$$

$$\mathscr{C}_u(\mathbf{u}) = \mathbf{f}_u \quad \text{on} \quad S_u \tag{3.24}$$

The stationarity principle can be stated as follows:

— Amongst all the admissible functions $\mathbf{u}$ (i.e. those satisfying derivability conditions and boundary conditions on S_u), one that satisfies (3.21a and 3.21b) renders the functional π stationary.

3.4.4 Lagrange Multipliers and Additional Functionals

In the functional π, the only unknown variables are functions $\mathbf{u}$ and they must satisfy the boundary conditions on S_u. With Lagrange multipliers it is possible to construct other functionals π^* for which the stationarity conditions give other integral formulations that could have the following characteristics:

— introduction of additional physical variables as unknown functions;
— relaxation of the continuity conditions for admissible functions $\mathbf{u}$.

Example 3.12 *Lagrange multiplier*

The extremum of functions:

$$\pi_1(u, v) = u^2 + v^2$$

is defined by:

$$\delta\pi_1 = 2u\,\delta u + 2v\,\delta v = 0 \quad \text{for all } \delta u\, \delta v$$

Hence

$$u = v = 0 \quad \pi_1(0,0) = 0$$

Presume that we look for a minimum of π_1, with condition:

$$g(u, v) = u - v + 2 = 0$$

A first method consists in using $g(u,v)=0$ to eliminate v from the expression for π_1:

$$\pi(u) = 2u^2 + 4u + 4$$
$$\delta\pi = 4(u+1)\delta u = 0$$
$$u = -1 \quad v = 1 \quad \pi(-1,1) = 2$$

A second method requires us to study the stationarity of a modified functional:

$$\pi^*(u,v,\lambda) = \pi_1(u,v) + \lambda g(u,v) = u^2 + v^2 + \lambda(u - v + 2)$$

where λ is a Lagrange multiplier corresponding to the condition $g = 0$.
The first variation of π^* is:

$$\delta\pi^* = \frac{\partial \pi^*}{\partial u}\delta u + \frac{\partial \pi^*}{\partial v}\delta v + \frac{\partial \pi^*}{\partial \lambda}\delta\lambda = 0 \quad \text{for all } \delta u, \delta v, \delta\lambda$$

Thus

$$\delta\pi^* = (2u + \lambda)\delta u + (2v - \lambda)\delta v + (u - v + 2)\delta\lambda = 0$$

hence

$$\begin{aligned} 2u + \lambda &= 0 & u &= -1 \\ 2v - \lambda &= 0 &\text{hence}\quad v &= 1 \\ u - v + 2 &= 0 & \lambda &= 2 \end{aligned}$$

The method avoids the elimination of one variable but leads to a functional π^* in 3 variables u, v, λ, instead of π in one variable u.

To generalize the results of the previous example we look for functions $\mathbf{u}$ that will make a functional $(\mathbf{u}, \mathbf{q})$ stationary, subject to the conditions:

$$\begin{aligned} g_1(\mathbf{u},\mathbf{q}) &= 0 \\ g_2(\mathbf{u},\mathbf{q}) &= 0 \\ &\cdots\cdots\cdots \quad \text{on } V \\ g_m(\mathbf{u},\mathbf{q}) &= 0 \end{aligned} \qquad (3.25)$$

where $\mathbf{q}$ are supplementary physical variables. A first method consists in eliminating m variables between $\mathbf{u}$ and $\mathbf{q}$ using (3.25). A second method introduces m Lagrange multipliers $(\lambda_1, \lambda_2, \ldots, \lambda_m)$ in the generalized functional π^*:

$$\pi^*(\mathbf{u},\mathbf{q},\lambda) = \pi_1(\mathbf{u},\mathbf{q}) + \int_V (\lambda_1 g_1(\mathbf{u},\mathbf{q}) + \lambda_2 g_2(\mathbf{u},\mathbf{q}) + \ldots + \lambda_m g_m(\mathbf{u},\mathbf{q}))dV \qquad (3.26)$$

The stationarity conditions of π^* include conditions (3.25)

$$\frac{\partial \pi^*}{\partial \mathbf{u}} = 0$$

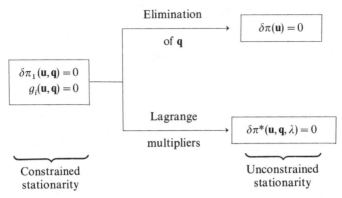

Figure 3.4 Transformation of a constrained functional into an unconstrained functional.

$$\frac{\partial \pi^*}{\partial \mathbf{q}} = 0 \tag{3.27}$$

$$\frac{\partial \pi^*}{\partial \lambda_i} = g_i = 0 \quad j = 1, 2, \ldots, m$$

The unknown functions pass from $\mathbf{u}, \mathbf{q}$ to $\mathbf{u}, \mathbf{q}$ and λ. Functional π^* is no more positive definite even if π is. Figure 3.4 shows the relations between π, π_1, and π^*.

Example 3.13 *Generalized functional for Poisson's equations*

Consider the functional of Example 3.9:

$$\pi(u) = \int_V \left(\frac{1}{2}\left(\frac{\partial u}{\partial x}\right)^2 + \frac{1}{2}\left(\frac{\partial u}{\partial y}\right)^2 - u f_V\right) dV + \int_{S_f} \left(\frac{1}{2}\alpha u^2 - u f_S\right) dS$$

It can also be written with the two variables q_x and q_y (see Example 3.6)

$$\pi_1(u, q_x, q_y) = \int_V \left(\frac{1}{2}(q_x^2 + q_y^2) - u f_V\right) dV + \int_{S_f} \left(\frac{1}{2}\alpha u^2 - u f_S\right) dS$$

with the following conditions:

$$q_x + \frac{\partial u}{\partial x} = 0$$

$$q_y + \frac{\partial u}{\partial y} = 0$$

The modified functional π^* with two Lagrange multipliers is:

$$\pi^*(u, q_x, q_y, \lambda_1, \lambda_2) = \int_V \left(\frac{1}{2}(q_x^2 + q_y^2) - u f_V + \lambda_1\left(q_x + \frac{\partial u}{\partial x}\right)\right.$$
$$\left. + \lambda_2\left(q_y + \frac{\partial u}{\partial y}\right)\right) dV + \int_{S_f} \left(\frac{1}{2}\alpha u^2 - u f_S\right) dS$$

The stationarity condition of π^* is:

$$\delta\pi^* = \frac{\partial\pi^*}{\partial u}\delta u + \frac{\partial\pi^*}{\partial q_x}\delta q_x + \frac{\partial\pi^*}{\partial q_y}\delta q_y + \frac{\partial\pi^*}{\partial \lambda_1}\delta\lambda_1 + \frac{\partial\pi^*}{\partial \lambda_2}\delta\lambda_2 = 0$$

Writing the first term explicitly:

$$\frac{\partial\pi^*}{\partial u}\delta u = \int_V \left(-f_V\delta u + \lambda_1\frac{\partial\delta u}{\partial x} + \lambda_2\frac{\partial\delta u}{\partial y}\right)dV + \int_{S_f} (\alpha u - f_S)\delta u\, dS = 0$$

or after integration by parts:

$$\frac{\partial\pi^*}{\partial u} = -\int_V \left(\frac{\partial\lambda_1}{\partial x} + \frac{\partial\lambda_2}{\partial y} + f_V\right)dV + \int_{S_f} (\alpha u - f_S + \lambda_1 m + \lambda_2 l)dS = 0$$

assuming that:

$$\delta u = 0 \quad \text{on} \quad S_u$$

l, m are the direction cosines of the normal to S_f. The remaining conditions of stationarity are:

$$\frac{\partial\pi^*}{\partial q_x} = \int_V (q_x + \lambda_1)dV = 0$$

$$\frac{\partial\pi^*}{\partial q_y} = \int_V (q_y + \lambda_2)dV = 0$$

$$\frac{\partial\pi^*}{\partial \lambda_1} = \int_V \left(q_x + \frac{\partial u}{\partial x}\right)dV = 0$$

$$\frac{\partial\pi^*}{\partial \lambda_2} = \int_V \left(q_y + \frac{\partial u}{\partial y}\right)dV = 0$$

Lagrange multipliers often have physical meaning; for example, rates of flow, fluxes, stresses. It is possible to construct mixed functionals π_r by eliminating the Lagrange multipliers from π^* with the help of equations (3.27). In structural mechanics, the functionals of Hellinger–Reissner [6] are mixed.

Example 3.14 *Mixed functional for Example 3.12*

Let us use one of the stationarity conditions of π^* from Example 3.12 to define λ in one of the following three forms:

$$\lambda = \begin{cases} -2u \\ 2v \\ v - u \end{cases}$$

Mixed functions are obtained after replacing λ from above into the functional

π^* of (3.12):

$$\pi_r(u,v) = \begin{cases} (-u^2 + v^2 + 2uv - 4u) \\ (u^2 - v^2 + 2uv + 4v) \\ (2uv - 2u + 2v) \end{cases}$$

The stationarity conditions of π_r give, in each case:

$$u = -1$$
$$v = 1$$
$$\pi_r(-1, 1) = 2$$

Example 3.15 Mixed functional for Poisson's equation

Using $\partial\pi^*/\partial q_x = 0$ and $\partial\pi^*/\partial q_y$ of Example 3.13 to eliminate λ_1 and λ_2 in π^*:

$$\lambda_1 = -q_x$$
$$\lambda_2 = -q_y$$

Functional π^* becomes mixed functional π_r:

$$\pi_r(u, q_x, q_y) = \int_V \left(\frac{1}{2}(q_x^2 + q_y^2) + q_x \frac{\partial u}{\partial x} - q_y \frac{\partial u}{\partial y} + u f_V \right) dV$$

$$+ \int_{S_f} (\tfrac{1}{2}\alpha u^2 - u f_s) dS$$

The stationarity conditions, after integration by parts of $(\partial \pi_r/\partial u)\delta u$, are:

$$\frac{\partial \pi_r}{\partial q_x} = -\int_V \left(q_x + \frac{\partial u}{\partial x} \right) dV = 0$$

$$\frac{\partial \pi_r}{\partial q_y} = -\int_V \left(q_y + \frac{\partial y}{\partial y} \right) dV = 0$$

$$\frac{\partial \pi_r}{\partial u} = \int_V \left(\frac{\partial q_x}{\partial x} + \frac{\partial q_y}{\partial y} - f_V \right) dV + \int_{S_f} (\alpha u - f_s - q_n) dS = 0$$

where
$$q_n = q_x l + q_y m$$

$$\left. \begin{array}{l} u = u_s \\ \delta u = 0 \end{array} \right\} \text{ on } S_u$$

Another form of π_r is obtained after integrating terms $q_x(\partial u/\partial x)$ and $q_y(\partial u/\partial y)$ by parts:

$$\pi_r^*(u, q_x, q_y) = \int_V \left(\frac{1}{2}(q_x^2 + q_y^2) - \left(\frac{\partial q_x}{\partial x} + \frac{\partial q_y}{\partial y} - f_V \right) u \right) dV$$

$$+ \int_{S_f} (\tfrac{1}{2}\alpha u^2 - u f_s - u q_n) dS - \int_{S_u} q_n u \, dS$$

In functional π of Example 3.9 the only variable is u. On the other hand, π_r and π_r^* have three independent variables u, q_x, q_y. A complementary functional π_c can also be obtained after eliminating $\mathbf{u}$ in π_r by explicitly satisfying constraints in q_x and q_y.

Example 3.16 *Complementary functional for Poisson's equation, when* $\alpha = 0$

If q_x and q_y are chosen to satisfy the following conditions:

$$\frac{\partial q_x}{\partial x} + \frac{\partial q_y}{\partial y} - f_V = 0 \quad \text{on} \quad V$$

$$q_n + f_S = 0 \quad \text{on} \quad S_f$$

functional π_r^* of Example 3.15 becomes the complementary functional

$$\pi_c = -\int_V \frac{1}{2}(q_x^2 + q_y^2)\,dV - \int_{S_u} q_n u\,dS$$

with values of u on S_u being known. This last functional depends only on the two variables q_x and q_y. Its stationarity conditions are:

$$\frac{\partial \pi_c}{\partial q_x} = -\int_V q_x\,dV - \int_{S_u} lu\,dS = 0$$

$$\frac{\partial \pi_c}{\partial q_y} = -\int_V q_y\,dV - \int_{S_u} mu\,dS = 0$$

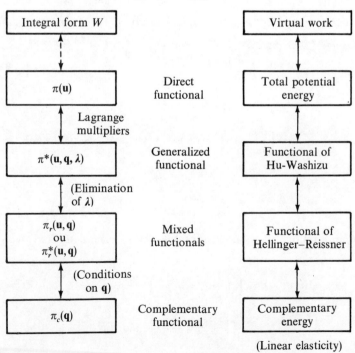

Figure 3.5 Various functionals

Figure 3.5 illustrates the relationships between π, π^*, π_r, π_c and their interpretation in linear elasticity.

3.5 Discretization of Integral Forms

3.5.1 *Discretization of Function W.*

In paragraphs 3.2 and 3.3 we have shown that the direct solution of partial differential equation (3.5) can be replaced by the search of a function **u** that nullifies the integral form (3.7):

$$W = \int_V \psi . R(\mathbf{u}) dV = \int_V \psi . (\mathscr{L}(\mathbf{u}) + f_V) dV = 0 \qquad (3.28)$$

for any weighting function ψ.

To obtain an approximation solution u we may transport (3.28) through discretization into an algebraic expression which may then be solved for unknowns. This may be devised in two steps.

— Unknown functions **u** are approximated by a 'n' parameter representation. The approximation parameters can be nodal values of **u** or not; they can be taken over the entire domain or restricted to a sub-domain (see paragraph 1.1). The method of undetermined parameters [1] uses non-nodal approximations (1.3). The finite element method uses nodal approximations over the subdomain (see paragraph (1.1.2). In all cases **u** can be written as:

$$\mathbf{u} = \mathbf{u}(a_1 a_2, \ldots, a_n) \qquad (3.29)$$

Expression (3.28) becomes:

$$W = \int_V \psi . (\mathscr{L}(u(a_1, a_2, \ldots, a_n)) + f_V)) dV = 0 \qquad (3.30)$$

for any ψ.

— n independent weighting functions $\psi_1, \psi_2, \ldots, \psi_n$ are chosen arbitrarily but the number n must be the same as the number of parameters chosen for **u** (3.29). The choice of the weighting functions leads to a wide variety of methods the most popular being: the collocation method, the Galerkin method, the least squares method, etc. Equations (3.30) can be written as:

$$\begin{aligned} W_1 &= \int_V \psi_1(\mathscr{L}(\mathbf{u}(a_1, a_2, \ldots, a_n)) + f_V)) dV = 0 \\ W_2 &= \int_V \psi_2(\mathscr{L}(\mathbf{u}(a_1, a_2, \ldots, a_n)) + f_V)) dV = 0 \\ &\cdots \\ W_n &= \int_V \psi_n(\mathscr{L}(\mathbf{u}(a_1, a_2, \ldots, a_n)) + f_V)) dV = 0 \end{aligned} \qquad (3.31)$$

After integration of (3.31) we get a system of algebraic equations for the

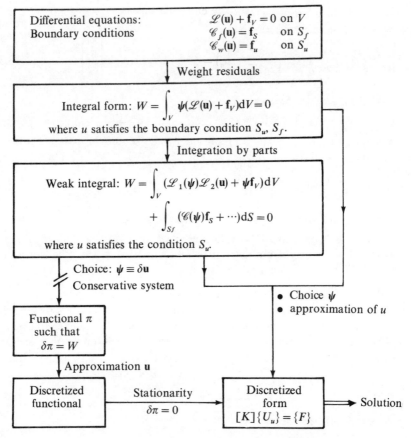

Figure 3.6 Relations between systems of defferential equations, integral forms and functionals.

determination of the parameters a_i in the approximation of **u**:

$$[K]\{a\} = \{F\} \quad (3.32)$$

Figure 3.6 summarizes various operation necessary for obtaining an approximate solution using the weighted residual method.

3.5.2 *Approximation of Functions u*

Functions **u** are approximated by the methods described in Chapter 1. The continuity requirements and boundary conditions of the integral form selected must be respected.

— Non-nodal approximation over the entire domain V (equation (1.3)).

$$\mathbf{u} = \mathbf{u}(\mathbf{x}, a_1, a_2, \ldots, a_n) = \langle P_1 \quad P_2 \ldots P_n \rangle \begin{Bmatrix} a_1 \\ a_2 \\ \vdots \\ a_n \end{Bmatrix} \quad (3.33)$$

— Nodal approximation over the entire domain V (equation (1.5)).

$$\mathbf{u} = \mathbf{u}(x, u_1, u_2, \ldots, u_n) = \langle N_1 \; N_2 \ldots N_n \rangle \begin{Bmatrix} u_1 \\ u_2 \\ \vdots \\ u_n \end{Bmatrix} \quad (3.34)$$

— Finite element approximations (paragraph 1.1.2) The following examples use non-nodal approximations over the entire domain (equation (3.33)) in order to obtain different approximate solutions depending on the choice of weighting functions. However, in the rest of the book we shall be employing only the finite element approximations coupled with the Galerkin-type weighting functions to obtain desired solutions.

Example 3.17 Non-nodal approximation of $\mathbf{u}$ over a square region

Consider Poisson's equation defined over a square region ($f_V = f = $ constant):

$$\mathcal{L}(u) + f_V = \frac{\partial^2 u}{\partial x^2} + \frac{\partial^2 u}{\partial y^2} + f = 0 \text{ on the square}$$

$$u = 0 \text{ on } S \begin{cases} x = \pm 1 \\ y = \pm 1 \end{cases}$$

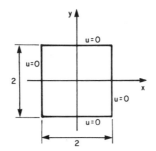

An approximation of u that satisfies the boundary conditions and the symmetry of the problem is:

$$u = \langle P_1 \; P_2 \rangle \begin{Bmatrix} a_1 \\ a_2 \end{Bmatrix} = \langle P \rangle \{a\}$$

where:
$$P_1 = (x^2 - 1)(y^2 - 1)$$
$$P_2 = (x^2 - 1)(y^2 - 1)(x^2 + y^2) = P_1(x^2 + y^2)$$

then:
$$\mathcal{L}(u) = \mathcal{L}(\langle P \rangle \{a\}) = \mathcal{L}(P_1)a_1 + \mathcal{L}(P_2)a_2$$

with:
$$\mathcal{L}(P_1) = 2(x^2 + y^2 - 2)$$
$$\mathcal{L}(P_2) = 2(6x^2 - 1)(y^2 - 1) + 2(6y^2 - 1)(x^2 - 1) + $$
$$+ 2(x^4 - x^2) + 2(y^4 - y^2)$$

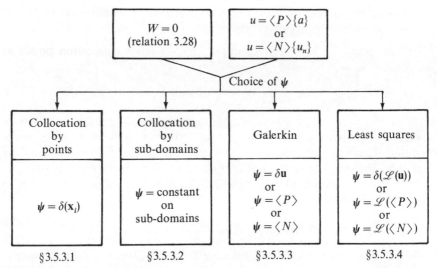

Figure 3.7 Various methods of undetermined parameters depending on the choice of ψ.

3.5.3 Choice of the Weighting Functions ψ

Depending on the choice of ψ_i, equation (3.31) leads to the various method illustrated in Figure 3.7

3.5.3.1 Collocation by points

Functions $\psi_i(x)$ are replaced by Dirac delta functions $\delta(x_i)$ at a number of collocation points equal to the number of parameters n in the approximation for **u**. Integral form (3.7) becomes:

$$W = \int_V \delta(\mathbf{x}_i) R(\mathbf{x}, \mathbf{u}) \, dV = R(\mathbf{x}_i, \mathbf{u}) = 0 \qquad (3.35)$$

Equation (3.31) becomes:

$$W_i(\mathbf{a}) = (\mathscr{L}(\mathbf{u}(\mathbf{x}, a_1, a_2, \ldots, a_n)) + \mathbf{f}_V)_{\mathbf{x}=\mathbf{x}_i} = 0 \quad i = 1, 2, \ldots, n$$

and upon substitution of (3.33) for **u**, we get:

$$\begin{aligned} W_i(\mathbf{a}) &= (\mathscr{L}(\langle P \rangle \{a\}) + f_V)_{\mathbf{x}=\mathbf{x}_i} = 0 \\ &= (\langle \mathscr{L}(P) \rangle \{a\} + f_V)_{\mathbf{x}=\mathbf{x}_i} = 0 \end{aligned} \qquad (3.36)$$

The precision of the solution depends on the number and position of collocation points. Symmetry of solution should be employed for the location of points. The method avoids the volume integration but, in general, leads to non-symmetrical systems of equations. The method may offer an interest for solving certain non-linear problems. In order to obtain symmetrical matrices one may employ the least squares technique to the discretized expression. This technique may also allow us to use more than n collocation points.

Example 3.18 Numerical application of the collocation method by points for Poisson's equation over a square domain

Using the problem defined in Example 3.17 we select collocation points as follows:

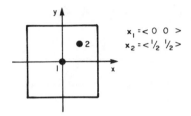

Weighting functions are: $\psi_1 = \delta(x_1)$ and $\psi_2 = \delta(x_2)$. Then equations (3.36) are:

$$W_1 = \langle \mathcal{L}(P_1) \quad \mathcal{L}(P_2) \rangle_{\mathbf{x}=\mathbf{x}_1} \begin{Bmatrix} a_1 \\ a_2 \end{Bmatrix} + f(\mathbf{x}_1) = 0$$

$$W_2 = \langle \mathcal{L}(P_1) \quad \mathcal{L}(P_2) \rangle_{\mathbf{x}=\mathbf{x}_2} \begin{Bmatrix} a_1 \\ a_2 \end{Bmatrix} + f(\mathbf{x}_2) = 0$$

The results from the previous example are:

$$\begin{matrix} W_1 = -4a_1 + 4a_2 + f = 0 \\ W_2 = -3a_1 - \tfrac{9}{4}a_2 + f = 0 \end{matrix} \begin{cases} a_1 = 0.2976 f \\ a_2 = 0.0476 f \end{cases}$$

The value of u at the centre of the square is:

$$u(\mathbf{x}_1) = \langle P_1(\mathbf{x}_1) \quad P_2(\mathbf{x}_1) \rangle \begin{Bmatrix} a_1 \\ a_2 \end{Bmatrix} = a_1$$

$$u_c = u(\mathbf{x}_1) = 0.2976 f$$

The exact value from a 14-term Fourier series solution is:

$$u_c \cong 0.2947 f$$

Using only one collocation point at x_1 and a one-parameter solution, we would get:

— with the collocation point at x_1: $u_c = 0.25 f$;
— with the collocation point at x_2: $u_c \approx 0.333 f$.

3.5.3.2 *Collocation by sub-domains*

For a subdivision of V into n sub-domains V^i we define the weighting functions as follows:

$$\psi_i = \begin{cases} 1 \text{ if } x \text{ belongs to } V^i \\ 0 \text{ otherwise} \end{cases} \quad (3.37)$$

Equation (3.31) leads to n equations of the type:

$$W_i(\mathbf{a}) = \int_{V^i} (\langle \mathscr{L}(P) \rangle \{a\} + f_V) dV = 0 \qquad (3.38)$$

The number of sub-domains must match the number of parameters in the approximation of **u**. We note that the sub-domains need not map the entire domain, they could even overlap.

Example 3.19 *Collocation by sub-domain for Poisson's equation*

We again refer to the problem described in Example 3.17. The following sub-domains are selected:

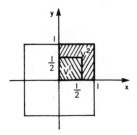

System (3.38) becomes:

$$W_1 = \int_{V^1} \langle \mathscr{L}(P_1) \quad \mathscr{L}(P_2) \rangle dV \begin{Bmatrix} a_1 \\ a_2 \end{Bmatrix} + \int_{V^1} f dV$$

$$= -0.9167 a_1 + 0.3875 a_2 + 0.25 f = 0$$

$$W_2 = \int_{V^2} \langle \mathscr{L}(P_1) \quad \mathscr{L}(P_2) \rangle dV \begin{Bmatrix} a_1 \\ a_2 \end{Bmatrix} + \int_{V^2} f dV$$

$$= -1.75 a_1 - 3.5875 a_2 + 0.75 f = 0$$

Hence:
$$a_1 = 0.2994 f$$
$$a_2 = 0.0630 f$$

The value of u at the centre is:

$$u_c = a_1 = 0.2994 f$$

For a one-parameter solution and the entire domain, we get: $u_c = 0.375 f$.

3.5.3.3 Galerkin Method

The weighting functions are identical to the functional bases of u. Thus we often represent ψ by a set of first variations δu of approximation u with respect to $\{a\}$.

$$\psi = \delta \mathbf{u} = \langle P \rangle \{\delta a\} \quad \text{for all } \{\delta a\} \qquad (3.39)$$

where $\{\delta a\}$ are the first variations of parameters $\{a\}$.

Equation (3.31) becomes:

$$W = \int_V \delta u(\mathscr{L}(\mathbf{u}) + f_V) dV = 0 \qquad (3.40)$$

$$W = \langle \delta a \rangle \int_V \{P\}(\mathscr{L}(\langle P \rangle \{a\}) + f_V) dV = 0 \qquad (3.41)$$

Since W must vanish for any value of $\{\delta a\}$, the previous expression is equivalent to the following system of algebraic equations:

$$\begin{aligned} W_1(\mathbf{a}) &= \int_V P_1(\langle \mathscr{L}(P) \rangle \{a\} + f_V) dV = 0 \\ &\vdots \\ W_n(\mathbf{a}) &= \int_V P_n(\langle \mathscr{L}(P) \rangle \{a\} + f_V) dV = 0 \end{aligned} \qquad (3.42)$$

The system is symmetrical if the operator $\mathscr{L}$ is self-adjoint.

Example 3.20 Solution of Poisson's equation by Galerkin method (without integration by parts)

Using functions P_1 and P_2 of Example 3.17 into the system (3.42) we get:

$$W_1 = \int_V \langle P_1.\mathscr{L}(P_1) \quad P_1.\mathscr{L}(P_2) \rangle dV \begin{Bmatrix} a_1 \\ a_2 \end{Bmatrix} + \int_V P_1 f dV = 0$$

$$W_2 = \int_V \langle P_2.\mathscr{L}(P_1) \quad P_2.\mathscr{L}(P_1) \rangle dV \begin{Bmatrix} a_1 \\ a_2 \end{Bmatrix} + \int_V P_2 f dV = 0$$

from which the following symmetrical system of equations is obtained:

$$5.689 a_1 + 1.9505 a_2 = 1.7778 f$$
$$1.9505 a_1 + 2.3839 a_2 = 0.7111 f$$
$$a_1 = 0.2922 f \quad a_2 = 0.0592 f$$
$$u_c = 0.2922 f$$

If we had used a one-parameter solution we would have obtained:

$$u_c = 0.3125 f$$

For a three-parameter solution, using the same functions P_1, P_2, and $P_3 = P_1.(x^2 y^2)$, we would have found:

$$a_1 = 0.2949 f \quad a_2 = 0.0401 f \quad a_3 = 0.1230 f$$
$$u_c = 0.2949 f$$

Integration by parts, as explained in paragraph 3.3, allows a transformation of system (3.42) into the following system:

$$W_1(\mathbf{a}) = \int_V \mathscr{L}_1(P_1)(\langle \mathscr{L}_2(P)\rangle\{a\})dV - \int_V P_1 f_V dV - \int_{S_f} P_1 f_S dS = 0$$
$$\vdots \qquad \vdots \qquad \qquad \vdots \qquad \vdots \qquad (3.43)$$
$$W_n(\mathbf{a}) = \int_V \mathscr{L}_1(P_n)(\langle \mathscr{L}_2(P)\rangle\{a\})dV - \int_V P_n f_V dV - \int_{S_f} P_n f_S dS = 0$$

The solutions of (3.42) and (3.43) are identical if the functions $\langle P \rangle$ are identical and all the conditions required by (3.42) are met. However, the admissibility conditions required by (3.43) being less restrictive it is advantageous to use a simpler set of functions $\langle P \rangle$ for (3.43).

Amongst all the weighted residual methods, this last variant of the Galerkin method is by far the most popular and is readily applicable to finite element approximations.

Example 3.21 *Solution of Poisson's equation by Galerkin method with integration by parts*

Using the form of W obtained in Example 3.5, we find:

$$\mathscr{L}_1 = \left\langle \frac{\partial}{\partial x} \; \frac{\partial}{\partial y} \right\rangle; \quad \mathscr{L}_2 = \begin{Bmatrix} \dfrac{\partial}{\partial x} \\ \dfrac{\partial}{\partial y} \end{Bmatrix}$$

Expression (3.43) becomes ($f_S = \alpha = 0$)

$$W_1 = \int_V \left\langle \frac{\partial P_1}{\partial x}\frac{\partial P_1}{\partial x} + \frac{\partial P_1}{\partial y}\frac{\partial P_1}{\partial y} ; \frac{\partial P_1}{\partial x}\frac{\partial P_2}{\partial x} + \frac{\partial P_1}{\partial y}\frac{\partial P_2}{\partial y} \right\rangle dV \begin{Bmatrix} a_1 \\ a_2 \end{Bmatrix}$$
$$- \int_V P_1 f\, dV = 0$$

$$W_2 = \int_V \left\langle \frac{\partial P_2}{\partial x}\frac{\partial P_1}{\partial x} + \frac{\partial P_2}{\partial y}\frac{\partial P_1}{\partial y} ; \frac{\partial P_2}{\partial x}\frac{\partial P_2}{\partial x} + \frac{\partial P_2}{\partial y}\frac{\partial P_2}{\partial y} \right\rangle dV \begin{Bmatrix} a_1 \\ a_2 \end{Bmatrix}$$
$$- \int_V P_2 f\, dV = 0$$

This last system has an identical solution to the one of Example 3.20.

3.5.3.4 Least Squares Method

The least squares method consists in minimizing the expression

$$\pi_m = \int_V R \cdot R\, dV \qquad (3.44)$$

with respect to parameters $a_1, a_2, \ldots, a_n$, where R is the residual function:

$$R = \mathscr{L}(u) + f_V = \langle \mathscr{L}(P) \rangle \{a\} + f_V \tag{3.45}$$

The stationarity conditions of (3.44) are:

$$W = \delta\pi_m(a_1, a_2, \ldots, a_n) = 0$$

$$W_1(\mathbf{a}) = \int_V \mathscr{L}(P_1)(\langle \mathscr{L}(P) \rangle \{a\} + f_V) \, dV = 0$$

$$\vdots \tag{3.46}$$

$$W_n(\mathbf{a}) = \int_V \mathscr{L}(P_n)(\langle \mathscr{L}(P) \rangle \{a\} + f_V) \, dV = 0$$

This method leads to a positive definite system of equations for the unknowns, no matter what operator $\mathscr{L}$ may be.

Example 3.22   Solution of Poisson's equation by the least squares method

Using functions P_1 and P_2 of Example 3.17, equation (3.46) gives the following results for the problem defined in Example 3.18:

$$W_1 = \int_V \langle \mathscr{L}(P_1) \cdot \mathscr{L}(P_1) \quad \mathscr{L}(P_1) \cdot \mathscr{L}(P_2) \rangle dV \begin{Bmatrix} a_1 \\ a_2 \end{Bmatrix} + \int_V \mathscr{L}(P_1) f \, dV = 0$$

$$W_2 = \int_V \langle \mathscr{L}(P_2) \cdot \mathscr{L}(P_1) \quad \mathscr{L}(P_2) \cdot \mathscr{L}(P_2) \rangle dV \begin{Bmatrix} a_1 \\ a_2 \end{Bmatrix} + \int_V \mathscr{L}(P_2) f \, dV = 0$$

$$31.2889 a_1 + 25.1937 a_2 = 10.6667 f$$
$$25.1937 a_1 + 87.4463 a_2 = 12.8000 f$$
$$a_1 = 0.2904 f \quad a_2 = 0.0627 f$$
$$u_c = 0.2904 f$$

With a one-parameter approximation $u = P_1(x, y) a_1$, the result is:

$$u_c = 0.03409 f$$

For a three-parameter solution with functions P_1, P_2, and $P_3 = P_1 \cdot x_2 y_2$ the result is:

$$a_1 = 0.2949 f \quad a_2 = 0.0385 f \quad a_3 = 0.1562 f$$
$$u_c = 0.2949 f$$

3.5.4 Discretization of a Functional (Ritz method)

In the Ritz method, a functional π is first discretized with an approximation of u like the one of (3.33). Then, the conditions of stationarity of π with respect to

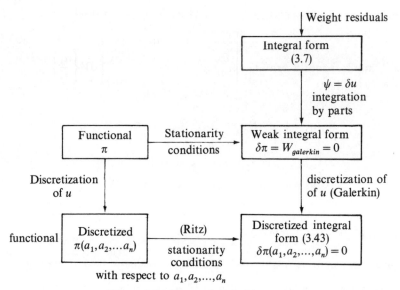

Figure 3.8 Ritz and Galerkin methods.

parameters $a_1, a_2, \ldots, a_n$ are obtained

$$\pi(\mathbf{u}) = \pi(\mathbf{u}(a_1, a_2, \ldots, a_n))$$

$$\delta\pi(a_1, a_2, \ldots, a_n) = \frac{\partial \pi}{\partial a_1}\delta a_1 + \frac{\partial \pi}{\partial a_2}\delta a_2 + \cdots + \frac{\partial \pi}{\partial a_n}\delta a_n = 0$$

This leads to the following n equations:

$$\frac{\partial \pi}{\partial a_1} = 0; \quad \frac{\partial \pi}{\partial a_2} = 0; \quad \ldots; \quad \frac{\partial \pi}{\partial a_n} = 0 \tag{3.47}$$

Functions **u** are still subject to the conditions imposed by the functional. If the functional π exists, its first variation is identical to the Galerkin integral form (3.43)

$$\delta\pi = W = 0 \tag{3.48}$$

The solution thus obtained by the Ritz method is identical to the Galerkin solution. Figure 3.8 shows that indeed the Ritz and Galerkin methods are identical.

Example 3.23 Solution of Poisson's equation by the Ritz method

The functional π is given in Example 3.11 ($\alpha = f_S = 0$). Using the approximation of

u given in Example 3.17, we get:

$$\pi(a_1, a_2) = \tfrac{1}{2} \langle a_1 \ a_2 \rangle \int_V \begin{bmatrix} \dfrac{\partial P_1}{\partial x} & \dfrac{\partial P_1}{\partial y} \\ \dfrac{\partial P_2}{\partial x} & \dfrac{\partial P_2}{\partial y} \end{bmatrix} \begin{bmatrix} \dfrac{\partial P_1}{\partial x} & \dfrac{\partial P_2}{\partial x} \\ \dfrac{\partial P_1}{\partial y} & \dfrac{\partial P_2}{\partial y} \end{bmatrix} dV \begin{Bmatrix} a_1 \\ a_2 \end{Bmatrix}$$

$$- \langle a_1 \ a_2 \rangle \int_V \begin{Bmatrix} P_1 \\ P_2 \end{Bmatrix} f \, dV$$

Then $\delta(\pi) = 0$ gives:

$$\int_V \begin{bmatrix} \dfrac{\partial P_1}{\partial x}\dfrac{\partial P_1}{\partial x} + \dfrac{\partial P_1}{\partial y}\dfrac{\partial P_1}{\partial y} & \dfrac{\partial P_1}{\partial x}\dfrac{\partial P_2}{\partial x} + \dfrac{\partial P_1}{\partial y}\dfrac{\partial P_2}{\partial y} \\ \text{Sym.} & \dfrac{\partial P_2}{\partial x}\dfrac{\partial P_2}{\partial x} + \dfrac{\partial P_2}{\partial y}\dfrac{\partial P_2}{\partial y} \end{bmatrix} dV \begin{Bmatrix} a_1 \\ a_2 \end{Bmatrix}$$

$$- \int_V \begin{Bmatrix} P_1 \ f \\ P_2 \ f \end{Bmatrix} dV = 0$$

This last system is identical to the system (3.21) obtained by the Galerkin method.

Example 3.24 *Comparison of results from the various methods of solution*

The values of $(u_{c/f})$ obtained from the various methods of solution for Poisson's equation of Example 3.17

$$P_1 = (x^2 - 1)(y^2 - 1); \quad P_2 = P_1(x^2 + y^2); \quad P_3 = P_1 \cdot x^2 y^2$$

Functions	Collocation by points Example 3.18	Collocation by sub-domains Example 3.19	Galerkin and Ritz Example 3.20 of 3.23	Least squares Example 3.22	Exact (Fourier series) (14 terms)
P_1	0.2500	0.3750	0.3125	0.3409	
P_1, P_2	0.2976	0.2994	0.2922	0.2904	0.2947
P_1, P_2, P_3			0.2949	0.2949	

3.5.5 Properties of the System of Equations

All the methods of solution lead to a system of equations like:

$$[K]\{a\} = \{F\} \tag{3.49}$$

Figure 3.9 summarizes the properties of such a system for each method of solution.

Methods	Terms K_{ij}	F_i	Conditions $u = \langle P \rangle \{a\}$ $\langle P \rangle$	Properties of $[K]$
Collocation by points	$\mathscr{L}(P_j)$ in $\mathbf{x} = \mathbf{x}_i$	$-f_V(\mathbf{x}_i)$	on S_u on S_f	non-symmetrical
Collocation by sub-domains	$\int_{V^i} \mathscr{L}(P_j) dV$	$-\int_V f_V dV$	on S_u on S_f	non-symmetrical
Galerkin	$\int_V P_i \mathscr{L}(P_j) dV$	$-\int_V P_i f_V dV$	on S_u on S_f	symmetrical if self-adjoint
Galerkin after integration by parts	$\int_V \langle \mathscr{L}_1(P_i) \rangle \{\mathscr{L}_2(P_j)\} dV$	$\int_V P_i f_V dV + \int_{S_f} P_i f_s dS$	S_u	symmetrical if self-adjoint $\mathscr{L}_1 = \mathscr{L}_2$
Least squares	$\int_V \mathscr{L}(P_i) \mathscr{L}(P_j) dV$	$-\int_V \mathscr{L}(P_i) f_V dV$	on S_u on S_f	symmetrical and positive definite
Ritz (if the functional exists)	$\int_V \langle \mathscr{L}_1(P_i) \rangle \{\mathscr{L}_1(P_j)\} dV$	$\int_V P_i f_V dV + \int_{S_f} P_i f_s dS$	on S_u	symmetrical

Figure 3.9 Properties of the system of equations coming from the method of undetermined parameters.

Important Results

Integral form of the weighted residual method:

$$W = \int_V \psi R(\mathbf{u}) dV = \int_V \psi(\mathscr{L}(\mathbf{u}) - f_V) dV = 0 \tag{3.7}$$

First variation of a functional:

$$\delta \pi = \frac{\partial \pi}{\partial \mathbf{u}} \delta \mathbf{u} + \frac{\delta \pi}{\partial \left(\frac{\partial \mathbf{u}}{\partial x}\right)} \delta \left(\frac{\partial \mathbf{u}}{\partial x}\right) + \cdots \tag{3.13b}$$

Functional associated with an integral form:

$$\pi \text{ such that } \delta\pi \equiv W = 0 \qquad (3.15a)$$

$$W = \int_V \delta u (\mathscr{L}(\mathbf{u}) + f_V) dV \qquad (3.15b)$$

Discretized integral form:

$$W_i = \int_V \psi_i(\mathscr{L}(\mathbf{u}(a_1, a_2, \ldots, a_n)) + f_V) dV = 0 \qquad i = 1, 2, \ldots, n \qquad (3.31)$$

Collocation by points:

$$W_i(\mathbf{a}) = (\langle \mathscr{L}(P) \rangle \{a\} + f_V)_{\mathbf{x} = \mathbf{x}_i} = 0 \qquad (3.36)$$

Collocation by sub-domains:

$$W_i(\mathbf{a}) = \int_{V^i} (\langle \mathscr{L}(P) \rangle \{a\} + f_V) dV = 0 \qquad (3.38)$$

Galerkin (after integration by parts) or Ritz:

$$W_i(\mathbf{a}) = \int_V \mathscr{L}_1(P_i) \langle \mathscr{L}_2(P) \rangle \{a\} dV$$

$$- \int_V P_i f_V dV - \int_{S_f} P_i f_S dS = 0 \qquad (3.42)$$

Least squares:

$$W_i(\mathbf{a}) = \int_V \mathscr{L}(P_i)(\langle \mathscr{L}(P) \rangle \{a\} + f_V) dV = 0 \qquad (3.46)$$

Notation

$\{a\}, a_1, a_2, \ldots$	parameters for the approximation of u
c	damping coefficient
$[C]$	damping matrix for a discrete system
E_u	ensemble of admissible functions u
E_ψ	ensemble of weighting functions
$\mathbf{f}_V, \mathbf{f}_S$	vector of body forces and surface tractions
$\{F\}$	vector of concentrated forces for a discrete system
$[K]$	global system (stiffness) matrix for a discrete system
l, m, n	components of a unit vector perpendicular to the boundary of a domain
$\mathscr{L}(u), \mathscr{C}(u)$	differential operators defining the governing equations and boundary conditions of a physical system
m	mass per unit volume
$[M]$	mass matrix of a discrete system
q	various physical variables like rate of flow, stress etc.

R	residual corresponding to a partial differential equation
S_u, S_f	parts of the boundary of a domain where u and f are specified
u	unknown variables of a physical system
$W, W_1, W_2, \ldots$	integral forms
$\delta u, \delta \pi$	first variation of a function or functional
$\delta^2 u, \delta^2 \pi$	second variation of a function or functional
$\delta(\mathbf{x})$	Dirac delta function at point x
Δ	Laplacian operator
$\lambda, \lambda_1, \lambda_2, \ldots$	Eigenvalues or Lagrange multipliers
$\psi, \psi_1, \psi_2, \ldots$	weighting functions
$\pi, \pi_1, \pi^*, \pi_r, \pi_c$	functionals

References

[1] S. H. Crandall, *Engineering Analysis*, McGraw-Hill, New York, 1956.
[2] L. Collatz, *The Numerical Treatment of Differential Equations*, Springer-Verlag, 1966.
[3] B. A. Finlayson, *The Method of Weighted Residuals and Variational Principles*, Academic Press, 1972.
[4] S. C. Mikhlin, *Variational Methods in Mathematical Physics*, Macmillan, 1964.
[5] S. C. Mikhlin, *The Numerical Performance of Variational Methods*, Wolters-Noordhoff, 1971.
[6] K. Washizu, *Variational Methods in Elasticity and Plasticity*, 2nd edn, Pergamon, Oxford, 1975.

4

MATRIX FORMULATION OF THE FINITE ELEMENT METHOD

4.0 Introduction

This chapter describes the various stages necessary for a finite element analysis. Matrix algebra is used to write the fundamental equations because such expressions are very easy to transform in computer codes.

First we define the finite element method as a discretization process for the integral forms of Galerkin; it replaces the global integral form W by a summation of element integral forms W^e. Then, every integral form is discretized using finite element approximations; this leads to elementary matrices to be assembled in a global matrix. Convergence conditions (to exact results) are discussed and the patch test is described.

Matrix algebra is used to write the discretized integral forms of the finite element method. Two examples, one for harmonic problems and one for elasticity problems, are used throughout to illustrate the various steps. FORTRAN codes are used to show how to implement the construction of elementary matrices and vectors.

Assembly techniques allowing to superpose (or merge) the elementary matrices into global system matrices are described. Various compacting schemes are examined for the sparse global matrix of coefficients. Special attention is paid to the skyline method of compaction. Finally, various techniques are examined to introduce the boundary conditions into the system of equations. The chapter ends with a detailed example of the method applied to the solution of Poisson's equation.

4.1 The Finite Element Method

4.1.1 *Definition*

The finite element method consists in discretizing an integral form W (paragraph 3.2) using a finite element approximation (paragraph 1.1.2) for the unknown variables **u** and solving the resulting system of algebraic equations. The various stages of the solution process are briefly described in this paragraph.

We start from Galerkin integral forms (paragraph 3.5.3.3) in which the weighting functions are given by $\psi \equiv \delta u$:

$$W = \int_V \delta u(\mathcal{L}(u) + f_V) dV = 0 \tag{4.1}$$

The integral over the entire domain is replaced by a summation of sub-domain (or element domain) integrals:

$$W = \sum_{e=1}^{n_{el}} W^e = \sum_{e=1}^{n_{el}} \int_{V^e} \delta u^e(\mathcal{L}(u) + f_V) dV = 0 \tag{4.2a}$$

For each term W^e (called element integral form) u and δu are replaced by a finite element approximation restricted by the element domain V^e:

$$\begin{aligned} u^e &= \langle N \rangle \{u_n\} \\ \delta u^e &= \langle N \rangle \{\delta u_n\} \end{aligned} \tag{4.2b}$$

Since $\langle N \rangle$ is defined null outside of region V^e and depends only on nodal values $\langle u_n \rangle$ belonging to V^e, the entire computational process is confined to the element domain. It is the repetitive nature of the elementary matrices computation that contributed to the enormous success of the method.

Using (4.2b), W^e becomes:

$$\begin{aligned} W^e &= \int_{V^e} \delta u^e(\mathcal{L}(u^e) + f_V) dV \\ W^e &= \langle \delta u_n \rangle \left(\int_{V^e} \{N\} \mathcal{L}(\langle N \rangle) dV \{u_n\} + \int_{V^e} \{N\} f_V dV \right) \end{aligned} \tag{4.2c}$$

Integration by parts is frequently used in equation (4.1) to reduce the order of the highest derivative contained in the expression (see paragraph 3.3). The integral form may then contain derivatives of δu and some supplementary contour integrals. The elementary integral forms W^e are then written in matrix notation as (for steady state systems):

$$W^e = \int_{V^e} (\langle \delta(\partial u^e) \rangle [D] \{\partial u^e\} - \delta u^e . f_V) dV - \int_{S_f^e} du^e . f_S dS \tag{4.3}$$

where

$$\langle \partial u^e \rangle = \left\langle u^e \frac{\partial u^e}{\partial x} \cdots \frac{\partial^2 u^e}{\partial x^2} \cdots \right\rangle$$

$$\langle \delta(\partial u^e) \rangle = \left\langle \delta u^e \delta\left(\frac{\partial u^e}{\partial x}\right) \cdots \delta\left(\frac{\partial^2 u^e}{\partial x^2}\right) \cdots \right\rangle$$

[D] is a matrix that is independent of u^e and its derivatives for linear operators $\mathcal{L}$. For non-linear operators $\mathcal{L}$, [D] is generally a matrix dependent on functions u^e and its derivatives;

f_V, f_S are body and surface tractions;

171

V^e is the volume of the element;
S_f^e is the part of the boundary of V^e over which integration by parts gives additional contour integral forms.

Example 4.1 Matrix expression for a Poisson's equation element integral form W^e.

For Poisson's equation (of Example 3.1), the element integral form W^e can be written as (note that superscript e is suppressed for variables u and δu in the rest of this example):

— before integration by parts (see Example 3.3)

$$W = \sum_e W^e = \sum_e \int_{V^e} \delta u \left(\frac{\partial^2 u}{\partial x^2} + \frac{\partial^2 u}{\partial y^2} + f_V \right) dV = 0$$

— after integration by parts (see Example 3.5)

$$W = \sum_e W^e = \sum_e \left(\int_{V^e} (\langle \delta(\partial u) \rangle [D] \{\partial u\} - \delta u f_V) dV \right.$$
$$\left. - \int_{S_f^e} du(f_S - \alpha u) dS \right) = 0$$

where

$$\langle \delta(\partial u) \rangle = \left\langle \delta\left(\frac{\partial u}{\partial x}\right) \delta\left(\frac{\partial u}{\partial y}\right) \right\rangle$$

$$\langle \partial u \rangle = \left\langle \frac{\partial u}{\partial x} \frac{\partial u}{\partial y} \right\rangle$$

$$[D] = \begin{bmatrix} 1 & 0 \\ 0 & 1 \end{bmatrix}$$

Consider a rectangular domain V divided into rectangular elements V^e. The boundary S of the domain V is divided into two parts S_u and S_f:

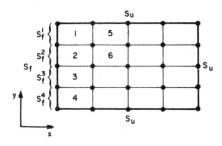

A contour integral on S_f exists only for one side of elements 1, 2, 3, and 4; for element 1 it is:

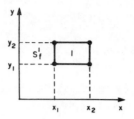

$$\int_{y_1}^{y_2} \delta u(x_1, y)(f_S - \alpha u(x_1, y)) \, dy$$

Finally, using expressions (4.2b) for u^e and δu^e as well as similar expressions for ∂u^e and $\delta(\partial u^e)$ into (4.3), we get the following matrix expression for the discretized element integral form W^e:

$$W^e = \langle \delta u_n \rangle ([k]\{u_n\} - \{f\}) \tag{4.4}$$

where $[k]$ is the element matrix; it is independent of u_n if $\mathscr{L}$ is linear; $\{f\}$ is the consistent element load vector; $\{u_n\}$ is the element vector of nodal values; and $\{\delta u_n\}$ is the first variation of the element nodal values.

The integral form (4.2a) is built by a merging process of the element integral forms (4.4):

$$W = \sum_e W^e = \sum_e \langle \delta u_n \rangle ([k]\{u_n\} - \{f\}) = 0 \tag{4.5a}$$

The result of the merging process is the following overall global matrix form:

$$W = \langle \delta U_n \rangle ([K]\{U_n\} - \{F\}) = 0 \tag{4.5b}$$

where $[K]$ is the global system matrix; $\{F\}$ is the global right-hand side load vector; $\{U_n\}$ is the global vector of all nodal values of the unknown function; and $\{\delta U_n\}$ is an arbitrary variation of U_n.

To pass from (4.5a) and (4.5b), a merging process respecting the topology of the elements must be used; this process, called assemblage, allows us to construct global matrix $[K]$ and vector $\{F\}$ from elementary matrices $[k]$ and vectors $\{f\}$. It will be examined in paragraph 4.4.

Since (4.5b) is null for arbitrary variations $\langle \delta U_n \rangle$, we must have:

$$[K]\{U_n\} = \{F\} \tag{4.5c}$$

For unsteady problems, terms like $\partial u/\partial t$ and $\partial^2 u/\partial t^2$ may appear and introduce additional corresponding expressions in the element integral form:

$$W^e = \int_{V^e} \delta u^e \frac{\partial u^e}{\partial t} dV \quad \text{and} \quad W^e = \int_{V^e} \delta u^e \frac{\partial^2 u^e}{\partial t^2} dV \tag{4.6a}$$

After discretization with (4.2b), these supplementary terms become:

$$W^e = \langle \delta u_n \rangle [c] \left\{ \frac{du_n}{dt} \right\} \quad \text{and} \quad W^e = \langle \delta u_n \rangle [m] \left\{ \frac{d^2 u_n}{dt^2} \right\}$$

$$[c] = [m] = \int_{V_e} \{N\} \langle N \rangle dV \tag{4.6b}$$

where $[m]$ is called the element mass matrix.

The element residual function is defined by:

$$\{r\} = \{f\} - [k]\{u_n\} \tag{4.6c}$$

The global residual function obtained after the assembly process is:

$$\{R\} = \sum_e \{r\} = \{F\} - [K]\{U_n\} \tag{4.6d}$$

This last residuals vanish when $\{U_n\}$ happens to be the exact solution of (4.5c).

4.1.2 Conditions to be Met for the Convergence to an Exact Result

The finite element method is an approximate method of solution, for boundary value problems, that guarantees convergence to an exact result by element size reduction provided the approximation of **u** fulfils the following two conditions.

— Complete polynomial basis (see paragraph 1.3.2 and 1.7). If the approximate solution is to approach the exact solution when h tends towards a value of zero, the truncation error of all the terms of W^e must be of order h^n with $n \geq 1$. In paragraph 1.7 we have seen that the approximation of **u** must have a basis that is at least complete up to order m if the highest derivative of **u** in W^e is of order m. For example, in a one-dimensional problem, if:

$$W^e = \int_{V_e} \delta\left(\frac{\partial^2 u^e}{\partial x^2}\right)\left(\frac{\partial^2 u^e}{\partial x^2}\right) dx \tag{4.7a}$$

the approximation of u^e must, at a minimum, contain a complete quadratic basis: $1, x, x^2$ or $1, \xi, \xi^2$ if the element is isoparametric.

— Continuity. To the previous condition which must be met everywhere within the element, we must add a certain degree of inter-element continuity so as to be able to write:

$$W \equiv \sum W^e$$

The approximate value of **u** in the entire domain V must satisfy the differentiability conditions of the integral form W. The function u and all its derivatives appearing in W must be bounded. If the highest derivative in W is of order m, when u and its derivatives up to order $m - 1$ are continuous across their common boundaries with other element, then, the continuity condition is satisfied and the element is said to be conformable. In the case of integral form (4.7a) a conformable element must be of class C^1.

When the continuity conditions are not all met, the element is said to be non-conformable. In such a case:

$$W = \sum_e W^e + W^d \qquad (4.7b)$$

where W^d is a supplementary term arising from inter-element discontinuities. Clearly, the solution will converge to an exact result if, and only if, W^d tends to zero with element size reduction. The patch test was devised to verify if this last minimum requirement is met by non-conformable elements. Unless a non-conformable passes all patch tests it should never be used.

4.1.3 Patch Tests

Different techniques have evolved in using Patch tests:

— an ad hoc numerical test [1] and
— an analytical test based on a variational method [2].

Numerical test

In the previous paragraph it was shown that the approximation for u in each element should have a polynomial basis complete to at least order m. We must verify that the condition is satisfied by the approximation over the entire domain V. A variety of numerical tests can be derived for such a verification. The following strategy is proposed:

— Assume that an arbitrary polynomial $P_m(\mathbf{x})$ of order m is a solution of the problem at hand. A typical mesh configuration is illustrated below:

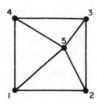

— At least one interior node must be included in the region, and a regular mesh should be avoided to prevent fortuitous self-correction by symmetry.

— Nodal values of u are calculated for all exterior nodes (1, 2, 3, 4) from $\mathbf{u} = P_m(\mathbf{x})$. With values of u at (1, 2, 3, 4) imposed, the finite element solution for the problem is obtained and the value of u_5 thus found compared to the value of u_5 obtained from $u = P_m(x_5)$. If $W^d = 0$, the two values of u at node 5 should be identical.

The test should be repeated for several configurations because, if W^d is zero for one configuration it is not necessarily so for all of them.

Example 4.2   Numerical patch test for the Laplace equation

Consider the non-conformable three-node element described in paragraph 2.3.3.6 for the following mesh:

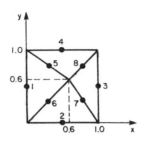

For the Laplace equation $m = 1$.

The integral form W is given in Example 4.1 where $f_V = f_S = \alpha = 0$. Let $P_m(x, y) = a_0 + a_1 x + a_2 y$ with $a_0 = a_1 = 1$ and $a_2 = 0$.

The nodal values of u are:

$$u_1 = 1.0, \quad u_2 = 1.5, \quad u_3 = 2.0, \quad u_4 = 1.5$$

The finite element solution for values of u at interior nodes 5, 6, 7, 8 is to be obtained from a code and should be identical to the following nodal values obtained from $P_m(x)$:

$$u_5 = 1.3 = P_m(0.3, \ 0.8)$$
$$u_6 = 1.3 = P_m(0.3, \ 0.3)$$
$$u_7 = 1.8 = P_m(0.8, \ 0.3)$$
$$u_8 = 1.8 = P_m(0.8, \ 0.8)$$

Variational Method

For integral forms of Galerkin type (3.43), the technique consists in verifying that:

$$W(P_m) = \sum_e W^e(P_m) = 0 \tag{4.8a}$$

Since $P_m(x)$ is of order m, the elementary expression W^e, in most cases, may be transformed to a contour integral through integration by parts:

$$\sum_e \int_{S^e} \ldots = 0 \tag{4.8b}$$

For each element, the contour integral can be split into two parts:

— the conformable portion, involving only nodal variables belonging to each side;
— the non-conformable portion, involving nodal variables not belonging to the side where integration is performed.

After assembly, the conformable portions of the contour integrations cancel each other and it remains to verify that:

$$\int_{S^e} (\text{non-conformable part}) = 0$$

for each element. Then (4.8b) is satisfied for any mesh. This last strategy is more general than the numerical tests.

Example 4.3   *Variational patch test for the Laplace equation*

The integral form for an element obtained in Example 4.1 is:

$$(f_S = f_V = \alpha = 0)$$

$$W^r = \int_{V^e} \left(\delta\left(\frac{\partial u}{\partial x}\right) \frac{\partial u}{\partial x} + \delta\left(\frac{\partial u}{\partial y}\right) \frac{\partial u}{\partial y} \right) dV$$

Integrating by parts:

$$W^e = -\int_{V^e} \delta u \left(\frac{\partial^2 u}{\partial x^2} + \frac{\partial^2 u}{\partial y^2} \right) dV + \int_{S^e} \delta u \frac{\partial u}{\partial n} dS$$

Using the polynomial P_m defined in Example 4.2, the Laplace equation is satisfied and $\partial P_m / \partial n = l$, therefore:

$$W^e = \int_{S^e} \delta u \, l \, dS$$

where l is the direction cosine of an outward unit normal vector to any side of an element. Consider a typical element:

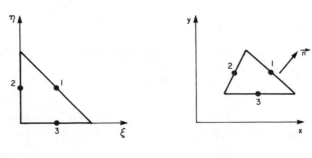

$$\delta u = \langle -1 + 2\xi + 2\eta \quad 1 - 2\xi \quad 1 - 2\eta \rangle \begin{Bmatrix} \delta u_1 \\ \delta u_2 \\ \delta u_3 \end{Bmatrix} = \langle N \rangle \{\delta u_m\}$$

The value of W^e on a typical side (say side 1; $\eta = 1 - \xi$) is:

$$W^e = \langle \delta u_n \rangle \int_0^1 \begin{Bmatrix} 1 \\ 1 - 2\xi \\ -1 + 2\xi \end{Bmatrix} lh \, d\xi$$

where h is the length of side 1:

$$W^e = \langle \delta u_n \rangle \begin{Bmatrix} lh \\ 0 \\ 0 \end{Bmatrix} = \delta u_1 . lh$$

The only non-zero term is the one involving nodal value δu_1 which is conformable, i.e. δh_1 is the same for an adjacent element but l would change sign, therefore, $\sum W^e = 0$ for an assembly of such elements.

4.2 Discretized Integral Forms W^e

4.2.1 *Matrix Expression for W^e*

To get discretized form (4.4) for W^e we use finite element approximations for u, δu, and their derivatives in the domain of an element e (see Chapters 1 and 2).

$$\begin{aligned} u &= \langle N \rangle \{u_n\} \\ \frac{\partial u}{\partial x} &= \left\langle \frac{\partial N}{\partial x} \right\rangle \{u_n\} \\ &\vdots \\ \delta u &= \langle N \rangle \{\delta u_n\} \\ \delta\left(\frac{\partial u}{\partial x}\right) &= \left\langle \frac{\partial N}{\partial x} \right\rangle \{\delta u_n\} \\ &\vdots \end{aligned} \quad (4.9a)$$

Then:

$$\{\partial u\} = \begin{Bmatrix} u \\ \frac{\partial u}{\partial x} \\ \vdots \end{Bmatrix} = \begin{bmatrix} \langle N \rangle \\ \left\langle \frac{\partial N}{\partial x} \right\rangle \\ \cdots \end{bmatrix} \{u_n\} = [B]\{u_n\}$$

$$\{\delta(\partial u)\} = \begin{Bmatrix} \delta u \\ \delta\left(\frac{\partial u}{\partial x}\right) \\ \vdots \end{Bmatrix} = \begin{bmatrix} \langle N \rangle \\ \left\langle \frac{\partial N}{\partial x} \right\rangle \\ \cdots \end{bmatrix} \{\delta u_n\} = [B_\delta]\{\delta u_n\} \quad (4.9b)$$

For self-adjoint operators $\mathscr{L}$:

$$\{\delta(\partial u)\} \equiv \delta(\{\partial u\}); \quad [B_\delta] \equiv [B]$$

Substitution of (4.9a) and (4.9b) into (4.3) gives:

$$W^e = \langle \delta u_n \rangle \left(\int_{V^e} [B_\delta]^T [D][B] \mathrm{d}V \{u_n\} - \int_{V^e} \{N\} f_V \mathrm{d}V - \int_{S_f^e} \{N\} f_S \mathrm{d}S \right) \quad (4.10a)$$

Thus, comparing with (4.4):

$$[k] = \int_{V^e} [B_\delta]^T [D][B] \mathrm{d}V \quad (4.10b)$$

$$\{f\} = \int_{V^e} \{N\} f_V \mathrm{d}V + \int_{S_f^e} \{N\} f_S \mathrm{d}S \quad (4.10c)$$

Example 4.4 Discretized form W^e for the problem defined in Example 4.1 (Poisson's equation)

The approximation of u on the element is:

$$u = \langle N \rangle \{u_n\}$$

$$\{\partial u\} = \begin{Bmatrix} \dfrac{\partial u}{\partial x} \\ \dfrac{\partial u}{\partial y} \end{Bmatrix} = \begin{bmatrix} \left\langle \dfrac{\partial N}{\partial x} \right\rangle \\ \left\langle \dfrac{\partial N}{\partial y} \right\rangle \end{bmatrix} \{u_n\} = [B]\{u_n\}$$

$$\{\delta(\partial u)\} = \begin{Bmatrix} \delta\left(\dfrac{\partial u}{\partial x}\right) \\ \delta\left(\dfrac{\partial u}{\partial y}\right) \end{Bmatrix} = \begin{bmatrix} \left\langle \dfrac{\partial N}{\partial x} \right\rangle \\ \left\langle \dfrac{\partial N}{\partial y} \right\rangle \end{bmatrix} \{\delta u_n\} = [B_\delta]\{\delta u_n\} = [B]\{\delta u_n\}$$

Note that in this case $[B_\delta] \equiv [B]$ because the Laplacian is a self-adjoint operator:

$$[K] = \int_{V^e} [B]^T [D][B] \mathrm{d}V + \int_{S_f^e} \alpha \{N\} \langle N \rangle \mathrm{d}S$$

The matrix $[k]$ is this symmetrical.

$$\{f\} = \int_{V^e} \{N\} f_V \mathrm{d}V + \int_{S_f^e} \{N\} f_S \mathrm{d}S$$

For a concentrated force f_i at point $x = x_i$ of S_f^e, function f_S becomes a Dirac delta function $\delta(x_i)$:

$$f_S(\mathbf{x}_i) = f_i \delta(\mathbf{x}_i)$$

Corresponding vector $\{f\}$ is:

$$\{f\} = \{N(\mathbf{x}_i)\} f_i$$

4.2.2 Case of a Non-Linear Differential Operator

For non-linear problems, the element matrix $[k]$ has terms depending on $\mathbf{u}_n$. The element matrix can be split into a linear part and a non-linear part:

$$[k(\mathbf{u}_n)] = [k_l] + [k_{nl}(\mathbf{u}_n)] \tag{4.11}$$

Example 4.5 Beam subject to large displacements

The integral form corresponding to a straight beam with rectangular cross-section $h \times b$, subject to large transverse displacements (but small strains and moderate rotations) is:

$$W^e = Ehb \int_{V^e} \delta\varepsilon \cdot \varepsilon \, dx + \frac{Eh^3 b}{12} \int_{V^e} \delta\kappa \cdot \kappa \, dx - \int_{V^e} \delta w \cdot f_V \, dx$$

where:

$$\varepsilon = u_{,x} + \tfrac{1}{2} w_{,x}^2 \qquad \delta\varepsilon = \delta(u_{,x}) + w_{,x} \delta(w_{,x})$$
$$\kappa = -w_{,xx} \qquad \delta\kappa = -\delta(w_{,xx})$$

$$W^e = Ehb \int_{V^e} (\delta(u_{,x})(u_{,x} + \tfrac{1}{2} w_{,x}^2) + \delta(w_{,x})(w_{,x} u_{,x} + \tfrac{1}{2} w_{,x}^3)) \, dx$$
$$+ \frac{Eh^3 b}{12} \int_{V^e} \delta(w_{,xx}) w_{,xx} \, dx - \int_{V^e} \delta w f_V \, dx$$

Note that W^e is the first variation of the total potential energy in this problem.

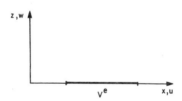

— u, w are axial and transverse displacement components of a point in the beam;
— E, h, b are Young's modulus, height, and thickness of the cross-section;
— f_V is the load per unit length applied to the beam.

We write W^e in matrix form splitting it in linear W_l^e and non-linear W_{nl}^e portions:

$$W_l^e = \int_{V^e} (\langle \delta(\partial u) \rangle [D_l] \{\partial u\} - \delta w . f_V) \, dx$$

where

$$\{\partial u\} = \begin{Bmatrix} u_{,x} \\ w_{,x} \\ w_{,xx} \end{Bmatrix} \quad \{\delta(\partial u)\} = \begin{Bmatrix} \delta(u_{,x}) \\ \delta(w_{,x}) \\ \delta(w_{,xx}) \end{Bmatrix}$$

$$[D_l] = \begin{bmatrix} Ehb & 0 & 0 \\ 0 & 0 & 0 \\ 0 & 0 & \dfrac{Eh^3 b}{12} \end{bmatrix}$$

$$W_{nl}^e = \int_{V^e} \langle \delta(\partial u) \rangle [D_{nl}] \{\partial u\}\, dx$$

$$[D_{nl}] = Ehb \begin{bmatrix} 0 & \tfrac{1}{2} w_{,x} & 0 \\ w_{,x} & \tfrac{1}{2} w_{,x}^2 & 0 \\ 0 & 0 & 0 \end{bmatrix} \quad \text{(non-symmetrical matrix)}$$

Let us use different approximations for u and w:

$$u = \langle N_u \rangle \{u_n\}$$
$$w = \langle N_w \rangle \{w_n\}$$

Hence:

$$\{\partial u\} = \begin{bmatrix} \langle N_{u,x} \rangle & 0 \\ 0 & \langle N_{w,x} \rangle \\ 0 & \langle N_{w,xx} \rangle \end{bmatrix} \begin{Bmatrix} u_n \\ w_n \end{Bmatrix} = [B]\{u_n\}$$

$$\{\delta(\partial u)\} = [B]\{\delta u_n\}$$

Linear and non-linear element matrices are:

$$[k_l] = \int_{V^e} [B]^T [D_l][B]\, dx$$

$$[k_{nl}] = \int_{V^e} [B]^T [D_{nl}][B]\, dx \quad \text{(non-symmetrical matrix)}$$

where $[D_{nl}]$ is written as a function of $\{W_n\}$:

$$[D_{nl}] = Ehb \begin{bmatrix} 0 & \tfrac{1}{2}\langle N_{w,x} \rangle \{w_n\} & 0 \\ \langle N_{w,x} \rangle \{w_n\} & \tfrac{1}{2}(\langle N_{w,x} \rangle \{w_n\})^2 & 0 \\ 0 & 0 & 0 \end{bmatrix}$$

Matrix $[k_{nl}]$ is symmetrical if we use another expression for $[D_{nl}]$ while keeping the same matrix $[B]$.

$$[D_{nl}] = Ehb \begin{bmatrix} 0 & \tfrac{1}{2} w_{,x} & 0 \\ \tfrac{1}{2} w_{,x} & \tfrac{1}{2} w_{,x}^2 + \tfrac{1}{2} u_{,x} & 0 \\ 0 & 0 & 0 \end{bmatrix}$$

4.2.3 Integral Form W^e in the Space of the Element of Reference

In Chapters 1 and 2, interpolation functions $N(\xi)$ were written in the space of the element of reference. In this chapter, expressions (4.3) and (4.10) for W^e contain:

— derivatives of **u** and $\delta\mathbf{u}$ with respect to **x**;
— nodal variables $\mathbf{u}_n$ and $\delta\mathbf{u}_n$ that may include derivatives of **u** and $\delta\mathbf{u}$ with respect to **x** at the nodes;
— integrations in the space of the real element V^e.

All derivatives and integrations in space **x** must be transformed in the space ξ.

4.2.3.1 Transformation of derivatives with respect to x

Derivatives $u_{,x}, u_{,y}, u_{,z}, u_{,xx}, \ldots$ are transformed into $u_{,\xi}, u_{,\eta}, u_{,\zeta}, u_{,\xi\xi}$, with the help of the Jacobian matrix of the geometrical transformation from paragraph 1.5. For example, in one dimension we have:

$$u(\xi) = \langle N(\xi) \rangle \{u_n\}$$

$$\frac{du}{dx} = \frac{d\xi}{dx}\frac{du}{d\xi} = \frac{d\xi}{dx}\left\langle \frac{dN(\xi)}{d\xi} \right\rangle \{u_n\}$$

Expression (4.9b) of matrix [B] is rewritten as:

$$[B] = [Q][B_\xi] \tag{4.12}$$

where [Q] is a matrix containing certain terms of $[j] = [J]^{-1}$; and $[B_\xi]$ is a matrix similar to [B] that contains derivatives of functions $N(\xi)$ in terms of ξ instead of derivatives of functions $N(\mathbf{x})$ with respect to **x**.

Example 4.6 *Transformation of matrix [B] for the integral form of Poisson's equation*

$$[B] = \begin{bmatrix} \left\langle \frac{\partial N}{\partial x} \right\rangle \\ \left\langle \frac{\partial N}{\partial y} \right\rangle \end{bmatrix} = \begin{bmatrix} \frac{\partial \xi}{\partial x} & \frac{\partial \eta}{\partial x} \\ \frac{\partial \xi}{\partial y} & \frac{\partial \eta}{\partial y} \end{bmatrix} \begin{bmatrix} \left\langle \frac{\partial N}{\partial \xi} \right\rangle \\ \left\langle \frac{\partial N}{\partial \eta} \right\rangle \end{bmatrix} = [Q][B_\xi]$$

In this case:

$$[Q] = [j] = [J]^{-1}$$

Then:

$$[k] = \int_{V^e} [B_\xi]^T [Q]^T [D][Q][B_\xi] dV$$

4.2.3.2 Transformation of Nodal Variables

For an element using Hermite polynomial approximations, nodal variables $\{u_n\}_\xi$ in the element of reference are related to the nodal variables $\{u_n\}_x$ in the real element by:

$$\{u_n\}_\xi = [T]\{u_n\}_x$$
$$\{\delta u_n\}_\xi = [T]\{\delta u_n\}_x$$

Transformation matrix $[T]$ contains terms of matrix $[J]$ evaluated at the nodes (see paragraph 2.3.4.1)

4.2.3.3 *Transformation of the domain of integration*

Integration over the volume of the real element V^e is replaced by an integration over the volume of the element of reference V^r (see paragraph 1.6.1) and (1.44):

$$\int_{V^e} \ldots dV = \int_{V^r} \ldots \det(J) d\xi \, d\eta \, d\zeta \qquad (4.13)$$

The limits of integration for all the classical elements of reference described previously are:

— One dimension

$$\int_{\xi=-1}^{\xi=1} \ldots \det(J) d\xi$$

— Two dimensions

Triangle $$\int_{\xi=0}^{\xi=1} \int_{\eta=0}^{\eta=1-\xi} \ldots \det(J) \, d\eta \, d\xi$$

Quadrilateral $$\int_{\xi=-1}^{\xi=1} \int_{\eta=-1}^{\eta=1} \ldots \det(J) \, d\eta \, d\xi$$

— Three dimensions

Tetrahedron $$\int_{\xi=0}^{\xi=1} \int_{\eta=0}^{\eta=1-\xi} \int_{\zeta=0}^{\zeta=1-\xi-\eta} \ldots \det(J) \, d\zeta \, d\eta \, d\xi$$

Hexahedron $$\int_{\xi=-1}^{\xi=1} \int_{\eta=-1}^{\eta=1} \int_{\zeta=-1}^{\zeta=1} \ldots \det(J) \, d\zeta \, d\eta \, d\xi$$

Prism $$\int_{\xi=0}^{\xi=1} \int_{\eta=0}^{\eta=1-\xi} \int_{\zeta=-1}^{\zeta=1} \ldots \det(J) \, d\zeta \, d\eta \, d\xi$$

4.2.3.4 *Transformation of the contour integrations*

(a) *Line integration in two or three dimensions.*

Integral

$$l = \int_S \ldots dS$$

is written in terms of a curvilinear abscissa s on the curve S

$$l = \int_{s_1}^{s_2} \ldots J_S \, dS \tag{4.14a}$$

In general, the abscissa s is one of the variables ξ, η or ζ. Curve S generally corresponds to one of the edges of an element of reference on which parameter S is defined:

$$x = \langle N(s) \rangle \{x_n\} \text{ etc.}$$
$$J_S = \sqrt{x_{,s}^2 + y_{,s}^2 + z_{,s}^2} \tag{4.14b}$$

Example 4.7 Contour integral for a four-node element

For side $\xi = 1$ of the quadrilateral element used in Example 1.16, we have:

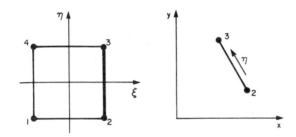

$$s \equiv \eta \quad ds \equiv d\eta$$

$$\langle N(s) \rangle = \langle N(\xi = 1, \eta) \rangle = \left\langle 0 \, \frac{1-\eta}{2} \, \frac{1+\eta}{2} \, 0 \right\rangle$$

$$x_{,\eta} = \langle N_{,\eta}(\xi = 1, \eta) \rangle \{x_n\} = \tfrac{1}{2}(x_3 - x_2)$$

$$y_{,\eta} = \tfrac{1}{2}(y_3 - y_2)$$

$$J_S = \sqrt{x_{,\eta}^2 + y_{,\eta}^2} = \sqrt{\left(\frac{x_3 - x_2}{2}\right)^2 + \left(\frac{y_3 - y_2}{2}\right)^2}$$

$$l = \int_{-1}^{1} \ldots J_S \, d\eta$$

(b) Surface integration in three dimensions

Integral

$$\int_S \ldots dS$$

is written in terms of surface coordinates s_1 and s_2 that are generally (ξ, η) or (ξ, ζ)

or (η, ζ):

$$\int_S \ldots J_s \, ds_1 \, ds_2 \tag{4.15a}$$

Surface S is one of the faces of the element of reference. In this face, a point is given in terms of the two parameters s_1 and s_2

$$J_s = \frac{x = \langle N(s_1, s_2) \rangle \{x_n\} \quad etc.}{\sqrt{(y_{,s_1} z_{,s_2} - z_{,s_1} y_{,s_2})^2 + (z_{,s_1} x_{,s_2} - x_{,s_1} z_{,s_2})^2 + (x_{,s_1} y_{,s_2} - y_{,s_1} x_{,s_2})^2}} \tag{4.15b}$$

Example 4.8 Surface integration for an eight-node element

On face $\zeta = 1$ of the eight-node hexahedronal element described in paragraph 2.6.1:

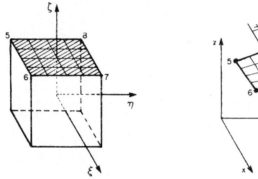

$$s_1 \equiv \xi \quad s_2 \equiv \eta$$
$$ds_1 \equiv d\xi \quad ds_2 \equiv d\eta$$

$$N(s_1, s_2) = N(\xi, \eta, \zeta = 1)$$
$$= \tfrac{1}{4} \langle 0 \quad 0 \quad 0 \quad 0 \quad (1-\xi)(1-\eta) \quad (1+\xi)(1-\eta)$$
$$(1+\xi)(1+\eta) \quad (1-\xi)(1+\eta) \rangle$$
$$N_{,\xi}(\xi, \eta, \zeta = 1) = \tfrac{1}{4} \langle 0; 0; 0; 0; -(1-\eta); (1-\eta); (1+\eta); -(1+\eta) \rangle$$
$$N_{,\eta}(\xi, \eta, \zeta = 1) = \tfrac{1}{4} \langle 0; 0; 0; 0; -(1-\xi); -(1+\xi); (1+\xi); (1-\xi) \rangle$$
$$\langle x_{,\xi} \quad y_{,\xi} \quad z_{,\xi} \rangle = \langle N_{,\xi} \rangle [\{x_n\} \{y_n\} \{z_n\}]$$
$$\langle x_{,\eta} \quad y_{,\eta} \quad z_{,\eta} \rangle = \langle N_{,\eta} \rangle [\{x_n\} \{y_n\} \{z_n\}]$$

J_s is given by (4.15b) and

$$I = \int_{-1}^{1} \int_{-1}^{1} \ldots J_s(\xi, \eta) \, d\xi \, d\eta$$

4.2.3.5 Expression of [k] and {f} in the space of the element of reference

Expressions (4.10b) and (4.10c) for matrix [k] and vector {f} transformed in the reference space V^r, are:

$$\{k\} = \int_{V^r} [B_{\delta\xi}]^T [Q_\delta]^T [D][Q][B_\xi] \det(J) d\xi \, d\eta \, d\zeta \qquad (4.16a)$$

$$\{f\} = \int_{V^r} \{N\} f_V \det(J) d\xi \, d\eta \, d\zeta + \int_{S^r} \{N\} f_S J_S ds_1 \, ds_2 \qquad (4.16b)$$

where $[Q]$ and $[B_\xi]$ are defined by (4.12); $[Q_\delta]$ and $[B_{\delta\xi}]$ are analogous to $[Q]$ and $[B_\xi]$, except for non-self-adjoint operators; and J_S is defined by (4.14b) or (4.15b).

Integrations (4.16a) and (4.16b) are generally evaluated numerically using the methods described in paragraph 5.1. (In particular, see (5.4) and Example 5.3.)

4.2.4 Some Classical Forms of W^e and Element Matrices

Integral forms W^e are in general a sum of several terms; for example:

$$W^e = \int_{V_e} \left(\delta\left(\frac{\partial u}{\partial x}\right) \cdot \frac{\partial u}{\partial x} + \delta\left(\frac{\partial u}{\partial y}\right) \cdot \frac{\partial u}{\partial y} \right) dV$$

Figure 4.1 describes the most frequently encountered terms and their corresponding matrices [B] and [D]. Figure 4.2 shows two one-dimensional elements explicitly for classical forms of W^e.

4.3 Computational Techniques for Element Matrices

4.3.1 Explicit Computations for a Triangular Element and Poisson's Equation

Let us apply the results of the preceding paragraphs to compute the element matrices when the integration can be performed explicitly. We use the three-node linear triangle of paragraph 2.3.2 and evaluate matrices [k], [m] and the load vector {f} formulated in Examples 4.4 and 4.6.

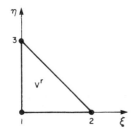
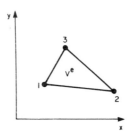

Terms	$[B_\delta]^T$	$[D]$	$[B]$	Property of $[k]$	
Symmetrical quadratic					
$\int_{V_e} \delta u.u \, dV$	$\{N\}$	1	$\langle N \rangle$	constant	symmetrical
$\int_{V_e} \delta\left(\frac{\partial u}{\partial x}\right) \cdot \frac{\partial u}{\partial x} dV$	$\left\{\dfrac{\partial N}{\partial x}\right\}$	1	$\left\langle \dfrac{\partial N}{\partial x} \right\rangle$	constant	symmetrical
$\int_{V_e} \delta\left(\frac{\partial^2 u}{\partial x^2}\right) \cdot \frac{\partial^2 u}{\partial x^2} dV$	$\left\{\dfrac{\partial^2 N}{\partial x^2}\right\}$	1	$\left\langle \dfrac{\partial^2 N}{\partial x^2} \right\rangle$	constant	symmetrical
$\int_{V_e} \delta\left(\frac{\partial^m u}{\partial x^m}\right) \cdot \frac{\partial^m u}{\partial x^m} dV$	$\left\{\dfrac{\partial^m N}{\partial x^m}\right\}$	1	$\left\langle \dfrac{\partial^m N}{\partial x^m} \right\rangle$	constant	symmetrical
Non-symmetrical quadratic					
$\int_{V_e} \delta u \cdot \frac{\partial u}{\partial x} dV$	$\{N\}$	1	$\left\langle \dfrac{\partial N}{\partial x} \right\rangle$	constant	non-symmetrical

$\int_{V_e} \delta\left(\dfrac{\partial^m u}{\partial x^m}\right) \dfrac{\partial^n u}{\partial x^n} dV$	$\left\{\dfrac{\partial^m N}{\partial x^m}\right\}$	1	$\left\langle \dfrac{\partial^n N}{\partial x^n} \right\rangle$	constant	non-symmetrical if $m \neq n$
Non-linear					
$\int_{V_e} \delta u \cdot u \cdot \dfrac{\partial u}{\partial x} dV$	$\{N\}$	$\left\langle \dfrac{\partial N}{\partial x} \right\rangle \{u_n\}$	$\langle N \rangle$	function of $\{u_n\}$	symmetrical
	$\{N\}$	$\langle N \rangle \{u_n\}$	$\left\langle \dfrac{\partial N}{\partial x} \right\rangle$	function of $\{u_n\}$	non-symmetrical
$\int_{V_e} \delta\left(\dfrac{\partial u}{\partial x}\right) \cdot \dfrac{\partial u}{\partial x} \dfrac{\partial u}{\partial x} dV$	$\left\{\dfrac{\partial N}{\partial x}\right\}$	$\left\langle \dfrac{\partial N}{\partial x} \right\rangle \{u_n\}$	$\left\langle \dfrac{\partial N}{\partial x} \right\rangle$	function of $\{u_n\}$	symmetrical
$\int_{V_e} \delta\left(\dfrac{\partial^m u}{\partial x^m}\right) \cdot D\left(u, \dfrac{\partial u}{\partial x}, \ldots\right) \cdot \dfrac{\partial^n u}{\partial x^n} dV$	$\left\{\dfrac{\partial^m N}{\partial x^m}\right\}$	$D(\{u_n\})$	$\left\langle \dfrac{\partial^n N}{\partial x^n} \right\rangle$	function of $\{u_n\}$	non-symmetrical if $m \neq n$
Quadratic contour integral terms					
$\int_{S_e} \delta u \cdot u \, dS$	$\{N\}$	1	$\langle N \rangle$	constant	symmetrical

Figure 4.1 (*Contd.*)

				Remarks
Linear volume integral terms $\int_{v_e} \delta u . f_V dV$	$\{N\}$	1	f_V	$W^e = \langle \delta u_n \rangle \{f\}$ $\{f\} = \int_{v_e} \{N\} f_V dV$
Linear surface integral terms $\int_{s_{e_f}} \delta u . f_S dS$	$\{N\}$	1	f_S	$W^e = \langle \delta u_n \rangle \{f\}$ $\{f\} = \int_{s_e^g} \{N\} f_S dS$
Non-stationary terms $\int_{v_e} \delta u . \dfrac{\partial u}{\partial t} dV$	$\{N\}$	1	$\langle N \rangle$	$W^e = \langle \delta u_n \rangle [c] \left\{\dfrac{du_n}{dt}\right\}$ $[c] = \int_{v_e} \{N\}\langle N \rangle dV$
$\int_{v_e} \delta u . \dfrac{\partial^2 u}{\partial t^2} dV$	$\{N\}$	1	$\langle N \rangle$	$W^e = \langle \delta u_n \rangle [m] \left\{\dfrac{d^2 u_n}{dt^2}\right\}$ $[m] = \int_{v_e} \{N\} dV$

2-noded linear element (paragraph 2.2.1)

$$\int_{x_1}^{x_2} \delta u \cdot u \, dx = \langle \delta u_1 \; \delta u_2 \rangle \frac{l}{6} \begin{bmatrix} 2 & 1 \\ 1 & 2 \end{bmatrix} \begin{Bmatrix} u_1 \\ u_2 \end{Bmatrix} = \langle \delta u_n \rangle [m] \{u_n\}$$

$$\int_{x_1}^{x_2} \delta\left(\frac{\partial u}{\partial x}\right) \cdot \frac{\partial u}{\partial x} dx = \langle \delta u_1 \; \delta u_2 \rangle \frac{1}{l} \begin{bmatrix} 1 & -1 \\ -1 & 1 \end{bmatrix} \begin{Bmatrix} u_1 \\ u_2 \end{Bmatrix} = \langle \delta u_n \rangle [k] \{u_n\}$$

$$l = x_2 - x_1.$$

Remark: m and k are used to define the mass and stiffness matrices of a truss element

2-noded cubic element (Hermite type) (paragraph 2.2.3.1)

$$\int_{x_1}^{x_2} \delta u \cdot u \, dx =$$

$$= \langle \delta u_n \rangle \frac{l}{420} \begin{bmatrix} 156 & 22l & 54 & -13l \\ & 4l^2 & 13l & -3l^2 \\ & & 156 & -22l \\ \text{Sym.} & & & 4l^2 \end{bmatrix} \{u_n\} = \langle \delta u_n \rangle [m] \{u_n\}$$

$$\int_{x_1}^{x_2} \delta\left(\frac{\partial u}{\partial x}\right) \cdot \frac{\partial u}{\partial x} dx =$$

$$= \langle \delta u_n \rangle \frac{1}{30l} \begin{bmatrix} 36 & 3l & -36 & 3l \\ & 4l^2 & -3l & -l^2 \\ & & 36 & -3l \\ \text{Sym.} & & & 4l^2 \end{bmatrix} \{u_n\} = \langle \delta u_n \rangle [k_m] \{u_n\}$$

$$\int_{x_1}^{x_2} \delta\left(\frac{\partial^2 u}{\partial x^2}\right) \frac{\partial^2 u}{\partial x^2} dx =$$

$$= \langle \delta u_n \rangle \frac{1}{l^3} \begin{bmatrix} 12 & 6l & -12 & 6l \\ & 4l^2 & -6l & 6l^2 \\ & & 12 & -6l \\ \text{Sym.} & & & 4l^2 \end{bmatrix} \{u_n\} = \langle \delta u_n \rangle [k_f] \{u_n\}$$

$$\langle \delta u_n \rangle = \langle \delta u_1 \; \delta u_{1,x} \; \delta u_2 \; \delta u_{2,x} \rangle ; \quad \langle u_n \rangle = \langle u_1 \; u_{1,x} \; u_2 \; u_{2,x} \rangle$$

$$l = x_2 - x_1.$$

Remark: m, k_m and k_f are used to define the mass, axial and bending stiffness matrices for a beam element

Figure 4.2 Explicit element matrices for a truss and a beam.

$$\langle N \rangle = \langle 1-\xi-\eta \quad \xi \quad \eta \rangle$$

$$[j] = \frac{1}{2A}\begin{bmatrix} y_3 - y_1 & -(y_2 - y_1) \\ -(x_3 - x_1) & x_2 - x_1 \end{bmatrix}$$

$$\det(J) = 2A = (x_2 - x_1)(y_3 - y_1) - (x_3 - x_1)(y_2 - y_1) \quad (4.17a)$$

$$[B_\xi] = \begin{bmatrix} -1 & 1 & 0 \\ -1 & 0 & 1 \end{bmatrix}; \quad [B] = [j][B_\xi] = \frac{1}{2A}\begin{bmatrix} y_2 - y_3 & y_3 - y_1 & y_1 - y_2 \\ x_3 - x_2 & x_1 - x_3 & x_2 - x_1 \end{bmatrix}$$

$$[D] = d\begin{bmatrix} 1 & 0 \\ 0 & 1 \end{bmatrix} \quad (4.17b)$$

where d is the isotropic conductivity coefficient which is equal to one for the Laplace equation.

$$[k] = \int_0^1 \int_0^{1-\xi} d[B]^T[B]\det(J)\,d\eta\,d\xi$$

Since matrix $[B]$ is constant: $[k] = A \cdot d \cdot [B]^T[B]$

$$[k] = \frac{d}{4A}\begin{bmatrix} (y_3-y_2)^2 + (x_3-x_2)^2 & (y_3-y_2)(y_1-y_3) + (x_3-x_2)(x_1-x_3) & (y_2-y_1)(y_3-y_2) + (x_2-x_1)(x_3-x_2) \\ & (y_3-y_1)^2 + (x_3-x_1)^2 & (y_1-y_3)(y_2-y_1) + (x_1-x_3)(x_2-x_1) \\ \text{Sym.} & & (y_2-y_1)^2 + (x_2-x_1)^2 \end{bmatrix}$$

(4.18)

For the case where:

$$x_1 = y_1 = 0; \quad x_2 = a; \quad y_2 = 0; \quad x_3 = 0; \quad y_3 = a$$

$$[k] = \frac{d}{2}\begin{bmatrix} 2 & -1 & -1 \\ & 1 & 0 \\ \text{Sym.} & & 1 \end{bmatrix} \quad (4.19)$$

Mass matrix (4.6b) is:

$$[m] = \int_0^1 \int_0^{1-\xi} \{N\}\langle N \rangle \det(J)\,d\eta\,d\xi \quad (4.20a)$$

$$[m] = \frac{A}{12}\begin{bmatrix} 2 & 1 & 1 \\ & 2 & 1 \\ \text{Sym.} & & 2 \end{bmatrix} \quad (4.20b)$$

Vector $\{f\}$ from (4.16b) for the case where f_V is constant and f_S zero is:

$$\{f\} = \frac{Af_V}{3}\begin{Bmatrix} 1 \\ 1 \\ 1 \end{Bmatrix} \quad (4.21)$$

```
      SUBROUTINE ELEM00(VCORE,VPREE,VKE,VFE,NDIM,NNEL,NDLE)        ELOO   1
C================================================================ELOO   2
C                                                                  ELOO   3
C     ELEMENT MATRIX AND VECTOR FOR A 3 NODES TRIANGLE,             ELOO   4
C     ISOTROPIC POISSON EQUATION                                    ELOO   5
C     (VPREE(1) = DX = DY = D)                                      ELOO   6
C                                                                  ELOO   7
C================================================================ELOO   8
      IMPLICIT REAL*8(A-H,O-Z)                                      ELOO   9
      DIMENSION VCORE(NDIM,NNEL),VPREE(2),VKE(NDLE,NDLE),VFE(NDLE)  ELOO  10
C-------  ELEMENT MATRIX                                            ELOO  11
      X32=VCORE(1,3)-VCORE(1,2)                                     ELOO  12
      X13=VCORE(1,1)-VCORE(1,3)                                     ELOO  13
      X21=VCORE(1,2)-VCORE(1,1)                                     ELOO  14
      Y23=VCORE(2,2)-VCORE(2,3)                                     ELOO  15
      Y31=VCORE(2,3)-VCORE(2,1)                                     ELOO  16
      Y12=VCORE(2,1)-VCORE(2,2)                                     ELOO  17
      C2A=X21*Y31-X13*Y12                                           ELOO  18
      C=VPREE(1)/(C2A*2.0D0)                                        ELOO  19
      VKE(1,1)=(Y23*Y23+X32*X32)*C                                  ELOO  20
      VKE(2,2)=(Y31*Y31+X13*X13)*C                                  ELOO  21
      VKE(3,3)=(Y12*Y12+X21*X21)*C                                  ELOO  22
      VKE(1,2)=(Y23*Y31+X32*X13)*C                                  ELOO  23
      VKE(1,3)=(Y12*Y23+X21*X32)*C                                  ELOO  24
      VKE(2,3)=(Y31*Y12+X13*X21)*C                                  ELOO  25
      VKE(2,1)=VKE(1,2)                                             ELOO  26
      VKE(3,1)=VKE(1,3)                                             ELOO  27
      VKE(3,2)=VKE(2,3)                                             ELOO  28
C-------  ELEMENT VECTOR                                            ELOO  29
      C=VPREE(2)*C2A/6.                                             ELOO  30
      VFE(1)=C                                                      ELOO  31
      VFE(2)=C                                                      ELOO  32
      VFE(3)=C                                                      ELOO  33
      RETURN                                                        ELOO  34
      END                                                           ELOO  35
```

Figure 4.3 Listing of subroutine ELEM00 used in program BBMEF of paragraph 6.2.2.

Figure 4.3 contains the listing of a subroutine for the computation of $[k]$ and $\{f\}$. The subroutine is used in program BBMEF of paragraph 6.2.2.

4.3.2 *Matrix Organization for the Numerical Computations of Element Matrices*

For the majority of elements, the fundamental matrices and vectors must be numerically integrated using methods described in paragraph 5.1. The computations can be carried out by the following steps.

(a) Operations common to all elements of the same type (i.e. having identical reference element):

— compute the weights w_r and coordinates ξ_r of integration points;
— construct the functions $N, \bar{N}$ and their derivatives with respect to ξ at the points of integration (for isoparametric elements $N \equiv \bar{N}$).

(b) Operations necessary for the computations of matrix $[k]$ of each element (4.16a):

— initialize the matrix $[k]$;
— for each point of integration ξ_r:

- construct the Jacobian matrix $[J]$ from the derivatives with respect to ξ of functions $\bar{N}$ and the nodal coordinates of the element (1.43). The inverse of $[J]$ and its determinant must also be found (see (1.39) to (1.41));
- construct the derivatives of functions N with respect to $\mathbf{x}$ starting from the derivatives with respect to $\boldsymbol{\xi}$ (1.37b);
- construct matrices $[B]$ and $[D]$;
- accumulate into $[k]$ the values of $[B]^T [D] [B] \det(J) w_r$ calculated for each integration point.

(c) Operations required to compute mass matrix $[m]$ (4.6b):

— initialize $[m]$;
— for each integration point ξ_r:
  - compute Jacobian matrix and its determinant;
  - accumulate the values of $\{N\} \langle N \rangle \det(J) w_r$ into matrix $[m]$.

(d) Operations necessary to compute consistent load vectors $\{f\}$ corresponding to f_V (4.16b):

— initialize $\{f\}$;
— for each integration point ξ_r:
  - compute the Jacobian matrix and its determinant;
  - accumulate the values of $\{N\} f_V \det(J) w_r$ into vector $\{f\}$.

(e) Operations required to compute the residue $\{r\}$ from the solution $\{u_n\}$ (4.6c):

— initialize vector $\{r\}$ with the value of $\{f\}$ from (d)
— for each integration point ξ_r:
  - compute matrices $[B]$, $[D]$, $[J]$ as in (b) above;
  - accumulate the product: $-[B]^T [D] [B] \{u_n\} w_r \det(J)$ into $\{r\}$.

(f) Operations necessary to compute gradients $\{\partial u\}$ at points of integration from the solution $\{u_n\}$ (4.9b):

— for each integration point ξ_r:
  - construct matrix $[B]$ as in (b) above;
  - compute and print gradient $\{\partial u\} = [B] \{u_n\}$.

4.3.3 *Control Subroutines for the Computation of Element Matrices*

Multi-purpose program MEF described in Chapter 6 is organized to contain a library of one-, two-, and three-dimensional elements for the solution of problems from a wide variety of disciplines. Problems from the mechanics of solids and fluids, problems governed by quasi-harmonic equations, etc. have been solved. For each element type '*nn*', one subroutine ELEMnn controls the computation of all the matrices and vectors described in paragraph 4.3.2. Control variable

ICODE specifies which element operation is desired; for example:

ICODE = 1   initialization of the characteristic parameters of an element (number of nodes, number of degrees of liberty).
ICODE = 2   operations required by a given reference element which are independent of the real geometry; construction of interpolation functions N and their derivatives with respect to ξ at the points of integration (see paragraphs 1.6.1 and 5.1).
ICODE = 3   construction of matrix $[k]$ in array VKE.
ICODE = 4   construction of matrix $[k_t]$ (tangent matrix needed for non-linear problems) in array VKE (see paragraph 5.3).
ICODE = 5   construction of mass matrix $[m]$ in array VKE.
ICODE = 6   computation of residual vector $\{r\}$ in array VFE.
ICODE = 7   computation of load vector $\{f\}$ in array VFE.
ICODE = 8   computation and printing of gradients $\{\partial u\}$.

Note that subroutine ELEMnn executes one operation at a time depending on the value of ICODE. For example, to compute matrices $[k]$ and $[m]$ one must successively use the following operations:

— Computation of $[k]$

  ICODE = 3
  CALL ELEMnn (...., VKE)
  array VKE must be saved in another location.

— Computation of $[m]$

  ICODE = 5
  CALL ELEMnn (...., VKE)
  again VKE must be saved in another location.

4.3.4 Subroutine ELEM01 (for quasi-harmonic problems)

Figure 4.4 contains the listing of programs ELEM01, NI01 for the element matrices of a quadratic isoparametric element suitable for the solution of problems governed by:

$$\frac{\partial}{\partial x}\left(d_x \frac{\partial u}{\partial x}\right) + \frac{\partial}{\partial y}\left(d_y \frac{\partial u}{\partial y}\right) + \frac{\partial}{\partial z}\left(d_z \frac{\partial u}{\partial z}\right) + f_V = 0 \quad (4.22)$$

The corresponding integral form is similar to the one of Example 4.4 with:

$$D = \begin{bmatrix} d_x & 0 & 0 \\ 0 & d_y & 0 \\ 0 & 0 & d_z \end{bmatrix} \quad \text{and} \quad \{\partial u\} = \begin{Bmatrix} \dfrac{\partial u}{\partial x} \\ \dfrac{\partial u}{\partial y} \\ \dfrac{\partial u}{\partial z} \end{Bmatrix}$$

Service programs used by ELEM01 and NI01

Name	Called by:	Listing in figure:	Function
GAUSS	ELEM01	5.1	Gauss point coordinates and weights
JACOB	ELEM01	1.9	Computation of $[j]$, $[J]$ and det (J)
DNIDX	ELEM01	1.10	Computation of $\langle N_{,x} \rangle$
PNINV	NI01	1.6	Computation of $[P_n^{-1}]$
NI	NI01	1.6	Computation of $\langle N \rangle$ and $\langle N_{,\xi} \rangle$
BASEP	NI, PNINV	1.6	Computation of a polynomial basis $\langle P \rangle$
INVERS	PNINV	1.6	Inversion of a full matrix

```
      SUBROUTINE ELEM01(VCORE,VPRNE,VPREE,VDLE,VKE,VFE)            EL01   1
C=======================================================================EL01   2
C     QUADRATIC ELEMENT FOR ANISOTROPIC HARMONIC PROBLEMS          EL01   3
C     IN 1,2 OR 3 DIMENSIONS :                                     EL01   4
C        1 DIMENSION:  3 NODES ELEMENT                             EL01   5
C        2 DIMENSIONS: 8 NODES ISOPARAMETRIC ELEMENT                EL01   6
C        3 DIMENSIONS: 20 NODES ISOPARAMETRIC ELEMENT               EL01   7
C     NUMBER OF INTEGRATION POINTS : 2 IN EACH DIRECTION           EL01   8
C     NUMBER OF DEGREES OF FREEDOM PER NODE : 1                    EL01   9
C     ELEMENT MATRIX OR VECTOR FORMED BY THIS SUBPROGRAM           EL01  10
C        ACCORDING TO ICODE VALUE :                                EL01  11
C           ICODE.EQ.1  RETURN OF PARAMETERS                       EL01  12
C           ICODE.EQ.2  EVALUATE INTERPOLATION FUNCTIONS AND       EL01  13
C                       NUMERICAL INTEGRATION COEFFICIENTS         EL01  14
C           ICODE.EQ.3  ELEMENT MATRIX (VKE)                       EL01  15
C           ICODE.EQ.4  TANGENT MATRIX (VKE)....NOT WRITTEN....    EL01  16
C           ICODE.EQ.5  MASS MATRIX (VKE)                          EL01  17
C           ICODE.EQ.6  K.U PRODUCT (VFE)                          EL01  18
C           ICODE.EQ.7  ELEMENT LOAD (VFE)....NOT WRITTEN....      EL01  19
C           ICODE.EQ.8  PRINT GRADIENTS                            EL01  20
C     ELEMENT PROPERTIES                                           EL01  21
C        VPREE(1)    COEFFICIENT DX                                EL01  22
C        VPREE(2)    COEFFICIENT DY                                EL01  23
C        VPREE(3)    COEFFICIENT DZ                                EL01  24
C        VPREE(4)    SPECIFIC HEAT CAPACITY C                      EL01  25
C=======================================================================EL01  26
      IMPLICIT REAL*8(A-H,O-Z)                                     EL01  27
      COMMON/COOR/NDIM                                             EL01  28
      COMMON/RGDT/IEL,ITPE,ITPE1,IGRE,IDLE,ICE,IPRNE,IPREE,INEL,IDEG,IPGEL01  29
     1      ,ICODE,IDLE0,INEL0,IPG0                                EL01  30
      COMMON/ES/M,MR,MP                                            EL01  31
      DIMENSION VCORE(1),VPRNE(1),VPREE(1),VDLE(1),VKE(1),VFE(1)   EL01  32
C.......  CHARACTERISTIC DIMENSIONS OF THE ELEMENT                 EL01  33
C              (VALID UP TO 3 DIMENSIONS)                          EL01  34
C     DIMENSION VCPG(IPG),VKPG(NDIM*IPG),XYZ(NDIM)                 EL01  35
      DIMENSION VCPG(  9),VKPG(     27),XYZ(  3)                   EL01  36
C     DIMENSION VJ (NDIM*NDIM),VJ1(NDIM*NDIM)                      EL01  37
      DIMENSION VJ (     9),VJ1(     9)                            EL01  38
C     DIMENSION VNIX( INEL*NDIM),VNI ((1+NDIM)*INEL*IPG),IPGKED(NDIM) EL01  39
      DIMENSION VNIX(       60),VNI (          2160),IPGKED(  3)   EL01  40
C        NUMBER OF G.P. IN KSI,ETA,DZETA DIRECTION                 EL01  41
      DATA IPGKED/3,3,3/                                           EL01  42
C.......                                                           EL01  43
      DATA ZERO/0.D0/,EPS/1.D-6/                                   EL01  44
      IKE=IDLE*(IDLE+1)/2                                          EL01  45
C                                                                  EL01  46
```

Figure 4.4 (*Contd.*)

```
C------- CHOOSE FUNCTION TO BE EXECUTED                      EL01  47
C                                                            EL01  48
      GO TO (100,200,300,400,500,600,700,800),ICODE          EL01  49
C                                                            EL01  50
C------- RETURN ELEMENT PARAMETERS IN COMMON 'RGDT'          EL01  51
C                                                            EL01  52
100   GO TO (110,120,130),NDIM                               EL01  53
110   IDLE0=3                                                EL01  54
      INEL0=3                                                EL01  55
      IPG0=3                                                 EL01  56
      RETURN                                                 EL01  57
120   IDLE0=8                                                EL01  58
      INEL0=8                                                EL01  59
      IPG0=9                                                 EL01  60
      RETURN                                                 EL01  61
130   IDLE0=20                                               EL01  62
      INEL0=20                                               EL01  63
      IPG0=27                                                EL01  64
      RETURN                                                 EL01  65
C                                                            EL01  66
C------- EVALUATE COORDINATES, WEIGHTS, FUNCTIONS N AND      EL01  67
C-------     THEIR DERIVATIVES AT G.P.                       EL01  68
C                                                            EL01  69
200   CALL GAUSS(IPGKED,NDIM,VKPG,VCPG,IPG)                  EL01  70
      CALL NI01(VKPG,VNI)                                    EL01  71
      RETURN                                                 EL01  72
C                                                            EL01  73
C------- COMPUTE ELEMENT STIFFNESS MATRIX                    EL01  74
C                                                            EL01  75
C------- INITIALIZE VKE                                      EL01  76
300   DO 310 I=1,IKE                                         EL01  77
310   VKE(I)=ZERO                                            EL01  78
C------- LOOP OVER THE INTEGRATION POINTS                    EL01  79
      INI=1+INEL                                             EL01  80
      DO 330 IG=1,IPG                                        EL01  81
C------- EVALUATE THE JACOBIAN MATRIX, ITS INVERSE AND ITS DETERMINANT EL01 82
      CALL JACOB(VNI(INI),VCORE,NDIM,INEL,VJ,VJ1,DETJ)       EL01  83
      IF(DETJ.LT.EPS) WRITE(MP,2000) IEL,IG,DETJ             EL01  84
2000  FORMAT(' *** ELEM ',I5,' P.G. ',I3,' DET(J)=',E12.5)   EL01  85
C------- PERFORM DETJ*WEIGHT                                 EL01  86
      COEF=VCPG(IG)*DETJ                                     EL01  87
C------- EVALUATE FUNCTIONS D(NI)/D(X)                       EL01  88
      CALL DNIDX(VNI(INI),VJ1,NDIM,INEL,VNIX)                EL01  89
C------- ACCUMULATE TERMS OF THE ELEMENT MATRIX              EL01  90
      IK=0                                                   EL01  91
      DO 320 J=1,IDLE                                        EL01  92
      DO 320 I=1,J                                           EL01  93
      I1=I                                                   EL01  94
      I2=J                                                   EL01  95
      C=ZERO                                                 EL01  96
      DO 315 IJ=1,NDIM                                       EL01  97
      C=C+VNIX(I1)*VNIX(I2)*VPREE(IJ)                        EL01  98
      I1=I1+IDLE                                             EL01  99
315   I2=I2+IDLE                                             EL01 100
      IK=IK+1                                                EL01 101
320   VKE(IK)=VKE(IK)+C*COEF                                 EL01 102
C------- NEXT G.P.                                           EL01 103
330   INI=INI+(NDIM+1)*INEL                                  EL01 104
      RETURN                                                 EL01 105
C                                                            EL01 106
C------- EVALUATE ELEMENT TANGENT MATRIX                     EL01 107
C                                                            EL01 108
400   CONTINUE                                               EL01 109
      RETURN                                                 EL01 110
```

Figure 4.4 (*Contd.*)

```
C                                                               EL01 111
C------ MASS MATRIX                                             EL01 112
C                                                               EL01 113
500     DO 510 I=1,IKE                                          EL01 114
510     VKE(I)=ZERO                                             EL01 115
        IF(VPREE(4).EQ.ZERO)RETURN                              EL01 116
        INI=0                                                   EL01 117
        DO 530 IG=1,IPG                                         EL01 118
C------ EVALUATE THE JACOBIAN MATRIX                            EL01 119
        I1=INI+INEL+1                                           EL01 120
        CALL JACOB(VNI(I1),VCORE,NDIM,INEL,VJ,VJ1,DETJ)         EL01 121
C------ COMPUTE THE WEIGHT                                      EL01 122
        COEF=VCPG(IG)*DETJ*VPREE(4)                             EL01 123
C------ TERMS OF THE MASS MATRIX                                EL01 124
        IK=0                                                    EL01 125
        DO 520 J=1,IDLE                                         EL01 126
        DO 520 I=1,J                                            EL01 127
        IK=IK+1                                                 EL01 128
        I1=INI+I                                                EL01 129
        I2=INI+J                                                EL01 130
520     VKE(IK)=VKE(IK)+VNI(I1)*VNI(I2)*COEF                    EL01 131
530     INI=INI+(NDIM+1)*INEL                                   EL01 132
        RETURN                                                  EL01 133
C                                                               EL01 134
C------ EVALUATE THE ELEMENT RESIDUAL                           EL01 135
C                                                               EL01 136
600     DO 605 I=1,INEL                                         EL01 137
605     VFE(I)=ZERO                                             EL01 138
        INI=1+INEL                                              EL01 139
        DO 640 IG=1,IPG                                         EL01 140
C------ EVALUATE THE JACOBIAN MATRIX AND THE DERIVATIVES OF N IN X,Y,Z EL01 141
        CALL JACOB(VNI(INI),VCORE,NDIM,INEL,VJ,VJ1,DETJ)        EL01 142
        CALL DNIDX(VNI(INI),VJ1,NDIM,INEL,VNIX)                 EL01 143
C------ COMPUTE THE COMMON COEFFICIENT                          EL01 144
        COEF=VCPG(IG)*DETJ                                      EL01 145
C------ VPREE*B*VDLE PRODUCT                                    EL01 146
        I1=0                                                    EL01 147
        DO 620 I=1,NDIM                                         EL01 148
        C=ZERO                                                  EL01 149
        DO 610 J=1,INEL                                         EL01 150
        I1=I1+1                                                 EL01 151
610     C=C+VNIX(I1)*VDLE(J)                                    EL01 152
620     VJ(I)=C*COEF*VPREE(I)                                   EL01 153
C------ (BT)*VJ PRODUCT                                         EL01 154
        DO 630 I=1,INEL                                         EL01 155
        I1=I-INEL                                               EL01 156
        DO 630 J=1,NDIM                                         EL01 157
        I1=I1+INEL                                              EL01 158
630     VFE(I)=VFE(I)+VNIX(I1)*VJ(J)                            EL01 159
640     INI=INI+(NDIM+1)*INEL                                   EL01 160
        RETURN                                                  EL01 161
C                                                               EL01 162
C------ EVALUATE FE                                             EL01 163
C                                                               EL01 164
700     CONTINUE                                                EL01 165
        RETURN                                                  EL01 166
C                                                               EL01 167
C------ EVALUATE AND PRINT GRADIENTS AT G.P.                    EL01 168
C                                                               EL01 169
800     WRITE(MP,2010) IEL                                      EL01 170
2010    FORMAT(//' GRADIENTS IN ELEMENT :',I4//)                EL01 171
        IDECL=(NDIM+1)*INEL                                     EL01 172
        INI0=1                                                  EL01 173
        INI=1+INEL                                              EL01 174
```

Figure 4.4 (*Contd.*)

```
              DO 830 IG=1,IPG                                          EL01 175
              CALL JACOB(VNI(INI),VCORE,NDIM,INEL,VJ,VJ1,DETJ)         EL01 176
              CALL DNIDX(VNI(INI),VJ1,NDIM,INEL,VNIX)                  EL01 177
      C------  EVALUATE THE COORDINATES OF THE G.P.                    EL01 178
              DO 803 I=1,NDIM                                          EL01 179
      803     XYZ(I)=ZERO                                              EL01 180
              IC=1                                                     EL01 181
              IO=INIO                                                  EL01 182
              DO 807 IN=1,INEL                                         EL01 183
              C=VNI(IO)                                                EL01 184
              DO 805 I=1,NDIM                                          EL01 185
              XYZ(I)=XYZ(I)+C*VCORE(IC)                                EL01 186
      805     IC=IC+1                                                  EL01 187
      807     IO=IO+1                                                  EL01 188
      C------  EVALUATE THE GRADIENT                                   EL01 189
              I1=0                                                     EL01 190
              DO 820 I=1,NDIM                                          EL01 191
              C=ZERO                                                   EL01 192
              DO 810 J=1,IDLE                                          EL01 193
              I1=I1+1                                                  EL01 194
      810     C=C+VNIX(I1)*VDLE(J)                                     EL01 195
      820     VJ(I)=C*VPREE(I)                                         EL01 196
      C------  PRINT THE GRADIENT                                      EL01 197
              WRITE(MP,2020) IG,(XYZ(I),I=1,NDIM)                      EL01 198
      2020    FORMAT(5X,'P.G. :',I3,' COORDINATES :',3E12.5)           EL01 199
              WRITE(MP,2025)(VJ(I),I=1,NDIM)                           EL01 200
      2025    FORMAT(15X,'GRADIENTS    :',3E12.5)                      EL01 201
              INIO=INIO+IDECL                                          EL01 202
      830     INI=INI+IDECL                                            EL01 203
              WRITE(MP,2030)                                           EL01 204
      2030    FORMAT(//)                                               EL01 205
              RETURN                                                   EL01 206
              END                                                      EL01 207
              SUBROUTINE NI01(VKPG,VNI)                                NI01   1
      C======================================================================NI01  2
      C       TO EVALUATE THE INTERPOLATION FUNCTIONS AND THEIR DERIVATIVES NI01 3
      C       D(N)/D(KSI) D(N)/D(ETA) BY THE GENERAL PN-INVERSE METHOD  NI01   4
      C       FOR 1,2 OR 3 DIMENSIONAL QUADRATIC ELEMENTS               NI01   5
      C          INPUT                                                  NI01   6
      C             VKPG    COORDINATES AT WHICH N IS TO BE EVALUATED   NI01   7
      C             IPG     NUMBER OF POINTS                            NI01   8
      C             INEL    NUMBER OF FUNCTIONS N (OR OF NODES) INEL.LE.20 NI01 9
      C             NDIM    NUMBER OF DIMENSIONS                NDIM.LE.3 NI01 10
      C          OUTPUT                                                 NI01  11
      C             VNI     FUNCTIONS N AND DERIVATIVES                 NI01  12
      C======================================================================NI01 13
              IMPLICIT REAL*8(A-H,O-Z)                                 NI01  14
              COMMON/COOR/NDIM                                         NI01  15
              COMMON/RGDT/IEL,ITPE,ITPE1,IGRE,IDLE,ICE,IPRNE,IPREE,INEL,IDEG,IPGNI01 16
              COMMON/TRVL/VKSI,VPN,VP,KEXP,KDER,K1                     NI01  17
              DIMENSION VKPG(1),VNI(1)                                 NI01  18
              DIMENSION VKSI1(3),KEXP1(3),VKSI2(16),KEXP2(16),VKSI3(60), NI01 19
             1 KEXP3(60)                                               NI01  20
      C                                                                NI01  21
      C.......  INFORMATION TO DEFINE THE 3 REFERENCE ELEMENTS         NI01  22
      C             (INEL.LE.20  NDIM.LE.3)                            NI01  23
      C       DIMENSION VKSI(NDIM*INEL),KEXP(NDIM*INEL),KDER(NDIM)     NI01  24
      C       DIMENSION VKSI(      60),KEXP(       60),KDER(    3)    NI01  25
      C       DIMENSION VPN (INEL*INEL),VP(INEL)                       NI01  26
      C       DIMENSION VPN (     400),VP(   20)                       NI01  27
      C       DIMENSION K1(INEL)                                       NI01  28
      C       DIMENSION K1(    20)                                     NI01  29
      C          CHARACTERISTICS FOR 1,2 AND 3 DIMENSIONAL REFERENCE ELEMENTS NI01 30
```

Figure 4.4 (*Contd.*)

```
      DATA VKSI1/-1.D0,0.D0,1.D0/,KEXP1/0,1,2/                         NI01 31
      DATA VKSI2/-1.D0,-1.D0, +0.D0,-1.D0, +1.D0,-1.D0, +1.D0,+0.D0,    NI01 32
     1              +1.D0,+1.D0, +0.D0,+1.D0, -1.D0,+1.D0, -1.D0,+0.D0/  NI01 33
      DATA KEXP2/0,0, 1,0, 0,1, 2,0, 1,1, 0,2, 2,1, 1,2/,IDEGR/2/        NI01 34
      DATA VKSI3/-1.D0,-1.D0,-1.D0, +0.D0,-1.D0,-1.D0,                   NI01 35
     1            +1.D0,-1.D0,-1.D0, +1.D0,+0.D0,-1.D0,                  NI01 36
     2            +1.D0,+1.D0,-1.D0, +0.D0,+1.D0,-1.D0,                  NI01 37
     3            -1.D0,+1.D0,-1.D0, -1.D0,+0.D0,-1.D0,                  NI01 38
     4            -1.D0,-1.D0,+0.D0, +1.D0,-1.D0,+0.D0,                  NI01 39
     5            +1.D0,+1.D0,+0.D0, -1.D0,+1.D0,+0.D0,                  NI01 40
     6            -1.D0,-1.D0,+1.D0, +0.D0,-1.D0,+1.D0,                  NI01 41
     7            +1.D0,-1.D0,+1.D0, +1.D0,+0.D0,+1.D0,                  NI01 42
     8            +1.D0,+1.D0,+1.D0, +0.D0,+1.D0,+1.D0,                  NI01 43
     9            -1.D0,+1.D0,+1.D0, -1.D0,+0.D0,+1.D0/                  NI01 44
      DATA KEXP3/0,0,0, 1,0,0, 0,1,0, 0,0,1, 1,1,1,                      NI01 45
     1 1,1,0, 0,1,1, 1,0,1, 2,0,0, 0,2,0, 0,0,2,                         NI01 46
     2 2,1,0, 2,0,1, 2,1,1, 1,2,0, 0,2,1, 1,2,1,                         NI01 47
     3 1,0,2, 0,1,2, 1,1,2/                                              NI01 48
C                                                                        NI01 49
C.......                                                                 NI01 50
      IDEG=IDEGR                                                         NI01 51
C------- SELECT TABLES VKSI AND KEXP ACCORDING TO NDIM                   NI01 52
      I1=NDIM*INEL                                                       NI01 53
      DO 5 I=1,I1                                                        NI01 54
      GO TO (1,2,3),NDIM                                                 NI01 55
1     VKSI(I)=VKSI1(I)                                                   NI01 56
      KEXP(I)=KEXP1(I)                                                   NI01 57
      GO TO 5                                                            NI01 58
2     VKSI(I)=VKSI2(I)                                                   NI01 59
      KEXP(I)=KEXP2(I)                                                   NI01 60
      GO TO 5                                                            NI01 61
3     VKSI(I)=VKSI3(I)                                                   NI01 62
      KEXP(I)=KEXP3(I)                                                   NI01 63
5     CONTINUE                                                           NI01 64
C------- EVALUATE THE PN-INVERSE MATRIX                                  NI01 65
      CALL PNINV(VKSI,KEXP,VP,K1,VPN)                                    NI01 66
C------- EVALUATE N,D(N)/D(KSI),D(N)/D(ETA) AT G.P.                      NI01 67
      I1=1                                                               NI01 68
      I2=1                                                               NI01 69
      DO 10 IG=1,IPG                                                     NI01 70
      KDER(1)=0                                                          NI01 71
      KDER(2)=0                                                          NI01 72
      KDER(3)=0                                                          NI01 73
      CALL NI(VKPG(I1),KEXP,KDER,VP,VPN,VNI(I2))                         NI01 74
      I2=I2+INEL                                                         NI01 75
      KDER(1)=1                                                          NI01 76
      CALL NI(VKPG(I1),KEXP,KDER,VP,VPN,VNI(I2))                         NI01 77
      I2=I2+INEL                                                         NI01 78
      IF(NDIM.EQ.1) GO TO 10                                             NI01 79
      KDER(1)=0                                                          NI01 80
      KDER(2)=1                                                          NI01 81
      CALL NI(VKPG(I1),KEXP,KDER,VP,VPN,VNI(I2))                         NI01 82
      I2=I2+INEL                                                         NI01 83
      IF(NDIM.EQ.2) GO TO 10                                             NI01 84
      KDER(2)=0                                                          NI01 85
      KDER(3)=1                                                          NI01 86
      CALL NI(VKPG(I1),KEXP,KDER,VP,VPN,VNI(I2))                         NI01 87
      I2=I2+INEL                                                         NI01 88
10    I1=I1+NDIM                                                         NI01 89
      RETURN                                                             NI01 90
      END                                                                NI01 91
```

Figure 4.4 Subroutines ELEM01 and NI01, for the computation of element matrices of quasi-harmonic problems, used in program MEF of Chapter 6.

These two subroutines give one-, two-, and three-dimensional elements (depending on the value of variable NDIM) which have one degree of freedom per node (the value of the unknown function):

NDIM	Number of nodes	Number of degrees of freedom	Element described in paragraph:
1	3	3	2.2.2.1 (rectilinear)
2	8	8	2.4.3.2 (quadrilateral)
3	20	20	2.6.2.2 (hexahedronal)

4.3.5 Subroutine ELEM02 (for plane elasticity)

From reference [3] we get the integral form required for two-dimensional elasticity problem:

$$W^e = \int_{V^e} \langle \delta\varepsilon \rangle [D] \{\varepsilon\} dV - \int_{V^e} \langle \delta u \rangle \begin{Bmatrix} f_{V_x} \\ f_{V_y} \end{Bmatrix} dV$$

$$- \int_{S_f^e} \langle \delta u \rangle \begin{Bmatrix} f_{S_x} \\ f_{S_y} \end{Bmatrix} dS \quad (4.23a)$$

where:

$\langle u \rangle = \langle u \quad v \rangle$ are the displacements of a point
$\langle \delta u \rangle = \langle \delta u \quad \delta y \rangle$ are the variations of the displacements
$\langle \varepsilon \rangle = \langle \varepsilon_x \quad \varepsilon_y \quad \gamma_{xy} \rangle$ are the infinitesimal strains
$= \left\langle \dfrac{\partial u}{\partial x}; \dfrac{\partial v}{\partial y}; \dfrac{\partial u}{\partial y} + \dfrac{\partial v}{\partial x} \right\rangle$

f_{Vx}, f_{Vy} are the forces per unit volume in directions x and y

f_{Sx}, f_{Sy} are the forces per unit area on S_f^e

$[D] = \begin{bmatrix} d_1 & d_2 & 0 \\ d_2 & d_1 & 0 \\ 0 & 0 & d_3 \end{bmatrix}$ is the matrix of elastic constants for an isotropic material

$$d_1 = \frac{E(1 - \alpha v)}{(1 + v)(1 - v - \alpha v)}$$

$$d_2 = \frac{v d_1}{(1 - \alpha v)}$$

$$d_3 = \frac{E}{2(1 + v)}$$

$\{\sigma\} = [D]\{\varepsilon\}$ are stresses

E is Young's modulus
v is Poisson's ratio
α is: 0 for plane stress
1 for plane strain

Using the eight-node element of paragraph 2.4.3.2 for the approximation of u and v (two degrees of freedom per node):

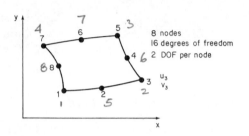

8 nodes
16 degrees of freedom
2 DOF per node

$$\{u\} = \begin{Bmatrix} u \\ v \end{Bmatrix} = [N]\{u_n\}$$

$$\{\delta u\} = \begin{Bmatrix} \delta u \\ \delta v \end{Bmatrix} = [N]\{\delta u_n\}$$

$$[N] = \begin{bmatrix} N_1 & 0 & N_2 & 0 & \cdots & N_8 & 0 \\ 0 & N_1 & 0 & N_2 & \cdots & 0 & N_8 \end{bmatrix}$$

where: $N_1 \ldots N_8$ are the functions of paragraph 2.4.3.2

$$\langle u_n \rangle = \langle u_1 \quad v_1 \quad u_2 \quad v_2 \quad \cdots \quad u_8 \quad v_8 \rangle$$
$$\langle \delta u_n \rangle = \langle \delta u_1 \quad \delta v_1 \quad \delta u_2 \quad \delta u_2 \quad \cdots \quad \delta u_8 \quad \delta v_8 \rangle$$
$$W^e = \langle \delta u_n \rangle [k]\{u_n\} - \langle \delta u_n \rangle \{f\}$$

$$\underset{(16 \times 16)}{[k]} = \int_{-1}^{1} \int_{-1}^{1} [B]^T [D][B] \det(J) d\xi d\eta \qquad (4.23b)$$

where:

$$\underset{(3 \times 16)}{[B]} = \begin{bmatrix} N_{1,x} & 0 & N_{2,x} & 0 & \cdots & N_{8,x} & 0 \\ 0 & N_{1,y} & 0 & N_{2,y} & \cdots & 0 & N_{8,y} \\ N_{1,y} & N_{1,x} & N_{2,y} & N_{2,x} & \cdots & N_{8,y} & N_{8,x} \end{bmatrix}$$

$$N_{i,x} = j_{11} N_{i,\xi} + j_{12} N_{i,\eta} \qquad i = 1, 2, \ldots, 8.$$
$$N_{i,y} = j_{21} N_{i,\xi} + j_{22} N_{i,\eta}$$

The vector $\{f\}$ is:

$$\underset{(16 \times 1)}{\{f\}} = \int_{-1}^{1} \int_{-1}^{1} [N]^T \begin{Bmatrix} f_{Vx} \\ f_{Vy} \end{Bmatrix} \det(J) d\xi d\eta \qquad (4.23c)$$

Mass matrix is:

$$[m]_{(16 \times 16)} = \int_{-1}^{1} \int_{-1}^{1} [N]^T [N] \det(J) \, d\xi \, d\eta \qquad (4.23\text{d})$$

Figure 4.5 contains the listing of subroutines ELEM02, NI02, D02, B02 and BTDB using results previously acquired.

```
      SUBROUTINE ELEM02(VCORE,VPRNE,VPREE,VDLE,VKE,VFE)             EL02    1
C===================================================================EL02    2
C     8 NODES QUADRATIC ELEMENT FOR 2 DIMENSIONAL ELASTICITY         EL02    3
C        EVALUATE ELEMENT INFORMATIONS ACCORDING TO ICODE VALUE      EL02    4
C        ICODE=1  ELEMENT PARAMETERS                                 EL02    5
C        ICODE=2  INTERPOLATION FUNCTIONS AND GAUSS COEFFICIENTS     EL02    6
C        ICODE=3  STIFFNESS MATRIX                                   EL02    7
C        ICODE=4  TANGENT MATRIX  ... NOT WRITTEN ...                EL02    8
C        ICODE=5  MASS MATRIX                                        EL02    9
C        ICODE=6  RESIDUALS                                          EL02   10
C        ICODE=7  SECOND MEMBER                                      EL02   11
C        ICODE=8  EVALUATE AND PRINT STRESSES                        EL02   12
C     ELEMENT PROPERTIES                                             EL02   13
C        VPREE(1)  YOUNG'S MODULUS                                   EL02   14
C        VPREE(2)  POISSON'S COEFFICIENT                             EL02   15
C        VPREE(3)  .EQ.0  PLANE STRESS                               EL02   16
C                  .EQ.1  PLANE STRAIN                               EL02   17
C        VPREE(4)  SPECIFIC MASS                                     EL02   18
C===================================================================EL02   19
      IMPLICIT REAL*8(A-H,O-Z)                                       EL02   20
      COMMON/COOR/NDIM                                               EL02   21
      COMMON/ASSE/NSYM                                               EL02   22
      COMMON/RGDT/IEL,ITPE,ITPE1,IGRE,IDLE,ICE,IPRNE,IPREE,INEL,IDEG,IPGEL02 23
     1 ,ICODE,IDLEO,INELO,IPGO                                       EL02   24
      COMMON/ES/M,MR,MP                                              EL02   25
      DIMENSION VCORE(1),VPRNE(1),VPREE(1),VDLE(1),VKE(1),VFE(1)     EL02   26
C....... CHARACTERISTIC DIMENSIONS OF THE ELEMENT                    EL02   27
C     DIMENSION VCPG(         IPG),VKPG(NDIM*IPG),VDE1(IMATD**2)     EL02   28
      DIMENSION VCPG(         9),VKPG(        18),VDE1(         9)   EL02   29
C     DIMENSION VBE (IMATD*IDLE),VDE (IMATD**2),VJ (NDIM*NDIM),VJ1(NDIM*EL02 30
      DIMENSION VBE (        48),VDE (        9),VJ (        4),VJ1(4) EL02 31
C     DIMENSION VNIX( INEL*NDIM),VNI ((1+NDIM)*INEL*IPG),IPGKED(NDIM) EL02   32
      DIMENSION VNIX(        16),VNI (       216),IPGKED(   2)       EL02   33
C        DIMENSION OF MATRIX D, NUMBER OF G.P.                       EL02   34
      DATA IMATD/3/,IPGKED/3,3/                                      EL02   35
C.......                                                             EL02   36
      DATA ZERO/0.D0/,DEUX/2.D0/,X05/0.5D0/,RADN/.5729577951308230D2/ EL02   37
      DATA EPS/1.D-6/                                                EL02   38
      SQRT(X)=DSQRT(X)                                               EL02   39
      ATAN2(X,Y)=DATAN2(X,Y)                                         EL02   40
C                                                                    EL02   41
C------ CHOOSE FUNCTION TO BE EXECUTED                               EL02   42
C                                                                    EL02   43
      GO TO (100,200,300,400,500,600,700,800),ICODE                  EL02   44
C                                                                    EL02   45
C------ RETURN ELEMENT PARAMETERS IN COMMON 'RGDT'                   EL02   46
C                                                                    EL02   47
  100 IDLEO=16                                                       EL02   48
      INELO=8                                                        EL02   49
      IPGO=9                                                         EL02   50
C     RETURN                                                         EL02   51
C                                                                    EL02   52
C------ EVALUATE COORDINATES, WEIGHTS, FUNCTIONS N AND THEIR         EL02   53
C------ DERIVATIVES AT G.P.                                          EL02   54
C                                                                    EL02   55
  200 CALL GAUSS(IPGKED,NDIM,VKPG,VCPG,IPG)                          EL02   56
```

Figure 4.5 (*Contd.*)

```
              IF(M.LT.2) GO TO 220                                  EL02  57
              WRITE(MP,2000) IPG                                    EL02  58
 2000   FORMAT(/I5,' GAUSS POINTS'/10X,'VCPG',25X,'VKPG')           EL02  59
              IO=1                                                  EL02  60
              DO 210 IG=1,IPG                                       EL02  61
              I1=IO+NDIM-1                                          EL02  62
              WRITE(MP,2010) VCPG(IG),(VKPG(I),I=IO,I1)             EL02  63
  210         IO=IO+NDIM                                            EL02  64
 2010   FORMAT(1X,F20.15,5X,3F20.15)                                EL02  65
  220   CALL NI02(VKPG,VNI)                                         EL02  66
              IF(M.LT.2) RETURN                                     EL02  67
              I1=3*INEL*IPG                                         EL02  68
              WRITE(MP,2020) (VNI(I),I=1,I1)                        EL02  69
 2020   FORMAT(/' FUNCTIONS N AND DERIVATIVES'/ (1X,8E12.5))        EL02  70
              RETURN                                                EL02  71
C                                                                   EL02  72
C------- EVALUATE ELEMENT STIFFNESS MATRIX                          EL02  73
C                                                                   EL02  74
C------- INITIALIZE VKE                                             EL02  75
  300   DO 310 I=1,136                                              EL02  76
  310   VKE(I)=ZERO                                                 EL02  77
C------- FORM MATRIX D                                              EL02  78
        CALL D02(VPREE,VDE)                                         EL02  79
        IF(M.GE.2) WRITE(MP,2030) (VDE(I),I=1,9)                    EL02  80
 2030   FORMAT(/' MATRIX D'/1X,9E12.5)                              EL02  81
C------- LOOP OVER THE G.P.                                         EL02  82
        I1=1+INEL                                                   EL02  83
        DO 330 IG=1,IPG                                             EL02  84
C------- EVALUATE THE JACOBIAN, ITS INVERSE AND ITS DETERMINANT     EL02  85
        CALL JACOB(VNI(I1),VCORE,NDIM,INEL,VJ,VJ1,DETJ)             EL02  86
        IF(DETJ.LT.EPS) WRITE(MP,2040) IEL,IG,DETJ                  EL02  87
 2040   FORMAT(' *** ELEM ',I5,' G.P. ',I3,' DET(J)=',E12.5)        EL02  88
        IF(M.GE.2) WRITE(MP,2050) VJ,VJ1,DETJ                       EL02  89
 2050   FORMAT(/' JACOBIAN=',4E12.5 / ' J INVERS=',4E12.5/' DETJ=',E12.5) EL02 90
C------- PERFORM D*COEF                                             EL02  91
        C=VCPG(IG)*DETJ                                             EL02  92
        DO 320 I=1,9                                                EL02  93
  320   VDE1(I)=VDE(I)*C                                            EL02  94
C------- FORM MATRIX B                                              EL02  95
        CALL DNIDX(VNI(I1),VJ1,NDIM,INEL,VNIX)                      EL02  96
        IF(M.GE.2) WRITE(MP,2060) (VNIX(I),I=1,16)                  EL02  97
 2060   FORMAT(/' VNIX'/(1X,8E12.5))                                EL02  98
        CALL B02(VNIX,INEL,VBE)                                     EL02  99
        IF(M.GE.2) WRITE(MP,2070) (VBE(I),I=1,48)                   EL02 100
 2070   FORMAT(/' MATRIX B'/(1X,10E12.5))                           EL02 101
        CALL BTDB(VKE,VBE,VDE1,IDLE,IMATD,NSYM)                     EL02 102
  330   I1=I1+3*INEL                                                EL02 103
        RETURN                                                      EL02 104
C                                                                   EL02 105
C------- EVALUATE THE ELEMENT TANGENT MATRIX                        EL02 106
C                                                                   EL02 107
  400   CONTINUE                                                    EL02 108
        RETURN                                                      EL02 109
C                                                                   EL02 110
C------- EVALUATE THE MASS MATRIX                                   EL02 111
C                                                                   EL02 112
  500   DO 510 I=1,136                                              EL02 113
  510   VKE(I)=ZERO                                                 EL02 114
C------- LOOP OVER THE G.P.                                         EL02 115
        IDIM1=NDIM-1                                                EL02 116
        IDECL=(NDIM+1)*INEL                                         EL02 117
        I1=1+INEL                                                   EL02 118
        I2=0                                                        EL02 119
        DO 550 IG=1,IPG                                             EL02 120
```

Figure 4.5 (*Contd.*)

```
      CALL JACOB(VNI(I1),VCORE,NDIM,INEL,VJ,VJ1,DETJ)          EL02 121
      D=VCPG(IG)*DETJ*VPREE(4)                                 EL02 122
C------ ACCUMULATE MASS TERMS                                  EL02 123
      IDL=0                                                    EL02 124
      DO 540 J=1,INEL                                          EL02 125
      JJ=I2+J                                                  EL02 126
      J0=1+IDL*(IDL+1)/2                                       EL02 127
      DO 530 I=1,J                                             EL02 128
      II=I2+I                                                  EL02 129
      C=VNI(II)*VNI(JJ)*D                                      EL02 130
      VKE(J0)=VKE(J0)+C                                        EL02 131
      IF(NDIM.EQ.1) GO TO 530                                  EL02 132
      J1=J0+IDL+2                                              EL02 133
      DO 520 II=1,IDIM1                                        EL02 134
      VKE(J1)=VKE(J1)+C                                        EL02 135
  520 J1=J1+J1+1                                               EL02 136
  530 J0=J0+NDIM                                               EL02 137
  540 IDL=IDL+NDIM                                             EL02 138
      I1=I1+IDECL                                              EL02 139
  550 I2=I2+IDECL                                              EL02 140
      RETURN                                                   EL02 141
C                                                              EL02 142
C------ EVALUATE THE ELEMENT RESIDUAL                          EL02 143
C                                                              EL02 144
C------ FORM MATRIX D                                          EL02 145
  600 CALL D02(VPREE,VDE)                                      EL02 146
C------ INITIALIZE THE RESIDUAL VECTOR                         EL02 147
      DO 610 ID=1,IDLE                                         EL02 148
  610 VFE(ID)=ZERO                                             EL02 149
C------ LOOP OVER THE G.P.                                     EL02 150
      I1=1+INEL                                                EL02 151
      DO 640 IG=1,IPG                                          EL02 152
C------ EVALUATE THE JACOBIAN                                  EL02 153
      CALL JACOB(VNI(I1),VCORE,NDIM,INEL,VJ,VJ1,DETJ)          EL02 154
C------ EVALUATE FUNCTIONS D(NI)/D(X)                          EL02 155
      CALL DNIDX(VNI(I1),VJ1,NDIM,INEL,VNIX)                   EL02 156
C------ EVALUATE STRAINS AND STRESSES                          EL02 157
      EPSX=ZERO                                                EL02 158
      EPSY=ZERO                                                EL02 159
      GAMXY=ZERO                                               EL02 160
      ID=1                                                     EL02 161
      DO 620 IN=1,INEL                                         EL02 162
      UN=VDLE(ID)                                              EL02 163
      VN=VDLE(ID+1)                                            EL02 164
      C1=VNIX(IN)                                              EL02 165
      IN1=IN+INEL                                              EL02 166
      C2=VNIX(IN1)                                             EL02 167
      EPSX=EPSX+C1*UN                                          EL02 168
      EPSY=EPSY+C2*VN                                          EL02 169
      GAMXY=GAMXY+C1*VN+C2*UN                                  EL02 170
  620 ID=ID+2                                                  EL02 171
      C1=VCPG(IG)*DETJ                                         EL02 172
      C2=VDE(2)*C1                                             EL02 173
      C3=VDE(9)*C1                                             EL02 174
      C1=VDE(1)*C1                                             EL02 175
      SIGX=C1*EPSX+C2*EPSY                                     EL02 176
      SIGY=C2*EPSX+C1*EPSY                                     EL02 177
      TAUXY=C3*GAMXY                                           EL02 178
C------ FORM THE RESIDUAL                                      EL02 179
      ID=1                                                     EL02 180
      DO 630 IN=1,INEL                                         EL02 181
      C1=VNIX(IN)                                              EL02 182
      IN1=IN+INEL                                              EL02 183
      C2=VNIX(IN1)                                             EL02 184
```

Figure 4.5 (*Contd.*)

```
              VFE(ID)=VFE(ID)+C1*SIGX+C2*TAUXY                   EL02 185
              VFE(ID+1)=VFE(ID+1)+C2*SIGY+C1*TAUXY               EL02 186
  630     ID=ID+2                                                EL02 187
  640     I1=I1+3*INEL                                           EL02 188
          RETURN                                                 EL02 189
C                                                                EL02 190
C------   EVALUATE VOLUMIC FORCES, FX,FY PER UNIT VOLUME         EL02 191
C         ( FOR GRAVITY FX=0 FY=-VPREE(4) )                      EL02 192
C                                                                EL02 193
  700     FX=ZERO                                                EL02 194
          FY=-VPREE(4)                                           EL02 195
          DO 710 I=1,16                                          EL02 196
  710     VFE(I)=ZERO                                            EL02 197
          I1=1                                                   EL02 198
          IDECL=(NDIM+1)*INEL                                    EL02 199
          DO 730 IG=1,IPG                                        EL02 200
          CALL JACOB(VNI(I1+INEL),VCORE,NDIM,INEL,VJ,VJ1,DETJ)   EL02 201
          DX=VCPG(IG)*DETJ                                       EL02 202
          DY=DX*FY                                               EL02 203
          DX=DX*FX                                               EL02 204
          I2=I1                                                  EL02 205
          I3=1                                                   EL02 206
          DO 720 IN=1,INEL                                       EL02 207
          VFE(I3)=VFE(I3)+DX*VNI(I2)                             EL02 208
          VFE(I3+1)=VFE(I3+1)+DY*VNI(I2)                         EL02 209
          I2=I2+1                                                EL02 210
  720     I3=I3+2                                                EL02 211
  730     I1=I1+IDECL                                            EL02 212
          RETURN                                                 EL02 213
C                                                                EL02 214
C------   EVALUATE AND PRINT STRESSES AT G.P.                    EL02 215
C                                                                EL02 216
  800     WRITE(MP,2080) IEL                                     EL02 217
 2080     FORMAT(//' STRESSES IN ELEMENT ',I5/                   EL02 218
         1 ' P.G.',7X,'X',11X,'Y',9X,'EPSX',8X,'EPSY',7X,'GAMXY',8X,'SIGX',EL02 219
         2 8X,'SIGY',7X,'TAUXY',8X,'TETA'/ 71X ,'SIG1',8X,'SIG2',7X,'TAUMAX'EL02 220
         3 /)                                                    EL02 221
C------   FORM THE MATRIX D                                      EL02 222
          CALL D02(VPREE,VDE)                                    EL02 223
C------   LOOP OVER THE G.P.                                     EL02 224
          I1=1+INEL                                              EL02 225
          I2=0                                                   EL02 226
          DO 820 IG=1,IPG                                        EL02 227
C------   EVALUATE THE JACOBIAN                                  EL02 228
          CALL JACOB(VNI(I1),VCORE,NDIM,INEL,VJ,VJ1,DETJ)        EL02 229
C------   EVALUATE FUNCTIONS D(NI)/D(X)                          EL02 230
          CALL DNIDX(VNI(I1),VJ1,NDIM,INEL,VNIX)                 EL02 231
C------   COMPUTE STRAINS AND COORDINATES AT G.P.                EL02 232
          EPSX=ZERO                                              EL02 233
          EPSY=ZERO                                              EL02 234
          GAMXY=ZERO                                             EL02 235
          X=ZERO                                                 EL02 236
          Y=ZERO                                                 EL02 237
          ID=1                                                   EL02 238
          DO 810 IN=1,INEL                                       EL02 239
          UN=VDLE(ID)                                            EL02 240
          VN=VDLE(ID+1)                                          EL02 241
          XN=VCORE(ID)                                           EL02 242
          YN=VCORE(ID+1)                                         EL02 243
          C1=VNIX(IN)                                            EL02 244
          IN1=IN+INEL                                            EL02 245
          C2=VNIX(IN1)                                           EL02 246
          IN1=IN+I2                                              EL02 247
          C3=VNI(IN1)                                            EL02 248
```

Figure 4.5 (*Contd.*)

```
      EPSX=EPSX+C1*UN                                           EL02 249
      EPSY=EPSY+C2*VN                                           EL02 250
      GAMXY=GAMXY+C1*VN+C2*UN                                   EL02 251
      X=X+C3*XN                                                 EL02 252
      Y=Y+C3*YN                                                 EL02 253
  810 ID=ID+2                                                   EL02 254
C------- COMPUTE THE STRESSES                                   EL02 255
      SIGX=VDE(1)*EPSX+VDE(2)*EPSY                              EL02 256
      SIGY=VDE(2)*EPSX+VDE(1)*EPSY                              EL02 257
      TAUXY=VDE(9)*GAMXY                                        EL02 258
C------- COMPUTE THE PRINCIPAL STRESSES                         EL02 259
      TETA=ATAN2(DEUX*TAUXY,SIGX-SIGY)*X05                      EL02 260
      TETA=TETA*RADN                                            EL02 261
      C1=(SIGX+SIGY)*X05                                        EL02 262
      C2=(SIGX-SIGY)*X05                                        EL02 263
      TAUMAX=SQRT(C2*C2+TAUXY*TAUXY)                            EL02 264
      SIG1=C1+TAUMAX                                            EL02 265
      SIG2=C1-TAUMAX                                            EL02 266
      WRITE(MP,2090) IG,X,Y,EPSX,EPSY,GAMXY,SIGX,SIGY,TAUXY,    EL02 267
     1 TETA,SIG1,SIG2,TAUMAX                                    EL02 268
 2090 FORMAT(1X,I5,8E12.5,5X,F5.1/66X,3E12.5)                   EL02 269
      I2=I2+3*INEL                                              EL02 270
  820 I1=I1+3*INEL                                              EL02 271
      RETURN                                                    EL02 272
      END                                                       EL02 273

      SUBROUTINE NI02(VKPG,VNI)                                 NI02   1
C=========================================================NI02   2
C     TO EVALUATE THE INTERPOLATION FUNCTIONS N AND THEIR DERIVATIVES  NI02   3
C     D(N)/D(KSI) AND D(N)/D(ETA) BY GENERAL PN-INVERSE METHOD  NI02   4
C       INPUT                                                   NI02   5
C         VKPG    COORDINATES AT WHICH N IS TO BE EVALUATED     NI02   6
C         IPG     NUMBER OF POINTS                              NI02   7
C         INEL    NUMBER OF FUNCTIONS N (OR OF NODES) INEL.EQ.8 NI02   8
C         NDIM    NUMBER OF DIMENSIONS              NDIM.EQ.2   NI02   9
C       OUTPUT                                                  NI02  10
C         VNI     FUNCTIONS N AND DERIVATIVES                   NI02  11
C=========================================================NI02  12
      IMPLICIT REAL*8(A-H,O-Z)                                  NI02  13
      COMMON/COOR/NDIM                                          NI02  14
      COMMON/RGDT/IEL,ITPE,ITPE1,IGRE,IDLE,ICE,IPRNE,IPREE,INEL,IDEG,IPGNI02 15
      DIMENSION VKPG(1),VNI(1)                                  NI02  16
C                                                               NI02  17
C....... INFORMATIONS RELATED TO THE 8 NODES REFERENCE SQUARE ELEMENT NI02 18
C         (INEL.EQ.8  NDIM.EQ.2)                                NI02  19
C     DIMENSION VKSI(NDIM*INEL),KEXP(NDIM*INEL),KDER(NDIM)      NI02  20
      DIMENSION VKSI(    16),KEXP(    16),KDER(   2)            NI02  21
C     DIMENSION VPN (INEL*INEL),VP(INEL),K1(INEL)               NI02  22
      DIMENSION VPN (    64),VP(   8),K1(   8)                  NI02  23
C       NODAL COORDINATES OF THE REFERENCE ELEMENT              NI02  24
      DATA VKSI/-1.D0,-1.D0, +0.D0,-1.D0, +1.D0,-1.D0, +1.D0,+0.D0, NI02 25
     1          +1.D0,+1.D0, +0.D0,+1.D0, -1.D0,+1.D0, -1.D0,+0.D0/ NI02 26
C       MONOMIAL EXPONENTS OF THE POLYNOMIAL BASIS, MAX-DEGREE  NI02  27
      DATA KEXP/0,0, 1,0, 0,1, 2,0, 1,1, 0,2, 2,1, 1,2/,IDEGR/2/ NI02 28
C                                                               NI02  29
C.......                                                        NI02  30
      IDEG=IDEGR                                                NI02  31
C------- EVALUATE THE PN-INVERSE MATRIX                         NI02  32
      CALL PNINV(VKSI,KEXP,VP,K1,VPN)                           NI02  33
C------- EVALUATE N,D(N)/D(KSI),D(N)/D(ETA) AT G.P.             NI02  34
      I1=1                                                      NI02  35
      I2=1                                                      NI02  36
      DO 10 IG=1,IPG                                            NI02  37
      KDER(1)=0                                                 NI02  38
```

Figure 4.5 (*Contd.*)

```
              KDER(2)=0                                           NI02  39
              CALL NI(VKPG(I1),KEXP,KDER,VP,VPN,VNI(I2))          NI02  40
              I2=I2+INEL                                          NI02  41
              KDER(1)=1                                           NI02  42
              CALL NI(VKPG(I1),KEXP,KDER,VP,VPN,VNI(I2))          NI02  43
              I2=I2+INEL                                          NI02  44
              KDER(1)=0                                           NI02  45
              KDER(2)=1                                           NI02  46
              CALL NI(VKPG(I1),KEXP,KDER,VP,VPN,VNI(I2))          NI02  47
              I2=I2+INEL                                          NI02  48
       10     I1=I1+NDIM                                          NI02  49
              RETURN                                              NI02  50
              END                                                 NI02  51

              SUBROUTINE D02(VPREE,VDE)                           D02   1
       C=====================================================================D02   2
       C      TO FORM MATRIX D (2 DIMENSIONAL ELASTICITY)         D02   3
       C         INPUT                                            D02   4
       C            VPREE    ELEMENT PROPERTIES                   D02   5
       C                     VPREE(1)    YOUNG'S MODULUS          D02   6
       C                     VPREE(2)    POISSON'S COEFFICIENT    D02   7
       C                     VPREE(3)   .EQ.0 PLANE STRESSES      D02   8
       C                                .EQ.1 PLANE STRAINS       D02   9
       C         OUTPUT                                           D02   10
       C            VDE      MATRIX D (FULL)                      D02   11
       C=====================================================================D02   12
              IMPLICIT REAL*8(A-H,O-Z)                            D02   13
              DIMENSION VPREE(1),VDE(9)                           D02   14
              DATA ZERO/0.D0/,UN/1.D0/,DEUX/2.D0/                 D02   15
              E=VPREE(1)                                          D02   16
              X=VPREE(2)                                          D02   17
              A=VPREE(3)                                          D02   18
              C1=E*(UN-A*X)/((UN+X)*(UN-X-A*X))                   D02   19
              C2=C1*X/(UN-A*X)                                    D02   20
              C3=E/(DEUX*(UN+X))                                  D02   21
              VDE(1)=C1                                           D02   22
              VDE(2)=C2                                           D02   23
              VDE(3)=ZERO                                         D02   24
              VDE(4)=C2                                           D02   25
              VDE(5)=C1                                           D02   26
              VDE(6)=ZERO                                         D02   27
              VDE(7)=ZERO                                         D02   28
              VDE(8)=ZERO                                         D02   29
              VDE(9)=C3                                           D02   30
              RETURN                                              D02   31
              END                                                 D02   32

              SUBROUTINE B02(VNIX,INEL,VBE)                       B02   1
       C=====================================================================B02   2
       C      TO FORM MATRIX B (2 DIMENSIONAL ELASTICITY)         B02   3
       C         INPUT                                            B02   4
       C            VNIX     DERIVATIVES OF INTERPOLATION FUNCTIONS W.R.T. X,Y,Z  B02   5
       C            INEL     NUMBER OF INTERPOLATION FUNCTIONS    B02   6
       C         OUTPUT                                           B02   7
       C            VBE      MATRIX B                             B02   8
       C=====================================================================B02   9
              IMPLICIT REAL*8(A-H,O-Z)                            B02   10
              DIMENSION VNIX(INEL,1),VBE(3,1)                     B02   11
              DATA ZERO/0.D0/                                     B02   12
              J=1                                                 B02   13
              DO 10 I=1,INEL                                      B02   14
              C1=VNIX(I,1)                                        B02   15
              C2=VNIX(I,2)                                        B02   16
              VBE(1,J)=C1                                         B02   17
```

Figure 4.5 (*Contd.*)

```
            VBE(1,J+1)=ZERO                                         B02   18
            VBE(2,J)=ZERO                                           B02   19
            VBE(2,J+1)=C2                                           B02   20
            VBE(3,J)=C2                                             B02   21
            VBE(3,J+1)=C1                                           B02   22
   10       J=J+2                                                   B02   23
            RETURN                                                  B02   24
            END                                                     B02   25
            SUBROUTINE BTDB(VKE,VBE,VDE,IDLE,IMATD,NSYM)            BTDB   1
   C=====================================================================BTDB   2
   C        TO ADD THE PRODUCT B(T).D.B  TO  VKE                    BTDB   3
   C        INPUT                                                   BTDB   4
   C           VKE     ELEMENT MATRIX NON SYMMETRICAL  (NSYM.EQ.1)  BTDB   5
   C                                   SYMMETRICAL    (NSYM.EQ.0)   BTDB   6
   C           VBE     MATRIX B                                     BTDB   7
   C           VDE     MATRIX D (FULL)                              BTDB   8
   C           IDLE    TOTAL NUMBER OF D.O.F. PER ELEMENT           BTDB   9
   C           IMATD   DIMENSION OF MATRIX D   (MAX. 6)             BTDB  10
   C        OUTPUT                                                  BTDB  11
   C           VKE                                                  BTDB  12
   C=====================================================================BTDB  13
            IMPLICIT REAL*8(A-H,O-Z)                                BTDB  14
            DIMENSION VKE(1),VBE(IMATD,1),VDE(IMATD,1),T(6)         BTDB  15
            DATA ZERO/0.D0/                                         BTDB  16
   C---------------------------------------------------------------BTDB  17
            IJ=1                                                    BTDB  18
            IMAX=IDLE                                               BTDB  19
            DO 40 J=1,IDLE                                          BTDB  20
            DO 20 I1=1,IMATD                                        BTDB  21
            C=ZERO                                                  BTDB  22
            DO 10 J1=1,IMATD                                        BTDB  23
   10       C=C+VDE(I1,J1)*VBE(J1,J)                                BTDB  24
   20       T(I1)=C                                                 BTDB  25
            IF(NSYM.EQ.0) IMAX=J                                    BTDB  26
            DO 40 I=1,IMAX                                          BTDB  27
            C=ZERO                                                  BTDB  28
            DO 30 J1=1,IMATD                                        BTDB  29
   30       C=C+VBE(J1,I)*T(J1)                                     BTDB  30
            VKE(IJ)=VKE(IJ)+C                                       BTDB  31
   40       IJ=IJ+1                                                 BTDB  32
            RETURN                                                  BTDB  33
            END                                                     BTDB  34
```

Figure 4.5 Subroutines ELEM02, NI02, D02, and B02, for the computation of element matrices of two-dimensional problems of elasticity, used in program MEF of Chapter 6.

4.4 Assembly of the Global Discretized Form W

Operations required to construct global or system matrices and load vectors $[K]$ and $\{F\}$ starting from element matrices $[k]$ and $\{f\}$.

4.4.1 *Assembly by Expansion of Element Matrices*

Each element integral form W^e gives, after discretization (4.4):

$$W^e = \langle \delta u_n \rangle ([k]\{u_n\} - \{f\})$$

where $[k]$ is the element matrix for element e; and $\{f\}$ is the element load vector resulting from all the body forces and surface tractions.

Vectors $\langle \delta u_n \rangle$ and $\{u_n\}$ are different for each element. Vectors $\{\delta U_n\}$ and $\{U_n\}$, being the global vectors, contain values for all the nodes of the complete domain V appearing in (4.5b). Element vectors $\langle \delta u_n \rangle$ and $\{u_n\}$ are thus contained in the global vectors.

Global variables

$$\langle \delta U_n \rangle = \langle \delta u_1 \quad \ldots \quad \delta u_i \quad \ldots \quad \delta u_j \quad \ldots \quad \delta u_k \quad \ldots \quad \delta u_n \rangle$$

Element variables

$$\langle \delta u_n \rangle = \langle \delta u_i \quad \delta u_j \quad \delta u_k \rangle$$

where δu_i, δu_j, δu_k are the nodal variables of the element.

Example 4.9   Element and global vectors $\{u_n\}$ and $\{U_n\}$

Consider domain V subdivided into two three-node triangular elements having only one degree of freedom per node:

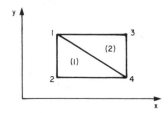

The global vectors are:

$$\langle \delta U_n \rangle = \langle \delta u_1 \quad \delta u_2 \quad \delta u_3 \quad \delta u_4 \rangle$$
$$\langle U_n \rangle = \langle u_1 \quad u_2 \quad u_3 \quad u_4 \rangle$$

Element vectors for element (1) are:

$$\langle \delta u_n^{(1)} \rangle = \langle \delta u_1 \quad \delta u_2 \quad \delta u_4 \rangle$$
$$\langle u_n^{(1)} \rangle = \langle u_1 \quad u_2 \quad u_4 \rangle$$

For element (2) we have:

$$\langle \delta u_n^{(2)} \rangle = \langle \delta u_1 \quad \delta u_4 \quad \delta u_3 \rangle$$
$$\langle u_n^{(2)} \rangle = \langle u_1 \quad u_4 \quad u_3 \rangle$$

The element integral forms are:

$$W^{(1)} = \langle \delta u_n^{(1)} \rangle ([k^{(1)}]\{u_n^{(1)}\} - \{f^{(1)}\})$$
$$W^{(2)} = \langle \delta u_n^{(2)} \rangle ([k^{(2)}]\{u_n^{(2)}\} - \{f^{(2)}\})$$

Note that a particular nodal variable u_n (or δu_n) may appear in many element vectors since a node may be common to many elements. Nodes 1 and 4 of Example 4.9 are shared by the two elements. Obviously, when the nodal variable is a vector, its components must be given in a common global system of reference.

Discretized global form W is the sum of discretized elementary forms W^e (4.5a). This operation is called the assembly process:

$$W = \sum_{\text{elements}} W^e$$

$$W = \sum_{\text{elements}} \langle \delta u_n \rangle ([k]\{u_n\} - \{f\})$$

We try to put this last expression in the following form (4.5b):

$$W = \langle \delta U_n \rangle ([K]\{U_n\} - \{F\})$$

To meet this objective, element forms W^e must be rewritten in terms of $\{U_n\}$ and $\langle \partial U_n \rangle$:

$$W^e = \langle \delta U_n \rangle ([K^e]\{U_n\} - \{F^e\}) \qquad (4.24)$$

Matrix $[K^e]$ is constructed by an expansion of matrix $[k]$ into an otherwise null matrix of the same dimensions as $[K]$.

Similarly, $\{F^e\}$ is constructed by an expansion of $\{f\}$ into an otherwise null vector of the same size as $\{F\}$.

(a) *Expansion of* $[k]$

This process is done in two stages; first, the post-factor $\{u_n\}$ is replaced by $\{U_n\}$; then, the pre-factor $\langle \delta u_n \rangle$ is similarly replaced by $\langle \delta U_n \rangle$. Consider the following example:

$$W^e = \langle \delta u_I \quad \delta u_J \rangle \begin{bmatrix} k_{11} & k_{12} \\ k_{21} & k_{22} \end{bmatrix} \begin{Bmatrix} u_I \\ u_J \end{Bmatrix} = \langle \delta u_n \rangle [k]\{u_n\} \qquad (4.25a)$$

The global vector of nodal variables is:

$$\langle U_n \rangle = \langle u_1 \quad u_2 \quad \ldots \quad \underline{u_I} \quad u_{I+1} \quad \ldots \quad \underline{u_J} \quad u_{J+1} \quad \ldots \quad u_n \rangle$$

— Replacement of $\{u_n\}$ by $\{U_n\}$

To keep W^e unchanged in numerical value when $\{u_n\}$ is replaced by $\{U_n\}$, matrix $[k]$ of size (2×2) must be replaced by a matrix $[k']$ of size $(2 \times n)$ in which column I is $\langle k_{11} \quad k_{21} \rangle^T$ and column J is $\langle k_{12} \quad k_{22} \rangle^T$ and all other terms are zero.

$$W^e = \langle \delta u_I \quad \delta u_J \rangle \underbrace{\begin{bmatrix} 0 \; 0 \ldots & \begin{Bmatrix} k_{11} \\ k_{21} \end{Bmatrix} & 0 \ldots & \begin{Bmatrix} k_{12} \\ k_{22} \end{Bmatrix} & 0 \ldots 0 \\ 0 \; 0 \ldots & & 0 \ldots & & 0 \ldots 0 \end{bmatrix}}_{(2 \times n)} \begin{Bmatrix} u_1 \\ \vdots \\ u_I \\ u_{I+1} \\ \vdots \\ u_J \\ u_{J+1} \\ \vdots \\ u_n \end{Bmatrix} \qquad (4.25b)$$

$$\text{column } I \qquad \text{column } J$$

— Note that if I is greater than J, the columns of $[k]$ are in reverse order in global matrix $[K]$ (4.25b).

— Replacement of $\langle \delta u_n \rangle$ by $\langle \delta U_n \rangle$

In this case, matrix $[K']$ must be replaced by matrix $[K^e]$ of the same size as $[K]$ in which row I is the first line of $[k']$ and row J is the second. All other terms are zero.

$$\langle \delta U_n \rangle = \langle \delta u_1 \quad \delta u_2 \quad \ldots \quad \underline{\delta u_I} \quad \delta u_{I+1} \quad \ldots \quad \underline{\delta u_J} \quad \delta u_{J+1} \quad \ldots \quad \delta u_n \rangle$$

$$W^e = \langle \delta U_n \rangle \begin{bmatrix} 0 & \cdots & 0 & \cdots & 0 & \cdots & 0 \\ \vdots & & \vdots & & \vdots & & \vdots \\ 0 & \cdots & k_{11} & \cdots & k_{12} & \cdots & 0 \\ \vdots & & \vdots & & \vdots & & \vdots \\ 0 & \cdots & 0 & \cdots & 0 & \cdots & 0 \\ \vdots & & \vdots & & \vdots & & \vdots \\ 0 & \cdots & k_{21} & \cdots & k_{22} & \cdots & 0 \\ \vdots & & \vdots & & \vdots & & \vdots \\ 0 & \cdots & 0 & \cdots & 0 & \cdots & 0 \end{bmatrix} \begin{matrix} \\ \\ \leftarrow \text{row } I \\ \\ \\ \\ \leftarrow \text{row } J \\ \\ \end{matrix} \{U_n\} = \langle \delta U_n \rangle [K^e] \{U_n\}$$

$$\underset{\text{column } I}{\uparrow} \quad \underset{\text{column } J}{\uparrow} \quad (n \times n) \qquad (4.25c)$$

(b) Expansion of $\{f\}$

Consider the following expression:

$$W^e = \langle \delta u_I \quad \delta u_J \rangle \begin{Bmatrix} f_1 \\ f_2 \end{Bmatrix} = \langle \delta u_n \rangle \{f\}. \qquad (4.26a)$$
$$\text{\small (2 × 1)}$$

Again, to keep W^e unchanged after replacement of $\langle \delta u_n \rangle$ by $\langle \delta U_n \rangle$ we must replace vector $\{f\}$ of size equal to 2 by vector $\{F^e\}$ of size n in which the term of row I is f_1 and the term of row J is f_2 with all the other terms equal to zero.

$$W^e = \langle \delta U_n \rangle \begin{Bmatrix} 0 \\ 0 \\ 0 \\ f_1 \\ 0 \\ \vdots \\ 0 \\ f_2 \\ 0 \\ \vdots \\ 0 \end{Bmatrix} = \langle \delta U_n \rangle \{F^e\} \qquad (4.26b)$$

Example 4.10 *Expansion of* $[k]$ *and* $\{f\}$ *of example* 4.9

In Example 4.9, the integral form W^e for element (1) is:

$$W^{(1)} = \langle \delta u_1 \quad \delta u_2 \quad \delta u_4 \rangle \left(\begin{bmatrix} k_{11} & k_{12} & k_{13} \\ k_{21} & k_{22} & k_{23} \\ k_{31} & k_{32} & k_{33} \end{bmatrix}^{(1)} \begin{Bmatrix} u_1 \\ u_2 \\ u_4 \end{Bmatrix} - \begin{Bmatrix} f_1 \\ f_2 \\ f_3 \end{Bmatrix}^{(1)} \right)$$

or using expanded matrix $[K^{(1)}]$ and expanded load vector $\{F^{(1)}\}$:

$$W^{(1)} = \langle \delta u_1 \quad \delta u_2 \quad \delta u_3 \quad \delta u_4 \rangle \left(\underbrace{\begin{bmatrix} k_{11} & k_{12} & 0 & k_{13} \\ k_{21} & k_{22} & 0 & k_{23} \\ 0 & 0 & 0 & 0 \\ k_{31} & k_{32} & 0 & k_{33} \end{bmatrix}}_{[K^{(1)}]}^{(1)} \begin{Bmatrix} u_1 \\ u_2 \\ u_3 \\ u_4 \end{Bmatrix} - \underbrace{\begin{Bmatrix} f_1 \\ f_2 \\ 0 \\ f_3 \end{Bmatrix}}_{\{F^{(1)}\}}^{(1)} \right)$$

For element (2) we get:

$$W^{(2)} = \langle \delta u_1 \quad \delta u_4 \quad \delta u_3 \rangle \left(\begin{bmatrix} k_{11} & k_{12} & k_{13} \\ k_{21} & k_{22} & k_{23} \\ k_{31} & k_{32} & k_{33} \end{bmatrix}^{(2)} \begin{Bmatrix} u_1 \\ u_4 \\ u_3 \end{Bmatrix} - \begin{Bmatrix} f_1 \\ f_2 \\ f_3 \end{Bmatrix}^{(2)} \right)$$

in expanded form:

$$W^{(2)} = \langle \delta u_1 \quad \delta u_2 \quad \delta u_3 \quad \delta u_4 \rangle \left(\underbrace{\begin{bmatrix} k_{11} & 0 & k_{13} & k_{12} \\ 0 & 0 & 0 & 0 \\ k_{31} & 0 & k_{33} & k_{32} \\ k_{21} & 0 & k_{23} & k_{22} \end{bmatrix}}_{[K^{(2)}]}^{(2)} \begin{Bmatrix} u_1 \\ u_2 \\ u_3 \\ u_4 \end{Bmatrix} - \underbrace{\begin{Bmatrix} f_1 \\ 0 \\ f_3 \\ f_2 \end{Bmatrix}}_{\{F^{(2)}\}}^{(2)} \right)$$

Note that the subscripts in $[K^e]$ and $\{F^e\}$ denote the position of each term in $[k]$ and $\{f\}$. On the other hand, the subscripts of $\langle \delta u_n \rangle$ and $\{u_n\}$ represent the node identification number corresponding to each nodal variable.

The global integral form W is obtained by a summation of expressions (4.24) where $\langle \delta U_n \rangle$ and $\{U_n\}$ are factored out:

$$W = \sum_e W^e = \sum_e \langle \delta U_n \rangle ([K^e]\{U_n\} - \{F^e\})$$

$$= \langle \delta U_n \rangle \left(\left[\sum_e [K^e] \right] \{U_n\} - \left\{ \sum_e \{F^e\} \right\} \right)$$

$$= \langle \delta U_n \rangle ([K]\{U_n\} - \{F\}) \qquad (4.27a)$$

where:

$$[K] = \sum_e [K^e]$$

$$\{F\} = \sum_e \{F^e\} \tag{4.27b}$$

Global matrix $[K]$ is then the sum of expanded element matrices $[K^e]$; the same is true for $\{F\}$ and $\{F^e\}$.

Example 4.11 *Global matrices for the problem of Example 4.10*

The global stiffness matrix is:

$$[K] = [K^{(1)}] + [K^{(2)}]$$

$$[K] = \begin{bmatrix} k_{11}^{(1)} + k_{11}^{(2)} & k_{12}^{(1)} & k_{13}^{(2)} & k_{13}^{(1)} + k_{12}^{(2)} \\ k_{21}^{(1)} & k_{22}^{(1)} & 0 & k_{23}^{(1)} \\ k_{31}^{(2)} & 0 & k_{33}^{(2)} & k_{32}^{(2)} \\ k_{31}^{(1)} + k_{21}^{(2)} & k_{32}^{(1)} & k_{23}^{(2)} & k_{33}^{(1)} + k_{22}^{(2)} \end{bmatrix}$$

The global load vector is:

$$\{F\} = \{F^{(1)}\} + \{F^{(2)}\} = \begin{Bmatrix} f_1^{(1)} + f_1^{(2)} \\ f_2^{(1)} \\ f_3^{(2)} \\ f_3^{(1)} + f_2^{(2)} \end{Bmatrix}$$

4.4.2 Assembly for Elements of Structural Mechanics

Historically, the assembly process of the finite element method was first used in structural mechanics for trusses and frames where elements were either bars or beams. The process is called the direct stiffness method. For each element e the discretized equation is:

$$[k]\{u_n\} - \{f\} = \{p\} \tag{4.28}$$

where $\{f\}$ are the external forces applied to the element (identical to the forces of (4.4)); $\{p\}$ are the internal reactions from the adjacent elements; and $[k]$ is the element stiffness matrix of (4.4).

The assembly process consists in obtaining the global system of equations:

$$[K]\{U_n\} = \{F\}$$

while respecting:

— the continuity of nodal displacements;
— the force equilibrium at every node, i.e. $\sum_e p_i^e = 0$.

Example 4.12 *Assembly of two elastic springs*

Consider the following two springs of moduli equal to 1:

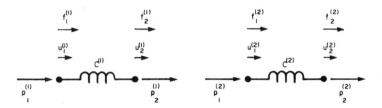

Equations (4.28) for each spring are:

spring 1: $\begin{bmatrix} 1 & -1 \\ -1 & 1 \end{bmatrix} \begin{Bmatrix} u_1^{(1)} \\ u_2^{(1)} \end{Bmatrix} - \begin{Bmatrix} f_1^{(1)} \\ f_2^{(1)} \end{Bmatrix} = \begin{Bmatrix} p_1^{(1)} \\ p_2^{(1)} \end{Bmatrix}$

spring 2: $\begin{bmatrix} 1 & -1 \\ -1 & 1 \end{bmatrix} \begin{Bmatrix} u_1^{(2)} \\ u_2^{(2)} \end{Bmatrix} - \begin{Bmatrix} f_1^{(2)} \\ f_2^{(2)} \end{Bmatrix} = \begin{Bmatrix} p_1^{(2)} \\ p_2^{(2)} \end{Bmatrix}$

The assembled structure is:

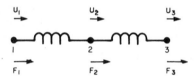

To preserve the continuity of displacements we must have:

$$u_1^{(1)} = U_1 \qquad u_1^{(2)} = U_2$$
$$u_2^{(1)} = U_2 \qquad u_2^{(2)} = U_3$$

Force equilibrium at the three nodes gives:

$$p_1^{(1)} = 0 \quad p_2^{(1)} + p_1^{(2)} = 0 \quad p_2^{(2)} = 0$$

which, after an expansion similar to (4.25c) and (4.26b), becomes:

$$\begin{bmatrix} 1 & -1 & 0 \\ -1 & 1 & 0 \\ 0 & 0 & 0 \end{bmatrix} \begin{Bmatrix} U_1 \\ U_2 \\ U_3 \end{Bmatrix} - \begin{Bmatrix} f_1^{(1)} \\ f_2^{(1)} \\ 0 \end{Bmatrix} + \begin{bmatrix} 0 & 0 & 0 \\ 0 & 1 & -1 \\ 0 & -1 & 1 \end{bmatrix}$$

$$\begin{Bmatrix} U_1 \\ U_2 \\ U_3 \end{Bmatrix} - \begin{Bmatrix} 0 \\ f_1^{(2)} \\ f_2^{(2)} \end{Bmatrix} = \begin{Bmatrix} 0 \\ 0 \\ 0 \end{Bmatrix}$$

$$[K^{(1)}]\{U_n\} - \{F^{(1)}\} + [K^{(2)}]\{U_n\} - \{F^{(2)}\} = 0$$

Thus:

$$\begin{bmatrix} 1 & -1 & \\ -1 & 2 & -1 \\ 0 & -1 & 1 \end{bmatrix} \begin{Bmatrix} U_1 \\ U_2 \\ U_3 \end{Bmatrix} = \begin{Bmatrix} f_1^{(1)} \\ f_2^{(1)} + f_1^{(2)} \\ f_2^{(2)} \end{Bmatrix},$$

$$[K]\{U_n\} = \{F\}$$

4.5 Assembly Techniques

4.5.1 *Phases of the Assembly Process*

In paragraph 4.4, two steps have been identified for the assembly process:

— construction of expanded matrices $[K^e]$ and $\{F^e\}$ for each element using (4.25c) and (4.26b);
— addition of expanded matrices (4.27b).

The two steps are executed simultaneously to avoid the explicit construction of $[K^e]$ and $\{F^e\}$.

4.5.2 *Rules for the Assembly Process Used in this Book*

To standardize the assembly operations we define a destination or location table (LOCE) for each element. The table gives the position of each term of $\{u_n\}$ and $\langle \delta u_n \rangle$ into $\{U_n\}$ and $\langle \delta U_n \rangle$ respectively. When the nodes have only one degree of freedom the table is identical to the connectivity table CONEC defined in paragraph 1.2.6. Array LOCE has a size equal to the total number of degrees of freedom n_{de} for the element.

Example 4.13 *Definition of the elements location table (or array)*

In the case of the following two triangular elements:

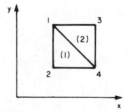

the connectivity matrix CONEC is:

elements	nodes		
1	1	2	4
2	1	4	3

(a) If there is only one degree of freedom per node (u):

$$\langle U_n \rangle = \langle u_1 \quad u_2 \quad u_3 \quad u_4 \rangle$$

— element 1

$$\langle u_n \rangle = \langle u_1 \quad u_2 \quad u_4 \rangle$$
$$\text{LOCE} = \langle 1 \quad 2 \quad 4 \rangle$$

— element 2

$$\langle u_n \rangle = \langle u_1 \quad u_4 \quad u_3 \rangle$$
$$\text{LOCE} = \langle 1 \quad 4 \quad 3 \rangle$$

(b) If there are two degrees of freedom per node (u and v):

$$\langle U_n \rangle = \langle u_1 v_1 \quad u_2 v_2 \quad u_3 v_3 \quad u_4 v_4 \rangle$$

— element 1

$$\langle u_n \rangle = \langle u_1 v_1 \quad u_2 v_2 \quad u_4 v_4 \rangle$$
$$\text{LOCE} = \langle 1 \quad 2 \quad 3 \quad 4 \quad 7 \quad 8 \rangle$$

— element 2

$$\langle u_n \rangle = \langle u_1 v_1 \quad u_4 v_4 \quad u_3 v_3 \rangle$$
$$\text{LOCE} = \langle 1 \quad 2 \quad 7 \quad 8 \quad 5 \quad 8 \rangle$$

Observe that the expansion process (4.25c) for any element matrix $[k]$ into $[K^e]$ using location table LOCE consists in transferring each term k_{ij} of $[k]$ to term K^e_{IJ} of $[K^e]$ where:

$$I = \text{LOCE}(i) \qquad i = 1, n_{de}$$
$$J = \text{LOCE}(j) \qquad j = 1, n_{de}$$
$$K^e_{IJ} \equiv K^e_{\text{LOCE}(i),\text{LOCE}(j)} \equiv k_{ij} \qquad (4.29a)$$

Similarly, each term f_i of $\{f\}$ is transferred to F^e_I of $\{F^e\}$ where:

$$F^e_I \equiv F^e_{\text{LOCE}(i)} \equiv f_i \qquad (4.29b)$$

A generalized algorithm for this two-step assembly process is:
— Initialize matrices $[K]$ and $\{F\}$.
— For each element e:
  • add each term k_{ij} of the element matrix to term K_{IJ} of the global matrix:

$$K_{IJ} = K_{IJ} + k_{ij} \quad i = 1, 2, \ldots, n_{de}$$
$$j = 1, 2, \ldots, n_{de}$$

where $I = \text{LOCE}(i)$
$J = \text{LOCE}(j)$

- add each term f_i of the element load vector to the term F_I of the global vector:

$$F_I = F_I + f_i \quad i = 1, 2, \ldots, n_{de}$$

where: $\quad I = \text{LOCE}(i)$

4.5.3 Example of Assembly Subroutine

Figure 4.6 contains a simple assembly subroutine for $[k]$ and $\{f\}$. The subroutine is used in program BBMEF of Chapter 6.

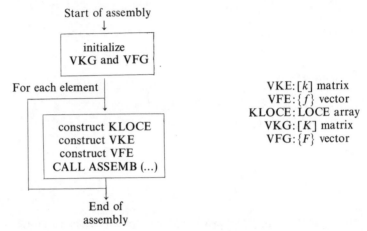

```
      SUBROUTINE ASSEMB(VKE,VFE,KLOCE,IDLE,NEQ,VKG,VFG)           ASMB   1
C=======================================================================ASMB   2
C                                                                 ASMB   3
C     SUBPROGRAM TO ASSEMBLE AN ELEMENT                           ASMB   4
C                                                                 ASMB   5
C     INPUT                                                       ASMB   6
C         VKE      ELEMENT MATRIX                                 ASMB   7
C         VFE      ELEMENT LOAD VECTOR                            ASMB   8
C         KLOCE    ELEMENT LOCALIZATION VECTOR                    ASMB   9
C         IDLE     NUMBER OF D.O.F. PER ELEMENT                   ASMB  10
C         NEQ      NUMBER OF EQUATIONS TO BE SOLVED               ASMB  11
C                                                                 ASMB  12
C     OUTPUT                                                      ASMB  13
C         VKG      GLOBAL MATRIX                                  ASMB  14
C         VFG      GLOBAL LOAD VECTOR                             ASMB  15
C                                                                 ASMB  16
C=======================================================================ASMB  17
      IMPLICIT REAL*8(A-H,O-Z)                                    ASMB  18
      DIMENSION VKE(IDLE,IDLE),VFE(IDLE),KLOCE(IDLE),             ASMB  19
     1          VKG(NEQ,NEQ),VFG(NEQ)                             ASMB  20
      DO 10 ID=1,IDLE                                             ASMB  21
      I=KLOCE(ID)                                                 ASMB  22
      VFG(I)=VFG(I)+VFE(ID)                                       ASMB  23
      DO 10 JD=1,IDLE                                             ASMB  24
      J=KLOCE(JD)                                                 ASMB  25
   10 VKG(I,J)=VKG(I,J)+VKE(ID,JD)                                ASMB  26
      RETURN                                                      ASMB  27
      END                                                         ASMB  28
```

Figure 4.6 Assembly subroutine ASSEMB, used in program BBMEF of Chapter 6.

4.5.4 Construction of Location Table LOCE

(a) *Case of one degree of freedom per node.*

The location array LOCE is identical to the connectivity array CONEC for an element (see paragraph 1.2.6).

(b) *Case of n_{dn} degrees of freedom per node $(u, v, \ldots)$.*

Presume that the nodal variables are organized in the following order:

$$\{u_n\}^T = \langle u_i \quad v_i \ldots; \quad u_j \quad v_j \ldots; \quad \ldots \rangle$$
$$\{U_n\}^T = \langle u_1 \quad v_1 \ldots; \quad u_2 \quad v_2 \ldots; \quad u_3 \quad v_3 \ldots; \quad \ldots; \quad u_n \quad v_n \rangle$$

where $i, j, \ldots$ are node identification numbers for the n_e nodes of element e, and n is the total number of nodes.

The total number of degrees of freedom for an element is:

$$n_{de} = n_e \times n_{dn}$$

Array LOCE is constructed in subroutine LOCEF (Figure 4.7) using the connectivity matrix CONEC. Subroutine LOCEF is used in program BBMEF of Chapter 6.

```
      SUBROUTINE LOCEF(KCONEC,NNEL,NDLN,KLOCE)               LOCF  1
C=====================================================================LOCF  2
C                                                            LOCF  3
C     SUBPROGRAM :                                           LOCF  4
C        TO FORM THE ELEMENT LOCALIZATION TABLE KLOCE (FIXED LOCF  5
C        NUMBER OF D.O.F.(NDLN) PER NODE)                    LOCF  6
C        THE ELEMENT AND GLOBAL D.O.F. ARE ORGANIZED ACCORDING TO : LOCF  7
C           U1 V1 .. U2 V2 .. U3 V3 ... ETC                  LOCF  8
C                                                            LOCF  9
C     INPUT                                                  LOCF 10
C        KCONEC   ELEMENT CONNECTIVITY ARRAY                 LOCF 11
C        NNEL     NUMBER OF NODES PER ELEMENT                LOCF 12
C        NDLN     NUMBER OF D.O.F. PER NODE                  LOCF 13
C                                                            LOCF 14
C     OUTPUT                                                 LOCF 15
C        KLOCE    LOCALIZATION TABLE LOCE                    LOCF 16
C                                                            LOCF 17
C=====================================================================LOCF 18
      IMPLICIT REAL*8 (A-H,O-Z)                              LOCF 19
      DIMENSION KCONEC(1),KLOCE(1)                           LOCF 20
      J=0                                                    LOCF 21
C------- LOOP OVER NNEL NODES OF THE ELEMENT                 LOCF 22
      DO 10 IN=1,NNEL                                        LOCF 23
      IDO=(KCONEC(IN)-1)*NDLN                                LOCF 24
C------- LOOP OVER NDLN DEGREES OF FREEDOM (D.O.F.)          LOCF 25
      DO 10 ID=1,NDLN                                        LOCF 26
      J=J+1                                                  LOCF 27
   10 KLOCE(J)=ID+IDO                                        LOCF 28
      RETURN                                                 LOCF 29
      END                                                    LOCF 30
```

Figure 4.7 Subroutine for the construction of the element location array LOCE (fixed number of degrees of freedom per node). This subroutine is used in program BBMEF of Chapter 6.

```
      SUBROUTINE LOCEV(KCONEC,KDLNC,NNEL,KLOCE)              LOCV  1
C=====================================================================LOCV  2
C                                                            LOCV  3
C     SUBPROGRAM :                                           LOCV  4
C        TO FORM THE ELEMENT LOCALIZATION TABLE KLOCE        LOCV  5
C        (VARIABLE D.O.F. AT EACH NODE)                      LOCV  6
C        THE ELEMENT AND GLOBAL D.O.F. ARE ORGANIZED ACCORDING TO :  LOCV  7
C           U1 V1 .. U2 V2 .. U3 V3 ... ETC                  LOCV  8
C                                                            LOCV  9
C     INPUT                                                  LOCV 10
C        KCONEC   ELEMENT CONNECTIVITY ARRAY                 LOCV 11
C        KDLNC    ARRAY OF NUMBERS OF DEGREES OF FREEDOM PER NODE  LOCV 12
C                 (CUMULATIVE)                               LOCV 13
C        NNEL     NUMBER OF NODES PER ELEMENT                LOCV 14
C                                                            LOCV 15
C     OUTPUT                                                 LOCV 16
C        KLOCE    LOCALIZATION TABLE LOCE                    LOCV 17
C                                                            LOCV 18
C=====================================================================LOCV 19
      IMPLICIT REAL*8(A-H,O-Z)                               LOCV 20
      DIMENSION KCONEC(1),KDLNC(1),KLOCE(1)                  LOCV 21
      J=0                                                    LOCV 22
C------ LOOP OVER THE NNEL NODES OF THE ELEMENT              LOCV 23
      DO 10 IN=1,NNEL                                        LOCV 24
      II=KCONEC(IN)                                          LOCV 25
      IDO=KDLNC(II)                                          LOCV 26
      IDLN=KDLNC(II+1)-IDO                                   LOCV 27
C------ LOOP OVER THE IDLN DEGREES OF FREEDOM OF NODE IN     LOCV 28
      DO 10 ID=1,IDLN                                        LOCV 29
      J=J+1                                                  LOCV 30
  10  KLOCE(J)=ID+IDO                                        LOCV 31
      RETURN                                                 LOCV 32
      END                                                    LOCV 33
```

Figure 4.8 Subroutine for the construction of the element location array LOCE (variable number of degrees of freedom per node).

(c) Case of a variable number of degrees of freedom per node

The number of degrees of freedom of each node must be stored in an array DLNC. For an increase in efficiency, the array is cumulative: DLNC $(I+1)$ stores the sum of all the degrees of freedom at nodes $1, 2, \ldots, I$. Array DLNC is dimensioned $(n+1)$. The number of degrees of freedom of node I is given by:

$$\text{DLNC}(I+1) - \text{DLNC}(I)$$

Table (or array) LOCE is constructed by subroutine LOCEF of Figure 4.8.

Example 4.14   *Location table for elements having a variable number of degrees of freedom per node*

Consider the following two three-node elements with a total of seven degrees of freedom (DOF) per element:

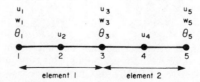

total number of nodes $\quad n = 5$
number of elements $\quad n_{el} = 2$
number of nodes per element $\quad n_e = 3$
number of DOF per element $\quad n_{de} = 7$
total number of DOF $\quad n_d = 11$

Connectivity table (CONEC)

elements	nodes		
1	1	2	3
2	3	4	5

Cumulative DOF per node in array DLNC:

$$\langle 0 \quad 3 \quad 4 \quad 7 \quad 8 \quad 11 \rangle$$

Location table LOCE:
— element 1

$$\langle 1 \quad 2 \quad 3 \quad 4 \quad 5 \quad 6 \quad 7 \rangle$$

— element 2

$$\langle 5 \quad 6 \quad 7 \quad 8 \quad 9 \quad 10 \quad 11 \rangle$$

4.6 Properties of Global Matrices

4.6.1 *Band Structure of Matrix* $[K]$

Global matrix $[K]$ is the result of the summation of very sparse expanded element matrices $[K^e]$:

$$[K] = \sum_{\text{elements}} [K^e]$$

According to the assembly rule of paragraph 4.5, non-zero terms of $[K^e]$ are such that:

$$K^e_{IJ} \equiv k^e_{ij} \tag{4.30}$$

where:

$$I = \text{LOCE}\,(i) \quad i = 1, 2, \ldots, n_{de}$$
$$J = \text{LOCE}\,(j) \quad j = 1, 2, \ldots, n_{de}$$

Consequently, term K_{IJ} is different from zero only when an element has nodal variables u_I and u_J.

The assembly rule is symmetrical in I and J: if there exists a non-zero term K_{IJ}, there also exists a non-zero term K_{JI}. We may thus study the topological structure (sparseness) of the upper half of $[K]$ for which $J \geq I$. Note that this does not imply that the matrix $[K]$ is symmetrical ($K_{IJ} = K_{JI}$).

For each non zero term K_{IJ}, horizontal distance b_{IJ} and vertical distance h_{IJ} measured from the diagonal of $[K]$ are defined as:

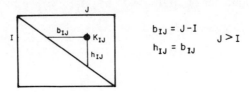

According to (4.30), b_{IJ} corresponding to term k_{ij} of element e is:

$$b_{IJ}^e = J - I = \text{LOCE}(j) - \text{LOCE}(i) \quad J > I$$

The element bandwidth b_I^e for row I of $[K^e]$ is the maximum value of b_{IJ}^e for all non-zero terms of that row:

$$\begin{array}{l} i = 1, 2, \ldots, n_{de} \\ \quad I = \text{LOCE}(i) \\ \quad\quad j = 1, 2, \ldots, n_{de} \\ \quad\quad\quad b_I^e = \underset{j}{\text{Max}}(\text{LOCE}(j) - I) = \underset{j}{\text{Max}}(\text{LOCE}(j)) - I \end{array} \quad (4.31a)$$

Similarly, the element bandheight h_J^e of column J of $[K^e]$ is the maximum value of h_{IJ}^e for all non-zero terms of that column:

$$\begin{array}{l} j = 1, 2, \ldots, n_{de} \\ \quad J = \text{LOCE}(j) \\ \quad\quad i = 1, 2, \ldots, n_{de} \\ \quad\quad\quad h_J^e = \underset{i}{\text{Max}}(J - \text{LOCE}(i)) = J - \underset{i}{\text{Min}}(\text{LOCE}(i)) \end{array} \quad (4.31b)$$

The bandwidth for row I of global matrix $[K]$ is:

$$b_I = \underset{e}{\text{Max}}(b_I^e) \quad (4.32a)$$

for all elements e. In row I, terms K_{IJ} are null for $J > b_I + I$. For column I, terms K_{JI} are also null for $J > b_I + I$.

Bandheight h_J for column J of global matrix $[K]$ is:

$$h_J = \underset{e}{\text{Max}}(h_J^e) \quad (4.32b)$$

for all elements e. Finally, the bandwidth b and bandheight h of global matrix $[K]$ are:

$$\begin{array}{ll} b = \underset{I}{\text{Max}}(b_I) & \text{for all rows } I \\ h = \underset{J}{\text{Max}}(h_J) & \text{for all columns } J \end{array} \quad (4.33)$$

Because of the symmetry in I and J

$$b = h$$

Observe that the bandwidth and bandheight defined above do not include a diagonal term. Therefore, for a diagonal matrix we would have $h = b = 0$.

Example 4.15 Bandwidths and bandheights for an assembly of three one-dimensional elements

Consider the following three elements having two nodes per element and one degree of freedom per node:

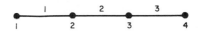

— Element 1

$$\text{LOCE} = \langle 1 \quad 2 \rangle$$
$$b_I^{(1)} = \langle 1 \quad 0 \quad 0 \quad 0 \rangle$$
$$h_J^{(1)} = \langle 0 \quad 1 \quad 0 \quad 0 \rangle$$

— Element 2

$$\text{LOCE} = \langle 2 \quad 3 \rangle$$
$$b_I^{(2)} = \langle 0 \quad 1 \quad 0 \quad 0 \rangle$$
$$h_J^{(2)} = \langle 0 \quad 0 \quad 1 \quad 0 \rangle$$

— Element 3

$$\text{LOCE} = \langle 3 \quad 4 \rangle$$
$$b_I^{(3)} = \langle 0 \quad 0 \quad 1 \quad 0 \rangle$$
$$h_J^{(3)} = \langle 0 \quad 0 \quad 0 \quad 1 \rangle$$

— For the assembled matrix

$$b_I = \langle 1 \quad 1 \quad 1 \quad 0 \rangle$$
$$h_J = \langle 0 \quad 1 \quad 1 \quad 1 \rangle$$
$$b = h = 1$$

$$[K] = \begin{bmatrix} x & x & 0 & 0 \\ x & x & x & 0 \\ 0 & x & x & x \\ 0 & 0 & x & x \end{bmatrix}$$

The banded structure of matrix $[K]$ is an important characteristic of the finite element method. It allows an important reduction of the storage needed for the matrix $[K]$ as well as an important reduction of the time needed for solution. Bandwidth b_I for each row of $[K]$ depends on the content of array LOCE which is derived from the connectivity of an element. Thus, the bandwidth is affected by the order in which the nodes are numbered. Even though the number of non-zero terms in [K] remains constant, the bandwidth depends strongly on the order in which nodes are numbered.

Example 4.16   Renumbering of the nodes of Example 4.15

$$[K] = \begin{bmatrix} x & 0 & 0 & x \\ 0 & x & x & x \\ 0 & x & x & 0 \\ x & x & 0 & x \end{bmatrix}$$

$$b = h = 3$$

Matrix [K] contains the same number of non-zero terms as in Example 4.15; however, the bandwidth has increased from 1 to 3.

A useful role to follow in order to reduce the bandwidth is to minimize the maximum difference between the integer identification of the various nodes of one element.

4.6.2 *Symmetry*

For many problems, matrices [k] and therefore also [K] are symmetrical. For such matrices important reductions of storage and solution time can be accomplished.

4.6.3 *Storage Strategies*

(a) *Full non-symmetrical matrix*

Such a matrix requires n^2 computer words for its storage.

(b) *Full symmetrical matrix.*

Only the upper triangular part of matrix [K] must be stored in an array VK; for example:

$$[K] = \begin{bmatrix} K_{11} & K_{12} & K_{13} \\ K_{12} & K_{22} & K_{23} \\ K_{13} & K_{23} & K_{33} \end{bmatrix} ; \quad VK = \langle K_{11}\ K_{12}\ K_{22}\ K_{13}\ K_{23}\ K_{33} \rangle$$

$$K_{IJ} \equiv VK_l \quad \text{if} \quad l = \frac{J(J-1)}{2} + I \tag{4.34}$$

$$J \geq I$$

Only $n(n+1)/2$ real computer words are needed for the storage in this case.

(c) Non-symmetrical band matrix

The matrix can be stored in a rectangular array VK of dimension $nx(2b+1)$ as shown below:

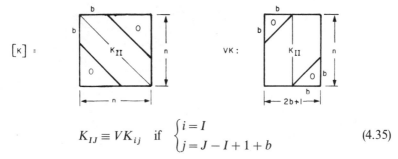

$$K_{IJ} \equiv VK_{ij} \quad \text{if} \quad \begin{cases} i = I \\ j = J - I + 1 + b \end{cases} \quad (4.35)$$

In this case, $n(2b+1)$ computer words are needed, including $b(b+1)$ zero terms.

(d) Symmetrical banded matrix

In this case:

$$K_{IJ} = VK_{ij} \quad \text{if} \quad \begin{cases} i = I \\ j = J - I + 1 \\ J \geq I \end{cases} \quad (4.36)$$

In this case, $n(b+1)$ computer words are needed, including $b(b+1)/2$ zero terms.

(e) Non-symmetrical skyline matrix

The most efficient strategy for the storage of global matrices is the skyline method. It consists in storing the terms of $[K]$ by rows and columns of variable length. Three arrays are used for convenience:

— VKGD contains the diagonal terms;
— VKGS contains the upper triangular part of $[K]$ stored in column order from the skyline profile down to, but without the main diagonal;
— VKGI contains the lower triangular matrix by rows from left to right.

For the following matrix:

Skyline profile

$$[K] = \begin{bmatrix} K_{11} & K_{12} & 0 & K_{14} & 0 \\ K_{21} & K_{22} & K_{23} & K_{24} & 0 \\ 0 & K_{32} & K_{33} & K_{34} & K_{35} \\ K_{41} & K_{42} & K_{43} & K_{44} & 0 \\ 0 & 0 & K_{55} & 0 & K_{55} \end{bmatrix} \quad (4.37)$$

$$[K] = \begin{bmatrix} 0 & K_{12} & 0 & K_{14} & 0 \\ 0 & 0 & K_{23} & K_{24} & 0 \\ 0 & 0 & 0 & K_{34} & K_{35} \\ 0 & 0 & 0 & 0 & 0 \\ 0 & 0 & 0 & 0 & 0 \end{bmatrix} + \begin{bmatrix} K_{11} & 0 & 0 & 0 & 0 \\ 0 & K_{22} & 0 & 0 & 0 \\ 0 & 0 & K_{33} & 0 & 0 \\ 0 & 0 & 0 & K_{44} & 0 \\ 0 & 0 & 0 & 0 & K_{55} \end{bmatrix}$$

Terms stored in VKGS   Terms stored in VKGD

$$+ \begin{bmatrix} 0 & 0 & 0 & 0 & 0 \\ K_{21} & 0 & 0 & 0 & 0 \\ 0 & K_{32} & 0 & 0 & 0 \\ K_{41} & K_{42} & K_{43} & 0 & 0 \\ 0 & 0 & K_{53} & 0 & 0 \end{bmatrix} \quad (4.38)$$

Terms stored in VKGI

$$\text{VKGS} = \langle K_{12}; K_{23}; K_{14} \quad K_{24} \quad K_{34}; K_{35} \quad 0 \rangle$$
$$\text{VKGI} = \langle K_{21}; K_{32}; K_{41} \quad K_{42} \quad K_{43}; K_{53} \quad 0 \rangle \quad (4.39)$$
$$\text{VKGD} = \langle K_{11} \quad K_{22} \quad K_{33} \quad K_{44} \quad K_{55} \rangle$$

The skyline is the envelope of the column of variable heights. The envelope is symmetrical even when $[K]$ is non-symmetrical. It is defined by the array of column heights h_J (4.32b); for matrix (4.37) we have:

$$h_J = \langle 0 \quad 1 \quad 1 \quad 3 \quad 2 \rangle \quad (4.40)$$

Zero terms outside the skyline envelope are not stored; however, zero terms within the skyline, like those of position 4, 5 and 5, 4 in matrix (4.37), must be stored because these terms are not invariant in a solution process.

To define the position of term K_{ij} in arrays VKGS and VKGI we must use the table locating the start of columns KLD of dimension $n+1$ defined by:

$$\text{KLD}(1) = 1; \quad \text{KLD}(2) = 1$$
$$\text{KLD}(I) = \text{KLD}(I-1) + h_J(I-1) \quad I = 3, 4, \ldots, n+1 \quad (4.41)$$

In the case of (4.40)

$$\text{KLD} = \langle 1 \quad 1 \quad 2 \quad 3 \quad 6 \quad 8 \rangle$$

Term K_{IJ} is then placed as follows:

— if $I = J$ in VKGD (I)
— if $I < J$ in VKGS (l) where $l = \text{KLD}(J+1) - J + 1$ (4.42)
— if $I > J$ in VKGI (l) where $l = \text{KLD}(I+1) - I + J$

The required amount of storage needed is:

— n real words for VKGD;
— $2(\text{KLD}(n+1) - 1)$ real words for VKGS or VKGI;

then, the grand total is:

$$n + 2(\text{KLD}(n+1) - 1) \text{ real words}$$

All the terms stored are needed, even the zeros inside the skyline because they can become non-zero during the solution process.

Example 4.17   *Skyline storage for a non-symmetrical matrix*

Matrix $[K]$ for Example 4.16 is:

$$\begin{bmatrix} K_{11} & 0 & 0 & K_{14} \\ 0 & K_{22} & K_{23} & K_{24} \\ 0 & K_{32} & K_{33} & 0 \\ K_{41} & K_{42} & 0 & K_{44} \end{bmatrix} \quad n = 4$$

In this case:

$$h_J = \langle 0 \; 0 \; 1 \; 3 \rangle$$
$$\text{KLD} = \langle 1 \; 1 \; 1 \; 2 \; 5 \rangle$$
$$\text{VKGS} = \langle K_{23} \; K_{14} \; K_{24} \; 0 \rangle$$
$$\text{VKGI} = \langle K_{32} \; K_{41} \; K_{42} \; 0 \rangle$$
$$\text{VKGD} = \langle K_{11} \; K_{22} \; K_{33} \; K_{44} \rangle$$

From (4.42) term K_{24} is at:

VKGS (l) where $l = \text{KLD}(4+1) - 4 + 2 = 3$

The required number of storage words is:

$$4 + 2(\text{KLD}(5) - 1) = 12 \text{ real words}$$

(f) Symmetrical matrix with skyline profile

The method of storage is identical to the one used for a non-symmetrical matrix except that only the diagonal terms and upper triangular part need to be stored. Array VKGI is not needed. The total storage needed (in computer words) is:

$$n + \text{KLD}(n+1) - 1$$

For a diagonal matrix, array VKGS is not used.

(g) *Out of core skyline matrix*

When the size of [K] is too big for the computer used, arrays VKGS and VKGI must be stored by blocks on a mass storage device such as a disk drive. The size of the blocks is defined by the amount of storage available in the computer memory. Each block contains a whole number of columns or rows that may vary from block to block. Blocks should not split the degrees of freedom of a given node.

Array VKGD containing the diagonal terms stays in memory. Two tables of pointers are used: table KLD, identical to that of previous sections (e) and (f), and table KEB defining the numerical identity of the first column or row in each block; it is dimensioned $n_b + 1$ (n_b is the number of blocks), and KEB $(n_b + 1) = n + 1$.

Example 4.18 *Segmentation of a matrix*

Consider the symmetrical skyline matrix shown below:

$$\begin{bmatrix} 1 & 2 & & 6 & & & & & & \\ & 3 & 4 & 7 & & & 15 & & & \\ & & 5 & 8 & 10 & & 16 & & & \\ & & & 9 & 11 & & 17 & & & \\ & & & & 12 & 13 & 18 & & & \\ & & & & & 14 & 19 & 21 & & \\ & & & & & & 20 & 22 & & \\ \text{Sym.} & & & & & & & 23 & & \end{bmatrix}$$

If the matrix is segmented in blocks of six terms, four blocks will be needed. Tables KLD and KEB are:

$$\text{KLD} = \langle 1 \quad 1 \quad 2 \quad 3 \quad 6 \quad 8 \quad 9 \quad 14 \quad 16 \rangle$$
$$\text{KEB} = \langle 1 \quad 5 \quad 7 \quad 8 \quad 9 \rangle$$

Successive blocks of the upper triangle part shall contain the following terms of the above matrix:

$$\text{block 1:} \langle 2 \quad 4 \quad 6 \quad 7 \quad 8 \quad 0 \rangle$$
$$\text{block 2:} \langle 10 \quad 11 \quad 13 \quad 0 \quad 0 \quad 0 \rangle$$
$$\text{block 3:} \langle 15 \quad 16 \quad 17 \quad 18 \quad 19 \quad 0 \rangle$$
$$\text{block 4:} \langle 21 \quad 22 \quad 0 \quad 0 \quad 0 \quad 0 \rangle$$

Array VKGD contains the following terms of the matrix:

$$\text{VKGD} = \langle 1 \quad 3 \quad 5 \quad 9 \quad 12 \quad 14 \quad 20 \quad 23 \rangle$$

For a non-symmetrical matrix with an analogous structure, tables KLD and KEB above remain unchanged as well as the blocks of the upper triangular part of the matrix:

$$\begin{bmatrix} 1 & 2 & & 6 & & & & & & \\ 2 & 3 & 4 & 7 & & & 15 & & & \\ & 4' & 5 & 8 & 10 & & 16 & & & \\ 6' & 7' & 8' & 9 & 11 & & 17 & & & \\ & & 10' & 11' & 12 & 13 & 18 & & & \\ & & & & 13' & 14 & 19 & 21 & \\ & & 15' & 16' & 17' & 18' & 19 & 20 & 22 \\ & & & & & & 21' & 22 & 23 \end{bmatrix}$$

Successive blocks of the lower triangular part shall contain:

$$\text{block 1:} \langle\ 2'\quad 4'\quad 6'\quad 7'\quad 8'\quad -\ \rangle$$
$$\text{block 2:} \langle 10'\quad 11'\quad 13'\quad -\quad -\quad -\ \rangle$$
$$\text{block 3:} \langle 15'\quad 16'\quad 17'\quad 18'\quad 19'\quad -\ \rangle$$
$$\text{block 4:} \langle 21'\quad 22'\quad -\quad -\quad -\quad -\ \rangle$$

In all cases, the diagonal terms are stored in VKGD as above:

$$\text{VKGD} = \langle 1\quad 3\quad 5\quad 9\quad 12\quad 14\quad 20\quad 23 \rangle$$

4.7 Global System of Equations

4.7.1 *Formulation of the System of Equations*

After assembly, the global discretized integral form (4.5b) is:

$$W = \langle \delta U_n \rangle ([K]\{U_n\} - \{F\}) = 0 \tag{4.43}$$

The problem consists in finding $\{U_n\}$ that makes W vanish no matter what is $\langle \delta U_n \rangle$ while satisfying boundary conditions defined on S_u in paragraph 3.3.2: $u = u_s$ and $\delta u = 0$. After discretization, these conditions are:

$$\delta U_i = 0 \tag{4.44a}$$

$$U_i = \bar{U}_i \tag{4.44b}$$

for all degrees of freedom U_i having a specified value of $\bar{U}_i$.

In summary, the algebraic system:

$$[K]\{U_n\} = \{F\} \tag{4.45}$$

must be solved for $\{U_n\}$ after some modifications required to satisfy conditions (4.44).

4.7.2 *Introduction of Boundary Conditions*

Conditions (4.44) can be introduced in system (4.45) by several methods:

(a) Using a large number on the diagonal term

Matrix [K] is assembled without paying attention to the boundary conditions; then, each specified value of the unknown $U_i = \bar{U}_i$ is introduced as follows:

— term K_{ii} is replaced by $K_{ii} + \alpha$ where α is a very large number with respect to all the other terms of K_{ij}; if α is chosen big enough, $K_{ii} + \alpha \cong \alpha$ (for example, in a six significant digit machine $100.343 + 1.00000 \times 10^9 = 1.00000 \times 10^9$); the corresponding unknown X_i in the same row would necessarily become negligibly small with respect to all the values of the other unknowns;

— all the terms of the right-hand side are replaced by their original values diminished by $F_j - K_{ji}\bar{U}_i = F_j^*$; the right-hand side of equation i is $-K_{ii}\bar{U}_i$; if we now compare any equation before and after modification, we get, for equation j:

- before: $K_{j1}U_1 + K_{j2}U_2 + \ldots K_{ji}\bar{U}_i + \ldots K_{jn}U_n = F_j$
- after: $K_{j1}U_1 + K_{j2}U_2 + \ldots K_{ji}X_i + \ldots K_{jn}U_n = F_j - K_{ji}\bar{U}_i$

(4.46)

since $K_{ji}X_i$ is negligibly small with respect to all the other terms, the second equation can be identical to the unmodified equation with a proper choice of α, for equation i:

- before: $K_{i1}U_1 + K_{i2}U_2 + \ldots + K_{ii}\bar{U}_i + \ldots + K_{in}U_n = R_i$
- after: $K_{i1}U_1 + K_{i2}U_2 + \ldots + \alpha X_i + \ldots + K_{in}U_n = -K_{ii}\bar{U}_i$

(4.47)

Subtracting the second from the unmodified equation, we get:

$$K_{ii}\bar{U}_i - \alpha X_i = R_i + K_{ii}\bar{U}_i$$

therefore $\qquad R_i = -\alpha X_i$

where R_i is the reaction corresponding to the specified value $\bar{U}_i$.

The method is very simple to code and gives, after solution, all the unknowns and reactions with no loss of accuracy. The matrix of coefficients retains its symmetry since only the diagonal term K_{ii} is modified.

(b) Wiping rows and columns in place

In this method, the load vector is modified as in the previous case for all specified value of the unknown $U_i = \bar{U}_i$, except for term F_i which is replaced by U_i:

$$F_j = F_j - K_{ji}\bar{U}_i \qquad j = 1, 2, \ldots, n \qquad j \neq i$$
$$F_i = \bar{U}_i$$
$$K_{ij} = K_{ji} = 0 \qquad j = 1, 2, \ldots, n \qquad j \neq i$$
$$K_{ii} = 1$$

Then, row and column intersecting on i are wiped out and term K_{ii} is replaced by

1. This effectively eliminates equation i from the system without changing the size and symmetry of $[K]$.

$$\begin{bmatrix} K_{11} & \cdots & K_{1,i-1} & 0 & K_{1,i+1} & \cdots & K_{1n} \\ \vdots & & \vdots & \vdots & \vdots & & \vdots \\ K_{i-1,1} & \cdots & K_{i-1,i-1} & 0 & K_{i-1,i+1} & \cdots & K_{i-1,n} \\ 0 & \cdots & 0 & 1 & 0 & \cdots & 0 \\ K_{i+1,1} & \cdots & K_{i+1,i-1} & 0 & K_{i+1,i+1} & \cdots & K_{i+1,n} \\ \vdots & & \vdots & \vdots & \vdots & & \vdots \\ K_{n1} & \cdots & K_{n,i-1} & 0 & K_{n,i+1} & \cdots & K_{nn} \end{bmatrix} \begin{Bmatrix} U_1 \\ \vdots \\ U_{i-1} \\ U_i \\ U_{i+1} \\ \vdots \\ U_n \end{Bmatrix} = \begin{Bmatrix} F_1 - K_{1i}\bar{U}_i \\ \vdots \\ F_{i-1} - K_{i-1,i}\bar{U}_i \\ \bar{U}_i \\ F_{i+1} - K_{i+1,i}\bar{U}_i \\ \vdots \\ F_n - K_{ni}\bar{U}_i \end{Bmatrix}$$

(4.48)

In this method, the reactions are not obtained directly.

(c) *Suppression of equations with specified value* $\bar{U}_i$

This effectively would require a complete reshuffling of the matrix of coefficients. The load vector is modified as in the two previous cases but row i and column i are both completely eliminated, reducing the size of matrix $[K]$. To avoid costly matrix manipulations, the assembly process is modified to prevent the construction of such equations altogether.

This method is used for program MEF of Chapter 6.

Example 4.19 *Application of boundary conditions*

The system of equations of Example 4.16 is:

$$\begin{bmatrix} K_{11} & 0 & 0 & K_{14} \\ 0 & K_{22} & K_{23} & K_{24} \\ 0 & K_{23} & K_{33} & 0 \\ K_{14} & K_{24} & 0 & K_{44} \end{bmatrix} \begin{Bmatrix} U_1 \\ U_2 \\ U_3 \\ U_4 \end{Bmatrix} = \begin{Bmatrix} F_1 \\ F_2 \\ F_3 \\ F_4 \end{Bmatrix}$$

Condition $U_1 = \bar{U}_1$ leads to the following three modified systems:

— dominating diagonal term ($\alpha = 10^{15}$)

$$\begin{bmatrix} 10^{15} & 0 & 0 & K_{14} \\ 0 & K_{22} & K_{23} & K_{24} \\ 0 & K_{23} & K_{33} & 0 \\ K_{14} & K_{24} & 0 & K_{44} \end{bmatrix} \begin{Bmatrix} X_1 \\ U_2 \\ U_3 \\ U_4 \end{Bmatrix} = \begin{Bmatrix} -K_{11}\bar{U}_1 \\ F_2 - 0 \\ F_3 - 0 \\ F_4 - K_{14}\bar{U}_1 \end{Bmatrix}$$

— wiping rows and columns

$$\begin{bmatrix} 1 & 0 & 0 & 0 \\ 0 & K_{22} & K_{23} & K_{24} \\ 0 & K_{23} & K_{33} & 0 \\ 0 & K_{24} & 0 & K_{44} \end{bmatrix} \begin{Bmatrix} U_1 \\ U_2 \\ U_3 \\ U_4 \end{Bmatrix} = \begin{Bmatrix} \bar{U}_1 \\ F_2 \\ F_3 \\ F_4 - K_{14}\bar{U}_1 \end{Bmatrix}$$

— elimination of equation 1

$$\begin{bmatrix} K_{22} & K_{23} & K_{24} \\ K_{23} & K_{33} & 0 \\ K_{24} & 0 & K_{44} \end{bmatrix} \begin{Bmatrix} U_2 \\ U_3 \\ U_4 \end{Bmatrix} = \begin{Bmatrix} F_2 \\ F_3 \\ F_4 - K_{14}\bar{U}_1 \end{Bmatrix}; \quad U_1 = \bar{U}_1$$

4.7.3 Reactions

For any equation i in which the unknown must take specified value $\bar{U}_i$, the right-hand side member becomes an unknown quantity called a reaction. The first method of application of the boundary conditions produced directly all the unknown values as well as all the reactions. The other two methods require *a posteriori* calculation from:

$$F_i = \sum_{j=1}^{n} K_{ij} U_j \tag{4.49}$$

Note that we can have another method for introducing boundary conditions and having reaction F_i explicitly as an unknown in the equations. This leads to:

$$\begin{bmatrix} K_{11} & \cdots & K_{1,i-1} & 0 & K_{1,i+1} & \cdots & K_{1n} \\ \vdots & & \vdots & & \vdots & & \vdots \\ K_{i-1,1} & \cdots & K_{i-1,i-1} & 0 & K_{i-1,i+1} & \cdots & K_{i-1,n} \\ K_{i1} & \cdots & K_{i,i-1} & -1 & K_{i,i+1} & \cdots & K_{in} \\ K_{i+1,1} & \cdots & K_{i+1,i-1} & 0 & K_{i+1,i+1} & \cdots & K_{i+1,n} \\ \vdots & & \vdots & & \vdots & & \vdots \\ K_{n1} & \cdots & K_{n,i-1} & 0 & K_{n,i+1} & \cdots & K_{nn} \end{bmatrix} \begin{Bmatrix} U_1 \\ \vdots \\ U_{i-1} \\ F_i \\ U_{i+1} \\ \vdots \\ U_n \end{Bmatrix} = \begin{Bmatrix} F_1 - K_{1i}\bar{U}_i \\ \vdots \\ F_{i-1} - K_{i-1,i}\bar{U}_i \\ -K_{ii}\bar{U}_i \\ F_{i+1} - K_{i+1,i}\bar{U}_i \\ \vdots \\ F_n - K_{ni}\bar{U}_i \end{Bmatrix}$$

$$(4.50)$$

However, in this case, the symmetry of K is lost even if $[K]$ was symmetrical *a priori*.

4.7.4 *Variable Transformations*

Presume that constraints must be imposed between some unknowns as follows:

$$\begin{aligned} \{\delta U_n\} &= [R]\{\delta U'_n\} \\ \{U_n\} &= [R]\{U'_n\} \end{aligned} \tag{4.51}$$

where $[R]$ could be a rectangular matrix of constants. Substituting (4.51) into (4.5b) gives:

$$W = \langle \delta U'_n \rangle ([K]\{U'_n\} - \{F'\}) = 0$$

where:
$$\begin{aligned} [K'] &= [R]^T [K][R] \\ \{F'\} &= [R]^T \{F\} \end{aligned} \tag{4.52}$$

Such a transformation may be needed for the following reasons:

— to change the reference system of some variables;
— to force a linear constraint between some variables.

Transformation (4.52) can also be performed at the element level.

Example 4.20   Rotation of a reference system at a node of a two-dimensional element mesh

Consider the assemblage of two triangular elements as shown. A change of reference system for node 2 is needed to impose a frictionless rolling boundary condition on the incline plane.

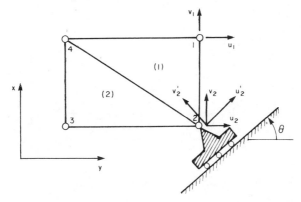

The global system of equations is:

$$\langle U_n \rangle = \langle u_1 \; v_1 \; u_2 \; v_2 \; u_3 \; v_3 \; u_4 \; v_4 \rangle$$

$$\underset{(8 \times 8)}{[K]} \underset{}{\{U_n\}} = \underset{(8 \times 1)}{\{F\}}$$

Boundary condition at node 2 is $v_2' = 0$, where v_2' is the component of displacement perpendicular to the rolling plane:

$$\begin{Bmatrix} u_2 \\ v_2 \end{Bmatrix} = \begin{bmatrix} c & -s \\ s & c \end{bmatrix} \begin{Bmatrix} u_2' \\ v_2' \end{Bmatrix} \qquad \begin{matrix} c = \cos\theta \\ s = \sin\theta \end{matrix}$$

The transformation matrix for the total vector of variables is:

$$[R] = \begin{bmatrix} 1 & & & & & & & \\ & 1 & & & & & & \\ & & c & -s & & & & \\ & & s & c & & & & \\ & & & & 1 & & & \\ & & & & & 1 & & \\ & & & & & & 1 & \\ & & & & & & & 1 \end{bmatrix}$$

$$\langle U_n' \rangle = \langle u_1 v_1 \; u_2' v_2' \; u_3 v_3 \; u_4 v_4 \rangle$$

Matrices $[K']$ and $\{F'\}$ are obtained from (4.52). The new system $[K']\{U'_n\} = \{F'\}$ can now be solved with condition $v'_2 = 0$.

4.7.5 Linear Constraints Between Variables

Transformation (4.51) permits the introduction of linear constraints between some variables:

$$a_i U_i + a_j U_j + a_k U_k + \ldots = U'_i = g \quad (4.53)$$

The transformation matrix $[R]$ between old $(U_i, U_j, U_k \ldots)$ and new $(U'_i, U_j, U_k \ldots)$ is:

$$[R]_{(n \times n)} = \begin{bmatrix} 1 & & & & & & \\ & 1 & & & & & \\ & & \ddots & & & & \\ & & & 1 & & & \\ \ldots & & -\dfrac{a_j}{a_i} & \ddots & \dfrac{1}{a_i} & \ldots & -\dfrac{a_k}{a_i} \\ & & & & & 1 & \\ & & & & & & \ddots \\ & & & & & & & 1 \end{bmatrix} \quad (4.54)$$

Then, variable U' is specified equal to g in the new system. Matrix $[K]$ is transformed in two steps:

— step 1:

$$[K''] = [K][R]$$

$$\text{column } i \text{ of } [K''] = \frac{1}{a_i} \times \text{column } i \text{ of } [K]$$

$$\text{column } j \text{ of } [K''] = \text{column } j \text{ of } [K] - \frac{a_j}{a_i} \times \text{column } i \text{ of } [K] \quad (4.55)$$

$$\text{column } k \text{ of } [K''] = \text{column } k \text{ of } [K] - \frac{a_k}{a_i} \times \text{column } i \text{ of } [K]$$

the other columns of $[K'']$ are identical to their corresponding column in $[K]$.

— step 2:

$$[K'] = [R]^T [K''] = [R]^T [K][R]$$

$$\text{row } i \text{ of } [K'] = \frac{1}{a_i} \times \text{row } i \text{ of } [K'']$$

$$\text{row } j \text{ of } [K'] = \text{row } j \text{ of } [K'] - \frac{a_j}{a_i} \times \text{row } i \text{ of } [K''] \quad (4.56)$$

$$\text{row } k \text{ of } [K'] = \text{row } k \text{ of } [K'] - \frac{a_k}{a_i} \times \text{row } i \text{ of } [K'']$$

Vector F is also modified as follows:

$$\{F'\} = [R]^T \{F\}$$

$$F'_i = \frac{1}{a_i} \cdot F_i$$

$$F'_j = F_j - \frac{a_j}{a_i} \cdot F_i \qquad (4.57)$$

$$F'_k = F_k - \frac{a_k}{a_i} \cdot F_i$$

The other terms of $\{F'\}$ are identical to their corresponding terms in $\{F\}$.

Example 4.21 Linear constraint between the variables of a one-dimensional problem

Consider the following assemblage of two bars:

$$\overset{U_1}{\bullet} \underset{(1)}{\rule{2cm}{0.4pt}} \overset{U_2}{\bullet} \underset{(2)}{\rule{3cm}{0.4pt}} \overset{U_3}{\bullet}$$

The global system of equations is:

$$\begin{bmatrix} K_{11} & K_{12} & 0 \\ K_{21} & K_{22} & K_{23} \\ 0 & K_{22} & K_{33} \end{bmatrix} \begin{Bmatrix} U_1 \\ U_2 \\ U_3 \end{Bmatrix} = \begin{Bmatrix} F_1 \\ F_2 \\ F_3 \end{Bmatrix}$$

To force the constraint $U_1 - U_3 = U'_3$ we use (4.54) in which $a_i = -1$, $a_j = 1$, $a_k = 0$:

$$\begin{Bmatrix} U_1 \\ U_2 \\ U_3 \end{Bmatrix} = [R] \begin{Bmatrix} U_1 \\ U_2 \\ U'_3 \end{Bmatrix}; \quad [R] = \begin{bmatrix} 1 & 0 & 0 \\ 0 & 1 & 0 \\ 1 & 0 & -1 \end{bmatrix}$$

Thus, from (4.55), (4.56), and (4.57)

$$[K'] = \begin{bmatrix} K_{11} + K_{33} & K_{12} + K_{32} & -K_{33} \\ K_{21} + K_{23} & K_{22} & -K_{23} \\ -K_{33} & -K_{32} & K_{33} \end{bmatrix}$$

$$\{F'\} = \begin{Bmatrix} F_1 + F_3 \\ F_2 \\ -F_3 \end{Bmatrix}$$

Equation (4.53) couples the variables U_i, U_j, U_k... and this can modify the structure of matrix $[K]$. To prevent this, coupling must be taken care of during the computation of column heights starting from the array LOCE. The column heights for i, j, k must be modified to make them of the same level as the highest one, as shown by the two examples below:

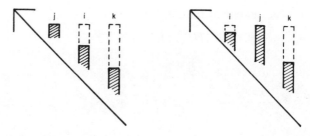

This can be obtained by adding a fictitious element for each condition like (4.53) such that:

$$\text{LOCE} = \langle i \quad j \quad k \ldots \rangle$$
$$[k] = [0]$$

Global transformation (4.55), (4.56), (4.57) can also be done at the element level. The transformation must be applied to all the elements sharing variable U_i. If an element e has U_i but not U_j or U_k, we must nevertheless include U_j and U_k in the list of variables for that element:

$$\text{LOCE} = \langle \langle \text{LOCE}^{(e)} \rangle, j, k \rangle,$$

$$[k] = \begin{bmatrix} & & & j & k \\ & [k^{(e)}] & & 0 & 0 \\ & & & 0 & 0 \\ & & & \vdots & \vdots \\ 0 & 0 & \ldots & 0 & \vdots \\ 0 & 0 & \ldots & & 0 \end{bmatrix} \qquad (4.58)$$

The transformation matrix $[r]$ at the element level is similar to (4.54) but transforms only the variables included in LOCE (see Example 4.22). With the modification of LOCE, matrix $[K]$ will automatically be constructed correctly.

Example 4.22   Linear constraint at the element level

In Example 4.21, destination arrays are:

$$\text{LOCE}^{(1)} = \langle 1 \quad 2 \rangle \quad \text{LOCE}^{(2)} = \langle 2 \quad 3 \rangle$$

Fictitious element 3 designed to apply the linear constraint $U_1 - U_3 = U'_3$ has the following destination array:

$$\text{LOCE}^{(3)} = \langle 1 \quad 3 \rangle$$

Only element 2 contains U_3; however, it does not contain U_1 so we must include it:

$$\text{LOCE}^{(2)} = \langle 2 \quad 3; \quad 1 \rangle$$

$$[k^{(2)}] = \begin{bmatrix} k_{11} & k_{12} & 0 \\ k_{21} & k_{22} & 0 \\ 0 & 0 & 0 \end{bmatrix}; \quad \{f^{(2)}\} = \begin{Bmatrix} f_1 \\ f_2 \\ 0 \end{Bmatrix}$$

The transformation matrix at the element level is:

$$\begin{Bmatrix} U_2 \\ U_3 \\ U_1 \end{Bmatrix} = [r] \begin{Bmatrix} U_2 \\ U_3' \\ U_1 \end{Bmatrix}$$

$$[r] = \begin{matrix} 2 & 3 & 1 \\ \begin{bmatrix} 1 & 0 & 0 \\ 0 & -1 & 1 \\ 0 & 0 & 1 \end{bmatrix} \end{matrix} ; \quad [k^{(2)\prime}] = [r]^T [k^{(2)}][r]; \quad \{f^{(2)\prime}\} = [r]^T \{f^{(2)}\}$$

$$[K^{(2)\prime}] = \begin{bmatrix} k_{11} & -k_{12} & k_{12} \\ -k_{21} & k_{22} & -k_{22} \\ k_{21} & -k_{22} & k_{22} \end{bmatrix}; \quad \{f^{(2)\prime}\} = \begin{Bmatrix} f_2 \\ f_1 \\ -f_2 \end{Bmatrix}$$

Matrices $[K']$ and $\{F'\}$ of Example (4.21) are then obtained by substituting $[K^{(2)\prime}]$ to $[K^{(2)}]$ and $\{f^{(2)\prime}\}$ to $\{f^{(2)}\}$ during the standard assembly process.

4.8 Example: Poisson's Equation

Consider the problem already studied in Example 3.17; equation:

$$\frac{\partial^2 u}{\partial x^2} + \frac{\partial^2 u}{\partial y^2} + f_V = 0$$

is defined in a square of side 2 with boundary condition $u = 0$ on all sides.

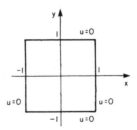

The element integral form from Example 4.1 is:

$$W^e = \int_{V^e} \langle \delta(\partial u) \rangle \{\partial u\} dV - \int_{V^e} \delta u\, f_V\, dV$$

$$\langle \delta(\partial u) \rangle = \left\langle \delta\left(\frac{\partial u}{\partial x}\right) \delta\left(\frac{\partial u}{\partial y}\right) \right\rangle$$

$$\langle \partial u \rangle = \left\langle \frac{\partial u}{\partial x}\, \frac{\partial u}{\partial y} \right\rangle$$

A finite element solution requires the following steps:
— Choice of element and mesh

Using the three node triangular element described in paragraph 4.3.1 and taking advantage of the double symmetry, we select the following mesh over $\frac{1}{8}$ of the total domain:

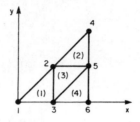

Nodal coordinates are:

	Nodes	1	2	3	4	5	6
CORG	x	0	0.5	0.5	1.0	1.0	1.0
	y	0	0.5	0	1.0	0.5	0

The connectivity table is:

	Elements	1	2	3	4
	node 1	3	5	2	6
CONEC	node 2	2	4	3	5
	node 3	1	2	5	3

— The boundary conditions are:
$$U_4 = U_5 = U_6 = 0$$

— Load vectors

We take $f_V = $ constant which could be the value for the following two specific situations:

- a membrane equilibrium problem with a uniform pressure;
- a heat conduction problem with a uniformly distributed heat source.
— Element matrices and vectors

All element matrices $[k]$ are identical to matrix (4.19) with $d = 1$, because the first node of the connectivity for each element is always the vertex at the right angle as in paragraph 4.31.

$$[k] = \frac{1}{2} \begin{bmatrix} 2 & -1 & -1 \\ -1 & 1 & 0 \\ -1 & 0 & 1 \end{bmatrix}$$

On the other hand, from (4.21):

$$\{f\} = \frac{f_V}{24}\begin{Bmatrix} 1 \\ 1 \\ 1 \end{Bmatrix}$$

— Assembly

After assembling the four elements and applying the boundary conditions on nodes 4, 5, and 6, the final system is:

$$\frac{1}{2}\begin{bmatrix} 1 & 0 & -1 \\ 0 & 4 & -2 \\ -1 & -2 & 4 \end{bmatrix}\begin{Bmatrix} U_1 \\ U_2 \\ U_3 \end{Bmatrix} = \frac{f_V}{24}\begin{Bmatrix} 1 \\ 2 \\ 3 \end{Bmatrix}$$

from which the solution is:

$$U_1 = \tfrac{30}{96}f_V = 0.3125 f_V$$
$$U_2 = \tfrac{17}{96}f_V = 0.1875 f_V$$
$$U_3 = \tfrac{22}{96}f_V = 0.2292 f_V$$

Consider the case of a concentrated load at node 1, of value f_c. This could be the load for the following two problems:

— equilibrium of a membrane under the action of a concentrated force at node 1;
— heat conduction with a point heat source at node 1.

The system matrix remains the same, but the load vector becomes:

$$\{F\} = \begin{Bmatrix} f_c \\ 0 \\ 0 \end{Bmatrix}$$

The corresponding solution is:

$$U_1 = 3.0 f_c$$
$$U_2 = 0.5 f_c$$
$$U_3 = 1.0 f_c$$

Important Results

Integral form:

$$W = \sum_{e=1}^{n_{el}} W^e = 0 \qquad (4.2a)$$

Discretized element integral form:

$$W^e = \langle \delta u_n \rangle ([k]\{u_n\} - \{f\}) \qquad (4.4)$$

Discretized global integral form:
$$W = \langle \delta U_n \rangle ([k]\{U_n\} - \{F\}) = 0 \tag{4.5b}$$

Mass matrix:
$$[m] = \int_{V^e} \{N\}\langle N \rangle dV \tag{4.6b}$$

Element matrix (stiffness matrix):
$$[k] = \int_{V^e} [B_\delta]^T [D][B] dV \quad \text{(see also (4.16a))} \tag{4.10b}$$

Element load vectors:
$$\{f\} = \int_{V^e} \{N\} f_V dV + \int_{S_f^e} \{N\} f_S dS \quad \text{(see also (4.16b))} \tag{4.10c}$$

Transformation of matrix $[B]$:
$$[B] = [Q][B_\xi] \tag{4.12}$$

Assembly rules:
$$K_{IJ}^e \equiv k_{ij} \quad \begin{array}{l} I = \text{LOCE}\,(i) \\ J = \text{LOCE}\,(j) \end{array} \tag{4.29a}$$

Transformation of variables:
$$[K'] = [R]^T [K][R] \quad \{F'\} = [R]^T \{F\} \tag{4.52}$$

Notation

b, h	width and height of the band of K
$[B]$	matrix relating the gradients in x and the nodal variables
$[B_\delta]$	matrix relating the variations of the gradients and the variations of the nodal variables
$[B_\xi]$	matrix relating the gradients in ξ and the nodal variables
$[D]$	matrix of material properties
e	index of an element
$\{f\}$	element load vectors
$\{F\}$	global load vector (or right-hand side)
$\{F^e\}$	expanded element load vector
$[k]$	element (or stiffness) matrix
$[K]$	global system matrix
$[K^e]$	expanded element (or stiffness) matrix
$[m]$	element mass matrix
n	total number of nodes
n_b	number of blocks stored for the global matrix

n_d	total number of degrees of freedom
n_{de}	number of degrees of freedom per element
n_{dn}	number of degrees of freedom per node
n_e	number of nodes per element
n_{el}	total number of elements
$[Q]$	transformation matrix for the gradients
$[T], [R], [r]$	transformation matrices for the nodal variables
$\{u_n\}$	nodal variables of an element
$\{U_n\}$	ensemble of all nodal variables
W	global integral form
W^e	element integral form
$\{\delta u_n\}$	variation of the nodal variables of an element
$\{\delta U_n\}$	variation of the ensemble of nodal variables
ξ_r, w_r	coordinates and weights of numerical integration points

References

[1]. B. M. Irons and A. A. Razzaque, Experience with the patch test, in *Mathematical Foundations of the Finite Element Method*, pp. 557–587, Academic Press, 1972.

[2] G. Strang and G. Fix, *An Analysis of Finite Element Method*, Prentice-Hall, New Jersey, 1973.

[3] K. Washizu, *Variational Methods in Elasticity and Plasticity*, 2nd edn, Pergamon Press, 1975.

5

NUMERICAL PROCEDURES

5.0 Introduction

An effective implementation of the finite element method in a computer code requires a variety of numerical methods to construct the necessary element matrices and to solve the resultant system of equations (Figure 5.1). This chapter describes the most frequently used numerical procedures for the construction of such codes.

We first review the numerical integration procedures most frequently used in the construction of element matrices defined in the space of the corresponding reference element. Various integration formulas are proposed for one, two-, and three-dimensional domains.

The second paragraph is devoted to the solution of linear systems of equation. Gaussian elimination techniques and some of its numerous variations are then described with an emphasis on decomposition algorithm for skyline matrices.

For non-linear systems, iteration methods, including the Newton–Raphson technique, are described.

Then, we describe the direct numerical integration methods (marching techniques) in the time domain used for solving time dependent problems of first and second order.

Finally, matrix eigenvalue and eigenvector iterative extraction methods are examined and we end this paragraph with the inverse iteration and the subspace iteration techniques.

5.1 Numerical Integration

5.1.1 *Introduction*

In the finite element method, element matrices $[k]$ and load vectors $\{f\}$ are expressed as integrals over lines, areas and volumes (4.10b) (4.10c) defined in the real element geometry V^e:

$$[k] = \int_{V^e} [B_\delta]^T [D][B] \, dV$$

$$\{f\} = \int_{V^e} \{N\} f_V \, dV + \int_{S_f^e} \{N\} f_S \, dS \qquad (5.1a)$$

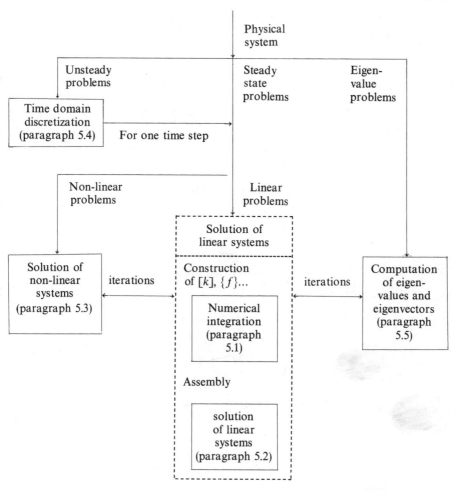

Figure 5.1 Numerical techniques used in the finite element method.

In the space of the reference element, these integrals become (4.16b) (4.16b):

$$[k] = \int_{V^r} [B_\delta(\xi)]^T [D(\xi)] [B(\xi)] \det(J(\xi)) \, dV^r$$

$$\{f\} = \int_{V^r} \{N(\xi)\} f_V \det(J(\xi)) \, dV^r + \int_{S_f^r} \{N(\xi_S)\} f_S \, dS \quad (5.1b)$$

where: V^r is the volume of the reference element;
S_f^r is the part of the boundary of the element of reference over which f_S is specified;
ξ_S represents the coordinates ξ on S_f^r;
$dS = J_S dS_1 dS_2$ (see paragraph 4.2.3.4);

[J] is the Jacobian matrix of the geometrical transformation (1.5.2) or, more compactly:

$$[k] = \int_{V^r} [k^*] dV^r$$

$$\{f\} = \int_{V^r} \{f_V^*\} dV^r + \int_{S_f^r} \{f_S^*\} dS \tag{5.2}$$

where:

$$[k^*] = [B_\delta(\xi)]^T [D(\xi)][B(\xi)] \det(J(\xi));$$
$$\{f_V^*\} = \{N(\xi)\} f_V \det(J(\xi));$$
$$\{f_S^*\} = \{N(\xi_s)\} f_S.$$

The terms of $[k^*]$, $\{f_V^*\}$ and $\{f_S^*\}$ are polynomials or complicated rational fractions. Explicit integration is possible for only the simplest of these expressions. We list some useful formulas for monomials:

— One dimension

$$\int_{-1}^{1} \xi^i d\xi = \begin{cases} 0 & \text{if } i \text{ is odd} \\ \dfrac{2}{i+1} & \text{if } i \text{ is even} \end{cases} \tag{5.3a}$$

— Two dimensions

  • Square element of reference

$$\int_{-1}^{1}\int_{-1}^{1} \xi^i \eta^j d\xi\, d\eta = \begin{cases} 0 & \text{if } i \text{ or } j \text{ is odd} \\ \dfrac{4}{(i+1)(j+1)} & \text{if } i \text{ and } j \text{ are even} \end{cases} \tag{5.3b}$$

  • Triangular element of reference

$$\int_{0}^{1}\int_{0}^{1-\xi} \xi^i \eta^j d\eta\, d\xi = \frac{i! j!}{(i+j+2)!} \tag{5.3c}$$

— Three dimensions

  • Cubic reference element

$$\int_{-1}^{1}\int_{-1}^{1}\int_{-1}^{1} \xi^i \eta^j \zeta^k d\xi\, d\eta\, d\zeta$$

$$= \begin{cases} 0 & \text{if } i,j, \text{ or } k \text{ is odd} \\ \dfrac{8}{(i+1)(j+1)(k+1)} & \text{if } i,j, \text{ and } k \text{ are even} \end{cases} \tag{5.3d}$$

- Tetrahedric element of reference

$$\int_0^1 \int_0^{1-\xi} \int_0^{1-\xi-\eta} \xi^i \eta^j \zeta^k \, d\zeta \, d\eta \, d\xi = \frac{i!j!k!}{(i+j+k+3)!} \qquad (5.3e)$$

For general cases where explicit integration is not possible, we then employ numerical integration formulas having the following generic form:

$$[k] = \sum_{i=1}^{r} w_i [k^*(\xi_i)]$$

$$\{f\} = \sum_{i=1}^{r} w_i \{f_v^*(\xi_i)\} \qquad (5.4)$$

where ξ_i are the coordinates of the r integration points; w_i are the weighting coefficients corresponding to each integration point.

In general, (5.4) gives an approximate value of the integrals; however, it has been shown that for many classes of problems the error of integration compensates the geometrical discretization error resulting in improved overall results.

5.1.2 One-Dimensional Numerical Integration [1, 2, 3]

5.1.2.1 Gauss method

In this method, r pairs of weights w_i and coordinates ξ_i are chosen so as to integrate exactly polynomials of degree $m \leq 2r - 1$.

Consider the following integration formula:

$$\int_{-1}^{1} y(\xi) \, d\xi = w_1 y(\xi_1) + w_2 y(\xi_2) + \cdots + w_i y(\xi_i) + \cdots + w_r y(\xi_r)$$

$$= \sum_{i=1}^{r} w_i y(\xi_i) \qquad (5.5)$$

The $2r$ coefficients $(w_1 \ldots w_r; \xi_1 \ldots \xi_r)$ are determined so as to integrate the following polynomial of degree $(2r - 1)$ exactly:

$$y(\xi) = a_1 + a_2 \xi + \cdots + a_{2r} \xi^{2r-1}$$

Substituting this last expression in (5.5):

$$a_1 \int_{-1}^{1} d\xi + a_2 \int_{-1}^{1} \xi \, d\xi + \cdots + a_{2r} \int_{-1}^{1} \xi^{2r-1} d\xi = a_1 (w_1 + w_2 + \cdots + w_r)$$

$$+ a_2 (w_1 \xi_1 + w_2 \xi_2 + \cdots + w_r \xi_r) + \cdots +$$

$$+ a_{2r} (w_1 \xi_1^{2r-1} + w_2 \xi_2^{2r-1} + \cdots + w_r \xi_r^{2r-1}) \qquad (5.6)$$

In order that the right-hand expression in (5.6) be equal to the explicit integration of the left-hand expression for arbitrary values of $a_1, a_2, \ldots, a_{2r}$, the coefficients w_i, ξ_i must satisfy the following $2r$ relations:

$$\int_{-1}^{1} \xi^\alpha \, d\xi = \frac{2}{\alpha+1} = \sum_{i=1}^{r} w_i \xi_i^\alpha \quad \alpha = 0, 2, 4, \ldots, 2r-2$$

$$\int_{-1}^{1} \xi^\alpha \, d\xi = 0 = \sum_{i=1}^{r} w_i \xi_i^\alpha \quad \alpha = 1, 3, 5, \ldots, 2r-1 \tag{5.7}$$

Then:

$$2 = w_1 + w_2 + \cdots + w_r$$
$$0 = w_1 \xi_1 + w_2 \xi_2 + \cdots + w_r \xi_r$$
$$\tfrac{2}{3} = w_1 \xi_1^2 + w_2 \xi_2^2 + \cdots + w_r \xi_r^2$$
$$\cdots$$
$$0 = w_1 \xi_1^{2r-1} + w_2 \xi_2^{2r-1} + \cdots + w_r \xi_r^{2r-1}$$

This system of $2r$ equations is linear in w_i but highly non linear ξ_i. The solution of the system is subject to the conditions:

$$\left. \begin{array}{c} w_i > 0 \\ -1 < \xi_i < 1 \end{array} \right\} \quad i = 1, 2, \ldots, r$$

Example 5.1   *Coefficients for the two Gauss point integration formula*

For this case $r = 2$ and (5.5) is:

$$\int_{-1}^{1} y(\xi) \, d\xi = w_1 y(\xi_1) + w_2 y(\xi_2)$$

If the formula is to give exact results for the integration of polynomials of degree $2r - 1 = 3$; equation (5.7) must be satisfied:

$$2 = w_1 + w_2$$
$$0 = w_1 \xi_1 + w_2 \xi_2$$
$$\tfrac{2}{3} = w_1 \xi_1^2 + w_2 \xi_2^2$$
$$0 = w_1 \xi_1^3 + w_2 \xi_2^3$$

The solution of the system is:

$$w_1 = w_2 = 1; \quad \xi_1 = -\xi_2 = \frac{1}{\sqrt{3}}$$

The solutions of (5.7) for ξ_i are equal to the roots of a Legendre polynomial of order r (see Davis and Rabinowitz [1] p. 88):

$$P_r(\xi) = 0$$

$$P_0(\xi) = 1$$
$$P_1(\xi) = \xi$$
$$\cdots$$
$$P_k(\xi) = \frac{2k-1}{k} \xi P_{k-1}(\xi) - \frac{k-1}{k} P_{k-2}(\xi); \quad k = 2, 3, \ldots, r$$
$$\tag{5.8}$$

The weights w_i are:

$$w_i = \frac{2(1-\xi_i^2)}{[rP_{r\,\delta 1}(\xi_i)]^2}; \quad i = 1, 2, \ldots, r \tag{5.9a}$$

The integration error is given by:

$$e = \frac{2^{2r+1}(r!)^4}{(2r+1)[(2r)!]^3} \frac{d^{2r}y}{d\xi^{2r}} \tag{5.9b}$$

Example 5.2   Numerical exercise

Let us compute the values of ξ_i and w_i already obtained in Example 5.1, using Legendre's polynomial (5.8):

$$P_0 = 1$$
$$P_1 = \xi$$
$$P_2 = \tfrac{3}{2}\xi^2 - \tfrac{1}{2}$$

The roots of $P_2 = 0$ are:

$$\xi_i = \pm \frac{1}{\sqrt{3}}$$

The weights from (5.9a) are:

$$w_i = \frac{2\left(1 - \dfrac{1}{3}\right)}{\left(2 \cdot \dfrac{1}{\sqrt{3}}\right)^2} = 1$$

Now let us use the two-point Gauss formula where the function is:

$$y = 1 + \xi^2 + \xi^3 + \xi^4$$

$$l_{\text{app}} = \left(1 + \frac{1}{3} + \frac{1}{3\sqrt{3}} + \frac{1}{9}\right) + \left(1 + \frac{1}{3} - \frac{1}{3\sqrt{3}} + \frac{1}{9}\right) = \frac{26}{9}$$

From (5.9b) the integration error is:

$$e = \frac{1}{135} \frac{d^4 y}{d\xi^4} = \frac{1}{135} \cdot 24 = \frac{8}{45}$$

Indeed:

$$\left| l_{\text{app}} - l_{\text{exact}} \right| = \left| \frac{26}{9} - \int_{-1}^{1} y(\xi) d\xi \right| = \left| \frac{130}{45} - \frac{138}{45} \right| = \frac{8}{45} \le e$$

Figure 5.2 gives all the coefficients w_i and ξ_i for the formulas having one to seven integration points [3]. Abscissas ξ_i are symmetric about the origin $\xi = 0$; the weights corresponding to two symmetric points are identical.

r	ξ_i	w_i	Error	Maximum polynomial degree for exact results
1	0	2	$\dfrac{1}{6}\dfrac{d^2y}{d\xi^2}$	1
2	$\pm 0.57735\,02691\,89626\,(\pm 1/\sqrt{3})$	1	$\approx 0.7 \times 10^{-2}\dfrac{d^4y}{d\xi^4}$	3
3	0 $\pm 0.77459\,66692\,41438\,(\pm\sqrt{3/5})$	$0.88888\,88888\,88889\,(8/9)$ $0.55555\,55555\,55556\,(5/9)$	$\approx 0.6 \times 10^{-4}\dfrac{d^6y}{d\xi^6}$	5
4	$\pm 0.33998\,10435\,84856\left(\pm\sqrt{\dfrac{3-2\sqrt{6/5}}{7}}\right)$ $\pm 0.86113\,63115\,94053\left(\pm\sqrt{\dfrac{3+2\sqrt{6/5}}{7}}\right)$	$0.65214\,51548\,62546\left(\dfrac{1}{2}+\dfrac{1}{6\sqrt{6/5}}\right)$ $0.34785\,48451\,37454\left(\dfrac{1}{2}-\dfrac{1}{6\sqrt{6/5}}\right)$	$\approx 0.3 \times 10^{-6}\dfrac{d^8y}{d\xi^8}$	7

$$\int_{-1}^{1} y(\xi) = \sum_{i=1}^{r} w_i y(\xi_i)$$

r	ξ_i	w_i	Error	Maximum polynomial degree for exact results
5	0 $\pm 0.53846\,93101\,05683$ $\pm 0.90617\,98459\,38664$	$0.56888\,88888\,88889\,(128/225)$ $\left(\dfrac{161}{450} + \dfrac{13}{180}\sqrt{5/14}\right)$ $\left(\dfrac{161}{450} - \dfrac{13}{180}\sqrt{5/14}\right)$	$\approx 0.8 \times 10^{-9} \dfrac{d^{10} y}{d\xi^{10}}$	9
6	$\pm 0.23861\,91860\,83197$ $\pm 0.66120\,93864\,66265$ $\pm 0.93246\,95142\,03152$	$0.46791\,39345\,72691$ $0.36076\,15730\,48139$ $0.17132\,44923\,79170$	$\approx 1.5 \times 10^{-12} \dfrac{d^{12} y}{d\xi^{12}}$	11
7	0 $\pm 0.40584\,51513\,77397$ $\pm 0.74153\,11855\,99394$ $\pm 0.94910\,79123\,42759$	$0.41795\,91836\,73469$ $0.38183\,00505\,05119$ $0.27970\,53914\,89277$ $0.12948\,49661\,68870$	$\approx 2.1 \times 10^{-15} \dfrac{d^{14} y}{d\xi^{14}}$	13

Figure 5.2 One-dimensional gauss quadrature.

Example 5.3 Integration of $[k]$ and $\{f\}$ for line elements using the Gauss integration scheme

Using a two-point Gauss formula, expressions (5.1b) become:

$$[k] \approx [B_\delta(\xi_1)]^T[[D].w_i.\det(J(\xi_1))][B(\xi_1)] +$$
$$+ [B_\delta(\xi_2)]^T[[D].w_2.\det(J(\xi_2))][B(\xi_2)]$$
$$\{f_V\} \approx \{N(\xi_1)\}(w_1.f_V.\det(J(\xi_1))) +$$
$$+ \{N(\xi_2)\}(w_2.f_V.\det(J(\xi_2)))$$

where:
$$\xi_1 = -\xi_2 = \frac{1}{\sqrt{3}}; \quad w_1 = w_2 = 1$$

It is possible to construct other numerical integration formulas for expressions containing a ponderation function $p(\xi)$:

$$\int_{-1}^{1} p(\xi)y(\xi)d\xi = \sum_{i=1}^{r} w_i y(\xi_i) \tag{5.10a}$$

For example, the Gauss–Jacobi method corresponds to a ponderation function equal to:

$$p(\xi) = (1 - \xi) \tag{5.10b}$$

If $p(\xi)$ is chosen equal to $1/(\xi - \xi_0)^\alpha$, we are then able to obtain integration formulas for functions that are singular at ξ_0.

5.1.2.2 Newton–Cotes method [1,2]

If the integration points are specified *a priori*, only 'r' weights $w_1, \ldots, w_r$ remain to be determined such that (5.5) integrates a polynomial of degree $r - 1$ exactly. In the Newton–Cotes method, points ξ_i are regularly spaced about the origin $\xi = 0$.

$$\xi_i = 2\frac{i-1}{r-1} - 1 \tag{5.11a}$$

To compute coefficients w_i, let $y(\xi)$ be given as a Lagrange polynomial of degree $(r - 1)$ taking values $y(\xi_i)$ at the 'r' points of integration ξ_i. Lagrange interpolation formulas have been obtained in paragraph 2.2.2.3.

$$y(\xi) = \sum_{i=1}^{r} N_i(\xi) y(\xi_i) \tag{5.11b}$$

Thus:

$$\int_{-1}^{1} y(\xi) = \sum_{i=1}^{r} \left(\int_{-1}^{1} N_i(\xi) d\xi . y(\xi_i) \right) = \sum_{i=1}^{r} w_i y(\xi_i)$$

The weighting coefficients w_i are then equal to the integrals of the Lagrange

r	ξ_i	w_i	Error	Maximum polynomial degree for exact results
2	± 1	1	$\dfrac{1}{6}\dfrac{d^2 y}{d\xi^2}$	1
3	0 ± 1	4/3 1/3	$\dfrac{1}{90}\dfrac{d^4 y}{d\xi^4}$	3
4	$\pm 1/3$ ± 1	3/4 1/4	$\dfrac{2}{405}\dfrac{d^4 y}{d\xi^4}$	3
5	0 $\pm 1/2$ ± 1	12/45 32/45 7/45	$\approx 0.6 \times 10^{-5}\dfrac{d^6 y}{d\xi^6}$	5
6	$\pm 1/5$ $\pm 3/5$ ± 1	50/144 75/144 19/144	$\approx 3.7 \times 10^{-5}\dfrac{d^6 y}{d\xi^6}$	5
7	0 $\pm 1/3$ $\pm 2/3$ ± 1	272/420 27/420 216/420 41/420	$\approx 3.2 \times 10^{-7}\dfrac{d^8 y}{d\xi^8}$	7
9	0 $\pm 1/4$ $\pm 1/2$ $\pm 3/4$ ± 1	$-4\,540/14175$ $10\,496/14175$ $-928/14175$ $5888/14175$ $989/14175$	$\approx 1.2 \times 10^{-9}\dfrac{d^{10} y}{d\xi^{10}}$	9

$$\int_{-1}^{1} y(\xi)d\xi = \sum_{i=1}^{r} w_i y(\xi_i)$$

Figure 5.3 One-dimensional Newton–Cotes formulas for numerical integration.

interpolation functions N_i:

$$w_i = \int_{-1}^{1} N_i(\xi) d\xi \tag{5.11c}$$

The weights of two symmetrical points about the origin $\xi = 0$ are equal. The integration error is of the form:

$$e = C_r \left(\frac{2}{r-1}\right)^{r+2} \frac{d^{r+1}y}{d\xi^{r+1}} \quad \text{if } r \text{ is odd}$$

$$e = C_r \left(\frac{2}{r-1}\right)^{r+1} \frac{d^r y}{d\xi^r} \quad \text{if } r \text{ is even} \tag{5.12}$$

An odd number of integration points is then to be preferred in this method. Coefficient C_r can be evaluated by integrating the approximation error of Lagrange given in (1.68).

Figure 5.3 contains the coefficients required by the Newton–Cotes formula with $2, 3, \ldots, 9$ points.

For a given number of integration points, the Newton–Cotes formulas does not perform as well as the Gauss method for polynomial functions (see comparisons in Figure 5.4). However, the formula allows integration points to coincide with the interpolation points. The integration of expressions containing interpolation functions N_i is thus greatly simplified since functions N_i are null at all nodes other than ξ_i. Such a technique can be used efficiently for mass matrices:

$$[m] = \int_{V_e} \{N\}\langle N \rangle dV$$

Mono-mials	Exact integration	Newton–Cotes				Gauss				
	Number of integration points r	2	3	4	5	1	2	3	4	5
1	$\int_{-1}^{1} 1.d\xi = 2$	2	2	2	2	2	2	2	2	2
ξ	0	0	0	0	0	0	0	0	0	0
ξ^2	2/3	2	2/3	2/3	2/3	0	2/3	2/3	2/3	2/3
ξ^3	0	0	0	0	0	0	0	0	0	0
ξ^4	2/5	2	2/3	14/27	2/5	0	2/9	2/5	2/5	2/5
ξ^5	0	0	0	0	0	0	0	0	0	0
ξ^6	2/7	2	2/3	122/243	1/3	0	2/27	6/25	2/7	2/7

Figure 5.4 Comparison of numerical integration formulas for monomials between -1 and 1, using Newton–Cotes and Gauss methods.

and for load vectors:

$$\{f\} = \int_{V^e} \{N\} f_V \, dV$$

Remarks

Numerical integration methods have been described for the space of the element of reference. Integration over the real element may be easily transformed to the reference element using (1.44).

$$\int_{x_1}^{x_2} y(x) \, dx = \frac{x_2 - x_1}{2} \int_{-1}^{1} y(\xi) \, d\xi \tag{5.13}$$

where:

$$x = \frac{1-\xi}{2} x_1 + \frac{1+\xi}{2} x_2$$

The Gauss method with r points integrates polynomial of degree $2r - 1$ exactly. The Newton–Cotes formula with r points integrates polynomials of degree $r - 1$ exactly. Since in practice the functions to integrate are very often polynomials of very high degree or non-polynomial functions (like rational polynomial fractions) the methods described will give approximate values of the integrals. The error of integration decreases as the number of integration points increases.

Example 5.4 Integration of a non-polynomial function using both Gauss and Newton–Cotes formulas

The exact value of

$$I = \int_{-1}^{1} \frac{1}{1+\xi^2} d\xi$$

is

$$I = \frac{\pi}{2} \approx 1.5708$$

The tabulated results shown below are obtained:

Number of points	Gauss		Newton–Cotes	
	value	relative error	value	relative error
1	2	27%	2	27%
2	1.5	4.5%	1	36%
3	1.5833	0.8%	1.6666	6%
4	1.5686	0.1%	1.6	2%

Instead of using a very large number of integration points over an entire domain, it is often advantageous to divide the domain into many sub-domains over which a smaller number of integration points may be used. The latter technique is essential in cases where functions $y(\xi)$ are sectionally discontinuous within the total domain.

5.1.3 *Numerical Integration in Two Dimensions* [1, 2]

(a) *Bidirectional methods*

In this method, a one-dimensional formula is selected for both ξ and η. If we have r_1 points in direction ξ and r_2 points in direction η, the Gauss method integrates the product of a $(2r_1 - 1)$ polynomial in ξ by a $(2r_2 - 1)$ polynomial in η, exactly. The method uses $r = r_1 . r_2$ points; it integrates all monomials:

$$\xi^i \eta^j \quad \text{such that} \quad 0 \leq i \leq 2r_1 - 1$$
$$0 \leq j \leq 2r_2 - 1$$

(b) *Direct methods*

The methods of paragraph 5.1.2 can also be directly extended in two dimensions:

$$\iint_{V^r} y(\xi, \eta) \, d\xi \, d\eta = \sum_{i=1}^{r} w_i y(\xi_i, \eta_i) \tag{5.14}$$

In particular, it is possible to construct formulas very similar to Gauss formulas that will integrate exactly all nomomials of order m:

$$\xi^i \eta^j \quad \text{such that} \quad i + j \leq m$$

Such methods are often more effective than bidirectional methods (see Stroud [2]). For quadrilateral elements, bidirectional formulas have less geometrical bias than direct methods; on the other hand, for triangular elements, direct methods have less geometrical bias than bidirectional methods.

5.1.3.1 *Quadrilateral elements*

The bidirectional formula is:

$$\int_{-1}^{1} \int_{-1}^{1} y(\xi, \eta) \, d\xi \, d\eta = \sum_{i=1}^{r_1} \sum_{j=1}^{r_2} w_i w_j y(\xi_i, \eta_j) \tag{5.15}$$

where w_i, w_j are the weighting coefficients given in Figure 5.2 (Gauss) or Figure 5.3 (Newton–Cotes); and ξ_i, η_j are the reference space coordinates of the corresponding points of integration.

Figure 5.5 shows a few direct formulas.

253

Example 5.5   Integration of [k] *by the Gauss bidirectional method*

For a 2 × 2 formula we get:

$$[k] \approx [B_\delta(\xi_1\eta_1)]^T[[D].w_1.w_1.\det(J(\xi_1\eta_1))][B(\xi_1\eta_1)]$$
$$+ [B_\delta(\xi_1\eta_2)]^T[[D].w_1.w_2.\det(J(\xi_1\eta_2))][B(\xi_1\eta_2)]$$
$$+ [B_\delta(\xi_2\eta_1)]^T[[D].w_2.w_1.\det(J(\xi_2\eta_1))][B(\xi_2\eta_1)]$$
$$+ [B_\delta(\xi_2\eta_2)]^T[[D].w_2.w_2.\det(J(\xi_2))][B(\xi_2\eta_2)]$$

	Order m	Number of points r	Coordinates ξ_i	η_i	Weights w_i
	2	3	$\sqrt{2/3}$ $-\dfrac{1}{\sqrt{6}}$	0 $\pm\dfrac{1}{\sqrt{2}}$	4/3 4/3
	2	3	1 $-5/9$ $1/3$	1 $2/9$ $-2/3$	4/7 27/14 3/2
	3	4 (2 × 2 points)	$\pm\dfrac{1}{\sqrt{3}}$	$\pm\dfrac{1}{\sqrt{3}}$	1
	3	4	± 1 0	0 $\pm\dfrac{1}{\sqrt{2}}$	2/3 4/3
	3	4	$\pm\sqrt{2/3}$ 0	0 $\pm\sqrt{2/3}$	1

	Order m	Number of points	Coordinates ξ_i	η_i	Weights w_i
	5	7	0 0 $\pm\sqrt{3/5}$	0 $\pm\sqrt{14/15}$ $\pm\sqrt{3/5}$	8/7 20/63 20/36
	5	7	0 $-r$ $+r$ $+s$ $-s$ $+t$ $-t$ $r=\sqrt{7/15}$ $s=\sqrt{\dfrac{7+\sqrt{24}}{15}}$ $t=\sqrt{\dfrac{7-\sqrt{24}}{15}}$	0 $-r$ $+r$ $-r$ $+t$ $-s$ $+s$ ≈ 0.683 ≈ 0.89 ≈ 0.374	8/7 $\left.\begin{array}{c}\\ \\ \end{array}\right\}$ 100/168 $\left.\begin{array}{c}\\ \\ \\ \end{array}\right\}$ 20/48

$$\int_{-1}^{1}\int_{-1}^{1} y(\xi,\eta)d\xi\, d\eta \approx \sum_{i=1}^{r} w_i y(\xi_i,\eta_i).$$

Formulas integrating exactly, polynomials of order m including terms like $\xi^i\eta^j$ such that —$(i+j\leq m)$.

Figure 5.5 Direct integration formulas over a square.

where $w_1 = w_2 = 1$, and

$$\xi_1 = \eta_1 = -\xi_2 = -\eta_2 = \frac{1}{\sqrt{3}}$$

Note that it is always preferable to multiply $[D]$ by the scalar factor $w_i w_j \det(J)$ rather than the triple matrix product $[B_\delta]^T [D][B]$ because $[D]$ contains far fewer terms the $[k]$.

Consider an eight-node element for Laplace equation. $[B]$ is a 2 × 8 matrix and $[D]$ is a 2 × 2 matrix and symmetric. For each point of integration, there will then be a number of multiplications equal to:

      2   for $c = w_1 . w_2 . \det(J)$
+ 3   for $[D.c]$
+ 32 for $[D.c][B]$
+ 72 for $[B]^T([D.c][B])$ (symmetrical)
_____
  109   Total

For four integration points we then have 436 multiplications without counting the operations required by the construction of $[B]$, $[D]$, and $\det(J)$. This number of multiplications can be significantly reduced if the zeros contained in $[B]$ and $[D]$ are taken into account.

Since the matrix $[D]$ is positive definite, it can be decomposed into a product of a lower triangular by an upper triangular matrix (5.38):

$$[D] = [L][L]^T$$

then, compute

$$([B]^T[L]) \quad ([L]^T[B])$$

for another reduction of the number of operations.

5.1.3.2 Triangular element

(a) Gauss–Radau method [4]

The bidirectional method requires that the integration over a triangle be transformed into an integration over a square (5.15) using a transformation of variables:

$$l = \int_0^1 \int_0^{1-\bar{\xi}} y(\bar{\xi}, \bar{\eta}) \, d\bar{\eta} \, d\bar{\xi} \tag{5.16}$$

Let us use a transformation that maps a square region onto a triangular region:

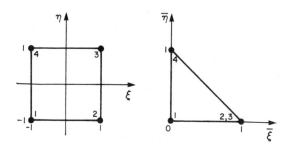

$$\bar{\xi} = \langle N \rangle \{\bar{\xi}_n\}$$
$$\bar{\eta} = \langle N \rangle \{\bar{\eta}_n\}$$

where

$$\langle \bar{\xi}_n \rangle = \langle 0 \quad 1 \quad 1 \quad 0 \rangle$$
$$\langle \bar{\eta}_n \rangle = \langle 0 \quad 0 \quad 0 \quad 1 \rangle$$

$\langle N \rangle$ are the expressions given in Example 1.16:

$$\bar{\xi} = \frac{1+\xi}{2}$$

$$\bar{\eta} = \frac{1-\xi}{2}\frac{1+\eta}{2}$$

$$d\bar{\xi}\,d\bar{\eta} = \det(J)\,d\xi\,d\eta = \frac{1-\xi}{2}d\xi\,d\eta \tag{5.17}$$

Integral (5.16) becomes:

$$I = \frac{1}{2}\int_{-1}^{1}(1-\xi)\int_{-1}^{1}y(\bar{\xi}(\xi),\bar{\eta}(\xi,\eta))\,d\eta\,d\xi$$

Using a Gauss integration formula in direction η gives:

$$I = \int_{-1}^{1}(1-\xi)\left[\sum_{j=1}^{r_2}\frac{w_j}{2}y(\bar{\xi}(\xi),\bar{\eta}(\xi,\eta_j))\right]d\xi \tag{5.18}$$

In direction ξ we use a Gauss–Jacobi formula with integration points ξ_i:

$$I = \sum_{i=1}^{r_1}\sum_{j=1}^{r_2}w_i\frac{w_j}{2}y(\bar{\xi}(\xi'_i),\bar{\eta}(\xi'_i,\eta_j)) \tag{5.19}$$

The Gauss–Radau method does not have geometrical isotropy and its use is generally restricted to functions having steep gradients in the neighbourhood of a vertex like A.

Figure 5.6 gives the weights and coordinates of the integration points for the Gauss–Radau method.

(b) *Direct method* [2, 5]

Figure 5.7 gives coefficients for direct formulas of order $m = 1, 2, \ldots, 6$ capable of exact integration of monomials $\xi^i\eta^j$ with $i+j \leq M$. These formulas are generally attributed to Hammer.

5.1.4 Numerical Integration in Three Dimensions [2,5,6,7]

5.1.4.1 Cubic (or brick) Elements

A tridirectional formula is obtained from a direct extension of the bidirectional formula:

$$\int_{-1}^{1}\int_{-1}^{1}\int_{-1}^{1}y(\xi,\eta,\zeta)\,d\xi\,d\eta\,d\zeta = \sum_{i=1}^{r_1}\sum_{j=1}^{r_2}\sum_{k=1}^{r_3}w_iw_jw_ky(\xi_i,\eta_j,\zeta_k) \tag{5.20}$$

Integration order in ξ^i or η^j	Number of points $r \times r$	RI	WI	SJ	AJ
1	1×1	0.5	1.0	0.3333333333	0.75
3	2×2	0.2113248654	0.5	0.1550510257	0.3764030627
		0.7886751346	0.5	0.6449489743	0.5124858262
5	3×3	0.1127016654	0.2777777778	0.0885879595	0.2204622112
		0.5	0.4444444444	0.4094668644	0.3881934688
		0.8872983346	0.2777777778	0.7876594618	0.3288443200
7	4×4	0.0694318442	0.1739274226	0.0571041961	0.1437135608
		0.3300094782	0.3260725774	0.2768430136	0.2813560151
		0.6699905218	0.3260725774	0.5835904324	0.3118265230
		0.9305681558	0.1739274226	0.8602401357	0.2231039011
9	5×5	0.0469100770	0.1184634425	0.0398098571	0.1007941926
		0.2307653449	0.2393143353	0.1980134179	0.2084506672
		0.5	0.2844444444	0.4379748102	0.2604633916
		0.7692346551	0.2393143353	0.6954642734	0.2426935942
		0.9530899230	0.1184634425	0.9014649142	0.1598203766

where

$$\int_1^1 \int_0^{1-\xi} y(\xi, \eta) d\eta d\xi = \sum_{j=1}^{r} \sum_{i=1}^{r} WJ(j).WI(i).y(\xi_j, \eta_{ij})$$

$$WJ = AJ(j)(1 - SJ(j))$$
$$\xi_j = SJ(j)$$
$$\eta_{ij} = RI(i)(1 - SJ(j)).$$

RI and WI are the coefficients of the Gauss quadrature formula over the span (0, 1)

$$\int_0^1 y(\xi) d\xi = \sum_{i=1}^{r} WI(i) y(RI)(i)).$$

Figure 5.6 Gauss–Radau formulas for a triangle.

where w_i, w_j, w_k are coefficients to be found in Figure 5.2 (Gauss) or Figure 5.3 (Newton–Cotes).

ξ_i, η_j, ζ_k are the coordinates of the integration points from Figure 5.2 or 5.3.

The total number of integration points is equal to $r_1.r_2.r_3$; it gives exact results for monomials $\xi^i \eta^j \zeta^k$ such that $i \leq 2r_1 - 1$, $j \leq 2r_2 - 1$, $k \leq 2r_3 - 1$. Figure 5.8 gives coefficients required for direct formulas of order $m = 2, 3, 5, 7$ capable of integrating monomials $\xi^i \eta^j \zeta^k$ exactly, provided $i + j + k \leq m$.

5.1.4.2 Tetrahedronal Elements

A tridirectional formula is rarely used for such geometry. Figure 5.9 gives the coefficients required for direct formulas of order $m = 1, 2, 3, 5$.

5.1.5 *Approximate Integration*

In general, for curved geometry, the functions to be integrated cannot be expressed in simple polynomials. In such cases, the numerical integration formulas will give approximate values to the integrals to be evaluated. In mechanics of solids, it has been observed that numerical integration errors generally compensate for the geometrical discretization error for an overall beneficial result. For rectangular and simple triangular elements with straight sides, however, it is generally possible to obtain exact results by numerical integration since the finite element process normally reduces the functions to be integrated, to simple polynomials.

	Order m	Number of points	Coordinates	η_i	Weights w_i
	1	1	1/3	1/3	1/2
	2	3	1/2 0 1/2	1/2 1/2 0	1/6
	2	3	1/6 2/3 1/6	1/6 1/6 2/3	1/6
	3	4	1/3 1/5 3/5 1/5	1/3 1/5 1/5 3/5	−27/96 25/96

Figure 5.7 (*contd.*)

259

	Order m	Number of points r	Coordinates ξ_i	η_i	Weights w_i
	4	6	a	a	
	$a = 0.445948490915965$		$1-2a$	a	$\Big\}\ 0.111690794839005$
			a	$1-2a$	
	$b = 0.091576213509771$		b	b	
			$1-2b$	b	$\Big\}\ 0.054975871827661$
			b	$1-2b$	
	5	7	$1/3$	$1/3$	$9/80$
	$a = \dfrac{6+\sqrt{15}}{21}$		a	a	$\Big\} A = \dfrac{155+\sqrt{15}}{2400}$
	$= 0.470142064105115$		$1-2a$	a	$= 0.0661970763942530$
	$b = \dfrac{4}{7} - a$		a	$1-2a$	$\Big\{ \dfrac{31}{240} - A =$
	$= 0.101286507323456$		b	b	$= 0.0629695902724135$
			$1-2b$	b	
			b	$1-2b$	
	6	12	a	a	
			$1-2a$	a	$\Big\}\ 0.025422453185103$
	$a = 0.063089014491502$		a	$1-2a$	
			b	b	
	$b = 0.249286745170910$		$1-2b$	b	$\Big\}\ 0.058393137863189$
			b	$1-2b$	
	$c = 0.310352451033785$		c	d	
			d	c	
	$d = 0.053145049844816$		$1-(c+d)$	c	$\Big\}\ 0.041425537809187$
			$1-(c+d)$	d	
			c	$1-(c+d)$	
			d	$1-(c+d)$	

$$\int_0^1 \int_0^{1-\xi} y(\xi,\eta) d\eta\, d\xi \cong \sum_{i=1}^r w_i y(\xi_i, \eta_i).$$

Formulas for the exact integration of polynomials of order m ($\xi^i \eta^j$ with $i+j \leq m$).

Figure 5.7 Direct integration formulas for a triangle (Hammer).

Example 5.6 Integration formulas for exact integration over an eight-node rectangular element valid for Laplace equation

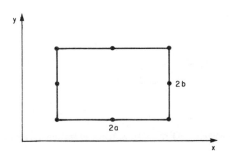

Order m	Number of points r	Coordinates ξ_i	η_i	ζ_i	Weights w_i
2	4	0 $\pm\sqrt{\frac{2}{3}}$	$\pm\sqrt{\frac{2}{3}}$ 0	$-\frac{1}{\sqrt{3}}$ $\frac{1}{\sqrt{3}}$	2
3	6	$\frac{1}{\sqrt{6}}$ $-\frac{1}{\sqrt{6}}$ $-\sqrt{\frac{2}{3}}$ $\sqrt{\frac{2}{3}}$	$\pm\frac{1}{\sqrt{2}}$ $\pm\frac{1}{\sqrt{2}}$ 0 0	$-\frac{1}{\sqrt{3}}$ $\frac{1}{\sqrt{3}}$ $-\frac{1}{\sqrt{3}}$ $\frac{1}{\sqrt{3}}$	$4/3$
3	6	± 1 0 0	0 ± 1 0	0 0 ± 1	$4/3$
5	14	$\pm a$ 0 0 $\pm b$	0 $\pm a$ 0 $\pm b$	0 0 $\pm a$ $\pm b$	$320/361$ $121/361$
		$a = \sqrt{\frac{19}{30}}$	$b = \sqrt{\frac{19}{33}}$		
7	34	$\pm a$ 0 0	0 $\pm a$ 0	0 0 $\pm a$	0.2957475994 51303
$a = 0.92582 00997 72552$		$\pm a$ 0 $\pm a$	$\pm a$ $\pm a$ 0	0 $\pm a$ $\pm a$	0.09410 15089 16324
$b = 0.33081 49636 99288$		$\pm b$	$\pm b$	$\pm b$	0.41233 38622 71436
$c = 0.73412 52878 52115$		$\pm c$	$\pm c$	$\pm c$	0.22470 31747 65601

$$\int_{-1}^{1}\int_{-1}^{1}\int_{-1}^{1} y(\xi,\eta,\zeta)\,\mathrm{d}\xi\,\mathrm{d}\eta\,\mathrm{d}\zeta = \sum_{i=1}^{r} w_i y(\xi_i,\eta_i,\zeta_i)$$

Figure 5.8 Direct integration formulas over a cube.

Order m	Number of points r	Coordinates $\xi_i \quad \eta_i \quad \zeta_i$	Weights w_i
1	1	$\frac{1}{4} \quad \frac{1}{4} \quad \frac{1}{4}$	$\frac{1}{6}$
2	4 $a = \frac{5-\sqrt{5}}{20}$ $b = \frac{5+3\sqrt{5}}{20}$	$a \quad a \quad a$ $a \quad a \quad b$ $a \quad b \quad a$ $b \quad a \quad a$	$\frac{1}{24}$
3	5 $a = \frac{1}{4}$ $b = \frac{1}{6}$ $c = \frac{1}{2}$	$a \quad a \quad a$ $b \quad b \quad b$ $b \quad b \quad c$ $b \quad c \quad b$ $c \quad b \quad b$	$-\frac{2}{15}$ $\frac{3}{40}$
5	15 $a = \frac{1}{4}$ $a = \frac{1}{4}$ $\frac{b_1}{b_2} = \frac{7-\sqrt{15}}{34}$ $\frac{c_1}{c_2} = \frac{13+3\sqrt{15}}{34}$ $d = \frac{5-\sqrt{15}}{20}$ $e = \frac{5+\sqrt{15}}{20}$	$a \quad a \quad a$ $b_i \quad b_i \quad b_i$ $b_i \quad b_i \quad b_i$ $b_i \quad b_i \quad c_i \quad i=1,2$ $b_i \quad c_i \quad b_i$ $c_i \quad b_i \quad b_i$ $d \quad d \quad e$ $d \quad e \quad d$ $e \quad d \quad d$ $d \quad e \quad e$ $e \quad d \quad e$ $e \quad e \quad d$	$\frac{8}{405}$ or $\frac{112}{5670}$ $\frac{w_1}{w_2} = \frac{2665 + 14\sqrt{15}}{226800}$ $\frac{5}{567}$

$$\int_0^1 \int_0^{1-\xi} \int_0^{1-\xi-\eta} y(\xi,\eta,\zeta)\,d\zeta\,d\eta\,d\xi = \sum_{i=1}^{r} w_i y(\xi_i, \eta_i, \zeta_i)$$

Figure 5.9 Direct integration formulas over a tetrahedron.

$$[k] = \int_{-1}^{1}\int_{-1}^{1} [B]^T[D][B] \det(J) \, d\xi \, d\eta$$

$$[B] = \frac{1}{\det(J)} \begin{bmatrix} b\left\langle \dfrac{\partial N}{\partial \xi} \right\rangle \\ a\left\langle \dfrac{\partial N}{\partial \eta} \right\rangle \end{bmatrix} ; \quad D = \begin{bmatrix} 1 & 0 \\ 0 & 1 \end{bmatrix} ; \quad \det(J) = ab$$

Functions $\langle N \rangle$ are those of paragraph 2.4.3.2; they contain the following eight monomials:

$$1 \quad \xi \quad \eta \quad \xi^2 \quad \xi\eta \quad \eta^2 \quad \xi^2\eta \quad \xi\eta^2$$

Functions $\partial N/\partial \xi$ and $\partial N/\partial \eta$ shall then contain the following monomials:

For $\dfrac{\partial N}{\partial \xi}$: 0 1 0 2ξ η 0 $2\xi\eta$ η^2

For $\dfrac{\partial N}{\partial \eta}$: 0 0 1 0 ξ 2η ξ^2 $2\xi\eta$

The matrix product $[B]^T[D][B]$ shall then contain monomials of no more than fourth order ($\xi^i\eta^j$; $i+j \leq 4$). Using a Gauss formula, only 3×3 points are required to produce exact results for $[k]$. Frequently, 2×2 point approximations produce improved overall results. In such cases, the formula is said to be a reduced integration formula.

Consider a load vector of the following type:

$$\{f\} = \int_{-1}^{1}\int_{-1}^{1} \{N\} f \det(J) \, d\xi \, d\eta$$

If f is constant, we must use 2×2 Gauss point or a third order direct formula to obtain exact results. For a load vector requiring integration in only one direction, for example, f constant and specified on side $\xi = 1$ of the element

$$\{f\} = \int_{-1}^{1} \{N(\xi = 1, \eta)\} f b \, d\eta$$

only two Gauss points are needed.

Example 5.7 *Formulas for an irregular eight-node quadrilateral element*

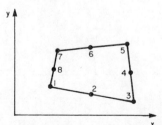

The Jacobian matrix given in Example 1.18 is:

$$[J] = \begin{bmatrix} J_{11} & J_{12} \\ J_{21} & J_{22} \end{bmatrix} = \begin{bmatrix} a_1 + b_1\eta & a_2 + b_2\eta \\ a_3 + b_3\xi & a_4 + b_4\xi \end{bmatrix}$$

$$\det(J) = A_0 + A_1\xi + A_2\eta$$

Then:

$$[B] = \begin{bmatrix} \left\langle \dfrac{\partial N}{\partial x} \right\rangle \\ \left\langle \dfrac{\partial N}{\partial y} \right\rangle \end{bmatrix} = \dfrac{1}{\det(J)} \begin{bmatrix} J_{22}\left\langle \dfrac{\partial N}{\partial \xi} \right\rangle - J_{12}\left\langle \dfrac{\partial N}{\partial \eta} \right\rangle \\ -J_{21}\left\langle \dfrac{\partial N}{\partial \xi} \right\rangle + J_{11}\left\langle \dfrac{\partial N}{\partial \eta} \right\rangle \end{bmatrix}$$

$$= \dfrac{1}{\det(J)}[B']$$

$$[k] = \int_{-1}^{1}\int_{-1}^{1} \dfrac{1}{\det(J)}[B']^T[D][B']\,d\xi\,d\eta$$

The terms of matrix $[k]$ are rational fractions: the numerator contains terms of degrees as high as four in both ξ and η; the denominator is linear in ξ and η. The term $1/\det(J)$ can be developed in series as follows:

$$\dfrac{1}{\det(J)} = \dfrac{1}{A_0 + A_1\xi + A_2\eta} = \dfrac{1}{A_0}\left(1 + \dfrac{A_1}{A_0}\xi + \dfrac{A_2}{A_0}\eta\right)^{-1}$$

$$= \dfrac{1}{A_0}\left(1 - \dfrac{A_1}{A_0}\xi - \dfrac{A_2}{A_0}\eta + \cdots\right)$$

The accuracy of the integration is affected by the ratios $A1/A_0$ and $A2/A_0$ characterizing the deformation of the element.

The following example shows the influence of the geometrical deformation of an element in the numerical integration formulas.

Example 5.8 Numerical integration over a four-node trapezoidal element

Consider the following element:

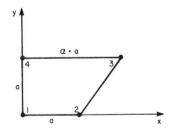

The Jacobian matrix of the geometrical transformation, given in Example 1.18, becomes:

$$J = A_0 + A_1\xi + A_2\eta$$

$$A_0 = a^2\frac{\alpha+1}{8}$$

$$A_1 = 0$$

$$A_2 = a^2\frac{\alpha-1}{8}$$

Integrating numerically:

$$l = \int_{-1}^{1}\int_{-1}^{1}\frac{1}{\det(J)}d\xi\,d\eta = \frac{16}{a^2}\int_{-1}^{1}\frac{1}{(\alpha+1)+(\alpha-1)\eta}d\eta$$

$$= \frac{16}{a^2}\frac{1}{\alpha-1}\operatorname{Log}\alpha$$

The values of $a^2 l/16$ are given below for various numerical values of α and a various number of integration points.

r \ α	2	3	5	10
1	0.666 66	0.50	0.333 33	0.181 82
2	0.692 31	0.545 45	0.391 30	0.234 04
3	0.693 12	0.549 02	0.400 67	0.249 62
exact	0.693 15	0.549 31	0.402 36	0.255 84

5.1.6 Choice of the Number of Integration Points [8,9]

The choice of the number of integration points depends upon the type of element used, the particular matrix to be constructed ($[k]$ or $[m]$) and the geometry of the real element. In general, since the number of operations required is equal to N^d, where N is the number of integration points and d an integer corresponding to the number of dimensions of the element, it is important to reduce N to its smallest acceptable value. It has been shown that approximations obtained by reduced integration [8] give more accurate overall results than exact integration under certain circumstances. Reduced integrations, however, may lead to spurious zero strain deformation patterns for non-rigid body motions. Under such conditions, the overall system matrix of coefficients may remain singular even after the application of boundary conditions. A minimum number of integration points is required to prevent the development of spurious non-straining nodes; for example, an eight-node isoparametric two-dimensional element needs at least 2×2 points for matrix $[k]$ and 3×3 points for $[m]$. Zienkiewicz [8] states that the minimum number of integration points is that which would integrate det(J) accurately. This last condition, although necessary, is not always sufficient; users should be aware of possible problems for a mesh consisting of a very small number of elements. At every integration point, one or more equations exist between the nodal variables of the element. The minimum number of integration points must always be such that there is a balance between the total number of equations established through these points and the total number of unknowns after boundary conditions are applied to prevent all rigid-body motion.

5.1.7 *Numerical Integration Code*

Codes for numerical integration should be able to accommodate either multilinear or direct integration formulas. Casting all integration formulas under the following form

$$I = \sum_{l=1}^{IPG} w_l y(\xi_l) \tag{5.21}$$

we have

— for one-dimensional formulas: $w_l = w_i$, IPG $= r$;
— for two-dimensional:
  • bilinear formulas: $w_l = w_i w_j$; IPG $= r_1, r_2$;
  • direct formulas: $w_l = w_i$, IPG $= r$;
— for three-dimensional:
  • trilinear formulas: $w_l = w_i w_j w_k$, IPG $= r_1.r_2.r_3$;
  • direct formulas: $w_l = w_i$, IPG $= r$;

where ξ_l are the coordinates of integration points 'l' corresponding to weight w_l; w_l is the weight corresponding to integration point number l; and IPG is the total number of integration points.

The computation process can be coded in a single loop according to equation (5.21).

In the code of Figure 5.10, the weights w_l are stored in the array of weighting coefficients VCPG of length IPG, and the coordinates ξ_i are stored in the array of coordinates of integration points VKPG of length IPG·NDIM where NDIM

```
      SUBROUTINE GAUSS(IPGKED,NDIM,VKPG,VCPG,IPG)            GAUS  1
C========================================================================GAUS  2
C     TO FORM ARRAYS OF COORDINATES AND WEIGHTS AT GAUSS POINTS    GAUS  3
C     (1,2 AND 3 DIMENSIONS)(1,2,3 OR 4 G.P. PER DIMENSION)        GAUS  4
C        INPUT                                                     GAUS  5
C           IPGKED  NUMBER OF POINTS IN KSI,ETA,ZETA DIRECTIONS    GAUS  6
C           NDIM    NUMBER OF DIMENSIONS (1,2 OR 3)                GAUS  7
C        OUTPUT                                                    GAUS  8
C           VKPG    COORDINATES OF GAUSS POINTS                    GAUS  9
C           VCPG    WEIGHTS AT GAUSS POINTS                        GAUS 10
C           IPG     TOTAL NUMBER OF GAUSS POINTS                   GAUS 11
C========================================================================GAUS 12
```

Figure 5.11 (*Contd*.)

```
      IMPLICIT REAL*8(A-H,O-Z)                                       GAUS 13
      DIMENSION IPGKED(1),VKPG(1),VCPG(1),G(10),P(10),INDIC(4)        GAUS 14
      DATA INDIC/1,2,4,7/                                             GAUS 15
      DATA G/0.0D0,-.577350269189626D0,.577350269189626D0,            GAUS 16
     1      -.774596669241483D0,0.0D0,.774596669241483D0,             GAUS 17
     2      -.861136311594050D0,-.339981043584860D0,                  GAUS 18
     3       .339981043584860D0,.861136311594050D0/                   GAUS 19
      DATA P/2.0D0,1.0D0,1.0D0,                                       GAUS 20
     1       0.555555555555556D0,0.888888888888889D0,0.555555555555556D0,GAUS 21
     2       .347854845137450D0,.652145154862550D0,                   GAUS 22
     3       .652145154862550D0,.347854845137450D0/                   GAUS 23
C-----------------------------------------------------------------GAUS 24
      II=IPGKED(1)                                                    GAUS 25
      IMIN=INDIC(II)                                                  GAUS 26
      IMAX=IMIN+II-1                                                  GAUS 27
      IF(NDIM-2) 10,20,30                                             GAUS 28
C-------  1 DIMENSION                                                 GAUS 29
   10 IPG=0                                                           GAUS 30
      DO 15 I=IMIN,IMAX                                               GAUS 31
      IPG=IPG+1                                                       GAUS 32
      VKPG(IPG)=G(I)                                                  GAUS 33
   15 VCPG(IPG)=P(I)                                                  GAUS 34
      RETURN                                                          GAUS 35
C-------  2 DIMENSIONS                                                GAUS 36
   20 II=IPGKED(2)                                                    GAUS 37
      JMIN=INDIC(II)                                                  GAUS 38
      JMAX=JMIN+II-1                                                  GAUS 39
      IPG=0                                                           GAUS 40
      L=1                                                             GAUS 41
      DO 25 I=IMIN,IMAX                                               GAUS 42
      DO 25 J=JMIN,JMAX                                               GAUS 43
      IPG=IPG+1                                                       GAUS 44
      VKPG(L)=G(I)                                                    GAUS 45
      VKPG(L+1)=G(J)                                                  GAUS 46
      L=L+2                                                           GAUS 47
   25 VCPG(IPG)=P(I)*P(J)                                             GAUS 48
      RETURN                                                          GAUS 49
C-------  3 DIMENSIONS                                                GAUS 50
   30 II=IPGKED(2)                                                    GAUS 51
      JMIN=INDIC(II)                                                  GAUS 52
      JMAX=JMIN+II-1                                                  GAUS 53
      II=IPGKED(3)                                                    GAUS 54
      KMIN=INDIC(II)                                                  GAUS 55
      KMAX=KMIN+II-1                                                  GAUS 56
      IPG=0                                                           GAUS 57
      L=1                                                             GAUS 58
      DO 35 I=IMIN,IMAX                                               GAUS 59
      DO 35 J=JMIN,JMAX                                               GAUS 60
      DO 35 K=KMIN,KMAX                                               GAUS 61
      IPG=IPG+1                                                       GAUS 62
      VKPG(L)=G(I)                                                    GAUS 63
      VKPG(L+1)=G(J)                                                  GAUS 64
      VKPG(L+2)=G(K)                                                  GAUS 65
      L=L+3                                                           GAUS 66
   35 VCPG(IPG)=P(I)*P(J)*P(K)                                        GAUS 67
      RETURN                                                          GAUS 68
      END                                                             GAUS 69
```

Figure 5.10 Subroutine GAUSS for the computation of coordinates and weights corresponding to integration points for one, two, or three dimensions. This subroutine is used in program MEF of Chapter 6.

```
      SUBROUTINE GAUSST(IPGKED,NDIM,VKPG,VCPG,IPG)                   GAUT   1
C=====================================================================GAUT   2
C     TO FORM ARRAYS OF COORDINATES AND WEIGHTS AT INTEGRATION        GAUT   3
C     POINTS FOR TRIANGULAR ELEMENTS                                  GAUT   4
C        INPUT                                                        GAUT   5
C           IPGKED(1)    NUMBER OF INTEGRATION POINTS                 GAUT   6
C           IPGKED(2)    .EQ.1 IF WE HAVE 3 POINTS AT MID SIDES       GAUT   7
C        OUTPUT                                                       GAUT   8
C           VKPG     COORDINATES OF INTEGRATION POINTS                GAUT   9
C           VCPG     WEIGHTS AT THE INTEGRATION POINTS                GAUT  10
C           IPG      TOTAL NUMBER OF INTEGRATION POINTS               GAUT  11
C=====================================================================GAUT  12
      IMPLICIT REAL*8(A-H,O-Z)                                        GAUT  13
      COMMON/ES/MR,MP                                                 GAUT  14
      DIMENSION IPGKED(1),VKPG(1),VCPG(1)                             GAUT  15
      DATA ZERO/0.D0/,UN/1.D0/,DEUX/2.D0/,TROIS/3.D0/,CINQ/5.D0/      GAUT  16
      DATA SIX/6.D0/                                                  GAUT  17
      SQRT(X)=DSQRT(X)                                                GAUT  18
      IPG=IPGKED(1)                                                   GAUT  19
C------- 1 POINT AT THE GRAVITY CENTER                                GAUT  20
      IF(IPG.NE.1) GO TO 10                                           GAUT  21
      VCPG(1)=UN/DEUX                                                 GAUT  22
      VKPG(1)=UN/TROIS                                                GAUT  23
      VKPG(2)=UN/TROIS                                                GAUT  24
      GO TO 100                                                       GAUT  25
C------- 3 POINTS                                                     GAUT  26
10    IF(IPG.NE.3) GO TO 20                                           GAUT  27
      C1=UN/SIX                                                       GAUT  28
      VCPG(1)=C1                                                      GAUT  29
      VCPG(2)=C1                                                      GAUT  30
      VCPG(3)=C1                                                      GAUT  31
C....... 3 POINTS AT MID SIDES                                        GAUT  32
      IF(IPGKED(2).NE.1) GO TO 11                                     GAUT  33
      VKPG(1)=UN/DEUX                                                 GAUT  34
      VKPG(2)=UN/DEUX                                                 GAUT  35
      VKPG(3)=ZERO                                                    GAUT  36
      VKPG(4)=UN/DEUX                                                 GAUT  37
      VKPG(5)=UN/DEUX                                                 GAUT  38
      VKPG(6)=ZERO                                                    GAUT  39
      GO TO 100                                                       GAUT  40
C....... 3 POINTS AT ONE THIRD OF THE MEDIANS                         GAUT  41
11    VKPG(1)=C1                                                      GAUT  42
      VKPG(2)=C1                                                      GAUT  43
      VKPG(3)=DEUX/TROIS                                              GAUT  44
      VKPG(4)=C1                                                      GAUT  45
      VKPG(5)=C1                                                      GAUT  46
      VKPG(6)=DEUX/TROIS                                              GAUT  47
      GO TO 100                                                       GAUT  48
C------- 4 POINTS AT ONE THIRD OF THE MEDIANS AND AT THEIR INTERSECTIONGAUT  49
20    IF(IPG.NE.4) GO TO 30                                           GAUT  50
      C2=-0.28125D0                                                   GAUT  51
      C3=0.260416666666667D0                                          GAUT  52
      VCPG(1)=C2                                                      GAUT  53
      VCPG(2)=C3                                                      GAUT  54
      VCPG(3)=C3                                                      GAUT  55
      VCPG(4)=C3                                                      GAUT  56
      VKPG(1)=UN/TROIS                                                GAUT  57
      VKPG(2)=UN/TROIS                                                GAUT  58
      VKPG(3)=UN/CINQ                                                 GAUT  59
      VKPG(4)=UN/CINQ                                                 GAUT  60
      VKPG(5)=TROIS/CINQ                                              GAUT  61
      VKPG(6)=UN/CINQ                                                 GAUT  62
      VKPG(7)=UN/CINQ                                                 GAUT  63
      VKPG(8)=TROIS/CINQ                                              GAUT  64
```

Figure 5.11 (*Contd.*)

```
              GO TO 100
C------- 6 POINTS
30     IF(IPG.NE.6) GO TO 40
       C1=0.111690794839005D0
       C2=0.054975871827661D0
       C3=0.445948490915965D0
       C4=0.091576213509771D0
       VCPG(1)=C1
       VCPG(2)=C1
       VCPG(3)=C1
       VCPG(4)=C2
       VCPG(5)=C2
       VCPG(6)=C2
       VKPG(1)=C3
       VKPG(2)=C3
       VKPG(3)=UN-DEUX*C3
       VKPG(4)=C3
       VKPG(5)=C3
       VKPG(6)=UN-DEUX*C3
       VKPG(7)=C4
       VKPG(8)=C4
       VKPG(9)=UN-DEUX*C4
       VKPG(10)=C4
       VKPG(11)=C4
       VKPG(12)=UN-DEUX*C4
       GO TO 100
C------- 7 POINTS
40     IF(IPG.NE.7) GO TO 99
       C1=9.D0/80.D0
       C2=(155.D0+SQRT(15.D0))/2400D0
       C3=31.D0/240.D0-C2
       C4=(6.D0+SQRT(15.D0))/21.D0
       C5=4.D0/7.D0-C4
       VCPG(1)=C1
       VCPG(2)=C2
       VCPG(3)=C2
       VCPG(4)=C2
       VCPG(5)=C3
       VCPG(6)=C3
       VCPG(7)=C3
       VKPG(1)=UN/TROIS
       VKPG(2)=UN/TROIS
       VKPG(3)=C4
       VKPG(4)=C4
       VKPG(5)=UN-DEUX*C4
       VKPG(6)=C4
       VKPG(7)=C4
       VKPG(8)=UN-DEUX*C4
       VKPG(9)=C5
       VKPG(10)=C5
       VKPG(11)=UN-DEUX*C5
       VKPG(12)=C5
       VKPG(13)=C5
       VKPG(14)=UN-DEUX*C5
100    RETURN
99     WRITE(MP,2000) IPG
2000   FORMAT('*** ERROR: NUMBER OF INTEGRATION POINTS (IPG) =',I5)
       RETURN
       END
```

Figure 5.11 Subroutine GAUSST for the computation of coordinates and weights corresponding to the integration points of a triangle (direct method).

stands for the number of dimensions.

$$VCPG = \langle w_1 \quad w_2 \ldots w_{IPG} \rangle$$
$$VKPG = \langle \xi_1 \quad \eta_1 \quad \zeta_1; \quad \xi_2 \quad \eta_2 \quad \zeta_2; \quad \ldots; \quad \xi_{IPG} \quad \eta_{IPG} \quad \zeta_{IPG} \rangle$$

Figures 5.10 and 5.11 contain listings of two subroutines: GAUSS for multilinear formulas, and GAUSST for direct formulas over triangular regions.

5.2 Solution of Systems of Linear Equations

5.2.1 Introduction

The solution of the system:

$$[K]\{U_n\} = \{F\} \tag{5.22}$$

is a most important step in the total finite element method of solution. The number n of unknowns U_n is directly proportional to the number of nodes and also to the number of degrees of freedom per node. The accuracy and range of application of the method is limited only by the number of simultaneous linear equations that can be solved economically in today's computers. Systems of a few thousand equations are routinely solved, but systems of more than 10 000 equations are still exceptional.

Methods of solutions are generally divided into two broad classes:

(a) direct methods, also called Gaussian elimination methods;
(b) iterative methods of which the Gauss–Seidel variation is the most popular.

During the early years of the finite element method, iterative techniques were thought to be more effective and easier to program than direct elimination method (see paragraph 5.3). However, further research in algorithmic solutions and computer programming resulted in a near-unanimous choice of elimination methods for the construction of codes because of their near-optimum accuracy and speed for general systems. Development of array processors may change the present state of the art because the hardware–software combination could make other strategies more economical.

5.2.2 Gaussian Elimination

The most common variations of the Gaussian elimination techniques divide the operations into two steps.

(a) Triangularization of the matrix of coefficients (also known as the deflation process)

A systematic series of linear transformations is applied to the system of equations in order to reduce it to a triangular system:

$$\begin{bmatrix} & S \\ 0 & \end{bmatrix} \{U_n\} = \{F'\} \tag{5.23}$$

(b) Back-substitution

For an upper triangular system (5.23), the last equation contains only one unknown and is readily solved. The previous equation contains two unknowns, one of which has just been calculated. After substitution, we find the remaining unknown. The process is repeated successively for each line of the system from bottom to top.

5.2.2.1 Triangularization process

$$\begin{bmatrix} K_{11} & K_{12} & \ldots & K_{1n} \\ K_{21} & K_{22} & \ldots & K_{2n} \\ \vdots & \vdots & \ldots & \vdots \\ K_{n1} & K_{n2} & \ldots & K_{nn} \end{bmatrix} \begin{Bmatrix} U_1 \\ U_2 \\ \vdots \\ U_n \end{Bmatrix} = \begin{Bmatrix} F_1 \\ F_2 \\ \vdots \\ F_n \end{Bmatrix} \qquad (5.24)$$

Unknowns U_s, $s = 1, 2, \ldots, n-1$ are eliminated from equations $s+1$ to n as follows:

— U_s is written in terms of $U_{s+1}, U_{s+2}, \ldots, U_n$ and F_s using equation s;
— then, the value of U_s is substituted in equations $s+1, s+2, \ldots, n$.

After eliminating U_s in equations $s+1$ to n, all terms of column s below the diagonal have become equal to zero. Therefore, after a similar elimination process for U_1 to U_{n-1}, matrix $[K]$ has been reduced to an upper triangular matrix.

The elimination of every unknown U_s modifies both $[K]$ and F:

ORIGINAL SYSTEM:

$[K^0]$ and $\{F^0\}$
↓   eliminate U_1 in equations 2 to n
$[K^1]$ and $\{F^1\}$
↓   eliminate U_2 in equations 3 to n
$[K^2]$ and $\{F^2\}$
↓
⋮   eliminate U_S in equations $s+1$ to n
$[K^S]$ and $\{F^S\}$
↓
⋮   eliminate U_{n-1} in equation n
↓
$[S]$ and $\{F^1\}$ UPPER TRIANGULAR SYSTEM

To eliminate variable U_1 from system (5.24), we solve the first equation for U_1:

$$U_1 = \frac{1}{K_{11}}(F_1 - K_{12}U_2 \ldots - K_{1n}U_n)$$

and substitute the expression of U_1 in equations $2, 3, \ldots, n$:

$$\begin{bmatrix} K_{11} & K_{12} & & K_{1n} \\ 0 & K_{22} - \dfrac{K_{21}}{K_{11}}K_{12} & \cdots & K_{2n} - \dfrac{K_{21}}{K_{11}}K_{1n} \\ \vdots & \vdots & & \vdots \\ 0 & K_{n2} - \dfrac{K_{n1}}{K_{11}}K_{12} & \cdots & K_{nn} - \dfrac{K_{n1}}{K_{11}}K_{1n} \end{bmatrix} \begin{Bmatrix} U_1 \\ U_2 \\ \vdots \\ U_n \end{Bmatrix} = \begin{Bmatrix} F_1 \\ F_2 - \dfrac{K_{21}}{K_{11}}F_1 \\ \vdots \\ F_n - \dfrac{K_{n1}}{K_{11}}F_1 \end{Bmatrix}$$

which we write:

$$\begin{bmatrix} K_{11} & K_{12} & \cdots & K_{1n} \\ 0 & K_{22}^1 & \cdots & K_{2n}^1 \\ \vdots & \vdots & & \vdots \\ 0 & K_{n2}^1 & \cdots & K_{nn}^1 \end{bmatrix} \begin{Bmatrix} U_1 \\ U_2 \\ \vdots \\ U_n \end{Bmatrix} = \begin{Bmatrix} F_1 \\ F_2^1 \\ \vdots \\ F_n^1 \end{Bmatrix}$$

where:

$$\begin{aligned} K_{ij}^1 &= K_{ij} - K_{i1} \cdot K_{11}^{-1} \cdot K_{1j} \\ F_i^1 &= F_i - K_{i1} \cdot K_{11}^{-1} \cdot F_1 \end{aligned} \quad i,j = 2, 3, \ldots, n$$

After elimination of $U_1 U_2 \ldots U_s$ $_{\delta 1}$:

$$\begin{bmatrix} K_{11} & K_{12} & \cdots & & K_{1s} & \cdots & K_{1n} \\ 0 & K_{22}^1 & \cdots & & K_{2s}^1 & \cdots & K_{2n}^1 \\ \vdots & \vdots & & & \vdots & & \vdots \\ 0 & 0 & \cdots & 0 & K_{ss}^{s-1} & \cdots & K_{sn}^{s-1} \\ 0 & 0 & & & \vdots & & \vdots \\ \vdots & \vdots & & & \vdots & & \vdots \\ 0 & 0 & \cdots & 0 & K_{ns}^{s-1} & \cdots & K_{nn}^{s-1} \end{bmatrix} \begin{Bmatrix} U_1 \\ U_2 \\ \vdots \\ U_s \\ \vdots \\ U_n \end{Bmatrix} = \begin{Bmatrix} F_1 \\ F_2^1 \\ \vdots \\ F_s^{s-1} \\ \vdots \\ F_n^{s-1} \end{Bmatrix}$$

$$[K^{s-1}]\{U_n\} = \{F^{s-1}\}$$

After elimination of U_s:

$$[K^s]\{U_n\} = \{F^s\}$$

The modified terms are:

$$\begin{aligned} K_{ij}^s &= K_{ij}^{s-1} - K_{is}^{s-1}(K_{ss}^{s-1})^{-1}K_{sj}^{s-1} \\ F_i^s &= F_i^{s-1} - K_{is}^{s-1}(K_{ss}^{s-1})^{-1}F_s^{s-1} \end{aligned} \quad i,j = s+1, s+2, \ldots, n \quad (5.25)$$

The final triangular system is:

$$\begin{bmatrix} K_{11} & \cdots & \cdots & \cdots & \cdots & \cdots & K_{1n} \\ 0 & K_{22}^1 & \cdots & \cdots & \cdots & \cdots & K_{2n}^1 \\ \vdots & 0 & K_{33}^2 & \cdots & \cdots & \cdots & K_{3n}^2 \\ \vdots & \vdots & 0 & \ddots & & & \vdots \\ \vdots & \vdots & \vdots & & K_{ss}^{s-1} & \cdots & K_{sn}^{s-1} \\ \vdots & \vdots & \vdots & & 0 & & \vdots \\ \vdots & \vdots & \vdots & & \vdots & & \vdots \\ 0 & 0 & 0 & \cdots & 0 & \cdots & K_{nn}^{n-1} \end{bmatrix} \begin{Bmatrix} U_1 \\ U_2 \\ U_3 \\ \vdots \\ U_s \\ \vdots \\ U_n \end{Bmatrix} = \begin{Bmatrix} F_1 \\ F_2^1 \\ F_3^2 \\ \vdots \\ F_s^{s-1} \\ \vdots \\ F_n^{n-1} \end{Bmatrix}$$

$$\begin{bmatrix} K^{n-1} \\ 0 \end{bmatrix} \{U_n\} = \{F^{n-1}\}$$

thus

$$\begin{bmatrix} S \\ 0 \end{bmatrix} \{U_n\} = \{F'\}$$

In practice, successive matrices $[K^1], [K^2], \ldots$ are constructed in the original space occupied by matrix K. The algorithm is:

$$\begin{aligned}
& s = 1, 2, \ldots, n-1 \\
& \quad i = s+1, s+2, \ldots, n \\
& \quad\quad c = K_{is} K_{ss}^{-1} \\
& \quad\quad F_i = F_i - c F_s \\
& \quad\quad j = s+1, s+2, \ldots, n \\
& \quad\quad\quad K_{ij} = K_{ij} - c K_{sj}
\end{aligned} \quad (5.26)$$

Remark

For a symmetric system, index j varies from i to n; also $K_{ij} = K_{ji}$.

Example 5.9 Triangularization of a non-symmetric system

$$\begin{bmatrix} 2 & 4 & 8 \\ 4 & 11 & 25 \\ 6 & 18 & 46 \end{bmatrix} \begin{Bmatrix} U_1 \\ U_2 \\ U_3 \end{Bmatrix} = \begin{Bmatrix} 34 \\ 101 \\ 180 \end{Bmatrix}$$

$$[K]\{U_n\} = \{F\}$$

After elimination of U_1:

$$\begin{bmatrix} 2 & 4 & 8 \\ 0 & 11 - \tfrac{4}{2}4 = 3 & 25 - \tfrac{4}{2}8 = 9 \\ 0 & 18 - \tfrac{6}{2}4 = 6 & 46 - \tfrac{6}{2}8 = 22 \end{bmatrix} \begin{Bmatrix} U_1 \\ U_2 \\ U_3 \end{Bmatrix} = \begin{Bmatrix} 34 \\ 101 - \tfrac{4}{2}34 = 33 \\ 180 - \tfrac{6}{2}34 = 78 \end{Bmatrix}$$

After elimination from U_1 and U_2:

$$\begin{bmatrix} 2 & 4 & 8 \\ 0 & 3 & 9 \\ 0 & 0 & 22 - \frac{60}{3}9 = 4 \end{bmatrix} \begin{Bmatrix} U_1 \\ U_2 \\ U_3 \end{Bmatrix} = \begin{Bmatrix} 34 \\ 33 \\ 78 - \frac{6}{3}33 = 12 \end{Bmatrix}$$

thus:

$$\begin{bmatrix} 2 & 4 & 8 \\ 0 & 3 & 9 \\ 0 & 0 & 4 \end{bmatrix} \begin{Bmatrix} U_1 \\ U_2 \\ U_3 \end{Bmatrix} = \begin{Bmatrix} 34 \\ 33 \\ 12 \end{Bmatrix}$$

$$[S]\{U_n\} = \{F'\}$$

Algorithm (5.26) fails if K_{ss} becomes equal to zero. (Diagonal term K_{ss} is called a pivot.) Then, row s must be interchanged with another row $i > s$ such that $K_{is} \neq 0$. For a symmetric system, symmetry will be preserved if the interchange is also performed for columns i and s. Note that column interchange modifies the order of the unknowns from s to $i - s$. Since diagonal term K_{ss} will be replaced by diagonal term K_{ii}, the latter has to be different than zero and the restriction on K_{is} is lifted.

The first process described is called row pivoting. For symmetrical matrices, it is in effect simultaneous row–column pivoting. A matrix is singular and cannot be inverted when for any s, terms K_{is} or K_{si} are equal to zero for $i \geq s$.

For a triangular matrix $[S]$, the determinant is equal to the product of the diagonal terms.

Example 5.10 Search for non-zero pivot

$$\begin{bmatrix} 0 & 4 & 8 \\ 4 & 0 & 6 \\ 8 & 6 & 0 \end{bmatrix} \begin{Bmatrix} U_1 \\ U_2 \\ U_3 \end{Bmatrix} = \begin{Bmatrix} 2 \\ 4 \\ 12 \end{Bmatrix}$$

The first pivot being null, interchange row 1 and 2:

$$\begin{bmatrix} 4 & 0 & 6 \\ 0 & 4 & 8 \\ 8 & 6 & 0 \end{bmatrix} \begin{Bmatrix} U_1 \\ U_2 \\ U_3 \end{Bmatrix} = \begin{Bmatrix} 4 \\ 2 \\ 12 \end{Bmatrix}$$

The matrix of coefficient is no more symmetrical and column interchange is prohibited because it would bring back a value of zero on the diagonal.

After elimination of U_i:

$$\begin{bmatrix} 4 & 0 & 6 \\ 0 & 4 & 8 \\ 0 & 6 & -12 \end{bmatrix} \begin{Bmatrix} U_1 \\ U_2 \\ U_3 \end{Bmatrix} = \begin{Bmatrix} 4 \\ 2 \\ 4 \end{Bmatrix}$$

After elimination of U_1 and U_2:

$$\begin{bmatrix} 4 & 0 & 6 \\ 0 & 4 & 8 \\ 0 & 0 & -24 \end{bmatrix} \begin{Bmatrix} U_1 \\ U_2 \\ U_3 \end{Bmatrix} = \begin{Bmatrix} 4 \\ 2 \\ 1 \end{Bmatrix}$$

The determinant of $[K]$ is:

$$4 \times 4 \times (-24) = -384$$

5.2.2.2 Solution of an upper triangular system

System (5.26) can be solved starting from the last equation for U_n, then next to last U_{n-1} and so on. This process is called back-substitution:

$$U_n = S_{nn}^{-1} F_n'$$
$$U_{n-1} = S_{n-1,n-1}^{-1}(F_{n-1}' - S_{n-1,n}U_n)$$
$$\ldots$$
$$U_1 = S_{11}^{-1}(F_1' - S_{12}U_2 - S_{13}U_3 - \ldots - S_{1n}U_n)$$

The algorithm uses both the triangularized matrix of coefficient $[K]$ and the modified load vector $\{F\}$:

$$\begin{aligned} &U_n = K_{nn}^{-1} F_n \\ &i = n-1, n-2, \ldots, 1 \\ &\quad U_i = K_{ii}^{-1}\left(F_i - \sum_{j=i+1}^{n} K_{ij} U_j\right) \end{aligned} \tag{5.27}$$

Example 5.11 *Solution of a triangular system*

Start from the triangular system obtained in Example 5.9:

$$\begin{bmatrix} 2 & 4 & 8 \\ 0 & 3 & 9 \\ 0 & 0 & 4 \end{bmatrix} \begin{Bmatrix} U_1 \\ U_2 \\ U_3 \end{Bmatrix} = \begin{Bmatrix} 34 \\ 33 \\ 12 \end{Bmatrix}$$

$$U_3 = \tfrac{12}{4} = 3$$
$$U_2 = \tfrac{1}{3}(33 - 9 \times 3) = 2$$
$$U_1 = \tfrac{1}{2}(34 - 4 \times 2 - 8 \times 3) = 1$$

5.2.2.3 Gaussian Elimination Computer Code

When matrix $[K]$ is stored in a two-dimensional array VKG, Gaussian elimination can be coded as shown in Figure 5.12. Array VFG contains the load vector $\{F\}$ upon entry into the subroutine and is replaced by the solution vector $\{U_n\}$ upon exit.

```
      SUBROUTINE RESOL(NSYM,NEQ,VKG,VFG)                              RESO   1
C======================================================================RESO   2
C                                                                     RESO   3
C        TO SOLVE A NON SYMMETRIC SYSTEM OF EQUATIONS                 RESO   4
C        BY GAUSS METHOD                                              RESO   5
C                                                                     RESO   6
C        INPUT                                                        RESO   7
C           NSYM       .EQ.1  NON SYMMETRIC SYSTEM                    RESO   8
C           NEQ        NUMBER OF EQUATIONS (.GE.2)                    RESO   9
C           VKG        MATRIX K STORED IN A 2 DIMENSIONAL ARRAY       RESO  10
C           VFG        SECOND MEMBER                                  RESO  11
C                                                                     RESO  12
C        OUTPUT                                                       RESO  13
C           VFG        SOLUTION                                       RESO  14
C                                                                     RESO  15
C======================================================================RESO  16
      IMPLICIT REAL*8(A-H,O-Z)                                        RESO  17
      DIMENSION VKG(NEQ,NEQ),VFG(NEQ)                                 RESO  18
C------- TRIANGULARIZE                                                RESO  19
      DATA ZERO/0.D0/                                                 RESO  20
      N1=NEQ-1                                                        RESO  21
      DO 50 IS=1,N1                                                   RESO  22
      PIV=VKG(IS,IS)                                                  RESO  23
      IF(PIV) 20,10,20                                                RESO  24
10    WRITE(6,2000) IS                                                RESO  25
2000  FORMAT(' ZERO PIVOT, EQUATION',I5)                              RESO  26
      STOP                                                            RESO  27
20    IS1=IS+1                                                        RESO  28
      DO 50 II=IS1,NEQ                                                RESO  29
      CL=VKG(II,IS)/PIV                                               RESO  30
      IF(CL.EQ.ZERO)GO TO 50                                          RESO  31
      VFG(II)=VFG(II)-CL*VFG(IS)                                      RESO  32
      IF(NSYM.NE.1) GO TO 32                                          RESO  33
      DO 30 IJ=IS1,NEQ                                                RESO  34
30    VKG(II,IJ)=VKG(II,IJ)-CL*VKG(IS,IJ)                             RESO  35
      GO TO 50                                                        RESO  36
32    DO 40 IJ=II,NEQ                                                 RESO  37
      VKG(II,IJ)=VKG(II,IJ)-CL*VKG(IS,IJ)                             RESO  38
40    VKG(IJ,II)=VKG(II,IJ)                                           RESO  39
50    CONTINUE                                                        RESO  40
C------- SOLVE THE TRIANGULAR SYSTEM                                  RESO  41
      VFG(NEQ)=VFG(NEQ)/VKG(NEQ,NEQ)                                  RESO  42
      DO 70 II=1,N1                                                   RESO  43
      IS1=IS1-1                                                       RESO  44
      CL=ZERO                                                         RESO  45
      IJ1=IS1+1                                                       RESO  46
      DO 60 IJ=IJ1,NEQ                                                RESO  47
60    CL=CL+VKG(IS1,IJ)*VFG(IJ)                                       RESO  48
70    VFG(IS1)=(VFG(IS1)-CL)/VKG(IS1,IS1)                             RESO  49
      RETURN                                                          RESO  50
      END                                                             RESO  51
```

Figure 5.12 Subroutine RESOL for the resolution of a system of linear equations, symmetric or non-symmetric, the matrix of which is stored in a two-dimensional array. This subroutine is used in program BBMEF of Chapter 6.

5.2.3 *Matrix Factorization*

5.2.3.1 *Introduction*

Gaussian elimination can be reformulated in a two phase process that does not require to modify $[K]$ and $\{F\}$ simultaneously. Such a procedure is preferred in the construction of finite element systems for the following reasons:

(a) matrix [K] is first decomposed into the product [L] [S]

$$[K] = \begin{bmatrix} & 0 \\ L & \end{bmatrix} \begin{bmatrix} & S \\ & 0 \end{bmatrix} = [L][S] \quad (5.28)$$

where: [L] is a lower unit triangular matrix; and [S] is the upper triangular matrix obtained by Gaussian elimination in paragraph 5.2.2.1.

(b) load vector $\{F\}$ is modified after the decomposition of [K]. This allows us to solve a system with multiple load vectors in a more convenient sequence of operations.

(c) we can develop other decomposition and solver algorithms which are more easily adapted to the skyline storage of matrix [K] described in paragraph 4.6.3. In the following section we give an algebraic representation of the Gauss process which then leads to different decomposition algorithms.

5.2.3.2 *Matrix formulation of Gaussian elimination*

The elimination of unknown U_s, (5.25) transforms $[K^{s-1}]$ into $[K^s]$. It can be written as follows:

$$[K^s] = \left([I] + \begin{bmatrix} 0 & & & & 0 \\ & \ddots & & & \\ & & 0 & & \\ & & -l_{s+1,s} & & \\ 0 & & \vdots & \ddots & \\ & & -l_{n,s} & & 0 \end{bmatrix} \right) [K^{s-1}] = [l^s][K^{s-1}] \quad (5.29)$$

where: [I] is an identity matrix

$$l_{is} = K_{is}^{s-1}(K_{ss}^{s-1})^{-1} \quad i = s+1, s+2, \ldots, n$$

Matrix $[L^s]$ is a lower unit triangular matrix with non-zero terms in column s only.

Elimination of unknowns $U_1, U_2, \ldots, U_{n-1}$ (algorithm 5.26) is done by successive applications of operation (5.29) with $s = 1, 2, \ldots, n-1$.

$$[K^{n-1}] = \begin{bmatrix} & S \\ 0 & \end{bmatrix} = [l^{n-1}][l^{n-2}]\ldots[l^1][K]$$

$$\begin{bmatrix} & S \\ 0 & \end{bmatrix} = \begin{bmatrix} & 0 \\ l & \end{bmatrix} [K] = [l][K] \quad (5.30)$$

Example 5.12 *Decomposition of the matrix of Example 5.9*

$$[K] = \begin{bmatrix} 2 & 4 & 8 \\ 4 & 11 & 25 \\ 6 & 18 & 46 \end{bmatrix}$$

Using expressions (5.29)

$s = 1$

$$[K^1] = [l^1][K]$$

$s = 2$

$$[K^2] = [l^2][K^1] = [l^2][l^1][K] = [l][K]$$

$$[l^1] = \begin{bmatrix} 1 & 0 & 0 \\ -2 & 1 & 0 \\ -3 & 0 & 1 \end{bmatrix}; \quad [K^1] = \begin{bmatrix} 2 & 4 & 8 \\ 0 & 3 & 9 \\ 0 & 6 & 22 \end{bmatrix}$$

$$[l^2] = \begin{bmatrix} 1 & 0 & 0 \\ 0 & 1 & 0 \\ 0 & -2 & 1 \end{bmatrix}; \quad \{K^2\} = \begin{bmatrix} 2 & 4 & 8 \\ 0 & 3 & 9 \\ 0 & 0 & 4 \end{bmatrix} = [S]$$

$$[l] = \begin{bmatrix} 1 & 0 & 0 \\ -2 & 1 & 0 \\ 1 & -2 & 1 \end{bmatrix}$$

Factorization of $[K]$ (5.28) is obtained by inverting matrix $[l]$ defined in (5.30)

$$\begin{aligned}[K] = [l]^{-1}[S] &= [l^1]^{-1}[l^2]^{-1}\ldots[l^{n-1}]^{-1}[S] \\ &= [L^1][L^2]\ldots[L^{n-1}][S] = \begin{bmatrix} \diagdown & 0 \\ L & \diagdown \end{bmatrix}[S] \quad (5.31) \\ &= [L][S]\end{aligned}$$

5.2.3.3 Properties of triangular matrices $[l^s]$

Triangular matrices $[l^s]$ and $[L^s]$ appearing in (5.30) and (5.31) have the following properties.

— The product of two lower triangular matrices is also a lower triangular matrix. The same is true for upper triangular matrices.
— The determinant of a triangular matrix is equal to the product of its diagonal terms.
— The inverse of a lower triangular matrix is also a lower triangular matrix. The same is true for upper matrices.
— The topology (bandwidth and profile) of $[S]$ is identical to the topology of the upper part of $[K]$, that of $[L]$ is the same as that of the lower part of $[K]$.
— The inverse of matrices $[l^s]$ (5.29) is:

$$[l^s]^{-1} = [L^s] = -[l^s] + 2[I] \quad (5.32)$$

— The product of two matrices $[L^i]$ and $[L^j]$, where $i \leq j$, is:

$$[L^i][L^j] = [L^i] + [L^j] - [I] \quad (5.33)$$

Example 5.13   Matrices [L] and [S] for Example 5.12

Applying (5.31) to the matrices obtained in Example 5.12 and using equations (5.32) and (5.33):

$$[K] = [L^1][L^2][S] = [L][S]$$

where:

$$[L^1] = [l^1]^{-1} = \begin{bmatrix} 1 & 0 & 0 \\ 2 & 1 & 0 \\ 3 & 0 & 1 \end{bmatrix}$$

$$[L^2] = [l^2]^{-1} = \begin{bmatrix} 1 & 0 & 0 \\ 0 & 1 & 0 \\ 0 & 2 & 1 \end{bmatrix}$$

$$[L] = [L^1][L^2] = \begin{bmatrix} 1 & 0 & 0 \\ 2 & 1 & 0 \\ 3 & 2 & 1 \end{bmatrix}$$

$$[S] = [K^2] = \begin{bmatrix} 2 & 4 & 8 \\ 0 & 3 & 9 \\ 0 & 0 & 4 \end{bmatrix}$$

Hence the final decomposition:

$$\begin{bmatrix} 2 & 4 & 8 \\ 4 & 11 & 25 \\ 6 & 18 & 46 \end{bmatrix} = \begin{bmatrix} 1 & 0 & 0 \\ 2 & 1 & 0 \\ 3 & 2 & 1 \end{bmatrix} \begin{bmatrix} 2 & 4 & 8 \\ 0 & 3 & 9 \\ 0 & 0 & 4 \end{bmatrix}$$

$$[K] \qquad\qquad [L] \qquad\quad [S]$$

5.2.3.4 *Representation of the Factorized Matrix [K] in Different Forms*

The basic decomposition algorithm (5.33):

$$[K] = [L][S] \tag{5.34}$$

is generally attributed to Doolittle.

Three other forms of the algorithm have been widely used in finite element codes:

(a)   LDU decomposition

Matrix [S] of (5.34) is further factored into a diagonal matrix [D] and an upper unit triangular matrix [U]:

$$[K] = \begin{bmatrix} & & \\ & D & \\ & & \end{bmatrix} \begin{bmatrix} & & U \\ & & \\ 0 & & \end{bmatrix}$$

$$U_{ii} = 1$$
$$U_{ij} = S_{ij}/S_{ii} \quad j > i$$
$$D_{ii} = S_{ii}$$

then: $\quad\quad\quad\quad [K] = [L][D][U] \quad\quad\quad\quad$ (5.35)

(b) *Crout decomposition*

$$[K] = [L'][U] \quad\quad\quad\quad (5.36)$$
where: $\quad\quad [L'] = [L][D]$

For symmetric matrices $[K] = [K]^T$, equation (5.35) becomes:

$$[K] = [L][D][U] = [K]^T = [U]^T[D][L]^T$$
hence: $\quad\quad [U] = [L]^T$
$$[K] = [L][D][L]^T \quad\quad\quad\quad (5.37)$$

(c) *Cholesky decomposition*

If $[K]$ is positive definite ($S_{ii} > 0$), equation (5.35) can be written as:

$$[K] = [L_c][L_c]^T \quad\quad\quad\quad (5.38)$$

where:

$$[L_c] = [L] \begin{bmatrix} \diagdown & & 0 \\ & \sqrt{S_{ii}} & \\ 0 & & \diagdown \end{bmatrix}$$

5.2.3.5 *Solution of a system of equations after decomposition of the matrix of coefficient*

The system to be solved:

$$[K]\{U_n\} = \{F\}$$

becomes:

$$[L][S]\{U_n\} = \{F\}$$

after replacing $[K]$ by its factors.

The solution proceeds in two steps:

$$[L]\{F'\} = \{F\} \text{ Lower triangular system}$$
$$\downarrow \quad\quad\quad\quad\quad\quad\quad\quad\quad\quad\quad\quad (5.39)$$
$$[S]\{U_n\} = \{F'\} \quad\quad \text{Upper triangular system}$$

Note that $\{F'\}$ is found in the same location as $\{F\}$ after the first elimination process (5.26).

Example 5.14 *Solution by decomposition*

Consider the system of Example 5.9 and the decomposed matrix $[K]$ obtained in Example 5.13:

— first step: lower triangular system

$$\underbrace{\begin{bmatrix} 1 & 0 & 0 \\ 2 & 1 & 0 \\ 3 & 2 & 1 \end{bmatrix}}_{[L]} \{F'\} = \underbrace{\begin{Bmatrix} 34 \\ 101 \\ 180 \end{Bmatrix}}_{\{F\}} \Rightarrow \{F'\} = \begin{Bmatrix} 34 \\ 33 \\ 12 \end{Bmatrix}$$

— second step: upper triangular system

$$\underbrace{\begin{bmatrix} 2 & 4 & 8 \\ 0 & 3 & 9 \\ 0 & 0 & 4 \end{bmatrix}}_{[S]} \{U_n\} = \underbrace{\begin{Bmatrix} 34 \\ 33 \\ 12 \end{Bmatrix}}_{\{F'\}} \Rightarrow \{U_n\} = \begin{Bmatrix} 1 \\ 2 \\ 3 \end{Bmatrix}$$

5.2.3.6 *Decomposition algorithms*

A decomposition algorithm can be developed to obtain all the terms of $[L]$ and $[S]$ from the original terms of $[K]$. All the terms of $[L]$ and $[S]$ are stored in the original matrix $[K]$ as follows:

$$\begin{bmatrix} S_{11} & S_{12} & \cdots & S_{1n} \\ L_{21} & & & \\ & & & S_{n-1,n} \\ L_{n1} & \cdots & L_{n,n-1} & S_{nn} \end{bmatrix} \qquad (5.40)$$

The Gaussian elimination algorithm (5.26) becomes a decomposition algorithm when terms L_{is} are saved in the lower triangular part of $[K]$:

$$\begin{aligned}
&s = 1, 2, \ldots, n-1 \\
&\quad i = s+1, s+2, \ldots, n \\
&\quad\quad K_{is} = K_{is} K_{ss}^{-1} \text{ (column } s \text{ of } L) \\
&\quad\quad (L_{is}) \\
&\quad\quad j = s+1, s+2, \ldots, n \\
&\quad\quad\quad K_{ij} = K_{ij} - K_{is}\ K_{sj} \\
&\quad\quad\quad\quad\quad\quad\ (L_{is})\ (S_{sj})
\end{aligned} \qquad (5.41)$$

For a given value of s, the algorithm produces the following results:

— creates column s of L (under the diagonal);
— creates line $s+1$ of S (to the right of the diagonal);
— modifies terms K_{ij}; $i, j > s$.

Other Algorithms can be obtained if we identify the terms of $[L][S]$ with the terms of $[K]$. For example, it is easy to construct a row of L and a column of S in succession. Such an algorithm is well adapted to a skyline matrix stored as described in paragraph 4.6.3 (see also paragraph 5.2.4.1).

$$\begin{bmatrix} 1 & & \\ L_{21} & 1 & \\ L_{31} & L_{32} & 1 \\ \hdots & & \end{bmatrix} \begin{bmatrix} S_{11} & S_{12} & S_{13} \\ & S_{22} & S_{23} & \cdots \\ & & S_{33} \\ & & & \end{bmatrix} = \begin{bmatrix} K_{11} & K_{12} & K_{13} & \cdots \\ K_{21} & K_{22} & K_{23} & \cdots \\ K_{31} & K_{32} & K_{33} & \cdots \\ \hdots & & & \end{bmatrix}$$

(5.42)

$s = 1$: $\quad 1 \cdot S_{11} = K_{11}$ from which: $S_{11} = K_{11}$

$s = 2$: $\quad L_{21} S_{11} = K_{21}$ $L_{21} = K_{21}(S_{11})^{-1}$

$\quad\quad\quad 1 \cdot S_{12} = K_{12}$ $S_{12} = K_{12}$

$\quad\quad\quad L_{21} S_{12} + S_{22} = K_{22}$ $S_{22} = K_{22} - L_{21} S_{12}$

$s = 3$: $\quad L_{31} S_{11} = K_{31}$ $L_{31} = K_{31}(S_{11})^{-1}$

$\quad\quad\quad L_{31} S_{12} + L_{32} S_{22} = K_{32}$ $L_{32} = (K_{32} - L_{31} S_{12})(S_{22})^{-1}$

$\quad\quad\quad 1 \cdot S_{13} = K_{13}$ $S_{13} = K_{13}$

$\quad\quad\quad L_{21} S_{13} + S_{23} = K_{23}$ $S_{23} = K_{23} - L_{21} S_{13}$

$\quad\quad\quad L_{31} S_{13} + L_{32} S_{23} + S_{33} = K_{33}$ $S_{33} = K_{33} - L_{31} S_{13} - L_{32} S_{23}$

For any value of s

$$L_{si} = \left(K_{si} - \sum_{m=1}^{i-1} L_{sm} S_{mi} \right) S_{ii}^{-1} \quad i = 1, 2, \ldots, s-1$$

$$S_{js} = K_{js} - \sum_{m=1}^{j-1} L_{jm} S_{ms} \quad j = 1, 2, \ldots, s$$

Matrices L and S being stored in $[K]$ as shown in (5.40), the previous algorithm becomes:

$s = 2, 3, \ldots, n$
$\quad i = 1, 2, \ldots, s-1$
$\quad\quad m = 1, 2, \ldots, i-1$
$\quad\quad\quad K_{si} = K_{si} - K_{sm} K_{mi}$ row of L
$\quad\quad\quad K_{is} = K_{is} - K_{im} K_{ms}$ column of S (5.43)
$\quad\quad K_{si} = K_{si} K_{ii}^{-1}$ normalize row L
$\quad m = 1, 2, \ldots, s-1$
$\quad\quad K_{ss} = K_{ss} - K_{sm} K_{ms}$ diagonal term

Example 5.15 *Decomposition of the matrix of Example 5.9 using algorithm* (5.43)

$$[K] = \begin{bmatrix} 2 & 4 & 8 \\ 4 & 11 & 25 \\ 6 & 18 & 46 \end{bmatrix}$$

After step $s=2$

$$[K] = \begin{bmatrix} 2 & 4 & 8 \\ 2 & 3 & 25 \\ 6 & 18 & 46 \end{bmatrix}$$

After step $s=3$

$$[K] = \begin{bmatrix} 2 & 4 & 8 \\ 2 & 3 & 9 \\ 3 & 2 & 4 \end{bmatrix} = \begin{bmatrix} S_{11} & S_{12} & S_{13} \\ L_{21} & S_{22} & S_{23} \\ L_{31} & L_{32} & S_{33} \end{bmatrix}$$

After decomposition by algorithm (5.43), matrix $[K]$ contains all the terms of L and S except the unit terms on the diagonal of L. The solution of the system proceeds in two steps as follows:

— step 1   forward substitution in lower triangular system

$$\begin{aligned} &i = 2, \ldots, n \\ &\quad F_i = F_i - \sum_{j=1}^{i-1} K_{ij} F_j \end{aligned} \qquad (5.44)$$

— step 2   back-substitution in upper triangular system

$$F_n = K_{nn}^{-1} F_n$$

$$\begin{aligned} &i = n-1, n-2, \ldots, 1 \\ &\quad F_i = K_{ii}^{-1} \left(F_i - \sum_{j=i+1}^{n} K_{ij} F_j \right) \end{aligned} \qquad (5.45)$$

$\{F\}$ contains solution $\{U_n\}$ upon exit from the algorithm.

5.2.4 *Algorithm for Skyline Matrices in Compact Storage*

5.2.4.1 *Case of a matrix $[K]$ stored in core*

To avoid operations by zero outside the skyline profile (see paragraph 4.6.3(e)), algorithm (5.43) must be modified as follows:

$$\begin{aligned} &s = 2, 3, \ldots, n \\ &\quad i = i_{0s}, \ldots, s-1 \\ &\quad\quad m = \text{Max}(i_{0i}, i_{0s}), \ldots, i-1 \\ &\quad\quad\quad K_{si} = K_{si} - K_{sm} K_{mi} \\ &\quad\quad\quad K_{is} = K_{is} - K_{im} K_{ms} \\ &\quad\quad K_{si} = K_{si} K_{ii}^{-1} \\ &\quad m = i_{0s}, \ldots, s-1 \\ &\quad\quad K_{ss} = K_{ss} - K_{sm} K_{ms} \end{aligned} \qquad (5.46)$$

where i_{0i} and i_{0s} are the integer identification of the high terms in columns i and s. The same value identify the left-most terms of rows i and s.

Taking into account (4.32b) and (4.41):

$$i_{0i} = i - h_J(i) = i - \text{KLD}(i+1) + \text{KLD}(i) \qquad (5.47)$$
$$i_{0s} = s - h_J(s) = s - \text{KLD}(s+1) + \text{KLD}(s)$$

Figure 5.13 shows the position of the various terms of $[K]$ appearing in algorithm (5.46). To use the compact storage method described in paragraph 4.6.3(e) for $[K]$, it remains simply to compute the positions in arrays VKGS, VKGD, VKGI various terms of $[K]$ appearing in the above algorithm. These computations use the array of pointers KLD (4.42).

In the case of symmetric systems, algorithm (5.46) is modified to avoid computations for the lower triangular terms of $[K]$ and their storage.

$$\begin{array}{l} s = 2, 3, \ldots, n \\ \quad i = i_{0s} + 1, \ldots, s-1 \\ \quad\quad m = \text{Max}(i_{0i}, i_{0s}), \ldots, i-1 \\ \quad\quad\quad K_{is} = K_{is} - K_{im} K_{ms} \\ \quad\quad c = 0 \\ \quad\quad m = i_{0s}, \ldots, i-1 \\ \quad\quad\quad c = c + K_{ms}^2 / K_{mm} \\ \quad\quad\quad K_{ms} = K_{ms} / K_{mm} \\ \quad\quad K_{ss} = K_{ss} - c \end{array} \qquad (5.48)$$

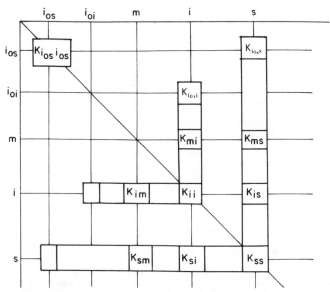

Figure 5.13 Position of the terms of K involved in algorithm (5.46).

Figure 5.14 defines the FORTRAN variables used in subroutines SOL that implements algorithms (5.46) and (5.48).

The subroutine listing of Figure 5.15 implements the solution of symmetrical (NSYM.EQ.0) and non-symmetrical (NSYM.EQ.1) systems for which skyline matrix storage information is contained in arrays VKGS, VKGD and VKGI (if NSYM.EQ.1).

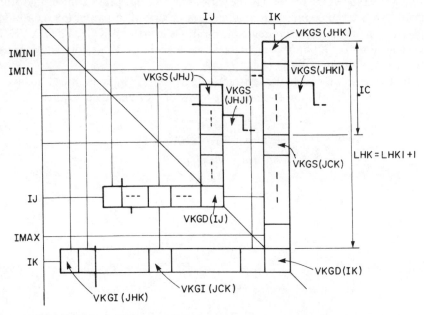

Figure 5.14 FORTRAN variables of subroutine SOL (Figure 5.15).

```
      SUBROUTINE SOL(VKGS,VKGD,VKGI,VFG,KLD,NEQ,MP,IFAC,ISOL,NSYM,ENERG)SOL   1
C========================================================================SOL   2
C        TO SOLVE A LINEAR SYSTEM (SYMMETRICAL OR NOT).                  SOL   3
C        THE MATRIX IS STORED IN CORE BY SKYLINES  IN ARRAYS             SOL   4
C        VKGS,VKGD,VKGI                                                  SOL   5
C          INPUT                                                         SOL   6
C            VKGS,VKGD,VKGI      SYSTEM MATRIX : UPPER, DIAGONAL AND     SOL   7
C                                LOWER PARTS                             SOL   8
C            VFG                 SECOND MEMBER                           SOL   9
C            KLD                 ADDRESSES OF COLUMN TOP TERMS           SOL  10
C            NEQ                 NUMBER OF EQUATIONS                     SOL  11
C            MP                  OUTPUT DEVICE NUMBER                    SOL  12
C            IFAC                IF IFAC.EQ.1 TRIANGULARIZE THE          SOL  13
C                                MATRIX                                  SOL  14
C            ISOL                IF ISOL.EQ.1 COMPUTE THE SOLUTION FROM  SOL  15
C                                TRIANGULARIZED MATRIX                   SOL  16
C            NSYM                INDEX FOR NONSYMMETRIC PROBLEM          SOL  17
C          OUTPUT                                                        SOL  18
C            VKGS,VKGD,VKGI      TRIANGULARIZED MATRIX (IF IFAC.EQ.1)    SOL  19
C            VFG                 SOLUTION (IF ISOL.EQ.1)                 SOL  20
C            ENERG               SYSTEM ENERGY (IF NSYM.EQ.0)            SOL  21
C========================================================================SOL  22
```

Figure 5.15 (*Contd.*)

```
        IMPLICIT REAL*8 (A-H,O-Z)                                  SOL  23
        DIMENSION VKGS(1),VKGD(1),VKGI(1),VFG(1),KLD(1)             SOL  24
        DATA ZERO/0.0D0/                                            SOL  25
C-----------------------------------------------------------------SOL  26
        IK=1                                                        SOL  27
        IF(VKGD(1).NE.ZERO) GO TO 10                                SOL  28
        WRITE(MP,2000) IK                                           SOL  29
        STOP                                                        SOL  30
10      ENERG=ZERO                                                  SOL  31
C                                                                   SOL  32
C-----  FOR EACH COLUMN IK TO BE MODIFIED                           SOL  33
C                                                                   SOL  34
        JHK=1                                                       SOL  35
        DO 100 IK=2,NEQ                                             SOL  36
C-----  ADDRESS OF THE NEXT COLUMN TOP TERM IK+1                    SOL  37
        JHK1=KLD(IK+1)                                              SOL  38
C-----  HEIGHT OF COLUMN IK (INCLUDE UPPER AND DIAGONAL TERMS)      SOL  39
        LHK=JHK1-JHK                                                SOL  40
        LHK1=LHK-1                                                  SOL  41
C-----  ROW OF FIRST TERM TO BE MODIFIED IN COLUMN IK               SOL  42
        IMIN=IK-LHK1                                                SOL  43
        IMIN1=IMIN-1                                                SOL  44
C-------  ROW OF LAST TERM TO BE MODIFIED IN COLUMN IK              SOL  45
        IMAX=IK-1                                                   SOL  46
        IF(LHK1.LT.0) GO TO 100                                     SOL  47
        IF(IFAC.NE.1) GO TO 90                                      SOL  48
        IF(NSYM.EQ.1) VKGI(JHK)=VKGI(JHK)/VKGD(IMIN1)               SOL  49
        IF(LHK1.EQ.0) GO TO 40                                      SOL  50
C                                                                   SOL  51
C-----  MODIFY NON-DIAGONAL TERM IN COLUMN IK                       SOL  52
C                                                                   SOL  53
        JCK=JHK+1                                                   SOL  54
        JHJ=KLD(IMIN)                                               SOL  55
C-----  FOR EACH TERM LOCATED AT JCK AND CORRESPONDING TO COLUMN IJ SOL  56
        DO 30 IJ=IMIN,IMAX                                          SOL  57
        JHJ1=KLD(IJ+1)                                              SOL  58
C-----  NUMBER OF MODIFICATIVE TERMS FOR COEFFICIENT LOCATED AT JCK SOL  59
        IC=MIN0(JCK-JHK,JHJ1-JHJ)                                   SOL  60
        IF(IC.LE.0.AND.NSYM.EQ.0) GO TO 20                          SOL  61
        C1=ZERO                                                     SOL  62
        IF(IC.LE.0) GO TO 17                                        SOL  63
        J1=JHJ1-IC                                                  SOL  64
        J2=JCK-IC                                                   SOL  65
        IF(NSYM.EQ.1) GO TO 15                                      SOL  66
        VKGS(JCK)=VKGS(JCK)-SCAL(VKGS(J1),VKGS(J2),IC)              SOL  67
        GO TO 20                                                    SOL  68
15      VKGS(JCK)=VKGS(JCK)-SCAL(VKGI(J1),VKGS(J2),IC)              SOL  69
        C1=SCAL(VKGS(J1),VKGI(J2),IC)                               SOL  70
17      VKGI(JCK)=(VKGI(JCK)-C1)/VKGD(IJ)                           SOL  71
20      JCK=JCK+1                                                   SOL  72
30      JHJ=JHJ1                                                    SOL  73
C                                                                   SOL  74
C-----  MODIFY DIAGONAL TERM                                        SOL  75
C                                                                   SOL  76
40      JCK=JHK                                                     SOL  77
        CDIAG=ZERO                                                  SOL  78
        DO 70 IJ=IMIN1,IMAX                                         SOL  79
        C1=VKGS(JCK)                                                SOL  80
        IF(NSYM.EQ.1) GO TO 50                                      SOL  81
        C2=C1/VKGD(IJ)                                              SOL  82
        VKGS(JCK)=C2                                                SOL  83
        GO TO 60                                                    SOL  84
50      C2=VKGI(JCK)                                                SOL  85
60      CDIAG=CDIAG+C1*C2                                           SOL  86
```

Figure 5.15 (*Contd.*)

```
70      JCK=JCK+1                                                       SOL    87
        VKGD(IK)=VKGD(IK)-CDIAG                                         SOL    88
        IF(VKGD(IK)) 90,80,90                                           SOL    89
80      WRITE(MP,2000) IK                                               SOL    90
2000    FORMAT(' *** ERROR,ZERO PIVOT EQUATION ',I5)                    SOL    91
        STOP                                                            SOL    92
C                                                                       SOL    93
C-----  SOLVE LOWER TRIANGULAR SYSTEM                                   SOL    94
C                                                                       SOL    95
   90   IF(ISOL.NE.1) GO TO 100                                         SOL    96
        IF(NSYM.NE.1) VFG(IK)=VFG(IK)-SCAL(VKGS(JHK),VFG(IMIN1),LHK)    SOL    97
        IF(NSYM.EQ.1) VFG(IK)=VFG(IK)-SCAL(VKGI(JHK),VFG(IMIN1),LHK)    SOL    98
100     JHK=JHK1                                                        SOL    99
        IF(ISOL.NE.1) RETURN                                            SOL   100
C                                                                       SOL   101
C-----  SOLVE DIAGONAL SYSTEM                                           SOL   102
C                                                                       SOL   103
        IF(NSYM.EQ.1) GO TO 120                                         SOL   104
        DO 110 IK=1,NEQ                                                 SOL   105
        C1=VKGD(IK)                                                     SOL   106
        C2=VFG(IK)/C1                                                   SOL   107
        VFG(IK)=C2                                                      SOL   108
110     ENERG=ENERG+C1*C2*C2                                            SOL   109
C                                                                       SOL   110
C-----  SOLVE DIAGONAL SYSTEM                                           SOL   111
C                                                                       SOL   112
120     IK=NEQ+1                                                        SOL   113
        JHK1=KLD(IK)                                                    SOL   114
130     IK=IK-1                                                         SOL   115
        IF(NSYM.EQ.1) VFG(IK)=VFG(IK)/VKGD(IK)                          SOL   116
        IF(IK.EQ.1) RETURN                                              SOL   117
        C1=VFG(IK)                                                      SOL   118
        JHK=KLD(IK)                                                     SOL   119
        JBK=JHK1-1                                                      SOL   120
        IF(JHK.GT.JBK)GO TO 150                                         SOL   121
        IJ=IK-JBK+JHK-1                                                 SOL   122
        DO 140 JCK=JHK,JBK                                              SOL   123
        VFG(IJ)=VFG(IJ)-VKGS(JCK)*C1                                    SOL   124
140     IJ=IJ+1                                                         SOL   125
150     JHK1=JHK                                                        SOL   126
        GO TO 130                                                       SOL   127
        END                                                             SOL   128

        FUNCTION SCAL(X,Y,N)                                            SCAL    1
C=================================================================SCAL   2
C       INNER PRODUCT OF VECTORS X AND Y OF LENGTH N                    SCAL    3
C         (FUNCTION TO BE WRITTEN EVENTUALLY IN ASSEMBLER)              SCAL    4
C=================================================================SCAL   5
        IMPLICIT REAL*8(A-H,O-Z)                                        SCAL    6
        DIMENSION X(1),Y(1)                                             SCAL    7
        DATA ZERO/0.0D0/                                                SCAL    8
C----------------------------------------------------------------SCAL    9
        SCAL=ZERO                                                       SCAL   10
        DO 10 I=1,N                                                     SCAL   11
10      SCAL=SCAL+X(I)*Y(I)                                             SCAL   12
        RETURN                                                          SCAL   13
        END                                                             SCAL   14
```

Figure 5.15 Subroutine SOL for the resolution of a symmetric or non-symmetric linear system with skyline matrix storage. This subroutine is used in program MEF of Chapter 6.

5.2.4.2 Skyline Matrix stored out of core by blocks

Algorithm (5.48) still applies when symmetrical matrix [K] resides out of core as described in paragraph 4.6.3(g). The algorithm successively modifies the terms in columns $s = 2, 3, \ldots, n$. Since the columns are stored by blocks, we alternatively read them from storage and write out blocks $1, 2, \ldots, n_b$ after modifications.

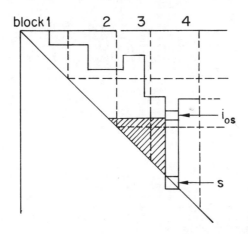

To modify the terms of a column s for which the high term is in row i_{0s}, the algorithm uses the terms of columns $i_{0s} + 1, \ldots s - 1$. Therefore, to modify all the columns in a given block, it is necessary to use the terms of the preceding blocks connected to that given block. Array KPB contains the integer identification of the first block connected to each block. For example, in the previous sketch, we have:

	Blocks
	1 2 3 4 ...
Number of the first connected block	1 1 1 2 ...

Hence: $\quad\quad\quad\quad KPB = \langle 1 \quad 1 \quad 1 \quad 2 \quad \ldots \rangle$

Array KPB is constructed from the information contained in arrays KLD and KEB defined in paragraph 4.6.3. A block I is connected to block J if:

$$KEB(I+1) > \min_{\text{block } J}(i_{0s})$$

which, after using (5.47), gives:

$$KEB(I+1) - 1 > \min_{i}(i - KLD(i+1) + KLD(i)) \quad\quad (5.49)$$

for i varying from $KEB(J)$ to $KEB(J+1) - 1$.

We use two blocks in core at all times in array VKGS: the block containing the

```
      SUBROUTINE SOLD(VKGS,VKGD,VKGI,VFG,KLD,NEQ,MP,IFAC,ISOL,NSYM,ENERGSOLD   1
     1 ,KEB,KPB)                                                       SOLD   2
C====================================================================SOLD   3
C        TO SOLVE A LINEAR SYSTEM (SYMMETRICAL OR NOT).                 SOLD   4
C        THE MATRIX IS STORED ON FILE M4 BY SKYLINES.                   SOLD   5
C        AFTER TRIANGULARIZATION IT IS STORED ON FILE M5                SOLD   6
C          INTPUT                                                       SOLD   7
C            VKGS,VKGD,VKGI     SYSTEM MATRIX : UPPER, DIAGONAL AND LOWER SOLD   8
C                               PARTS                                   SOLD   9
C            VFG                SECOND MEMBER                           SOLD  10
C            KLD                ADDRESSES OF COLUMN TOP TERMS           SOLD  11
C            NEQ                NUMBER OF EQUATIONS                     SOLD  12
C            MP                 OUTPUT DEVICE NUMBER                    SOLD  13
C            IFAC               IF IFAC.EQ.1 TRIANGULARIZATION OF       SOLD  14
C                               THE MATRIX                              SOLD  15
C            ISOL               IF ISOL.EQ.1 COMPUTE SOLUTION FROM THE  SOLD  16
C                               TRIANGULARIZED MATRIX                   SOLD  17
C            NSYM               INDEX FOR NON SYMMETRIC PROBLEM         SOLD  18
C            KEB                NUMBER OF FIRST EQUATION IN EACH        SOLD  19
C                               BLOCK                                   SOLD  20
C            KPB                NUMBER OF FIRST BLOCK CONNECTED TO EACH SOLD  21
C                               BLOCK                                   SOLD  22
C          OUTPUT                                                       SOLD  23
C            VKGS,VKGD,VKGI     TRIANGULARIZED MATRIX (IF IFAC.EQ.1)    SOLD  24
C            VFG                SOLUTION (IF ISOL.EQ.1)                 SOLD  25
C            ENERG              SYSTEM ENERGY (IF NSYM.EQ.0)            SOLD  26
C====================================================================SOLD  27
      IMPLICIT REAL*8 (A-H,O-Z)                                         SOLD  28
      COMMON/LIND/NLBL,NBLM                                             SOLD  29
      COMMON/ES/M,MR,MP1,M1,M2,M3,M4,M5                                 SOLD  30
      DIMENSION VKGS(1),VKGD(1),VKGI(1),VFG(1),KLD(1),KEB(1),KPB(1)     SOLD  31
      DATA ZERO/0.0D0/                                                  SOLD  32
C--------------------------------------------------------------------SOLD  33
      REWIND M4                                                         SOLD  34
      REWIND M5                                                         SOLD  35
      IK=1                                                              SOLD  36
      IF(VKGD(1).NE.ZERO) GO TO 5                                       SOLD  37
      WRITE(MP,2000) IK                                                 SOLD  38
      STOP                                                              SOLD  39
5     ENERG=ZERO                                                        SOLD  40
C                                                                       SOLD  41
C------- FOR EACH BLOCK TO BE TRIANGULARIZED                            SOLD  42
C                                                                       SOLD  43
      J1MIN=NLBL+1                                                      SOLD  44
      J1MAX=NLBL+NLBL                                                   SOLD  45
      DO 105 IB=1,NBLM                                                  SOLD  46
C------- READ A BLOCK TO BE TRIANGULARIZED                              SOLD  47
      READ(M4) (VKGS(I),I=1,NLBL)                                       SOLD  48
      IF(NSYM.EQ.1) READ(M4) (VKGI(I),I=1,NLBL)                         SOLD  49
C------- PARAMATERS FOR BLOCK IB                                        SOLD  50
      IK0=KEB(IB)                                                       SOLD  51
      IK1=KEB(IB+1)-1                                                   SOLD  52
      IB0=KPB(IB)                                                       SOLD  53
      J0=KLD(IK0)-1                                                     SOLD  54
      IF(IB0.EQ.IB) GO TO 11                                            SOLD  55
C------- BACKSPACE ON CONNECTED BLOCKS                                  SOLD  56
      I1=IB-IB0                                                         SOLD  57
      DO 10 I=1,I1                                                      SOLD  58
      BACKSPACE M5                                                      SOLD  59
      IF(NSYM.EQ.1) BACKSPACE M5                                        SOLD  60
10    CONTINUE                                                          SOLD  61
C------- FOR EACH CONNECTED BLOCK (INCLUDING BLOCK IB ITSELF)           SOLD  62
11    DO 103 IBC=IB0,IB                                                 SOLD  63
      IF(IBC.EQ.IB) GO TO 12                                            SOLD  64
```

Figure 5.16 (*Contd.*)

```
              READ(M5) (VKGS(I),I=J1MIN,J1MAX)                        SOLD  65
              IF(NSYM.EQ.1) READ(M5) (VKGI(I),I=J1MIN,J1MAX)           SOLD  66
C------- PARAMETERS OF CONNECTED BLOCK                                 SOLD  67
12            IIO=KEB(IBC)                                             SOLD  68
              II1=KEB(IBC+1)-1                                         SOLD  69
              JCO=KLD(IIO)-1                                           SOLD  70
              IF(IBC.NE.IB) JCO=JCO-NLBL                               SOLD  71
C                                                                     SOLD  72
C------- FOR EACH COLUMN OF BLOCK IB TO BE MODIFIED                    SOLD  73
C                                                                     SOLD  74
              DO 100 IK=IKO,IK1                                        SOLD  75
              JHK=KLD(IK)-JO                                           SOLD  76
C----- ADDRESS OF NEXT COLUMN TOP TERM IK+1                            SOLD  77
              JHK1=KLD(IK+1)-JO                                        SOLD  78
C----- HEIGHT OF COLUMN IK (INCLUDE UPPER AND DIAGONAL TERMS)          SOLD  79
              LHK=JHK1-JHK                                             SOLD  80
              LHK1=LHK-1                                               SOLD  81
C----- ROW OF FIRST TERM TO BE MODIFIED IN COLUMN IK                   SOLD  82
              IMIN=IK-LHK1                                             SOLD  83
              IMIN1=IMIN-1                                             SOLD  84
C------- ROW OF LAST TERM TO BE MODIFIED IN COLUMN IK                  SOLD  85
              IMAX=IK-1                                                SOLD  86
              IF(LHK1.LT.0) GO TO 100                                  SOLD  87
              IF(IFAC.NE.1) GO TO 90                                   SOLD  88
              IF(NSYM.EQ.0) GO TO 14                                   SOLD  89
              IB1=IB                                                   SOLD  90
              IF(IMIN1.LT.IKO) IB1=IBO                                 SOLD  91
              IF(IBC.EQ.IB1) VKGI(JHK)=VKGI(JHK)/VKGD(IMIN1)           SOLD  92
14            IF(IBC.EQ.IB.AND.IK.EQ.IKO) GO TO 40                     SOLD  93
              IF(LHK1.EQ.0) GO TO 40                                   SOLD  94
C------- FIND FIRST AND LAST ROW OF COLUMN IK AFFECTED                 SOLD  95
C              BY CONNECTED BLOCK IBC                                  SOLD  96
              IMINC=MAX0(IMIN,IIO)                                     SOLD  97
              IMAXC=MIN0(IMAX,II1)                                     SOLD  98
              IF(IMINC.GT.IMAXC) GO TO 40                              SOLD  99
C                                                                     SOLD 100
C----- MODIFY NON DIAGONAL TERMS OF COLUMN IK                          SOLD 101
C                                                                     SOLD 102
              JCK=JHK+IMINC-IMIN1                                      SOLD 103
              JHJ=KLD(IMINC)-JCO                                       SOLD 104
C------- FOR EACH TERM TO BE MODIFIED, LOCATED AT JCK                  SOLD 105
              DO 30 IJ=IMINC,IMAXC                                     SOLD 106
              JHJ1=KLD(IJ+1)-JCO                                       SOLD 107
C----- NUMBER OF MODIFICATIVE TERMS OF COEFFICIENT LOCATED AT JCK      SOLD 108
              IC=MIN0(JCK-JHK,JHJ1-JHJ)                                SOLD 109
              IF(IC.LE.0.AND.NSYM.EQ.0) GO TO 20                       SOLD 110
              C1=ZERO                                                  SOLD 111
              IF(IC.LE.0) GO TO 17                                     SOLD 112
              J1=JHJ1-IC                                               SOLD 113
              J2=JCK-IC                                                SOLD 114
              IF(NSYM.EQ.1) GO TO 15                                   SOLD 115
              VKGS(JCK)=VKGS(JCK)-SCAL(VKGS(J1),VKGS(J2),IC)           SOLD 116
              GO TO 20                                                 SOLD 117
15            VKGS(JCK)=VKGS(JCK)-SCAL(VKGI(J1),VKGS(J2),IC)           SOLD 118
              C1=SCAL(VKGS(J1),VKGI(J2),IC)                            SOLD 119
17            VKGI(JCK)=(VKGI(JCK)-C1)/VKGD(IJ)                        SOLD 120
20            JCK=JCK+1                                                SOLD 121
30            JHJ=JHJ1                                                 SOLD 122
C                                                                     SOLD 123
C----- MODIFY DIAGONAL TERM                                            SOLD 124
C                                                                     SOLD 125
40            IF(IBC.NE.IB) GO TO 90                                   SOLD 126
              JCK=JHK                                                  SOLD 127
              CDIAG=ZERO                                               SOLD 128
```

Figure 5.16 (*Contd.*)

```
      DO 70 IJ=IMIN1,IMAX                                      SOLD 129
      C1=VKGS(JCK)                                             SOLD 130
      IF(NSYM.EQ.1) GO TO 50                                   SOLD 131
      C2=C1/VKGD(IJ)                                           SOLD 132
      VKGS(JCK)=C2                                             SOLD 133
      GO TO 60                                                 SOLD 134
   50 C2=VKGI(JCK)                                             SOLD 135
   60 CDIAG=CDIAG+C1*C2                                        SOLD 136
   70 JCK=JCK+1                                                SOLD 137
      VKGD(IK)=VKGD(IK)-CDIAG                                  SOLD 138
      IF(VKGD(IK)) 90,80,90                                    SOLD 139
   80 WRITE(MP,2000) IK                                        SOLD 140
 2000 FORMAT(' *** ERROR, ZERO PIVOT EQUATION ',I5)            SOLD 141
      STOP                                                     SOLD 142
C                                                              SOLD 143
C----- SOLVE LOWER TRIANGULAR SYSTEM                           SOLD 144
C                                                              SOLD 145
   90 IF(ISOL.NE.1) GO TO 100                                  SOLD 146
      IF(IBC.NE.IB) GO TO 100                                  SOLD 147
      IF(NSYM.NE.1) VFG(IK)=VFG(IK)-SCAL(VKGS(JHK),VFG(IMIN1),LHK)  SOLD 148
      IF(NSYM.EQ.1) VFG(IK)=VFG(IK)-SCAL(VKGI(JHK),VFG(IMIN1),LHK)  SOLD 149
  100 CONTINUE                                                 SOLD 150
C------- NEXT CONNECTED BLOCK                                  SOLD 151
  103 CONTINUE                                                 SOLD 152
C-------  END OF ELIMINATION OF THIS BLOCK                     SOLD 153
      IF(IB.EQ.NBLM) GO TO 105                                 SOLD 154
      WRITE(M5) (VKGS(I),I=1,NLBL)                             SOLD 155
      IF(NSYM.EQ.1)  WRITE(M5) (VKGI(I),I=1,NLBL)              SOLD 156
  105 CONTINUE                                                 SOLD 157
      IF(ISOL.NE.1) RETURN                                     SOLD 158
C                                                              SOLD 159
C----- SOLVE DIAGONAL SYSTEM                                   SOLD 160
C                                                              SOLD 161
      IF(NSYM.EQ.1) GO TO 120                                  SOLD 162
      DO 110 IK=1,NEQ                                          SOLD 163
      C1=VKGD(IK)                                              SOLD 164
      C2=VFG(IK)/C1                                            SOLD 165
      VFG(IK)=C2                                               SOLD 166
  110 ENERG=ENERG+C1*C2*C2                                     SOLD 167
C                                                              SOLD 168
C----- SOLVE UPPER TRIANGULAR SYSTEM                           SOLD 169
C                                                              SOLD 170
  120 IB=NBLM                                                  SOLD 171
      IK0=KEB(IB)-1                                            SOLD 172
      J0=KLD(IK0+1)-1                                          SOLD 173
      IK=NEQ+1                                                 SOLD 174
      JHK1=KLD(IK)-J0                                          SOLD 175
C-------  FOR EVERY EQUATION FROM NEQ TO 1                     SOLD 176
  130 IK=IK-1                                                  SOLD 177
C-------  READ A BLOCK IF REQUIRED                             SOLD 178
      IF(IK.NE.IK0) GO TO 135                                  SOLD 179
      BACKSPACE M5                                             SOLD 180
      IF(NSYM.EQ.1) BACKSPACE M5                               SOLD 181
      READ(M5) (VKGS(I),I=1,NLBL)                              SOLD 182
      IF(NSYM.EQ.1) READ(M5) (VKGI(I),I=1,NLBL)                SOLD 183
      BACKSPACE M5                                             SOLD 184
      IF(NSYM.EQ.1) BACKSPACE M5                               SOLD 185
      IB=IB-1                                                  SOLD 186
      IK0=KEB(IB)-1                                            SOLD 187
      J0=KLD(IK0+1)-1                                          SOLD 188
      JHK1=KLD(IK+1)-J0                                        SOLD 189
C-------  MODIFY THE UNKNOWN VECTOR                            SOLD 190
  135 IF(NSYM.EQ.1) VFG(IK)=VFG(IK)/VKGD(IK)                   SOLD 191
      IF(IK.EQ.1) RETURN                                       SOLD 192
```

Figure 5.16 (*Contd.*)

```
            C1=VFG(IK)                              SOLD 193
            JHK=KLD(IK)-J0                          SOLD 194
            JBK=JHK1-1                              SOLD 195
            IF(JHK.GT.JBK)GO TO 150                 SOLD 196
            IJ=IK-JBK+JHK-1                         SOLD 197
            DO 140 JCK=JHK,JBK                      SOLD 198
            VFG(IJ)=VFG(IJ)-VKGS(JCK)*C1            SOLD 199
        140 IJ=IJ+1                                 SOLD 200
        150 JHK1=JHK                                SOLD 201
            GO TO 130                               SOLD 202
            END                                     SOLD 203
```

Figure 5.16 Subroutine SOLD for the resolution of a symmetric or non-symmetric linear system with skyline matrix storage by blocks out of core. This subroutine is used in program MEF of Chapter 6.

columns to be modified and one of its connected block. Diagonal terms and vector $\{F\}$ remain in core in arrays VKGD and VFG, respectively. For a non-symmetrical matrix (algorithm 5.46), two more blocks from the lower triangular part of $[K]$ must also reside in core in array VKGI. Figure 5.16 contains the listing for subroutine SOLD implementing the solution of symmetric and non-symmetric systems for which skyline matrix $[K]$ is stored by blocks out of core.

5.3 Solution of Non-linear Systems

5.3.1 *Introduction*

Two types of non-linearities appear in the formulation of physical problems:

— the material properties of the physical systems may become dependent on the value of the nodal variables U_n; this is the case for example, in plasticity, non-Newtonian flows, flow through unsaturated porous media, etc.;
— geometrical non-linearities may appear in the fundamental governing equation of the problem; for example, the Navier–Stokes equations (Example 3.2) contain non-linear terms like:

$$u\frac{\partial u}{\partial x} + v\frac{\partial u}{\partial y} + \cdots$$

and for large displacement elasticity (Example 4.5):

$$\varepsilon_x = \frac{\partial u}{\partial x} + \frac{1}{2}\left(\frac{\partial v}{\partial x}\right)^2$$

The discretization of non-linear problems can be written in the form of equation (4.5b):

$$W = \langle \delta U_n \rangle ([K(U_n)]\{U_n\} - \{F\}) = 0 \text{ for all } \langle \delta U_n \rangle$$

after dropping subscript n, it becomes:

$$[K(U)]\{U\} = \{F\} \text{ or } \{R(U)\} = \{F\} - [K(U)]\{U\} = 0 \qquad (5.50)$$

For certain problems, such as plasticity, the formulation must be recast in an incremental form (5.50):

$$[K(U)]\{\Delta U\} = \{\Delta F\} \qquad (5.51)$$

The non-linear solution process consists in searching vector $\{U\}$ that renders the residual $\{R(U)\}$ as small as possible. An exact solution would make the residual vanish. The search for a solution is done iteratively:

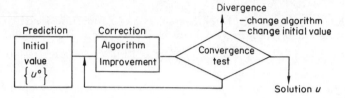

Most algorithms include the solution of systems of linear equations at each step of iteration. The choice of an algorithm must take many considerations into account:

— type of non-linearity and severity;
— spread of non-linearity, i.e. whether it is localized or not;
— existence of more than one solution;
— ease of construction of the code implementing the algorithm;
— precision of desired results;
— speed of convergence and risks of divergence.

No method is general enough to cover all types of non-linear problems; however, most methods are derived from the three following strategies singly or in combinations:

— substitution method;
— Newton–Raphson iteration;
— incremental method.

5.3.2 *Substitution Method*

The method consists in constructing a sequence of solutions $\{U^0\}, \{U^1\}, \ldots, \{U^i\}$, where $\{U^i\}$ is calculated from the previous value $\{U^{i-1}\}$ and solving the linear system:

$$[K(U^{i-1})]\{U^i\} = \{F\}; \quad i = 1, 2, 3 \ldots \qquad (5.52)$$

which can be written in a more convenient incremental form as:

$$\begin{aligned} \{R^i\} &= \{R(U^{i-1})\} = \{F\} - [K(U^{i-1})]\{U^{i-1}\} \\ [K(U^{i-1})]&\{\Delta U^i\} = \{R^i\} \\ \{U^i\} &= \{^{i-1}\} + \{\Delta U^i\} \end{aligned} \qquad (5.53)$$

Supposing the vector $\{U^{i-1}\}$ is known, we can then compute $[K(U^{i-1})]$ and the residue $\{R^i\}$. Using any of the methods discussed in paragraph 5.2, we can solve the linearized system (5.53) for $\{\Delta U^i\}$. Obviously, the starting value $\{U^0\}$ is not known but in many cases it is possible to assume it to be zero. The convergence of solution may depend upon the type of non-linearity, the choice of initial values and loading amplitude.

Algorithm for substitution method.
— Select initial approximation $\{U^0\}$ (very often zero).
— Construct $\{F\}$ from element values $\{f\}$.

$i = 1, 2, \ldots$ (for each iteration)

for each element
Extract $\{u^{i-1}\}$ from $\{U^{i-1}\}$
compute $[k(u^{i-1})]$
compute element residual

$$\{r\} = \{f\} - [k]\{u^{i-1}\}$$

assemble linearized system matrices
$[K]$ and $[R^i]$

solve linear system $[K]\{\Delta U^i\} = \{R^i\}$
update value of solution

$$U^i = U^{i-1} + \omega\{\Delta U^i\}$$

(ω is called an over-relaxation factor)
compute the norm $\|n\|$ of $\{\Delta U^i\}$ or
$\|m\|$ of $\{R^i\}$

— execute convergence test with $\|n\|$ or $\|m\|$

Remarks

(a) *Over-relaxation*

An over-relaxation factor ω, used in (5.54), will frequently improve the convergence rate but its numerical value is problem dependent and usually determined by numerical experimentation. For plasticity problems, values between 1.7 and 1.9 have often been used. If $\omega = 1$, over-relaxation disappears.

(b) *Norms*

Two different vector norms are commonly used in convergence tests:

— the maximum norm:

$$\|n\| = \underset{j}{\text{Max}} |\Delta U_j|^i \quad \text{or} \quad \|m\| = \underset{j}{\text{Max}} |R_j|^i \tag{5.55}$$

— the Euclidian norm:

$$\|n\| = \sqrt{\langle \Delta U^i \rangle \{\Delta U^i\}} \quad \text{or} \quad \|m\| = \sqrt{\langle R^i \rangle \{R^i\}} \tag{5.56}$$

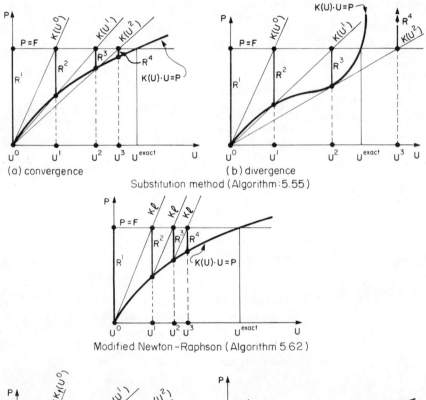

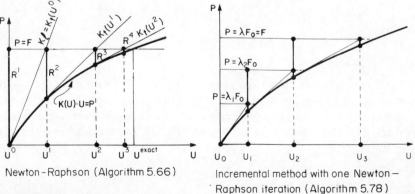

Figure 5.17 Graphical illustration of various algorithms for a one-variable problem.

Frequently, convergence tests use norms of relative values:

$$\|n\| = \underset{j}{\text{Max}} \left| \frac{\Delta U_i}{U_j} \right|^i \tag{5.57}$$

(if U_j is zero or so small that the norm becomes useless, the average of the absolute values of U_j is frequently substituted for U_j); for Euclidian norms we get:

$$\|n\| = \frac{\sqrt{\langle \Delta U^i \rangle \{\Delta U^i\}}}{\sqrt{\langle U^i \rangle \{U^i\}}} \tag{5.58}$$

The iterative process is terminated when the previous norms are sufficiently small:

$$\|n\| < \varepsilon$$

where ε is an arbitrarily selected small number, for example, $\varepsilon = 0.05$ for the norm of (5.58).

For a problem having one unknown, the algorithm (5.54) can be characterized by graphs of Figure 5.17; possibilities of divergence are evident on these graphs.

Example 5.16 *Non-linear spring, using substitution method*

$$k.U = F$$
thus $$(1 - U).U = F$$

Starting with an initial assumption $U^0 = 0$, the solution proceeds as illustrated in the table shown below:

Iteration i	U^{i-1}	$k = 1 - U^{i-1}$	$R^i = F - k(U^{i-1}).U^{i-1}$	$\Delta U^i = k^{-1} R^i$	$U^i = U^{i-1} + \Delta U^i$	$\|n\|$ (5.57)
1	0	1	0.2	0.2	0.2	1
2	0.2	0.8	0.04	0.05	0.25	0.2
3	0.25	0.75	0.0125	0.0167	0.2667	0.06
4	0.2667	0.733	0.0044	0.006	0.2727	0.02

The exact value of U is $0.5 \mp \sqrt{0.05}$, i.e. 0.2764 or 0.7236.

In equation (5.53), we split $[K]$ into a matrix of constants $[K_l]$ and a matrix of non-linear terms $[K_{nl}]$, we get:

$$([K_l] + [K_{nl}(U^{i-1})])\{\Delta U^i\} = \{R^i\} \tag{5.59}$$

In algorithm (5.54), $[K]$ must be assembled, then factored at every iteration step. Neglecting $[K_{nl}]$ in (5.59), we get:

$$[K_l]\{\Delta U^i\} = \{R^i\}$$
$$\{U^i\} = \{U^{i-1}\} + \{\Delta U^i\} \qquad (5.60)$$

Matrix $[K_l]$ being constant needs to be assembled and factored only once at the beginning of the process. The algorithm corresponding to (5.60) is often called the modified substitution or Newton–Raphson method.

Modified Newton–Raphson algorithm

Select or compute an approximate solution $\{U^0\}$; a null vector can be used in absence of any better initial value.
Assemble $\{F\}$ using element vectors $\{f\}$.
Assemble $[K_l]$ using element $[k_i]$.
Factorize $[K_l]$

$i = 1, 2, \ldots$ (for each iteration)
    For each element
        Extract $\{u^{i-1}\}$ from $\{U^{i-1}\}$;
        Assemble residual $[R^i]$ using elemental residual

$$\{r\} = \{f\} - [k]\{u^{i-1}\} \qquad (5.61)$$

    Solve $[K_l]\{\Delta U^i\} = \{R^i\}$
    Compute $\{U^i\} = \{U^{i-1}\} + \{\Delta U^i\}$
    Compute $\|n\|$
Execute convergence test.

This algorithm is graphically illustrated in Figure 5.17 for problems with one variable.

Example 5.17 *Non-linear spring using modified Newton–Raphson algorithm*

In the preceding example:

$$k = (1 - U) = k_l + k_{nl}(U)$$

where
$$k_l = 1$$
$$k_{nl} = -U$$

Using algorithm (5.61):

Iteration i	U^{i-1}	k_i	$k =$ $1 - U^{i-1}$	$R^i =$ $F - k.U^{i-1}$	$\Delta U^i =$ $k_i^{-1} R^i$	$U^i =$ $U^{i-1} + \Delta U^i$	$\|n\|$ (5.57)
1	0	1	1	0.2	0.2	0.2	1
2	0.2	1	0.8	0.04	0.04	0.24	0.166
3	0.24	1	0.76	0.0176	0.0176	0.2576	0.068
4	0.2576	1	0.7424	0.0087	0.0087	0.2663	0.032

In algorithm (6.61), matrix $[K_t]$ is assembled and decomposed only once while algorithm (5.54) necessitated an assembly and decomposition of $[K]$ for every iteration. Algorithm (5.61) is very often used for problems that are known to be only mildly non-linear; in general, when the severity of non-linearities is not known, the Newton–Raphson method explained in the next section will converge more rapidly.

5.3.3 Newton–Raphson Method

Presume that an approximation U^{i-1} has the following residual:

$$\{R(U^{i-1})\} = \{F\} - [K(U^{i-1})]\{U^{i-1}\} \neq 0 \quad (5.62)$$

For iteration i, we look for an approximation U^i such that:

$$\{R(U^i)\} = \{R(U^{i-1} + \Delta U^i)\} \approx 0 \quad (5.63)$$

If in a Taylor series the residual function is developed in the neighbourhood of U^{i-1}:

$$\{R(U^{i-1} + \Delta U^i)\} = \{R(U^{i-1})\} + \left[\frac{\partial R}{\partial U}\right]_{U=U^{i-1}}\{\Delta U^i\} + \ldots = 0 \quad (5.64)$$

Neglecting all terms of order higher than one:

$$-\left[\frac{\partial R}{\partial U}\right]\{\Delta U^i\} = \{R(U^{i-1})\}$$

or:

$$\begin{aligned}\{K_t(U^{i-1})\}\{\Delta U^i\} &= \{R(U^{i-1})\} \\ \{U^i\} &= \{U^{i-1}\} + \{\Delta U^i\}\end{aligned} \quad (5.65)$$

Matrix $[K_t(U^{i-1})]$ is obtained by differentiating expression (5.50):

$$[K_t(U)] = -\left[\frac{\partial R}{\partial U}\right] = -\left[\frac{\partial F}{\partial U}\right] + [K(U)] + \left[\frac{\partial [K(U)]}{\partial U}\{U\}\right] \quad (5.66)$$

When F is independent of U:

$$[K_t(U)] = [K(U)] + \left[\frac{\partial [K(U)]}{\partial U}\{U\}\right] \quad (5.67)$$

or, using individual terms ij of $[K_t]$ and $[K]$

$$(K_t)_{ij} = K_{ij} + \sum_l \frac{\partial K}{\partial U_j} U_l$$

Expression (5.65) leads to an algorithm that is similar to algorithm (5.54) where $[K]$ is replaced by $[K_t]$. It is illustrated graphically in Figure 5.17 for problems with one variable.

Example 5.18 Non-linear spring using Newton–Raphson

$$k_t = k + \frac{\partial k}{\partial U} U = (1 - U) + (-1)U = 1 - 2U$$

Iteration i	U^{i-1}	$k_t =$ $1-2U^{i-1}$	$k =$ $1-U^{i-1}$	$R^i =$ $F-kU^{i-1}$	$\Delta U^i =$ $k_t^{-1} R^i$	$U^i =$ $U^{i-1}+\Delta U^i$	$\|n\|$ (5.57)
1	0	1	1	0.2	0.2	0.2	1
2	0.2	0.6	0.8	0.04	0.0667	0.2667	0.25
3	0.2667	0.466	0.7333	0.0044	0.0095	0.2762	0.03

Construction of $[K_t]$

Global tangent matrix $[K_t]$ is obtained by assembling element tangent matrices $[k_t]$. However, it is impracticable to use (5.66) for calculating $[k_t(u)]$ since the derivation of $[k(u)]$ would require that its terms be defined explicitly which is something to be avoided. It is much easier to derive the expression for $[k_t]$ starting from the integral form before discretization (4.3):

$$W(\mathbf{u}) = \sum_e W^e = \sum_e \int_{V^e} \delta(\partial \mathbf{u}) R(\mathbf{u}) \, dV = 0 \qquad (5.68)$$

A Taylor series development of W in the neighbourhood of U^{i-1} gives:

$$W(\mathbf{u}^i) = W(\mathbf{u}^{i-1}) + \Delta(W)_{=\,i-1} + \ldots = 0 \qquad (5.69)$$

where $\Delta(W)$ is the first variation of W not to be confused in general with δ terms appearing in (5.68). Using a finite element discretization of u, we get:

$$W(U^{i-1}) = \langle \delta U_n \rangle ([K(U^{i-1})]\{U_n^{i-1}\} - \{F\})$$
$$= -\langle \delta U_n \rangle \{R(U^{i-1})\} \qquad (5.70)$$

$$\Delta W(U^{i-1}) = \langle \delta U_n \rangle [K_t(U^{i-1})]\{\Delta U^i\} \qquad (5.71)$$

where $\{\Delta U^i\}$ replaces variation $\{\Delta U_n\}$ of $\{U_n\}$ during one iteration. Equation

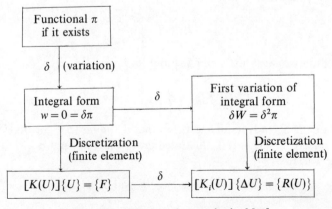

Figure 5.18 Construction method of $[K_t]$.

(5.69) becomes:

$$\langle \delta U_n \rangle ([K_t(U^{i-1})]\{\Delta U^i\} - \{R(U^{i-1})\}) = 0 \quad \text{for all} \quad \langle \delta U_n \rangle \quad (5.72)$$

which is identical to (5.65).

$[K_t]$ can thus be obtained by a direct discretization of ΔW. Figure 5.18 illustrates both methods of construction of $[K_t]$.

Example 5.19   Matrices $[k]$ and $[k_t]$ for a beam

The element integral form of a beam subject to large displacement was obtained in Example 4.5:

$$W^e = \int \left(\langle \delta\varepsilon \rangle [D]\{\varepsilon\} - \langle \delta u\, \delta w \rangle \begin{Bmatrix} f_u \\ f_w \end{Bmatrix} \right) dx$$

where:
$$\{\varepsilon\} = \begin{Bmatrix} \varepsilon_m \\ \kappa \end{Bmatrix} = \begin{Bmatrix} \dfrac{du}{dx} + \dfrac{1}{2}\left(\dfrac{dw}{dx}\right)^2 \\ -\dfrac{d^2 w}{dx^2} \end{Bmatrix}$$

$$[D] = \begin{bmatrix} EA & 0 \\ 0 & EI \end{bmatrix}$$

$$\delta(W^e) = \int (\langle \delta\varepsilon \rangle [D]\{\delta\varepsilon\} + \langle \delta^2\varepsilon \rangle [D]\{\varepsilon\}) dx$$

where:
$$\{\delta\varepsilon\} = \begin{Bmatrix} \delta\left(\dfrac{du}{dx}\right) + \dfrac{dw}{dx}\delta\left(\dfrac{dw}{dx}\right) \\ -\delta\left(\dfrac{d^2 w}{dx^2}\right) \end{Bmatrix}$$

$$\{\delta^2\varepsilon\} = \begin{Bmatrix} \delta\left(\dfrac{dw}{dx}\right)\delta\left(\dfrac{dw}{dx}\right) \\ 0 \end{Bmatrix}$$

since:
$$\delta^2 u = \delta u^2 w = 0$$

Discretization with finite elements:

$$u = \langle N_u \rangle \{u_n\}$$
$$w = \langle N_w \rangle \{w_n\}$$

$$\{\varepsilon\} = ([B_l] + [B_{nl}])\begin{Bmatrix} u_n \\ w_n \end{Bmatrix} = [B]\{u^e\}$$

$$[B_l] = \begin{bmatrix} \langle N_{u,x} \rangle & 0 \\ 0 & -\langle N_{w,xx} \rangle \end{bmatrix}$$

$$[B_{nl}] = \begin{bmatrix} 0 & \tfrac{1}{2}(\langle N_{w,x} \rangle\{w_n\})\langle N_{w,x} \rangle \\ 0 & 0 \end{bmatrix}$$

$$\{\delta\varepsilon\} = ([B_l] + 2[B_{nl}])\{\delta u^e\}$$

Then:

$$W^e = \langle \delta u^e \rangle ([k]\{u^e\} - \{f^e\})$$

$$[k] = \int ([B_l] + 2[B_{nl}])^T [D]([B_l] + [B_{nl}]) dx$$

This last matrix is non-symmetric; however, it has an equivalent symmetrical form (see Example 4.5):

$$[k] = [k_l] + [k_{nl}^1] + [k_{nl}^2] + [k_\sigma]$$

$$[k_l] = \int [B_l]^T [D][B_l] dx$$

$$[k_{nl}^1] = \int ([B_{nl}]^T [D][B_l] + [B_l]^T [D][B_{nl}]) dx$$

$$[k_{nl}^2] = \int [B_{nl}]^T [D][B_{nl}] dx$$

$$[k_\sigma] = \begin{bmatrix} 0 & 0 \\ 0 & \int \{N_{w,x}\}(EA\varepsilon_m)\langle N_{w,x}\rangle dx \end{bmatrix}$$

$$\delta W^e = \langle \delta u^e \rangle [k_t]\{\delta u^e\}$$

$$[k_t] = [k_l] + 2[k_{nl}^1] + 4[k_{nl}^2] + [k_\sigma]$$
$$= [k] + [k_{nl}^1] + 3[k_{nl}^2]$$

In this example, we represented $\Delta(W)$ by $\delta(W)$ in order to reduce unnecessary algebra for reorganizing various terms.

If a functional π exists, matrices $[k_t]$ and $[K_t]$ are symmetric, that is to say, if:

$$\pi = \pi(U_1, U_2, \ldots, U_n) \tag{5.73}$$

where $U_1, U_2, \ldots, U_n$ are the nodal variables

$$\delta W = \delta^2 \pi = \langle \delta U_n \rangle [K_t]\{\delta U_n\} \tag{5.74}$$

and

$$(K_t)_{ij} = \frac{\partial^2 \pi}{\partial U_i \partial U_j} = (K_t)_{ji} = \frac{\partial^2 \pi}{\partial U_j \partial U_i}$$

5.3.4 Incremental Method (Step by Step)

The initial value of the solution $\{U^0\}$, in the previous iteration methods, plays a primordial role. A bad choice of initial value may easily result in a diverging process.

The incremental method consists in replacing the solution of:

$$\{K(U)\}\{U\} = \lambda\{F_0\} = \{F\} \tag{5.75}$$

by successive solutions of:
$$\{K(U_j)\}\{U_j\} = \lambda_j\{F_0\} \tag{5.76}$$
where
$$\lambda_j = \lambda_1, \lambda_2, \ldots, \lambda$$

The initial value to be used in the calculation of U_j is the solution U_{j-1} obtained in the previous step. Each step is a non-linear problem that can be solved by one or more iterations of the Newton–Raphson or modified Newton–Raphson method. If one Newton–Raphson iteration is used in each step of the incremental method, we get:

$$\{R(U_{j-1})\} = \lambda_{j-1}\{F_0\} - [K(U_{j-1})]\{U_{j-1}\}$$
$$[K_t(U_{j-1})]\{\Delta U_j\} = \{R(U_{j-1})\} + (\lambda_j - \lambda_{j-1})\{F_0\} \tag{5.77}$$
$$\{U_j\} = \{U_{j-1}\} + \{\Delta U_j\}$$

This algorithm is shown graphically in Figure 5.17 for a one-variable problem. If more than one Newton–Raphson iteration is used, in each step we get:

$$\{K_t(U_j^{i-1})]\{\Delta U_j^i\} = \{R(U_j^{i-1})\} + (\lambda_j - \lambda_{j-1})\{F_0\}$$
$$\{U_j^i\} = \{U_j^{i-1}\} + \{\Delta U_j^i\} \quad i = 2, 3, \ldots \tag{5.78}$$

For $i = 1$, (5.77) is used.

Example 5.20 Non-linear spring: incremental method

Applying load $F = 0.2$ in two increments:
$$\lambda_1 = 0.5 \quad \lambda_2 = 1 \quad F_0 = 0.2$$

Using algorithm (5.78):

Step j	$\lambda_j F_0$	Iteration i	U_{j-1} or $(U_j^i \approx {}^1)$	$k(U_{j-1})$	R	$k_t(U_{j-1})$	ΔU	U_j^i
1	0.1	1	0	1	0.1	1	0.1	0.1
2	0.2	1	0.1	0.9	0.11	0.8	0.1375	0.2375
	0.2	2	0.2375	0.7625	0.019	0.525	0.0362	0.2737

5.3.5 Change of Independent Variables [12]

Until now, loads have been given and nodal variables $\{U_n\}$ were the unknowns.

For many problems more than one solution $\{U_n\}$ exists for a certain load level $\lambda\{F_0\}$:

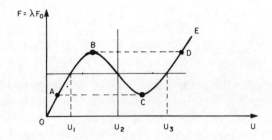

The methods presented so far could give solutions in regions OAB and EDC. To get a solution in domain BC, we chose a given value for a component u_l of the vector $\{U\}$, the load level parameter λ thus becomes unknown. Notice that between B and C, there corresponds only one value $\lambda\{F\}$ for a given U_l.

Algorithm (5.78) becomes:

$$[K_t]\{\Delta U\} = \{R\} + \Delta\lambda\{F_0\} \tag{5.79}$$

Moving $\Delta\lambda$ to the left-hand side in the vector of unknowns and imposing $\Delta U_l = 0$:

$$\begin{bmatrix} K_{t_{11}} & \cdots & & \cdots & K_{t_{1n}} \\ \vdots & & & & \vdots \\ & & -F_0 & & \\ \vdots & & & & \vdots \\ K_{t_{n1}} & \cdots & & \cdots & K_{t_{nn}} \end{bmatrix} \begin{Bmatrix} \Delta U_1 \\ \vdots \\ \Delta U_{l-1} \\ \Delta\lambda \\ \Delta U_{l+1} \\ \vdots \\ \Delta U_n \end{Bmatrix} = \{R\} \tag{5.80}$$

and:

$$U_l = \bar{U}_l$$

Replacing column l of $[K_t]$ by $\{F_0\}$ destroys its symmetry and changes its bandwidth. To avoid such an inconvenience, we can solve (5.79) twice with different load vectors but using the same matrix $[K_t]$.

$$[K_t]\{\Delta U^R\} = \{R\}$$
$$[K_t]\{\Delta U^F\} = \{F_0\}. \tag{5.81}$$

The solution of (5.79) is then:

$$\{\Delta U\} = \{\Delta U^R\} + \Delta\lambda\{\Delta U^F\} \tag{5.82}$$

Unknown $\Delta\lambda$ is obtained from:

$$\Delta U_l = 0 = \Delta U_l^R + \Delta\lambda \Delta U_l^F$$

$$\Delta\lambda = -\frac{\Delta U_l^R}{\Delta U_l^F} \tag{5.83}$$

The corresponding algorithm is:
> Choice of load increment j
>> Modify component l of U_{j-1}:
>> $$(U_{j-1})_l = \bar{U}_l$$
>>> Iteration i
>>> Compute residual $\{R\}$
>>> Compute matrix $[K_t]$
>>> Solve $[K_t]\{\Delta U^R\} = \{R\}$
>>> $$[K_t]\{\Delta U^F\} = \{F_0\}$$
>>> Compute $\Delta\lambda = -\dfrac{\Delta U_l^R}{\Delta U_l^F}$ (5.84)
>>> $\{U\} = \{U\} + \{\Delta U^R\} + \Delta\lambda\{\Delta U^F\}$
>>> $\lambda = \lambda + \Delta\lambda$
>> Convergence test

Different variations of this algorithm may be useful in developing automatic solution strategies, such as automatic choice of $\Delta\lambda$. That is, another criterion may be employed for selecting λ than that given by (5.83).

5.3.6 Solution Strategy

All the previous methods are contained in the following algorithm which, for a given load level, is:

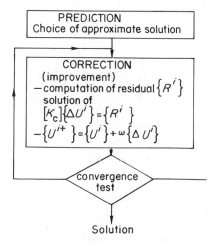

For any method of solution, the expression for the residual $\{R\}$ remains the same. However, the expression of $[K_c]$ varies from one method to the other and affects the rate of convergence:

$[K_c] \equiv [K]$ for the substitution method
$[K_c] \equiv [K_l]$ for the modified Newton–Raphson method
$[K_c] \equiv [K_t]$ for the Newton–Raphson method.

Other expressions of $[K_c]$ lead to iterative procedures used in solving systems of linear equations. An over-relaxation factor ω is often very effective in improving the convergence rate. For example:

$$[K_c] \equiv \begin{bmatrix} \ddots & & 0 \\ & K_{ii} & \\ 0 & & \ddots \end{bmatrix} \text{ for the Jacobi method}$$

$$[K_c] \equiv \begin{bmatrix} K_{11} & 0 & \cdots & 0 \\ K_{21} & K_{22} & & \vdots \\ \vdots & & \ddots & 0 \\ K_{n1} & \cdots & \cdots & K_{nn} \end{bmatrix} \text{ for the Gauss–Seidel method}$$

Figure 5.19 details an algorithm common to the various methods in which the following choices must be made:

```
                    Choice of ω, Δλ IMETH
                           IKT = 1
                           λ = 0
Load step j: λ = λ + Δλ
   Prediction: {U_j^0} = {U_{j-1}} or extrapolation of
              {U_{j-1}}, {U_{j-2}}, ...
j = 0
   Iteration: i = i + 1
      Elements:
         Computation and assemblage of residual {R(λ, F_0, U^{i-1})}
         If IKT equals 1: computation and assemblage of
         [K_c(U^{i-1})], depending on the value of IMETH

      If IKT equals 1: decomposition of [K_c]
      solution of [K_c]{ΔU} = {R}
      Correction of U: {U^i} = {U^{i-1}} + ω{ΔU}
      Relative norms of ‖n‖, {ΔU} or {R}
      Output of iteration i

      Strategy: choice of IKT, IMETH, ω, Δλ
             Convergence test

         Output of load level at iteration j
```

Figure 5.19 General algorithm for the solution of non-linear problems (Newton–Raphson type).

— number and size of load increment $\Delta\lambda$;
— maximum number of iterations per incremental step and convergence criteria (type of norm and desired accuracy);
— type of matrix to be used for each iteration leading to $\{\Delta U\}$; depending on the value of IMETH, we could select $[K_1]$ of (5.60), $[K(U^{i-1})]$ of (5.53), or $[K_t(U^{i-1})]$ of (5.67);
— a predictor formula to be used for the first iteration of each load increment;
— a value for the over-relaxation factor ω.

The various choices to be made depend upon the problem to be solved; certain problems require very small incremental loads and for such cases it conceivably could require very few iterations per step. Note that the algorithm of Figure 5.19 is almost identical to the one of Figure 5.21 for non-linear time dependent problems.

5.4 Solution of Time-dependent Problems 11, 13, 17

5.4.1 *Introduction*

Time-dependent problems are often called propagation problems. The finite element geometrical discretization of a propagation problem results in systems of ordinary differential equations such as:

— first order system

$$[C]\{\dot{U}\} + [K]\{U\} = \{F\} \quad \text{for} \quad t > t_0 \qquad (5.85)$$

and
$$\{U(t_0)\} = \{U_0\}$$

— second order system

$$[M]\{\ddot{U}\} + [C]\{\dot{U}\} + [K]\{U\} = \{F\} \quad \text{for} \quad t > t_0 \qquad (5.86)$$

and
$$\{U(t_0)\} = \{U_0\}; \quad \{\dot{U}(t_0)\} = \{\dot{U}_0\}$$

where:

$$\{\ddot{U}\} = \frac{\partial}{\partial t}\{\dot{U}\}; \quad \{\ddot{U}\} = \frac{\partial^2}{\partial t^2}\{U\} = \frac{\partial}{\partial t}\{\dot{U}\}$$

$[M]$ is the mass matrix;
$[C]$ is the damping or thermal capacitance matrix, etc.;
$[K]$ is the stiffness or conductivity matrix, etc.;
$\{F\}$ is a right-hand side load vector;
$\{U(t)\}$ is the vector of unknown values.

For a linear system, $[M]$, $[C]$, $[K]$, and $\{F\}$ are independent of $\{U\}$ and its derivatives. Moreover, in many physical problems, $[M]$, $[C]$, and $[K]$ are independent of the time.

In a non-linear system, $[K]$ and sometimes $[C]$ and $[M]$ are dependent on $\{U\}$ and its derivatives.

Note that it is always possible to transform a differential equation of order n into n simultaneous equations of first order. However, this is rarely done for the second order equations encountered in vibration problems because more efficient direct methods have been developed for such cases.

Example 5.21   Transformation of a second order system in two simultaneous equations of first order

Consider

$$[M]\{\ddot{U}\} + [C]\{\dot{U}\} + [K]\{U\} = \{F\}$$

let

$$\{\dot{U}\} = \{V\}$$
$$[M]\{\dot{V}\} + [C]\{V\} + [K]\{U\} = \{F\}$$
$$[I]\{\dot{U}\} - [I]\{V\} = 0$$
$$\begin{bmatrix} [M] & 0 \\ 0 & [I] \end{bmatrix} \begin{Bmatrix} \dot{V} \\ \dot{U} \end{Bmatrix} + \begin{bmatrix} [C] & [K] \\ -[I] & 0 \end{bmatrix} \begin{Bmatrix} V \\ U \end{Bmatrix} = \begin{Bmatrix} F \\ 0 \end{Bmatrix}$$
$$[C']\{\dot{U}'\} + [K']\{U'\} = \{F'\}$$

where $[I]$ is a unit matrix. In this new formula, the number of unknowns has doubled and the symmetry of the systems has been destroyed.

To solve systems (5.85) or (5.86) one must find time dependent functions $\{U(t)\}$ satisfying (5.85) or (5.86) at all times, including specified initial values at $t = t_0$. Two methods shall be described:

— direct integration, or step by step method;
— mode superposition methods.

Direct integration methods consist in a sequential construction of values of the solution from its initial value $\{U_0\}$ at times: $t_0 + \Delta t, t_0 + 2\Delta t, \ldots, t_0 + n\Delta t, \ldots$

$$\{U(t_0)\} \to \{U(t_0 + \Delta t)\} \to \{U(t_0 + 2\Delta t)\} \to \cdots \to \{U(t_0 + n\Delta t)\} \tag{5.87}$$

A rich variety of methods can be derived from finite difference approximation formulas of $\{U\}$ and $\{\ddot{U}\}$. Zienkiewicz has also proposed a finite element discretization of the time domain leading to new formulas [8].

In the modal superposition methods, the coupled systems (5.85) or (5.86) are first decoupled into modal equations. Each modal equation can then be solved numerically and the final results obtained by a linear recombination of the various modes.

5.4.2 Direct Integration Methods for First Order Systems

5.4.2.1 Explicit Euler's Method

Algorithm

Systems of differential equations of first order can be written as:
$$\{\dot{U}\} = \{f(\{U\},t)\} \quad \text{for} \quad t > t_0$$
$$\{U(t_0)\} = \{U_0\} \tag{5.88}$$

where, for system (5.85), we would have:
$$\{f\} = [C]^{-1}(\{F\} - [K]\{U\}) \tag{5.89}$$

Using the following left offset finite difference approximation for $\{\dot{U}\}$:
$$\{\dot{U}(t)\} = \{\dot{U}_t\} \approx \frac{1}{\Delta t}(\{U_{t+\Delta t}\} - \{U_t\}). \tag{5.90}$$

into equation (5.88) written at time t, we get:
$$\{U_{t+\Delta t}\} = \{U_t\} + \Delta t\{f(\{U_t\},t)\} \tag{5.91}$$

The last recurrence formula is said to be explicit because $\{U_{t+\Delta t}\}$ does not appear in the right-hand side of the equation. The formula is self-starting at $t = t_0$.

Example 5.22 *Solution of a one-variable first order equation by explicit Euler's formula*

Consider the following equation with only one variable:
$$\frac{du}{dt} + u = 0 \quad t > 0 \quad \text{thus} \quad \frac{du}{dt} = f(u,t) = -u$$
$$u_0 = 1 \qquad t = 0$$

Algorithm (5.91) for this case is:
$$u_{t+\Delta t} = u_t - \Delta t \cdot u_t = (1 - \Delta t)u_t$$

at instant $t = n\Delta t$:
$$u_{n\Delta t} = (1 - \Delta t)^n u_0$$

The value of $u_{n\Delta t}$ stays bounded when n tends to infinity if
$$|1 - \Delta t| \leq 1:$$
$$-1 \leq 1 - \Delta t \quad \text{thus} \quad \Delta t \leq 2$$

furthermore, to prevent numerical oscillations as n increase we must have:
$$0 \leq 1 - \Delta t \quad \text{or} \quad \Delta t \leq 1$$

t	0	0.1	0.2	0.3	0.4	0.5	0.6
$\Delta t = 0.2$	1	—	0.800	—	0.640	—	0.512
$\Delta t = 0.1$	1	0.900	0.810	0.729	0.656	0.590	0.531
Exact (e^{-t})	1	0.905	0.819	0.741	0.670	0.607	0.549

In practice, to solve (5.85), Euler's algorithm is reorganized using (5.89) in order to replace the inversion of $[C]$ by successive resolutions from the following linear systems:

$$[C]\{U_{t+\Delta t}\} = \Delta t\{F_t\} + ([C] - \Delta t[K])\{U_t\} \qquad (5.92)$$

Another variation can be written as:

$$[C]\{\Delta U\} = \Delta t(\{F_t\} - [K]\{U_t\}) = \{R_t\}$$
$$\{U_{t+\Delta t}\} = \{U_t\} + \{\Delta U\} \qquad (5.93)$$

An algorithm corresponding to (5.93) is contained in Figure 5.20. The numerical efficiency of (5.93) is greatly improved when $[C]$ is a diagonal matrix.

Stability

As we have seen, for a particular case in Example 5.22, explicit Euler's algorithm is subject to numerical stability if Δt becomes greater or equal to a critical value Δt_c.

Rewriting (5.92) as:

$$\{U_{t+\Delta t}\} = [B]\{F_t\} + [A]\{U_t\}$$
$$[B] = \Delta t[C]^{-1}; \quad [A] = [I] - \Delta t[C]^{-1}[K] \qquad (5.94)$$

we get, by recurrence:

$$\{U_{t_0+n\Delta t}\} = [A]^n\{U_0\} + [A]^{n-1}[B]\{F_0\} + [A]^{n-2}[B]\{F_{\Delta t}\} + \cdots +$$
$$+ [B]\{F_{t_0+(n-1)\Delta t}\} \qquad (5.95)$$

```
    t = t_0
      Define {U_0}, Δt
        Construct [C]
Triangularize [C]
For each time step
    t = t + Δt
    Construct {R_t} = Δt({F_t} − [K]{U_t})
    Solve using triangularized [C]
          [C]{ΔU} = {R_t}
    Compute: {U_{t+Δt}} = {U_t} + {ΔU}
```

Figure 5.20 Explicit Euler's algorithm: incremental form (5.93).

If $\{U_{t_0+n\Delta t}\}$ is to stay bounded as n tends towards infinity, the spectral radius $\rho(A)$ of matrix $[A]$ must be smaller or at most equal to one:

$$\rho(A) = \max |\lambda_i| \leq 1 \tag{5.96}$$

where λ_i are the eigenvalues of $[A]$.
If l_i are the eigenvalues of $[C]^{-1}[K]$, we get:

$$\lambda_i = 1 - \Delta t l_i \tag{5.97}$$

For the case when $[C]$ and $[K]$ are positive definite matrices, stability criteria (5.96) becomes:

$$-\leq 1 - \Delta t l_{max} \quad \text{thus} \quad \Delta t \leq \frac{2}{l_{max}} \tag{5.98}$$

$$l_{max} = \max(l_i)$$

Moreover, if oscillations of $\{U\}$ are to be prevented as n increases (see Example 5.22), we must have:

$$0 \leq 1 - \Delta t l_{max} \quad \text{thus} \quad \Delta t \leq \frac{1}{l_{max}} = \Delta t_c. \tag{5.99}$$

Note that the stability criteria has nothing to do with the precision of the solution. To get an acceptable result, the time step Δt must also be chosen to reduce the truncation error to an acceptable value. In summary, the choice of the time step must be guided simultaneously by a stability criteria and an acceptable truncation error.

Non-linear problems

Explicit Euler's formula can be applied directly to non-linear problems in which $[K]$ may be a function of $\{U\}$ but $[C]$ is constant. Since the equations are evaluated at time t, matrix $[K]$ can be calculated explicitly from $\{U_t\}$. All that is needed is to replace $[K]$ by $[K(U_t)]$ in (5.92) and (5.93); for example, (5.93) becomes:

$$[C]\{\Delta U\} = \Delta t(\{F_t\} - [K(U_t)]\{U_t\}) \tag{5.100}$$
$$\{U_{t+\Delta t}\} = \{U_t\} + \{\Delta U\}$$

Because of the non-linearity of $[K]$, a stability criterion is more difficult to define explicitly.

Example 5.23 *Solution of a two-variable first order system using the explicit Euler's method*

$$[C]\{\dot{U}\} + [K]\{U\} = \{F\} \quad \text{for} \quad t > 0; \quad \{U\} = \{0\} \quad \text{for} \quad t = 0$$

$$[C] = \begin{bmatrix} 1 & 1 \\ 1 & 2 \end{bmatrix}; \quad [K] = \begin{bmatrix} 1 & 1 \\ 1 & 3 \end{bmatrix}; \quad \{F\} = \begin{Bmatrix} 2 \\ 3 \end{Bmatrix}$$

Stability condition (5.99) is:

$$\Delta t \le \frac{1}{l_{max}}$$

where l_{max} is the largest eigenvalue of:

$$[C]^{-1}[K] = \begin{bmatrix} 1 & 1 \\ 1 & 2 \end{bmatrix}^{-1} \begin{bmatrix} 1 & 1 \\ 1 & 3 \end{bmatrix}$$

$$= \begin{bmatrix} 2 & -1 \\ -1 & 1 \end{bmatrix} \begin{bmatrix} 1 & 1 \\ 1 & 3 \end{bmatrix} = \begin{bmatrix} 1 & -1 \\ 0 & 2 \end{bmatrix}$$

thus $l_{max} = 2$. Therefore:

$$\Delta t \le \tfrac{1}{2}$$

Choosing $\Delta t = 0.1$ and applying the algorithm of Figure 5.20

$$t = 0$$
$$\{U_0\} = \{0\} \quad \Delta t = 0.1$$
$$[C] = \begin{bmatrix} 1 & 1 \\ 1 & 2 \end{bmatrix} = \begin{bmatrix} 1 & 0 \\ 1 & 1 \end{bmatrix} \begin{bmatrix} 1 & 1 \\ 0 & 1 \end{bmatrix}$$

Step number 1:

$$t = 0.1$$

$$\{R_t\} = 0.1 \left(\begin{Bmatrix} 2 \\ 3 \end{Bmatrix} - \begin{bmatrix} 1 & 1 \\ 1 & 3 \end{bmatrix} \begin{Bmatrix} 0 \\ 0 \end{Bmatrix} \right) = \begin{Bmatrix} 0.2 \\ 0.3 \end{Bmatrix}$$

$$[C]\{\Delta U\} = \begin{Bmatrix} 0.2 \\ 0.3 \end{Bmatrix} \to \{\Delta U\} = \begin{Bmatrix} 0.1 \\ 0.1 \end{Bmatrix}$$

$$\{U_1\} = \{U_0\} + \{\Delta U\} = \begin{Bmatrix} 0.1 \\ 0.1 \end{Bmatrix}$$

Step number 2:

$$t = 0.2$$

$$\{R_t\} = 0.1 \left(\begin{Bmatrix} 2 \\ 3 \end{Bmatrix} - \begin{bmatrix} 1 & 1 \\ 1 & 3 \end{bmatrix} \begin{Bmatrix} 0.1 \\ 0.1 \end{Bmatrix} \right) = \begin{Bmatrix} 0.18 \\ 0.26 \end{Bmatrix}$$

$$[C]\{\Delta U\} = \begin{Bmatrix} 0.18 \\ 0.26 \end{Bmatrix} \to \{\Delta U\} = \begin{Bmatrix} 0.10 \\ 0.08 \end{Bmatrix}$$

$$\{U_2\} = \{U_1\} + \{\Delta U\} = \begin{Bmatrix} 0.20 \\ 0.18 \end{Bmatrix}$$

Finally, the solution obtained is $\{U(t)\} = \begin{Bmatrix} u_1 \\ u_2 \end{Bmatrix}$:

t	u_1	u_2
0	0	0
0.1	0.1	0.1
0.2	0.20	0.18
0.3	0.298	0.244
0.4	0.393	0.295
0.5	0.483	0.336
0.6	0.568	0.369
0.7	0.648	0.395
0.8	0.723	0.416
0.9	0.792	0.433
1.0	0.856	0.446
2.0	1.263	0.494
3.0	1.416	0.499
4.0	1.471	0.5
5.0	1.499	0.5
6.0	1.5	0.5
10.0	1.5	0.5

Note that for t very large, $\{U\}$ tends towards the solution of: $[K]\{U\} = \{F\}$.

5.4.2.2 Implicit Euler's Method

Algorithm

In this method, equation (5.88) is written for time $t + \Delta t$ and derivatives are approximated by a left offset finite difference approximation:

$$\{\dot{U}_{t+\Delta t}\} \approx \frac{1}{\Delta t}(\{U_{t+\Delta t}\} - \{U_t\}) \tag{5.101}$$

gives the following recurrence formula for system (5.88):

$$\{U_{t+\Delta t}\} = \{U_t\} + \Delta t \{f(\{U_{t+\Delta t}\}, t + \Delta t)\} \tag{5.102}$$

In this case, the value of $\{f\}$ contains the unknowns $\{U_{t+\Delta t}\}$. To adapt (5.102) for systems of type (5.85), we write:

$$[\bar{K}]\{U_{t+\Delta t}\} = \{\bar{R}_{t+\Delta t}\} \tag{5.103}$$

where:
$$[\bar{K}] = [C] + \Delta t [K]$$
$$\{\bar{R}_{t+\Delta t}\} = \Delta t \{F_{t+\Delta t}\} + [C]\{U_t\}$$

or using $\{\Delta U\}$:

$$[\bar{K}]\{\Delta U\} = \{\bar{R}_{t+\Delta t}\} - [\bar{K}]\{U_t\} = \{R_{t+\Delta t}\} \tag{5.104}$$

where
$$\{R_{t+\Delta t}\} = \Delta t(\{F_{t+\Delta t}\} - [K]\{U_t\})$$
$$\{U_{t+\Delta t}\} = \{U_t\} + \{\Delta U\}$$

An algorithm corresponding to (5.104) is contained in Figure 5.21, if we take $\alpha = 1$.

Example 5.24 Solution of a non-variable first order system with implicit Euler's method

Implicit equation (5.103) corresponding to the equation used in Example 5.22 is:
$$(1 + \Delta t)u_{t+\Delta t} = u_t$$
or
$$u_{t+\Delta t} = \frac{1}{1+\Delta t}u_t$$
thus
$$u_{n\Delta t} = \left(\frac{1}{1+\Delta t}\right)^n u_0$$

Since $(1/1 + \Delta t) < 1$ for all $\Delta t > 0$, the method is unconditionally stable. The results are:

t	0	0.1	0.2	0.3	0.4	0.5	0.6
$\Delta t = 0.2$	1	—	0.833	—	0.694	—	0.578
$\Delta t = 0.1$	1	0.909	0.826	0.751	0.683	0.620	0.564
Exact (e^{-1})	1	0.905	0.819	0.741	0.670	0.607	0.549

Example 5.25 Solution of a two-variable first order system with implicit Euler's method

Consider the system of Example 5.23:
Using (5.104), we get:
$$[\bar{K}] = \begin{bmatrix} 1 & 1 \\ 1 & 2 \end{bmatrix} + 0.1\begin{bmatrix} 1 & 1 \\ 1 & 3 \end{bmatrix} = \begin{bmatrix} 1.1 & 1.1 \\ 1.1 & 2.3 \end{bmatrix}$$

Step number 1:
$$\{R_{t+\Delta t}\} = 0.1\left(\begin{Bmatrix} 2 \\ 3 \end{Bmatrix} - \begin{bmatrix} 1 & 1 \\ 1 & 3 \end{bmatrix}\begin{Bmatrix} 0 \\ 0 \end{Bmatrix}\right) = \begin{Bmatrix} 0.2 \\ 0.3 \end{Bmatrix}$$

$$\begin{bmatrix} 1.1 & 1.1 \\ 1.1 & 2.3 \end{bmatrix}\{\Delta U\} = \begin{Bmatrix} 0.2 \\ 0.3 \end{Bmatrix} \rightarrow \{\Delta U\} = \begin{Bmatrix} 0.098\,484\,8 \\ 0.083\,333\,3 \end{Bmatrix}$$

$$\{U_1\} = \begin{Bmatrix} 0.098\,484\,8 \\ 0.083\,333\,3 \end{Bmatrix}$$

Solution:

t	u_1	u_2
0	0	0
0.1	0.098	0.083
0.2	0.194	0.153
0.3	0.287	0.211
0.4	0.375	0.259
0.5	0.459	0.299
0.6	0.538	0.333
0.7	0.613	0.360
0.8	0.683	0.384
0.9	0.749	0.403
1.0	0.810	0.419
2.0	1.216	0.487
3.0	1.387	0.498
4.0	1.456	0.499
5.0	1.483	0.5
6.0	1.493	0.5
10.0	1.5	0.5

Stability

Expression (5.103) can be written as follows:

$$\{U_{t+\Delta t}\} = [B]\{F_{t+\Delta t}\} + [A]\{U_t\}$$
$$[B] = \Delta t([C] + \Delta t[K])^{-1} \qquad (5.105)$$
$$[A] = ([I] + \Delta t[C]^{-1}[K])^{-1}$$

The stability criteria based on the spectral radius (5.180) of $[A]$ is:

$$\rho(A) \leq 1 \qquad (5.106)$$

thus:

$$(1 + \Delta t . l_i) \geq 1 \qquad (5.107)$$

where l_i are the eigenvalues of $[C]^{-1}[K]$. When $[C]$ and $[K]$ are both positive definite, (5.106) is always verified; implicit Euler's method is thus unconditionally stable for such systems.

Non-linear problems

For non-linear problems in which $[C]$ is constant and $[K]$ a function of $\{U\}$, a non-linear expression (5.103) of the form of (5.50) must be solved for each time step. The correction of $\{\Delta U^i\}$ in $\{U_{t+\Delta t}\}$ for an iteration step is obtained by solving:

$$[K_{nl}]\{\Delta U^i\} = \{R_{nl}\} \tag{5.108}$$

where

$$\begin{aligned}\{R_{nl}\} &= \{\bar{R}_{t+\Delta t}\} - [\bar{K}(U_{t+\Delta t}^{i-1})]\{U_{t+\Delta t}^{i-1}\} \\ &= \Delta t(\{F_{t+\Delta t}\} - [K(U_{t+\Delta t}^{i-1})]\{U_{t+\Delta t}^{i-1}\}) + [C](\{U_t\} - \{U_{t+\Delta t}^{i-1}\}) \end{aligned} \tag{5.109}$$

In the substitution method (paragraph 5.3.2)

$$[K_{nl}] = [\bar{K}] = [C] + \Delta t[K(U_{t+\Delta t}^{i-1})] \tag{5.110}$$

In the Newton–Raphson method (paragraph 5.3.3)

$$[K_{nl}] = [K_t] = [\bar{K}] + \Delta t\left[\frac{\partial K}{\partial U} \cdot U\right]_{t+\Delta t}^{i-1} \tag{5.111}$$

In practice, the construction of $[K_t]$ by differentiation of $[K]$ is sometimes very complicated. It is preferable then to discretize the first variation $\Delta(W)$ of the integral form W (see paragraph 5.3.3).

5.4.2.3 Semi-implicit Euler's method

Algorithm

The method starts by writing equation (5.88) at time $t + \alpha\Delta t$, with $0 \le \alpha \le 1$. Euler's formula can thus be written:

$$\{U_{t+\Delta t}\} = \{U_t\} + \Delta t\{f(\{U_{t+\alpha\Delta t}\}, t+\alpha\Delta t)\} \tag{5.112}$$

where $\quad \{U_{t+\alpha\Delta t}\} = \alpha\{U_{t+\Delta t}\} + (1-\alpha)\{U_t\}$

With $\alpha = 0$, we get explicit Euler's formula and with $\alpha = 1$, we get implicit Euler's formula. An algorithm of type (5.103) is:

$$[\bar{K}]\{U_{t+\Delta t}\} = \{\bar{R}_{t+\Delta t}\} \tag{5.113}$$

where:

$$[\bar{K}] = [C] + \alpha\Delta t[K]$$
$$\{\bar{R}_{t+\Delta t}\} = \Delta t(\alpha\{F_{t+\Delta t}\} + (1-\alpha)\{F_t\} - (1-\alpha)[K]\{U_t\}) + [C]\{U_t\}$$

The algorithm of type (5.104) is:

$$[\bar{K}]\{\Delta U\} = \{R_{t+\Delta t}\} \tag{5.114}$$

where:

$$\{R_{t+\Delta t}\} = \Delta t(\alpha\{F_{t+\Delta t}\} + (1-\alpha)\{F_t\} - [K]\{U_t\})$$
$$\{U_{t+\Delta t}\} = \{U_t\} + \{\Delta U\}$$

The various steps required to implement the algorithm are described in Figure 5.21.

Stability

Expression (5.113) can be written in a form analogous to (5.94):

$$\{U_{t+\Delta t}\} = [B](\alpha\{F_{t+\Delta t}\} + (1-\alpha)\{F_t\}) + [A]\{U_t\} \quad (5.115)$$

$$[B] = [\bar{K}]^{-1} \quad (5.116)$$

$$[A] = [\bar{K}]^{-1}([C] - (1-\alpha)\Delta t[K]) \quad (5.117)$$

Stability criteria $\rho(A) \leq 1$, for the case where $[C]$ and $[K]$ are positive definite, is:

$$(1 - 2\alpha)\Delta t l_{max} \leq 2 \quad (5.118)$$

This is always true for $\alpha \geq 0.5$. For $0 \leq \alpha < 0.5$ (5.118) gives:

$$\Delta t \leq \frac{2}{(1-2\alpha)l_{max}} \equiv \Delta t_c \quad (5.119)$$

where l_{max} is the largest eigenvalue of $[C]^{-1}[K]$. For a stable solution without numerical oscillation, we must have $(1-\alpha)\Delta t l_{max} < 1$. Domains of stability are shown in the following diagram:

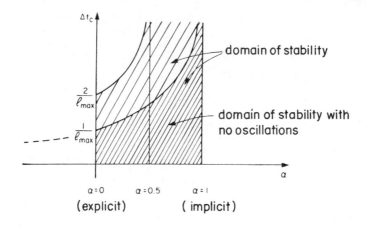

For $\alpha = 0.5$, the algorithm becomes the popular Crank–Nicholson formula.
The formula (5.113) gives the truncation error of the order $0(\Delta t^2)$ for $\alpha \neq 0.5$ and $0(\Delta t^3)$ for $\alpha = 0.5$.

Non-linear problems

There is an algorithm of the same type as (5.108) for non-linear problems. The residual corresponding to (5.114) is:

$$\{R_{nl}\} = \Delta t(\alpha\{F_{t+\Delta t}\} + (1-\alpha)\{F_t\} - (1-\alpha)[K(U_t)]\{U_t\}$$
$$- \alpha[K(U_{t+\Delta t}^{i-1})]\{U_{t+\Delta t}^{i-1}\}) + [C](\{U_t\} - \{U_{t+\Delta t}^{i-1}\}) \quad (5.120)$$

```
                    Choice of α, ω, Δt, IMETH
                              IKT = 1
                              t = t₀
 ┌─ Time step: t = t + Δt
 │  ┌─ Prediction: {U⁰_{t+Δt}} = {U_t} or extrapolation of
 │  │              {U_t}, {U_{t−Δt}}, {U_{t−2Δt}} ...
 │  │  i = 0
 │  │  Iteration: i = i + 1
 │  │  ┌─ Elements:
 │  │  │  Computation and assembly of residual {R_{n1}}
 │  │  └─ If IKT is 1: computation and assembly of [K_t] or [K̄]
 │  │              depending on the value of IMETH
 │  │     If KIT is 1: decomposition of [K_t]
 │  │     Solution of [K_t]{ΔUⁱ} = {R_{n1}}
 │  │     Correction of {U_{t+Δt}}:{Uⁱ_{t+Δt}} = {U^{i−1}_{t+Δt}} + ω{ΔUⁱ}
 │  │     Computation of relative norms ‖n‖ of {ΔUⁱ} or {R_{n1}}
 │  │     Output of iteration i
 │  └─ Strategy: choice of IKT, IMETH, ω, α, Δt,
 │                 convergence criteria
 │     Convergence test
 └─    Output of time step t
```

Figure 5.21 Semi-implicit Euler's algorithm, for unsteady and non-linear problems, using the Newton–Raphson method.

Matrix $[K_{nl}]$ is given below:

— for the substitution method:

$$[K_{nl}] = [\bar{K}] = [C] + \alpha\Delta t[K(U^{i-1}_{t+\Delta t})] \tag{5.121}$$

— for the Newton–Raphson method:

$$[K_{nl}] = [K_t] = [\bar{K}] + \alpha\Delta t\left[\frac{\partial K}{\partial U}.U\right]^{i-1}_{t+\Delta t} \tag{5.122}$$

where $[\bar{K}]$ is found in (5.113).

The various steps required to implement the previous algorithm are listed in Figure 5.21. The steps are analogous to the steps contained in Figure 5.19, if we interpret the load increment as a time step. Where $[C] = 0$, we rediscover the non-linear algorithm of Figure 5.19 for stationary problems.

Program MEF of Chapter 6 contains a functional block ('TEMP') implementing the algorithm of Figure 5.21.

5.4.2.4 *Predictor–Corrector Formulas* [13, 15]

In these methods, an explicit formula is used to predict the first value of $\{U_{t+\Delta t}\}$, then an implicit formula is used to improve that value by an iteration in place.

Explicit predictor formulas

For system like (5.88), we have:

$$\{U^0_{t+\Delta t}\} = \{U_t\} + \Delta t \sum_{j=1}^{n} b_j \{f(U_{t+\Delta t - j\Delta t}, t + \Delta t - j\Delta t)\} \quad (5.123)$$

For formulation (5.85), we have:

$$[C]\{\Delta U\} = \Delta t \sum_{j=1}^{n} b_j\{R_j\}; \quad \{U_{t+\Delta t}\} = \{U_t\} + \{\Delta U\} \quad (5.124)$$

$$\{R_j\} = \{F_{t+\Delta t - j\Delta t}\} - [K(U_{t+\Delta t - j\Delta t})]\{U_{t+\Delta t - j\Delta t}\} \quad (5.125)$$

The coefficients b_j depend on the selected formula:

	n	Error	b_1	b_2	b_3	b_4
Formula of order 2 (Euler explicit)	1	$O(\Delta t^2)$	1	0	0	0
Formula of order 3	2	$O(\Delta t^3)$	$\frac{3}{2}$	$-\frac{1}{2}$	0	0
Formula of order 4	3	$O(\Delta t^4)$	$\frac{23}{12}$	$-\frac{16}{12}$	$\frac{5}{12}$	0
Formula of order 5	4	$O(\Delta t^5)$	$\frac{55}{24}$	$-\frac{59}{24}$	$\frac{37}{24}$	$-\frac{9}{24}$

Corrector formulas

For systems like (5.88), we have:

$$\{U_{t+\Delta t}\} = \{U_t\} + \Delta t \sum_{j=0}^{n} b_j\{f(U_{t+\Delta t - j\Delta t}, t + \Delta t - j\Delta t)\} \quad (5.126)$$

For systems like (5.85), the formula is identical to (5.124) except that index j starts at a zero value. The corresponding coefficients b_j are:

	n	Error	b_0	b_1	b_2	b_3
Formula of order 3 (of Euler type with $\alpha = 0.5$)	1	$O(\Delta t^3)$	$\frac{1}{2}$	$\frac{1}{2}$	0	0
Formula of order 4	2	$O(\Delta t^4)$	$\frac{5}{12}$	$\frac{8}{12}$	$-\frac{1}{12}$	0
Formula of order 5	3	$O(\Delta t^5)$	$\frac{9}{24}$	$\frac{19}{24}$	$-\frac{5}{24}$	$\frac{1}{24}$

Since $\{U_{t+\Delta t}\}$ appears in $\{f\}$, one or more iterations are needed for each time step:

$$[C]\{\Delta U^i\} = \{\bar{R}\}; \quad \{U^i_{t+\Delta t}\} = \{U_t\} + \{\Delta U^i\} \quad (5.127)$$

$${\bar{R}} = \Delta t b_0 \{R_0\} + \Delta t \sum_{j=1}^{n} b_j \{R_j\} \qquad (5.128)$$

The values of $\{R_j\}$ are given by (5.125) and stay constant at each step.

$$\{R_0\} = \{F_{t+\Delta t}\} - [K(U_{t+\Delta t}^{i-1})]\{U_{t+\Delta t}^{i-1}\} \qquad (5.129)$$

The combined use of (5.124) and (5.127) is very effective if $[C]$ is diagonal; all operations are then carried on vectors. For linear problems, the corrector formula converges if the eigenvalues of $[C]^{-1}\Delta t b_0[K]$ are smaller than 1 in absolute value. If $[C]$ is diagonal, this criteria is verified when:

$$\Delta t \leq \frac{1}{b_0 \underset{i}{\text{Max}} \sum_j \left|\frac{K_{ij}}{C_{ii}}\right|} \equiv \Delta t_c. \qquad (5.130)$$

The same algorithm (5.127) can be used for non-linear problems. However, (5.130) may give a very small value to the time step Δt. The following modification of (5.127) will generally allow a greater value for Δt_c.

$$[C']\{\Delta U^i\} = \{\bar{R}'\} \qquad (5.131)$$

where:

$[C'] = [C] + \Delta t b_0 [K_l];$

$\{\bar{R}'\} = \Delta t b_0 \{R_0'\} + \Delta t \sum_{j=1}^{n} b_j \{R_j\};$

$\{R_0'\} = \{F_{t+\Delta t}\} - [K_{nl}(U_{t+\Delta t}^{i-1})]\{U_{t+\Delta t}^{i-1}\};$

$[K_l]$ and $[K_{nl}]$ are the linear and non-linear parts of $[K]$.

The convergence criteria becomes;

$$\rho([C']^{-1}\Delta t b_0[K_{nl}]) \leq 1 \qquad (5.132)$$

Note that matrices $[C]$ of (5.127) and $[C']$ of (5.131) are assembled and factored only once.

All methods, except explicit Euler methods, use results acquired in previous steps; they must therefore use special starting procedures to begin the process. Euler explicit formula or any Runge–Kutta formulas (described in next paragraph) could be used to start the step by step solution.

Example 5.26 *Solution of a two-variable first order problem using predictor–corrector formulas*

The problem to be solved is described in Example 5.23. We use starting values obtained in Example 5.27 (obtained by a Runge–Kutta formula).

A fourth order predictor formula (5.124) is:

$$\{\Delta U\} = \Delta t [C]^{-1}(\{F\} - \tfrac{1}{12}[K](23\{U_t\} - 16\{U_{t+\Delta t}\} + 5\{U_{t-2\Delta t}\}))$$
$$\{U_{t+\Delta t}^0\} = \{U_t\} + \{\Delta U\}$$

The corresponding corrector (5.126) is:

$$\{\Delta U^1\} = \Delta t[C]^{-1}(\{F\} - \tfrac{1}{12}[K](5\{U^0_{t+\Delta t}\} + 8\{U_t\} - \{U_{t-\Delta t}\}))$$
$$\{U^1_{t+\Delta t}\} = \{U_t\} + \{\Delta U^1\}$$

For $\Delta t = 0.1$ and 0.4, the tabulated results shown below are obtained:

t	$\Delta t = 0.1$		$\Delta t = 0.4$	
	u_1	u_2	u_1	u_2
0.0	0.0	0.0	0.0	0.0
0.1	0.099 690 5*	0.090 634 6*		
0.2	0.197 698 5*	0.164 840 0*		
0.3	0.293	0.226		
0.4	0.384	0.275	0.378 666 7*	0.282 666 7*
0.5	0.471	0.316		
0.6	0.553	0.349		
0.7	0.630	0.377		
0.8	0.702	0.399	0.698 453 3*	0.405 532 4*
1.2	0.943	0.455	0.950	0.446
1.6	1.112	0.480	1.125	0.469
2.0	1.238	0.491	1.242	0.485
4.0	1.464	0.500	1.463	0.500
6.0	1.495	0.500	1.495	0.500
8.0	1.499	0.500	1.499	0.500
10.0	1.500	0.500	1.500	0.500

* Solution obtained by Runge–Kutta.

5.4.2.5 *Explicit methods of Runge–Kutta type* [10, 13].

These methods solve (5.88) through multiple evaluations of $\{f(\{U\},t)\}$ within each time step Δt. They are all self-starting and pose no problems for step value changes. However, for a fixed value of Δt, the predictor–corrector formulas require a lesser amount of computations. A general expression for formulas of the Runge–Kutta type is:

$$\{U_{t+\Delta t}\} = \{U_t\} + \Delta t \sum_{j=1}^{n} b_j\{f(U_j, t_j)\} \qquad (5.133)$$

Order of the formula	n	Error	Formulas
3	1	$O(\Delta t^3)$	$b_1 = 1; t_1 = t + \dfrac{\Delta t}{2}; \quad \{U_1\} = \{U_t\} + \dfrac{\Delta t}{2}\{f(U_t, t)\}$
4	2	$O(\Delta t^4)$	$b_1 = \tfrac{1}{4}; t_1 = t; \quad \{U_1\} = \{U_t\}$
			$b_2 = \tfrac{3}{4}; t_2 = t + \dfrac{2\Delta t}{3}; \{U_2\} = \{U_t\} + \tfrac{2}{3}\Delta t$
			$\times \left\{ f\left(U^*, t + \dfrac{\Delta t}{3}\right)\right\}$
			$\{U^*\} = \{U_t\} + \dfrac{\Delta t}{3}\{f(U_t, t)\}$
5	4	$O(\Delta t^5)$	$b_1 = \tfrac{1}{6}; t_1 = t; \quad \{U_1\} = \{U_t\}$
			$b_2 = \tfrac{1}{3}; t_2 = t + \dfrac{\Delta t}{2}; \quad \{U_2\} = \{U_t\} + \tfrac{1}{2}\Delta t\{f(U_t, t)\}$
			$b_3 = \tfrac{1}{3}; t_3 = t + \dfrac{\Delta t}{2}; \quad \{U_3\} = \{U_t\} + \tfrac{1}{2}\Delta t\{f(U_2, t_2)\}$
			$b_4 = \tfrac{1}{6}; t_4 = t + \Delta t; \quad \{U_4\} = \{U_t\} + \Delta t\{f(U_3, t_3)\}$

Example 5.27   Solution of a two-variable first order problem using a Runge–Kutta formula

Consider the problem defined in Example 5.23 and a fourth order Runge–Kutta formula for its solution.

Formula (5.133) with its fourth order coefficients is:

$$\{U_{t+\Delta t}\} = \{U_t\} + \dfrac{\Delta t}{4}[C]^{-1}(4\{F\} - [K](\{U_t\} + 3\{U_2\}))$$

where

$$\{U_2\} = \{U_t\} + \tfrac{2}{3}\Delta t[C]^{-1}(\{F\} - [K]\{U^*\})$$

and

$$\{U^*\} = U_t + \dfrac{\Delta t}{3}[C]^{-1}(\{F\} - [K]\{U_t\})$$

For $\Delta t = 0.1$ and 0.4, the solution is shown in the following table:

t	$\Delta t = 0.1$		$\Delta t = 0.4$	
	u_1	u_2	u_1	u_2
0.0	0.0	0.0	0.0	0.0
0.1	0.996 905	0.906 346		
0.2	0.197 698 5	0.164 840 0		
0.3	0.293	0.226		
0.4	0.384	0.275	0.378 666 7	0.282 667
0.5	0.471	0.316		
0.6	0.553	0.349		
0.7	0.630	0.377		
0.8	0.702	0.399	0.698 453 3	0.405 532 4
1.2	0.943	0.455	0.941	0.459
1.6	1.117	0.480	1.116	0.482
2.0	1.238	0.491	1.239	0.492
4.0	1.464	0.500	1.464	0.500
6.0	1.495	0.500	1.495	0.500
8.0	1.499	0.500	1.499	0.500
10.0	1.500	0.500	1.500	0.500

5.4.3 *Modal Superposition for First Order System*

In this method, the coupled linear system (5.85) is first decoupled using the following transformation:

$$\{U(t)\} = [X]\{V(t)\} \tag{5.134}$$

where the transformation matrix $[X]$ is constituted of the n eigenvectors $\{X_i\}$ defined by (see paragraph 5.5):

$$([K] - \lambda_i[C])\{X_i\} = 0 \quad i = 1, 2, \ldots, n \tag{5.135}$$

$$[X] = [\{X_1\}\{X_2\}\ldots\{X_n\}] \tag{5.136}$$

Matrix $[X]$ satisfies the following orthogonality conditions, (5.175):

$$[X]^T[C][X] = [I]$$

$$[X]^T[K][X] = [\lambda] = \begin{bmatrix} \lambda_1 & & 0 \\ & \lambda_2 & \\ 0 & & \lambda_n \end{bmatrix} \tag{5.137}$$

System (5.85) transformed with (5.134) gives:

$$[X]^T[C][X]\{\dot{V}(t)\} + [X]^T[K][X]\{V(t)\} = [X]^T\{F(t)\} = \{\bar{F}(t)\}$$

which, after using (5.137), becomes:

$$\{\dot{V}(t)\} + [\lambda]\{V(t)\} = \{\bar{F}(t)\} \quad \text{for} \quad t > t_0 \tag{5.138}$$

This last system is completely decoupled and each equation can be integrated independently from all others. Initial conditions $\{V(t_0)\}$ are obtained by a premultiplication of (5.134) by $[X]^T[C]$:

$$[X]^T[C]\{U(t_0)\} = [X]^T[C][X]\{V(t_0)\} = \{V(t_0)\} \tag{5.139}$$

The general solution of the ith equation of (5.138) is:

$$V_i(t) = e^{-\lambda_i(t-t_0)}\left(V_i(t_0) + \int_{t_0}^{t} e^{\lambda_i(s-t_0)}\bar{F}_i(s)\,\mathrm{d}s\right) \tag{5.140}$$

Whenever expression $\{\bar{F}(t)\}$ allows it, expression (5.140) is integrated explicitly for $V_i(t)$, $i = 1, 2, \ldots, n$. The final solution $\{U(t)\}$ is a linear combination of eigenvectors $\{X_i\}$ obtained by substitution of $\{V_i\}$ in (5.134).

In practice, it is not necessary to consider all the modal components. The solution is usually dominated by a few modes and only the corresponding vectors are needed in $[X]$.

When $\{\bar{F}(t)\}$ is complicated, it may be more efficient to use a direct step by step integration of (5.138) as described in paragraph 5.4.2 instead of a numerical integration of (5.140).

In summary, modal superposition proceeds as follows:

— select the number of modes to be used, p;
— compute the smallest p eigenvalues and eigenvectors (see paragraph 5.5) of system (5.135);
— construct vector $\{\bar{F}\}$, then solve the p decoupled equations in (5.138);
— superpose the modes according to (5.134).

Example 5.28 *Solution of a first order system by modal superposition*

Consider the system defined in Example 5.23:

$$[C]\{\dot{U}\} + [K]\{U\} = \{F\} \quad (\text{for } t > 0);$$

$$[C] = \begin{bmatrix} 1 & 1 \\ 1 & 2 \end{bmatrix}; \quad [K] = \begin{bmatrix} 1 & 1 \\ 1 & 3 \end{bmatrix}; \quad \{F\} = \begin{Bmatrix} 2 \\ 3 \end{Bmatrix};$$

$$\{U(t=0)\} = \begin{Bmatrix} 0 \\ 0 \end{Bmatrix}$$

Eigenvalues and eigenvectors of:

$$([K] - \lambda[C])\{X\} = 0$$

are (see paragraph 5.5 and Example 5.31):

$$\lambda_1 = 1 \quad \{X_1\} = \begin{Bmatrix} 1 \\ 0 \end{Bmatrix}$$

$$\lambda_2 = 2 \quad \{X_2\} = \begin{Bmatrix} 1 \\ -1 \end{Bmatrix}$$

The decoupled system (5.138) is:

$$\{\dot{V}\} + \begin{bmatrix} 1 & 0 \\ 0 & 2 \end{bmatrix} \{V\} = \begin{bmatrix} 1 & 0 \\ 1 & -1 \end{bmatrix} \begin{Bmatrix} 2 \\ 3 \end{Bmatrix}$$

and

$$\{V_0\} = \begin{bmatrix} 1 & 1 \\ 0 & -1 \end{bmatrix} \begin{bmatrix} 1 & 1 \\ 1 & 2 \end{bmatrix} \begin{Bmatrix} 0 \\ 0 \end{Bmatrix} = \begin{Bmatrix} 0 \\ 0 \end{Bmatrix}$$

thus

$$\dot{V}_1 + V_1 = 2 \qquad V_1(t=0) = 0$$
$$\dot{V}_2 + 2V_2 = -1 \qquad V_2(t=0) = 0$$

The solutions of these two equations are (5.140):

$$V_1(t) = 2(1 - e^{-t})$$
$$V_2(t) = \tfrac{1}{2}(e^{-2t} - 1)$$

The solution of the coupled system is therefore (5.134):

$$\{U\} = \begin{bmatrix} 1 & 1 \\ 0 & -1 \end{bmatrix} \begin{Bmatrix} 2(1-e^{-t}) \\ \tfrac{1}{2}(e^{-2t}-1) \end{Bmatrix} = \begin{Bmatrix} \tfrac{3}{2} - 2e^{-t} + \tfrac{1}{2}e^{-2t} \\ \tfrac{1}{2} - \tfrac{1}{2}e^{-2t} \end{Bmatrix}$$

t	u_1	u_2
0	0	0
0.1	0.100	0.091
0.2	0.198	0.165
0.3	0.293	0.226
0.4	0.384	0.275
0.5	0.471	0.316
0.6	0.553	0.349
0.7	0.630	0.377
0.8	0.702	0.399
0.9	0.769	0.417
1.0	0.832	0.432
2.0	1.238	0.491
3.0	1.402	0.499
4.0	1.463	0.5
5.0	1.486	0.5
6.0	1.495	0.5
10.0	1.5	0.5
∞	1.5	0.5

5.4.4 Direct Integration of Second Order Systems [11, 16, 17]

5.4.4.1 Central Finite Difference Method

Recall the second order equation (5.86):

$$[M]\{\ddot{U}\} + [C]\{\dot{U}\} + [K]\{U\} = \{F(t)\} \quad t > t_0 \quad (5.141)$$

$\{U\}$ and $\{\dot{U}\}$ being given for $t = t_0$

The central finite difference method is explicit. It uses the governing equation evaluated at time t and the following central difference formulas for first and second order derivatives:

$$\{\ddot{U}_t\} \approx \frac{1}{\Delta t^2}(\{U_{t+\Delta t}\} - 2\{U_t\} + \{U_{t-\Delta t}\}) \quad (5.142a)$$

$$\{\dot{U}_t\} \approx \frac{1}{2\Delta t}(\{U_{t+\Delta t}\} - \{U_{t-\Delta t}\}) \quad (5.143b)$$

System (5.141) can then be written:

$$[\bar{K}]\{U_{t+\Delta t}\} = \{R_t\} \quad (5.143a)$$

where:
$$[\bar{K}] = [M] + \frac{\Delta t}{2}[C]$$

$$\{R_t\} = \Delta t^2 \{F_t\} + [M](2\{U_t\} - \{U_{t-\Delta t}\})$$
$$+ \frac{\Delta t}{2}[C]\{U_{t-\Delta t}\} - \Delta t^2[K]\{U_t\}$$

or in incremental form:

$$[\bar{K}]\{\Delta U\} = \{\bar{R}_t\}$$
$$\{\bar{R}_t\} = \Delta t^2\{F_t\} + [M](\{U_t\} - \{U_{t-\Delta t}\})$$
$$+ \frac{\Delta t}{2}[C](-\{U_t\} + \{U_{t-\Delta t}\}) - \Delta t^2[K]\{U_t\}$$

$$\{U_{t+\Delta t}\} = \{U_t\} + \{\Delta U\} \quad (5.143b)$$

At instant $t = t_0$, initial values of $\{U_0\}$ and $\{\dot{U}_0\}$ are known and $\{\ddot{U}_0\}$ can be calculated from the governing equation (5.141). Furthermore, $\{U_{t_0-\Delta t}\}$ is evaluated by eliminating $\{U_{t+\Delta t}\}$ from the two equations (5.142) written for time $t = t_0$:

$$\{U_{t_0-\Delta t}\} = \{U_0\} - \Delta t\{\dot{U}_0\} + \frac{\Delta t^2}{2}\{\ddot{U}_0\} \quad (5.144)$$

Figure 5.22 lists all the steps required to implement the algorithm described above.

The solution is stable provided the time step Δt is smaller than a threshold value depending on the smallest fundamental period $T_{\min}$ of the physical system under study. Furthermore, the truncation errors of (5.142a) and (5.142b) must be acceptable. The method being explicit, $[K]$ does not appear in the left-hand side of either (5.143a) nor (5.143b). For non-linear problems in which $[K]$ is a function of $\{U\}$, the method can still be applied directly using $K(\{U_t\})$. Non-linearities usually require smaller time step Δt.

5.4.4.2 Houbolt's method [18]

This implicit method uses expression (5.141) evaluated at time $t + \Delta t$ and the following offset finite difference approximations for derivatives:

$$\{\ddot{U}_{t+\Delta t}\} = \frac{1}{\Delta t^2}(2\{U_{t+\Delta t}\} - 5\{U_t\} + 4\{U_{t-\Delta t}\} - \{U_{t-2\Delta t}\}) \quad (5.145a)$$

$$\{\dot{U}_{t+\Delta t}\} = \frac{1}{6\Delta t}(11\{U_{t+\Delta t}\} - 18\{U_t\} + 9\{U_{t-\Delta t}\} - 2\{U_{t-2\Delta t}\}) \quad (5.145b)$$

The truncation error of these last two formulas is of order $(\Delta t)^2$. The corresponding algorithm is:

$$[\bar{K}]\{U_{t+\Delta t}\} = \{R_{t+\Delta t}\}$$

where: $\quad [\bar{K}] = 2[M] + \frac{11}{6}\Delta t[C] + \Delta t^2[K] \quad (5.146)$

$$\{R_{t+\Delta t}\} = \Delta t^2 \{F_{t+\Delta t}\} + [M](5\{U_t\} - 4\{U_{t-\Delta t}\} + \{U_{t-2\Delta t}\})$$
$$+ \Delta t[C](3\{U_t\} - \tfrac{3}{2}\{U_{t-\Delta t}\} + \tfrac{1}{3}\{U_{t-2\Delta t}\}) \quad (5.147)$$

Introducing $\{\Delta U\}$:

$$[\bar{K}]\{\Delta U\} = \{R_{t+\Delta t}\} - [\bar{K}]\{U_t\}$$
$$\{U_{t+\Delta t}\} = \{U_t\} + \{\Delta U\}$$

This algorithm is unconditionally stable; however, it is not self-starting. Other methods must be used to start a solution. Figure 5.22 describes the steps required to implement the formula in a code.

For non-linear problems in which $[M]$ and $[C]$ are constant but $[K]$ depends on $\{U\}$, the correction $\{U_{t+\Delta t}\}$ during each iteration within a time step is:

$$[K_{nl}]\{\Delta U^i\} = \{R_{t+\Delta t}\} - [\bar{K}(U_{t+\Delta t}^{i-1})]\{U_{t+\Delta t}^{i-1}\}$$
$$\{U_{t+\Delta t}^i\} = \{U_{t+\Delta t}^{i-1}\} + \{\Delta U^i\} \quad (5.148)$$

For the substitution method, we use:

$$[K_{nl}] = [\bar{K}] \quad (5.149)$$

For the Newton–Raphson method, we use:

$$[K_{nl}] = [K_t] = [\bar{K}] + \Delta t^2 \left[\frac{\partial K}{\partial U} \cdot U\right]_{t+\Delta t}^{i-1} \quad (5.150)$$

	Central finite difference	Houbolt	Newmark–Wilson
Preliminary operations Define the parameters Assemble and triangularize $[K] = a_0[M] + a_1[C] + a_2[K]$	Δt $a_0 = 1, a_1 = \dfrac{\Delta t}{2}, a_2 = 0$	Δt $a_0 = 2, a_1 = \tfrac{11}{6}\Delta t, a_2 = \Delta t^2$	$\Delta t, \theta, a, b$ $a_0 = 1, a_1 = \theta \Delta t a,$ $a_2 = \dfrac{(\theta \Delta t)^2 b}{2}$
Initialize vectors	$\{U_{t_0}\}$ known $\{U_{t_0-\Delta t}\}$ by (5.145)	$\{U_{t_0}\}$ known $\left.\begin{array}{l}\{U_{t_0-\Delta t}\}\\\{U_{t_0-\Delta t}\}\end{array}\right\}$ other methods	$\{U_{t_0}\}$ known $\{U_{t_0}\}$ known $\{U_{t_0}\}$ by (5.142)
Within each time step Compute residues Solve system in $\{U_{t+\Delta t}\}$	(5.144a)	(5.147)	(5.153)
Prepare next step	Transfer of 2 vectors	Transfer of 3 vectors	Computation of $\{U\}\,\{\dot U\}\,\{\ddot U\}$ by (5.155)–(5.158) in the case of Wilson ($\theta \ne 1$)

Figure 5.22 Operations corresponding to the various direct integration methods of second order systems

5.4.4.3 Methods of Newmark and Wilson [11, 19, 20]

These implicit methods use the governing equation evaluated at time $t+\Delta t$ and the following truncated expansions for $\{\dot{U}_{t+\tau}\}$ and $\{U_{t+\tau}\}$:

$$\{\dot{U}_{t+\tau}\} = \{\dot{U}_t\} + \tau((1-a)\{\ddot{U}_t\} + a\{\ddot{U}_{t+\tau}\}) \quad (5.151a)$$

$$\{U_{t+\tau}\} = \{U_t\} + \tau\{\dot{U}_t\} + \frac{\tau^2}{2}((1-b)\{\ddot{U}_t\} + b\{\ddot{U}_{t+\tau}\}) \quad (5.151b)$$

For $a = b = \frac{1}{2}$, these approximations are physically equivalent to assuming a constant acceleration equal to its average value during internal $(t, t+\tau)$.

For $a = \frac{1}{2}$ and $b = \frac{1}{3}$, the same approximations are equivalent to assuming an acceleration varying linearly over the interval $(t, t+\tau)$. Expression (5.141) written at time $(t+\tau)$ is:

$$[\bar{K}]\{U_{t+\tau}\} = \{\bar{R}_{t+\tau}\} \quad (5.152)$$

where:
$$[\bar{K}] = [M] + \tau a[C] + \frac{\tau^2}{2}b[K]$$

$$\{\bar{R}_{t+\tau}\} = \frac{\tau^2}{2}b\{F_{t+\tau}\} + [M]\left(\{U_t\} + \tau\{\dot{U}_t\} + \frac{\tau^2}{2}(1-b)\{\ddot{U}_t\}\right)$$
$$+ [C]\left(\tau a\{U_t\} + \frac{\tau^2}{2}(2a-b)\{\dot{U}_t\} + \frac{\tau^3}{2}(a-b)\{\ddot{U}_t\}\right)$$

Newmark's Method

When $\tau = \Delta t$, we get Newmark's method which is unconditionally stable if:

$$a \geq \tfrac{1}{2}; \quad b \geq \tfrac{1}{2}(a+\tfrac{1}{2})^2 \quad (5.153)$$

For $a = \frac{1}{2}$ a stability criteria can be written as [19]:

$$\Delta t \leq (l/c)\sqrt{\frac{1}{1-2b}}$$

where c is the speed of sound in the medium; for an elastic solid:

$$c^2 = \frac{E(1-v)}{\rho(1+v)(1-2v)},$$

with E equal to Young's modulus, v equal to Poisson's ratio; and l is the smallest dimension of an element.

For $b = 0$, the method becomes explicit. The values most often used are $a = b = \frac{1}{2}$. After solving (5.152), the values of $\{\ddot{U}_{t+\Delta t}\}$ and $\{\dot{U}_{t+\Delta t}\}$ are computed with (5.151).

Wilson's method

In Wilson's method, we use $\tau = \theta \Delta t$ with $\theta > 1$. Newmark's method is then used to construct solution $\{U_{t+\theta\Delta t}\}$ by solving (5.152). The solution must then be calculated at $t+\Delta t$ for the next time step. For this purpose the following computations are performed:

$$\{\ddot{U}_{t+\theta\Delta t}\} = \frac{2}{b(\theta\Delta t)^2}(\{U_{t+\theta\Delta t}\} - \{U_t\})$$

$$-\frac{2}{b\theta\Delta t}\{\dot{U}_t\} - \left(\frac{1}{b}-1\right)\{\ddot{U}_t\} \qquad (5.154)$$

— $\{\ddot{U}_{t+\Delta t}\}$ assuming linear variations of $\{\ddot{U}\}$:

$$\{\ddot{U}_{t+\Delta t}\} = \{\ddot{U}_t\} + \frac{1}{\theta}(\{\ddot{U}_{t+\theta\Delta t}\} - \{\ddot{U}_t\}) \qquad (5.155)$$

— $\{\dot{U}_{t+\Delta t}\}$ using (5.151a) with $\tau = \Delta t$:

$$\{\dot{U}_{t+\Delta t}\} = \{\dot{U}_t\} + \Delta t((1-a)\{\ddot{U}_t\} + a\{\ddot{U}_{t+\Delta t}\}) \qquad (5.156)$$

— $\{U_{t+\Delta t}\}$ using (5.151b) with $\tau = \Delta t$:

$$\{U_{t+\Delta t}\} = \{U_t\} + \Delta t\{\dot{U}_t\} + \frac{\Delta t^2}{2}((1-b)\{\ddot{U}_t\} + b\{\ddot{U}_{t+\Delta t}\}) \qquad (5.157)$$

An increase of θ attenuates the high frequency modes. An unconditionally stable formula results from choosing $a = \frac{1}{2}$, $b = \frac{1}{3}$ and $\theta = 1.4$.

For non-linear problems, expressions similar to (5.148), (5.149), (5.150) are constructed from matrix $[\bar{K}]$ and a residual defined by (5.152). Figure 5.22 summarizes the various steps of Newmark–Wilson methods.

Example 5.29 *Solution of a two-variable second order system using central difference and Newmark's methods*

Consider the system:

$$[M]\{\ddot{U}\} + [C]\{\dot{U}\} + [K]\{U\} = \{F\}; \quad \{U_0\} = \{\dot{U}_0\} = 0$$

where:

$$[M] = \begin{bmatrix} 1 & 1 \\ 1 & 2 \end{bmatrix}; \quad [C] = [K] = \begin{bmatrix} 1 & 1 \\ 1 & 3 \end{bmatrix}; \quad \{F\} = \begin{Bmatrix} 1 \\ 2 \end{Bmatrix}$$

Substituting $\{U_0\}$ and $\{\dot{U}_0\}$ into the system we get:

$$[M]\{\ddot{U}_0\} = \begin{Bmatrix} 1 \\ 2 \end{Bmatrix} \quad \{\ddot{U}_0\} = \begin{Bmatrix} 0 \\ 1 \end{Bmatrix}$$

Central finite difference method ($\Delta t = 0.5$, algorithm (5.143a))

$$[\bar{K}]\{U_{t+\Delta t}\} = \{R_t\}$$

$$\{R_t\} = 0.25\{F_t\} + [M](2\{U_t\} - \{U_{t-\Delta t}\})$$
$$+ 0.25[C]\{U_{t-\Delta t}\} - 0.25[K]\{U_t\}$$

From (5.143a):
$$[\bar{K}] = [M] + 0.25[C] = \begin{bmatrix} 1.25 & 1.25 \\ 1.25 & 2.75 \end{bmatrix}$$

From (5.144) (with $t_0 = 0$);
$$\{U_{-\Delta t}\} = \begin{Bmatrix} 0 \\ 0.125 \end{Bmatrix}$$

The results are:

t	u_1	u_2
0	0	0
0.5	0.000	0.125
1.0	0.083	0.292
1.5	0.233	0.417
2.0	0.399	0.486
2.5	0.535	0.514
3.0	0.619	0.519
4.0	0.638	0.508
5.0	0.561	0.501
6.0	0.495	0.499
7.0	0.474	0.500
8.0	0.482	0.500
9.0	0.496	0.500
10.0	0.503	0.500
15.0	0.499	0.500

Newmark: ($\Delta t = 0.5$, $\theta = 1$, $a = b = 0.5$, algorithm (5.152))

$$[\bar{K}] = [M] + 0.25[C] + 0.0625[K] = \begin{bmatrix} 1.3125 & 1.3125 \\ 1.3125 & 2.9375 \end{bmatrix}$$

$$\{R_{t+\Delta t}\} = 0.0625\{F_{t+\Delta t}\} + [M](\{U_t\} + 0.5\{\dot{U}_t\} + 0.0625\{\ddot{U}_t\}) + [C](0.25\{U_t\} + 0.0625\{\dot{U}_t\})$$

After each time step, speeds and accelerations are obtained from (5.151b) and (5.151a)

$$\{\ddot{U}_{t+\Delta t}\} = -\{\ddot{U}_t\} + 16(\{U_{t+\Delta t}\} - \{U_t\} - 0.5\{\dot{U}_t\})$$
$$\{\dot{U}_{t+\Delta t}\} = \{\dot{U}_t\} + 0.25(\{\ddot{U}_t\} + \{\ddot{U}_{t+\Delta t}\})$$

t	u_1	u_2	$\dot{u}_1$	$\dot{u}_2$	$\ddot{u}_1$	$\ddot{u}_2$
0	0	0	0	0	0	0
0.5	0.018	0.077	0.073	0.308	0.293	0.231
1.0	0.090	0.237	0.213	0.331	0.265	−0.136
1.5	0.219	0.379	0.303	0.239	0.096	−0.235
2.0	0.371	0.471	0.307	0.129	−0.078	−0.201
2.5	0.509	0.515	0.241	0.048	−0.187	−0.126
3.0	0.604	0.528	0.141	0.002	−0.215	−0.058
4.0	0.651	0.515	−0.035	−0.019	−0.120	0.007
5.0	0.581	0.502	−0.090	−0.007	0.003	0.011
6.0	0.504	0.498	−0.057	0.000	0.051	0.003
7.0	0.471	0.499	−0.010	0.001	0.039	0.000
8.0	0.475	0.500	0.014	0.000	0.011	−0.001
9.0	0.491	0.500	0.015	0.000	−0.007	0.000
10.0	0.503	0.500	0.007	0.000	−0.009	0.000
15.0	0.499	0.500	0.000	0.000	0.001	0.000

5.4.5 *Modal Superposition for Second Order Systems* [16]

This method is similar to the one described in paragraph 5.4.3 for first order systems. However, for second order system (5.86), we use the eigenvalues and eigenvectors of:

$$([K] - \lambda[M])\{X\} = 0 \tag{5.158}$$

After transformation (5.134), system (5.86) becomes:

$$\{\ddot{V}(t)\} + [X]^T[C][X]\{\dot{V}(t)\} + [\lambda]\{V(t)\} = \{\bar{F}(t)\} \tag{5.159}$$

This last system decouples only under the following hypothesis for the value of matrix $[C]$:

— $[C]$ is a null matrix, or
— $[C]$ is a linear combination of $[M]$ and $[K]$.

$$[C] = c_1[M] + c_2[K] \tag{5.160}$$

Then:

$$[X]^T[C][X] = c_1[I] + c_2[\lambda]$$

One often replaces matrix $[X]^T[C][X]$ by the following diagonal form:

$$\begin{bmatrix} 2\xi_1\omega_1 & & & 0 \\ & 2\xi_2\omega_2 & & \\ & & 2\xi_i\omega_i & \\ 0 & & & \end{bmatrix} ; \quad \omega_i = \sqrt{\lambda_i} \tag{5.161}$$

where coefficient ξ_i is the damping ratio of mode i over its critical damping $2\omega_i$.

Note that hypothesis (5.160) allows only two possible non-zero values for ξ_i. When equations (5.159) are decoupled, we get:

$$\ddot{V}_i(t) + 2\omega_i\xi_i\dot{V}_i(t) + \omega_i^2 V(t) = \bar{F}_i(t) \tag{5.162}$$

A general solution can be written with the help of the Duhamel integral for $t_0 = 0$:

$$V_i(t) = e^{-\xi_i\omega_i t}(\alpha_1 \sin \bar{\omega}_i t + \alpha_2 \cos \bar{\omega}_i t)$$

$$+ \frac{1}{\bar{\omega}_i} \int_0^i e^{-\xi_i\omega_i(t-s)} \sin \bar{\omega}_i(t-s) . \bar{F}_i(s) ds \tag{5.163}$$

$$\bar{\omega}_i = \omega_i \sqrt{1-\xi_i^2}$$

Vector $\{\bar{F}_i(t)\}$ is often defined as a ramp function,

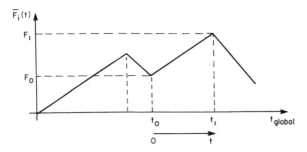

$$\bar{F}_i(t) = F_0 + \frac{F_1 - F_0}{t_1 - t_0} t = a + bt \quad \text{for} \quad 0 \le t \le t_1 - t_0.$$

In this case, the solution of (5.162) is explicit:

$$V_i(t_0 + t) = a_0 + a_1 t + e^{-\xi_i\omega_i t}(a_2 \cos \bar{\omega} t + a_3 \sin \bar{\omega} t) \tag{5.164}$$

$$a_0 = \frac{a}{\omega_i^2} - 2\frac{\xi_i b}{\omega_i^3}; \quad a_1 = \frac{b}{\omega_i^2}$$

$$a_2 = V_i(t_0) - a_0; \quad a_3 = \frac{1}{\bar{\omega}_i}(\dot{V}_i(t_0) + \xi_i\omega_i a_2 - a_1)$$

Examples 5.30 Solution of a two-variable second order problem by modal superposition

Consider the system defined in Example 5.29. Eigenvalues and eigenvectors of:

$$\left(\begin{bmatrix} 1 & 1 \\ 1 & 3 \end{bmatrix} - \lambda \begin{bmatrix} 1 & 1 \\ 1 & 2 \end{bmatrix} \right) \{X\} = 0$$

have been found in Example 5.29:

$$\lambda_1 = 1 \quad \{X_1\} = \begin{Bmatrix} 1 \\ 0 \end{Bmatrix} \quad [X] = \begin{bmatrix} 1 & 1 \\ 0 & -1 \end{bmatrix}$$

$$\lambda_2 = 2 \quad \{X_2\} = \begin{Bmatrix} 1 \\ -1 \end{Bmatrix}$$

System (5.159) decouples because $[C]$ is a linear combination of $[M]$ and $[K]$:

$$\{C\} = 0.\begin{bmatrix} 1 & 1 \\ 1 & 2 \end{bmatrix} + 1.\begin{bmatrix} 1 & 1 \\ 1 & 3 \end{bmatrix}$$

Hence

$$\{\ddot{V}\} + \begin{bmatrix} 1 & 0 \\ 0 & 2 \end{bmatrix}\{\dot{V}\} + \begin{bmatrix} 1 & 0 \\ 0 & 2 \end{bmatrix}\{V\} = \begin{Bmatrix} 1 \\ -1 \end{Bmatrix}$$

thus:

$$\ddot{V}_1 + \dot{V}_1 + V_1 = 1$$
$$\ddot{V}_2 + 2\dot{V}_2 + 2V_2 = -1$$

Initial conditions are obtained from (5.139):

$$\{V_0\} = [X]^T[M]\{U_0\} = \begin{Bmatrix} 0 \\ 0 \end{Bmatrix}; \quad \{\dot{V}_0\} = [X]^T[M]\{\dot{U}_0\} = \begin{Bmatrix} 0 \\ 0 \end{Bmatrix}$$

The solution of the first decoupled equation is obtained using (5.164):

$$a = 1, \quad b = 0, \quad \omega_1 = 1, \quad \xi_1 = \tfrac{1}{2}, \quad \bar{\omega}_1 = \frac{\sqrt{3}}{2}$$

$$a_0 = 1, \quad a_1 = 0, \quad a_2 = -1, \quad a_3 = -\frac{1}{\sqrt{3}}$$

$$V_1 = 1 - e^{-t/2}\left(\cos\frac{\sqrt{3}}{2}t + \frac{\sqrt{3}}{3}\sin\frac{\sqrt{3}}{2}t\right)$$

For the second decoupled equation:

$$a = -1, \quad b = 0, \quad \omega_2 = \sqrt{2}, \quad \xi_2 = \frac{\sqrt{2}}{2}, \quad \bar{\omega}_2 = 1$$

$$a_0 = -\tfrac{1}{2}, \quad a_1 = 0, \quad a_2 = \tfrac{1}{2}, \quad a_3 = \tfrac{1}{2}$$
$$V_2 = -\tfrac{1}{2} + e^{-t}(\tfrac{1}{2}\cos t + \tfrac{1}{2}\sin t)$$

The solution of the initial system is:

$$\{U\} = [X]\begin{Bmatrix} V_1 \\ V_2 \end{Bmatrix} = \begin{bmatrix} 1 & 1 \\ 0 & -1 \end{bmatrix}$$

$$\times \begin{Bmatrix} 1 - e^{-t/2}\left(\cos\frac{\sqrt{3}}{2}t + \frac{\sqrt{3}}{3}\sin\frac{\sqrt{3}}{2}t\right) \\ -\tfrac{1}{2} + e^{-t}(\tfrac{1}{2}\cos t + \tfrac{1}{2}\sin t) \end{Bmatrix}$$

$$\{U\} = \begin{Bmatrix} \tfrac{1}{2} - e^{-t/2}\left(\cos\frac{\sqrt{3}}{2}t + \frac{\sqrt{3}}{3}\sin\frac{\sqrt{3}}{2}t\right) + e^{-t}(\tfrac{1}{2}\cos t + \tfrac{1}{2}\sin t) \\ \tfrac{1}{2} - e^{-t}(\tfrac{1}{2}\cos t + \tfrac{1}{2}\sin t) \end{Bmatrix}$$

t	u_1	u_2
0	0	0
0.5	0.016	0.088
1.0	0.094	0.246
1.5	0.230	0.381
2.0	0.383	0.466
2.5	0.515	0.508
3.0	0.603	0.521
4.0	0.640	0.513
5.0	0.572	0.502
6.0	0.503	0.499
7.0	0.475	0.499
8.0	0.479	0.500
9.0	0.493	0.500
10.0	0.502	0.500
15.0	0.499	0.500

5.5 Solution of the Matrix Eigenvalue Problem [11, 13, 21]

5.5.1 Introduction

Solving a matrix eigenvalue problem consists in finding values of λ_i and $\{X_i\}$ satisfying the following equation:

$$[K]\{X_i\} = \lambda_i [M]\{X_i\} \tag{5.165}$$

and a normalization condition for the vectors:

$$\text{either } \langle X_i \rangle \{X_i\} = 1 \quad \text{or} \quad \langle X_i \rangle [M]\{X_i\} = 1$$

Such problems appear in various disciplines; for example structural vibrations, the buckling load of a structure, and non-stationary problems of first and second order:

Structural Vibrations

The determination of the fundamental modes of vibration of a structure leads to an equation similar to (5.165) where:

  $[K]$ is the stiffness matrix of the structure;
  $[M]$ is the mass matrix;
  $\{X_i\}$ is the displacement vector of the structure for the ith mode of vibration;
  $\lambda_i = \omega_i^2$ is the square of the corresponding frequency.

Buckling Load of a structure

The search for the buckling load of a structure also leads to an expression like (5.165) where:

[K] is again the stiffness matrix;
$[M] \equiv [K_G]$ is a geometrical or initial stress matrix of a structure;
$\{X_i\}$ is the displacement vector of the structure for its ith mode of buckling;
λ_i defines the amplitude of the critical load.

Solution of non-stationary problems of first and second order

We have already seen in paragraphs 5.4.3 and 5.4.5 that the mode superposition method begins with the computation of the eigenvectors and eigenvalues of a system like (5.165).

The literature on the matrix eigenvalue problem is abundant. The works of Wilkinson [21], and more recently Bathe and Wilson [11], contain detailed descriptions of the various strategies available for the solution of such problems. Here, we enumerate without proof the most important properties of (5.165) and describe three methods of solution:

— inverse iteration;
— Jacobi rotation;
— subspace iteration.

5.5.2 Fundamental Properties of Some Matrix Eigenvalue Problems

5.5.2.1 Simplified formulation

We restrict our discussion to problems in which [K] and [M] are real symmetrical matrices of which at least one is positive definite. It is then possible to reduce the general equation to:

$$[\bar{K}]\{\bar{X}\} = \lambda \{\bar{X}\} \tag{5.166}$$

If [M] is positive definite, it is decomposed as (5.38):

$$[M] = [L][L]^T \tag{5.167}$$

then, let

$$\{\bar{X}\} = [L]^T \{X\}$$
$$[\bar{K}] = [L]^{-1}[K][L]^{-1,T}$$

If [M] is not positive definite, then [K] will be; we modify (5.165) as follows:

$$[M]\{X\} = \frac{1}{\lambda}[K]\{X\} = \lambda'[K]\{X\} \tag{5.168}$$

which has the same form as (5.166).

5.5.2.2 Eigenvalues

There are n real eigenvalues λ_i, distinct or not, which satisfy (5.165). If the two matrices are positive definite, all the value of λ_i are positive:

$$\lambda_n \geq \lambda_{n-1} \geq \ldots \lambda_2 \geq \lambda_1$$

We rewrite (5.165) as follows:
$$([K] - \lambda[M])\{X\} = 0 \tag{5.169}$$
Non-trivial solutions $\{X\}$ exist for (5.169) only if $([K] - \lambda[M])$ is singular:
$$\det([K] - \lambda[M]) = P(\lambda) = 0 \tag{5.170}$$
This last expression is a characteristic polynomial of order n in λ. The search for eigenvalue solutions is equivalent to the search for the roots of the characteristic polynomial. There is no closed form solution for $n > 4$; iterative methods must be used. Furthermore, the formation of the characteristic polynomial of a large eigensystem is awkward and seldom used directly.

5.5.2.3 Eigenvectors

To each eigenvalue we have a corresponding eigenvector solution of:
$$([K] - \lambda_i[M])\{X_i\} = 0 \tag{5.171a}$$

To solve this singular homogeneous system, we need a supplementary equation to define the vector uniquely. We use a convenient normalization condition; for example:
$$\langle X_i \rangle [M]\{X_i\} = 1 \tag{5.171b}$$

The eigenvectors satisfy the following orthogonality conditions:
$$\langle X_i \rangle [K]\{X_j\} = \lambda_i \delta_{ij}; \quad \delta_{ij} = \begin{cases} 0 & \text{if } i \neq j \\ 1 & \text{if } i = j \end{cases} \tag{5.172a}$$
$$\langle X_i \rangle [M]\{X_j\} = \delta_{ij}. \tag{5.172b}$$

We then say that the eigenvectors are '$[K]$-orthogonal' and '$[M]$-orthonormal', respectively.

Regrouping all eigenvectors into a modal matrix:
$$[X] = [\{X_1\} \ldots \{X_i\} \ldots] \tag{5.173}$$

The complete solution of (5.165) is written as:
$$[K][X] = [M][X][\lambda] \tag{5.174}$$

$$[\lambda] = \begin{bmatrix} \lambda_1 & & & 0 \\ & \lambda_2 & & \\ & & \lambda_i & \\ 0 & & & \ddots \end{bmatrix}$$

Then, equations (5.172) become:
$$[X]^T[K][X] = [\lambda] \tag{5.175a}$$
$$[X]^T[M][X] = [I] \tag{5.175b}$$

Also if we premultiply (5.175b) by $[X]$ and post-multiply by $[X]^{-1}$, we get:
$$[X][X]^T[M] = [I] \tag{5.176}$$

5.5.2.4 Spectral Decomposition

Post-multiplying (5.174) by $[X]^T[M]$, and using (5.176), we get the following expression since $[M]$ is symmetrical:

$$[K] = [M][X].[\lambda].([M][X])^T \tag{5.177}$$

Or, if we let $\{Y_i\} = [M]\{X_i\}$:

$$[K] = \sum_{i=1}^{n} \lambda_i \{Y_i\}\langle Y_i \rangle \tag{5.178a}$$

The spectral decomposition of $[K]$ is defined by (5.178a) when $[M] \equiv [I]$

$$[K] = \sum_{i=1}^{n} \lambda_i \{X_i\}\langle X_i \rangle = \sum_{i=1}^{n} \lambda_i [P_i] \tag{5.178}$$

where symmetrical matrices $[P_i]$ are the projectors of $[K]$. This last expression permits us to evaluate $[K]^p$:

$$[K]^p = \sum_{i=1}^{n} \lambda_i^p [P_i] \quad (p \text{ positive or negative}) \tag{5.179}$$

The spectral radius of $[K]$ is defined by:

$$\rho(K) = \max |\lambda_i| \leq \|K\|_\infty \tag{5.180}$$

where:

$$\|K\|_\infty = \max_i \left(\sum_{j=1}^{n} |K_{ij}| \right)$$

Thus, $[K]^p$ will be bound as p tends towards infinity if:

$$\rho(K) < 1 \tag{5.181}$$

An arbitrary vector V in that space can be represented using the eigenvectors as a basis of that space (5.165):

$$\{V\} = \sum_{i=1}^{n} c_i \{X_i\} \tag{5.182}$$

The components c_i are given by:

$$c_i = \langle X_i \rangle [M]\{V\} \tag{5.183}$$

5.5.2.5 Transformation of $[K]$ and $[M]$

Let us transform $[K]$ and $[M]$ as follows:

$$[\bar{K}] = [Q]^T[K][Q]; \quad [\bar{M}] = [Q]^T[M][Q] \tag{5.184a}$$

The eigenvalues and eigenvectors of (5.184a) are defined by:

$$([\bar{K}] - \bar{\lambda}[\bar{M}])\{\bar{X}\} = 0 \tag{5.184b}$$

or equivalently:

$$[Q]^T([K] - \bar{\lambda}[M])[Q]\{\bar{X}\} = 0$$

Then:

$$\det([\bar{K}] - \bar{\lambda}[M]) = \det([Q]^T([K] - \bar{\lambda}[M])[Q])$$
$$= \det([Q]^T).\det([Q]).\det([K] - \bar{\lambda}[M]) = 0$$

This shows that the eigenvalues $\bar{\lambda}$ of the transformed system (5.184b) are identical to λ of system (5.169) if $\det([Q]) \neq 0$. This type of transformation was used in paragraph 5.5.2.1. The eigenvectors $\{\bar{X}\}$ of (5.184b) are related to those of (5.169) by:

$$\{X\} = [Q]\{\bar{X}\} \qquad (5.184c)$$

Example 5.31 Eigenvalues and eigenvectors

Consider the following eigenvalue problem:

$$[K]\{X_i\} = \lambda_i[M]\{X_i\}; \quad [K] = \begin{bmatrix} 1 & 1 \\ 1 & 3 \end{bmatrix}; \quad [M] = \begin{bmatrix} 1 & 1 \\ 1 & 2 \end{bmatrix}$$

Matrices $[K]$ and $[M]$ are positive definite; the characteristic polynomial of the system is:

$$\det\left(\begin{bmatrix} 1 & 1 \\ 1 & 3 \end{bmatrix} - \lambda \begin{bmatrix} 1 & 1 \\ 1 & 2 \end{bmatrix}\right) = (1 - \lambda)(2 - \lambda) = 0$$

Thus, the eigenvalues are:

$$\lambda_1 = 1$$
$$\lambda_2 = 2$$

The eigenvectors are solutions of the following singular systems:

$$\left(\begin{bmatrix} 1 & 1 \\ 1 & 3 \end{bmatrix} - 1\begin{bmatrix} 1 & 1 \\ 1 & 2 \end{bmatrix}\right)\{X_i\} = 0 \rightarrow \{X_1\} = \begin{Bmatrix} x_1 \\ 0 \end{Bmatrix}$$

$$\left(\begin{bmatrix} 1 & 1 \\ 1 & 3 \end{bmatrix} - 2\begin{bmatrix} 1 & 1 \\ 1 & 2 \end{bmatrix}\right)\{X_2\} = 0 \rightarrow \{X_2\} = \begin{Bmatrix} x_2 \\ -x_2 \end{Bmatrix}$$

Using the 'M-orthonormality' condition of $\{X_1\}$ and $\{X_2\}$ (5.172b), we get:

$$\{X_1\} = \begin{Bmatrix} 1 \\ 0 \end{Bmatrix} \quad \{X_2\} = \begin{Bmatrix} 1 \\ -1 \end{Bmatrix}$$

$$[X] = \begin{bmatrix} 1 & 1 \\ 0 & -1 \end{bmatrix}$$

Orthogonality conditions (5.175) are:

$$\begin{bmatrix} 1 & 0 \\ 1 & -1 \end{bmatrix}\begin{bmatrix} 1 & 1 \\ 1 & 3 \end{bmatrix}\begin{bmatrix} 1 & 1 \\ 0 & -1 \end{bmatrix} = \begin{bmatrix} 1 & 0 \\ 0 & 2 \end{bmatrix}$$

$$\begin{bmatrix} 1 & 0 \\ 1 & -1 \end{bmatrix}\begin{bmatrix} 1 & 1 \\ 1 & 2 \end{bmatrix}\begin{bmatrix} 1 & 1 \\ 0 & -1 \end{bmatrix} = \begin{bmatrix} 1 & 0 \\ 0 & 1 \end{bmatrix}$$

The decomposition of matrix $[K]$, according to (5.178a), is:

$$\{Y_1\} = \begin{bmatrix} 1 & 1 \\ 1 & 2 \end{bmatrix} \begin{Bmatrix} 1 \\ 0 \end{Bmatrix} = \begin{Bmatrix} 1 \\ 1 \end{Bmatrix}$$

$$\{Y_2\} = \begin{bmatrix} 1 & 1 \\ 1 & 2 \end{bmatrix} \begin{Bmatrix} 1 \\ -1 \end{Bmatrix} = \begin{Bmatrix} 0 \\ -1 \end{Bmatrix}$$

$$[K] = 1 \begin{bmatrix} 1 & 1 \\ 1 & 1 \end{bmatrix} + 2 \begin{bmatrix} 0 & 0 \\ 0 & 1 \end{bmatrix}$$

5.5.2.6 Rayleigh Quotient

The Rayleigh quotient for any arbitrary vector $\{V\}$ is defined by:

$$R(\{V\}) = \frac{\langle V \rangle [K] \{V\}}{\langle V \rangle [M] \{V\}} \tag{5.185}$$

If $[K]$ and $[M]$ are positive definite, $R(V) > 0$; moreover, the value of $R(V)$ is bound, above by the maximum eigenvalue and below by the minimum eigenvalue of (5.169)

$$\lambda_1 \leq R(\{V\}) \leq \lambda_n \tag{5.186}$$

When vector $\{V\}$ is equal to an eigenvector $\{X_i\}$ of (5.169), then the Rayleigh quotient is equal to the corresponding eigenvalue λ_i:

$$R(\{X_i\}) = \lambda_i \tag{5.187}$$

If $\{V\}$ is a linear combination of the first p eigenvectors, then Rayleigh quotient is bound by λ_p and λ_1:

$$\lambda_1 \leq R\left(\sum_{i=1}^{p} c_i \{X_i\} \right) \leq \lambda_p \tag{5.188}$$

and the maximum value of R is obtained for $c_1 = c_2 = \cdots = c_{p-1} = 0$; $c_p = 1$. Assuming that $\{V\}$ is in the neighbourhood of eigenvector $\{X_i\}$, it can be written as:

$$\{V\} = \{X_i\} + \varepsilon \{Z\} \tag{5.189}$$

where ε is a very small number and $\{Z\}$ an arbitrary perturbation vector. The Rayleigh quotient is then close to λ_i and of the form:

$$R(\{V\}) = \lambda_i + 0(\varepsilon^2) \tag{5.190}$$

This means that the Rayleigh quotient is stationary with respect to ε in the neighbourhood of an eigenvector $\{X_i\}$ since the term $0(\varepsilon)$ is missing in (5.190).

5.5.2.7 Separation of eigenvalues

Let $\lambda_1 \lambda_2 \ldots \lambda_n$ be the n eigenvalues of system (5.169) and $l_1 l_2 \ldots l_{n-1}$ and $n-1$ eigenvalues of the following modified system:

$$[K]^1 \{X\} = l[M]^1 \{X\} \tag{5.191}$$

$[K]^1$ and $[M]^1$ are obtained by eliminating the last row and column from $[K]$ and $[M]$, respectively. We may write the following inequalities:

$$\lambda_1 \leq l_1 \leq \lambda_2 \leq l_2 \leq \cdots \leq \lambda_{n-1} \leq l_{n-1} \leq \lambda_n \tag{5.192}$$

More generally, let us define $[K]^r$ and $[M]^r$ as the matrices obtained after suppressing the last r rows and columns from $[K]$ and $[M]$. If $p^r(\lambda)$ is the corresponding characteristic polynomial, the roots of $p^{r-1}(l)$ alternate with the roots of $p^r(\lambda)$ as follows:

$$\lambda_1 \leq l_1 \leq \lambda_2 \leq \cdots \leq l_{n-r-2} \leq \lambda_{n-r-1} \leq l_{n-r-1} \leq \lambda_{n-r} \tag{5.193}$$

The sequence of polynomial $p^r(\lambda)$ is called a Sturm sequence. Decomposing matrix $[K] - \mu[M]$, where μ is an arbitrary number different from an eigenvalue, according to (5.36), we get:

$$[K] - \mu[M] = [L][D][L]^T \tag{5.194}$$

Assuming that the eigenvalues of 5.169 are positive: the number of negative terms in matrix D is equal to the number of eigenvalues smaller than μ. The Sturm sequence is an effective test to verify if all the desired eigenvalues have been obtained in a certain range of values.

5.5.2.8 Shifting of Eigenvalues

If matrix $[K]$ of system (5.169) is replaced by:

$$[\bar{K}] = [K] - a[M] \tag{5.195}$$

we get the following modified eigensystem:

$$[\bar{K}]\{X\} = \bar{\lambda}[M]\{X\} \quad \text{or} \quad [K]\{X\} = \{\bar{\lambda} + a\}[M]\{X\} \tag{5.196}$$

The eigenvectors of that system are identical to those of (5.169) but the eigenvalues $\bar{\lambda}$ have shifted by an amount equal to a from those of (5.169):

$$\bar{\lambda} = \lambda - a \tag{5.197}$$

Example 5.32 Rayleigh quotient and separation of eigenvalues

Consider matrices $[K]$ and $[M]$ of Example 5.31. Rayleigh quotient (5.185) of an arbitrary vector $\{V\} = \langle 1, \ 1 \rangle^T$ is:

$$R(\{V\}) = \frac{\langle 1 \ \ 1 \rangle \begin{bmatrix} 1 & 1 \\ 1 & 3 \end{bmatrix} \begin{Bmatrix} 1 \\ 1 \end{Bmatrix}}{\langle 1 \ \ 1 \rangle \begin{bmatrix} 1 & 1 \\ 1 & 2 \end{bmatrix} \begin{Bmatrix} 1 \\ 1 \end{Bmatrix}} = \tfrac{6}{5} = 1.2$$

We verify that

$$\lambda_1 \leq R(\{V\}) \leq \lambda_2$$

with

$$\lambda_1 = 1, \quad \lambda_2 = 2$$

Eliminating the last row and column of $[K]$ and $[M]$:

$$[K]^1 = [1] \quad [M]^1 = [1]$$

The eigenvalue of

$$([1] - l[1])X = 0$$

is $l = 1$

We verify property (5.192)

$$\lambda_1 \leq l \leq \lambda_2$$

After decomposition (5.194) with $\mu = 3$

$$[K] - 3[M] = \begin{bmatrix} -2 & -2 \\ -2 & -3 \end{bmatrix} = \begin{bmatrix} 1 & 0 \\ 1 & 1 \end{bmatrix} \begin{bmatrix} -2 & 0 \\ 0 & -1 \end{bmatrix} \begin{bmatrix} 1 & 1 \\ 0 & 1 \end{bmatrix}$$

Since the two diagonal terms of D are negative, all the eigenvalues are smaller than three.

5.5.3 *Computations of Eigenvalues*

5.5.3.1 *Inverse iteration method*

Search for the smallest eigenvalue λ_1 in absolute value

Inverse iteration gives the smallest eigenvalue λ_1 and corresponding eigenvector $\{X_1\}$ for system (5.169). $[K]$ must be positive definite; if it is not, a shift (as described in (5.195) and of appropriate magnitude) transforms $[K]$ into a positive definite matrix $[K]$. The algorithm is elaborated in Figure 5.23.

Convergence

The inverse iteration is applicable, in general, to problems with positive or negative eigenvalues. ($[K]$ positive definite, $[M]$ symmetrical). However, in the following discussion, we assume for convenience, all the values to be positive. The convergence rate of inverse iteration is proportional to λ_2/λ_1 where λ_2 is the next higher value above λ_1. It can be very small if λ_1 and λ_2 are very close values. A shift of magnitude a improves the convergence rate if

$$\frac{\lambda_2 - a}{\lambda_1 - a} > \frac{\lambda_2}{\lambda_1} \quad (5.198)$$

The shift magnitude 'a' can be taken equal to Rayleigh quotient $R(\{V^{i+1}\})$. The convergence rate is thus greatly improved but matrix $([K] - a[M])$ must be factored at every iteration.

Triangularize $[K] = [L]^T[D][L]$.
Choice of initial vector $\{V^1\}$, not M-orthogonal to the desired vector ($d \neq 0$)
  Computation of : $\{F^1\} = [M]\{V^1\}$.
┌─ For each iteration $i = 1, 2, \ldots$
│ Solve $[L]^T[D][L]\{V^{i+1}\} = \{F^i\}$.
│   Compute: $\{\bar{F}\} = [M]\{V^{i+1}\}$.
│ Evaluate $d = \langle V^{i+1} \rangle \{F\}$.
│   Compute   Rayleigh   quotient   (5.186)

$$\lambda_1^{i+1} = R(\{V^{i+1}\}) = \frac{\langle V^{i+1} \rangle \{F^i\}}{d}.$$

Compute: $\{F^{i+1}\}$:

$$\{F^{i+1}\} = \frac{1}{\sqrt{|d|}}\{\bar{F}\}.$$

└─ Verify convergence λ_1^{i+1} : $|\lambda_1^{i+1} - \lambda_1^i| < \varepsilon$.
Compute eigenvector normalized with respect to M.

$$\{X_1\} = \frac{1}{\sqrt{|d|}}\{V^{i+1}\}.$$

Figure 5.23 Inverse iteration algorithm.

Search for the largest eigenvalue (absolute)

The largest eigenvalue can be obtained by direct iteration. The process consists in applying inverse iteration to system:

$$[M]\{X\} = \frac{1}{\lambda}[K]\{X\} = \lambda'[K]\{X\} \qquad (5.199)$$

$[M]$ must be positive definite. If it is not, a shift of sufficient magnitude can be performed to make $[\bar{M}] = [M] - a[K]$ positive definite.

Search for intermediary eigenvalues

Intermediary values λ_p included between λ_1 and λ_n can be obtained by inverse iteration if a shift $a \approx \lambda_p$ is used. However, it is often difficult to predict which eigenvalue will result from a given shift a.

Search of successive eigenvalues $\lambda_1 \lambda_2 \ldots$

A systematic technique to find successive eigenvalues $\lambda_1 \lambda_2 \ldots$ can be developed with the inverse iteration method if the previously found vectors are swept out of the current iteration using (5.182):

$$\{V^{i+1}\} = \{V^{i+1}\} - c_1\{X_1\}$$

where:
$$c_1 = \langle X_1 \rangle \{\bar{Y}\}$$
$$\{\bar{Y}\} = [M]\{V^{i+1}\} \tag{5.200}$$

For eigenvalue λ_p and eigenvector $\{X_p\}$, $\{V^{i+1}\}$ must be swept out of all the previously found vectors $\{X_1\}, \{X_2\}, \ldots, \{X_{p-1}\}$:

$$\{V^{i+1}\} = \{V^{i+1}\} - \sum_{j=1}^{p-1} (\langle X_j \rangle \{\bar{Y}\})\{X_j\} \tag{5.201}$$

Note that sweeping operations are numerically sensitive and may have to be repeated several times during a particular iteration sequence for the same eigenvalue.

5.5.3.2 Jacobi Rotation Method

Generalities

The method described here is restricted to cases where $[K]$ and $[M]$ are symmetrical and positive definite. It is often referred to as a diagonalization process. It consists in developing a sequence of similarity transformations that eventually results in diagonal matrices $[K^d]$ and $[M^d]$:

$$\begin{aligned}
[K^1] &= [K] & [M^1] &= [M] \\
[K^2] &= [Q^1]^T[K^1][Q^1] & [M^2] &= [Q^1]^T[M^1][Q^1] \\
&\ldots & &\ldots \\
[K^{k+1}] &= [Q^k]^T[K^k][Q^k] & [M^{k+1}] &= [Q^k]^T[M^k][Q^k]
\end{aligned} \tag{5.202}$$

Matrices $[K^{k+1}]$ and $[M^{k+1}]$ tend to become diagonal when the transformation matrices $[Q^i]$ are constructed as shown below. After diagonalization, we find the eigenvalues and eigenvectors as follows:

$$[\lambda] = [K^d][M^d]^{-1} \quad \text{or} \quad \lambda_i = K^d_{ii}/M^d_{ii} \tag{5.203}$$

$$[X] = [Q^1][Q^2]\ldots[Q^k][Q^{k+1}] \begin{bmatrix} \ddots & & 0 \\ & \dfrac{1}{\sqrt{M^d_{ii}}} & \\ 0 & & \ddots \end{bmatrix}$$

Transformation matrices

Matrix $[Q^k]$ is chosen to annihilate an off-diagonal non-zero term (i,j) of $[K^k]$ and $[M^k]$ after transformation (5.202). Matrix $[Q^k]$ has the following structure:

$$[Q^k] = \begin{bmatrix} 1 & & & & 0 \\ & \ddots & & & \\ & & 1 & a & \\ & & b & 1 & \\ & & & & \ddots \\ 0 & & & & 1 \end{bmatrix} \begin{matrix} \\ \\ \ldots \text{row } i \\ \ldots \text{row } j \\ \\ \end{matrix} \tag{5.204}$$

$$\begin{matrix} \vdots & \vdots \\ \text{column } i & \text{column } j \end{matrix}$$

The coefficients a and b are calculated from $K_{ij}^{k+1} = M_{ij}^{k+1} = 0$:

$$aK_{ii} + (1+ab)K_{ij} + bK_{jj} = 0$$
$$aM_{ii} + (1+ab)M_{ij} + bM_{jj} = 0 \tag{5.205}$$

For the particular case where:

$$\frac{K_{ii}}{M_{ii}} = \frac{K_{jj}}{M_{jj}} = \frac{K_{ij}}{M_{ij}}$$

the values of a and b are taken as:

$$a = 0 \quad b = -\frac{K_{ij}}{K_{jj}} \tag{5.206}$$

For the general case, define the following terms:

$$c_1 = K_{ii}M_{ij} - M_{ii}K_{ij}$$
$$c_2 = K_{jj}M_{ij} - M_{jj}K_{ij}$$
$$c_3 = K_{ii}M_{jj} - M_{ii}K_{jj}$$

$$d = \frac{c_3}{2} + \operatorname{sign}(c_3)\sqrt{\left(\frac{c_3}{2}\right)^2 + c_1 c_2}$$

Then:

$$a = \frac{c_2}{d} \quad b = -\frac{c_1}{d} \tag{5.207}$$

Note that when $[M]$ is positive definite, $c_1 c_2$ is always positive. When $d = 0$, a and b are given by (5.206).

Classical Jacobi Rotation Method

For the case where $[M]$ is a unit matrix (5.166), matrices $[Q^k]$ are orthogonal and can be written as rotation matrices:

$$[Q^k] = \begin{bmatrix} 1 & & & & \\ & \ddots & & & \\ & & \cos\theta & -\sin\theta & \\ & & \sin\theta & \cos\theta & \\ & & & & \ddots \\ & & & & & 1 \end{bmatrix} \begin{matrix} \\ \\ \ldots \text{row } i \\ \ldots \text{row } j \\ \\ \end{matrix} \tag{5.208}$$

where:

$$tg(2\theta) = \frac{2K_{ij}}{K_{ii} - K_{jj}} \quad \text{if } K_{ii} \neq K_{jj}$$

$$\theta = \frac{\pi}{4} \quad \text{if } K_{ii} = K_{jj}$$

The general formulation (5.207) gives the correct results when $[M] = [I]$.
An algorithm corresponding to the general formulation is described in Figure 5.24 and a corresponding FORTRAN code listing is found in Figure 5.25.

Set convergence accuracy (desired) ε.
For each cycle s
    Define dynamic tolerance $\varepsilon_s = 10^{-2s}$.
For each row $i = 1, 2, \ldots, n$.
    For each column $j = i+1, \ldots, n$.
    Compute coupling factors

$$f_K = \frac{|K_{ij}|}{\sqrt{K_{ii} K_{jj}}} \quad f_M = \frac{|M_{ij}|}{\sqrt{M_{ii} M_{ij}}}.$$

If f_K or $f_M > \varepsilon_s$:
- Compute a and b using (5.208)
- Transform matrices $[K]$ and $[M]$
    column $i =$ column $i + b$.column j
    column $j = a$.column $i +$ column j
    row i = row i + b.row j
    row j = a.row i + row j
- Modify eigenvectors $[X]$
    column $i =$ column $i + b$.column j
    column $j = a$.column $i +$ column j

Computation of eigenvalues

$$\lambda_i = \frac{K_{ii}}{M_{ii}} \quad \text{and} \quad F_i = \operatorname*{Max}_{i} \frac{|\lambda_i^s - \lambda_i^{s-1}|}{|\lambda_i^{s-1}|}.$$

Computation of coupling factors

$$F_K = \operatorname*{Max}_{i,j} \frac{|K_{ij}|}{\sqrt{K_{ii} K_{jj}}} \quad F_M = \operatorname*{Max}_{i,j} \frac{|M_{ij}|}{\sqrt{M_{ii} M_{jj}}}.$$

Convergence test: $F_K < \varepsilon$ and $F_M < \varepsilon$ or $F_\lambda < \varepsilon$.

Figure 5.24 Algorithm for the general Jacobi method.

```
      SUBROUTINE JACOBI(VK,VM,N,NCYM,EPS,VALPO,VALP,VECT)          JACI   1
C=====================================================================JACI   2
C     TO SOLVE THE EIGENPROBLEM K-LAMBDA.M BY THE GENERALIZED         JACI   3
C     JACOBI METHOD                                                   JACI   4
C     INPUT                                                           JACI   5
C        VK       MATRIX K (UPPER TRIANGLE BY DESCENDING              JACI   6
C                 COLUMNS)                                            JACI   7
C        VM       MATRIX M (UPPER TRIANGLE BY DESCENDING              JACI   8
C                 COLUMNS)                                            JACI   9
C        N        ORDER OF MATRICES K AND M                           JACI  10
C        NCYM     MAXIMUM NUMBER OF SWEEPS ALLOWED (15)               JACI  11
C        EPS      CONVERGENCE TOLERANCE (1.D-12)                      JACI  12
C     WORKSPACE                                                       JACI  13
C        VALPO    WORKING VECTOR (DIMENSION N)                        JACI  14
C     OUTPUT                                                          JACI  15
C        VALP     EIGENVALUES                                         JACI  16
C        VECT     EIGENVECTORS                                        JACI  17
C=====================================================================JACI  18
      IMPLICIT REAL*8(A-H,O-Z)                                        JACI  19
      COMMON/ES/M,MR,MP                                               JACI  20
```

Figure 5.25 (*Contd.*)

```
      DIMENSION VK(1),VM(1),VALPO(N),VALP(N),VECT(N,N)              JACI  21
      DATA EPSD0/1.D-4/,ZERO/0.D0/,UN/1.D0/,DEUX/2.D0/,QUATR/4.D0/   JACI  22
      SQRT(X)=DSQRT(X)                                               JACI  23
      ABS(X)=DABS(X)                                                 JACI  24
      EPS2=EPS*EPS                                                   JACI  25
      ITR=0                                                          JACI  26
C------ VERIFY IF DIAGONAL TERMS ARE POSITIVE                        JACI  27
C       AND INITIALIZE EIGENVALUES                                   JACI  28
      II=0                                                           JACI  29
      DO 20 I=1,N                                                    JACI  30
      II=II+I                                                        JACI  31
      IF(VK(II).GT.ZERO.AND.VM(II).GT.ZERO) GO TO 10                 JACI  32
      WRITE(MP,2000) I                                               JACI  33
 2000 FORMAT(' ** ERROR, NEGATIVE DIAGONAL TERM IN JACOBI, ROW ',    JACI  34
     1  I5)                                                          JACI  35
      STOP                                                           JACI  36
 10   VALP(I)=VK(II)/VM(II)                                          JACI  37
 20   VALPO(I)=VALP(I)                                               JACI  38
C------ INITIALIZE EIGENVECTORS                                      JACI  39
      DO 40 I=1,N                                                    JACI  40
      DO 30 J=1,N                                                    JACI  41
 30   VECT(I,J)=ZERO                                                 JACI  42
 40   VECT(I,I)=UN                                                   JACI  43
C------ FOR EACH SWEEP                                               JACI  44
      DO 250 IC=1,NCYM                                               JACI  45
C------ DYNAMIC TOLERANCE                                            JACI  46
      EPSD=EPSD0**IC                                                 JACI  47
C------ SWEEP ROWWISE OVER UPPER TRIANGLE                            JACI  48
      IMAX=N-1                                                       JACI  49
      II=0                                                           JACI  50
      DO 180 I=1,IMAX                                                JACI  51
      IO=II+1                                                        JACI  52
      II=II+I                                                        JACI  53
      IP1=I+1                                                        JACI  54
      IJ=II+I                                                        JACI  55
      JJ=II                                                          JACI  56
      DO 180 J=IP1,N                                                 JACI  57
      JP1=J+1                                                        JACI  58
      JM1=J-1                                                        JACI  59
      JO=JJ+1                                                        JACI  60
      JJ=JJ+J                                                        JACI  61
      J3=JJ-1                                                        JACI  62
C------ COMPUTE COUPLING FACTORS                                     JACI  63
      FK=(VK(IJ)*VK(IJ))/(VK(II)*VK(JJ))                             JACI  64
      FM=(VM(IJ)*VM(IJ))/(VM(II)*VM(JJ))                             JACI  65
      IF(FK.LT.EPSD.AND.FM.LT.EPSD) GO TO 180                        JACI  66
C------ COMPUTE THE TRANSFORMATION COEFFICIENTS                      JACI  67
      ITR=ITR+1                                                      JACI  68
      C1=VK(II)*VM(IJ)-VM(II)*VK(IJ)                                 JACI  69
      C2=VK(JJ)*VM(IJ)-VM(JJ)*VK(IJ)                                 JACI  70
      C3=VK(II)*VM(JJ)-VM(II)*VK(JJ)                                 JACI  71
      DET=(C3*C3/QUATR)+(C1*C2)                                      JACI  72
      IF(DET.GE.ZERO) GO TO 50                                       JACI  73
      WRITE(MP,2005) I,J                                             JACI  74
 2005 FORMAT(' **ERROR, SINGULAR JACOBI TRANSFORMATION I=',I5,       JACI  75
     1  ' J=',I5)                                                    JACI  76
      STOP                                                           JACI  77
 50   DET=SQRT(DET)                                                  JACI  78
      D1=C3/DEUX+DET                                                 JACI  79
      D2=C3/DEUX-DET                                                 JACI  80
      D=D1                                                           JACI  81
      IF(ABS(D2).GT.ABS(D1))D=D2                                     JACI  82
      IF(D.EQ.ZERO) GO TO 60                                         JACI  83
      A=C2/D                                                         JACI  84
```

Figure 5.25 (*Contd.*)

```
            B=-C1/D                                              JACI  85
            GO TO 65                                             JACI  86
60          A=ZERO                                               JACI  87
            B=-VK(IJ)/VK(JJ)                                     JACI  88
C------  MODIFY COLUMNS OF K AND M                               JACI  89
65          IF(I.EQ.1) GO TO 80                                  JACI  90
            IK=I0                                                JACI  91
            J1=IJ-1                                              JACI  92
            DO 70 JK=J0,J1                                       JACI  93
            C1=VK(IK)                                            JACI  94
            C2=VK(JK)                                            JACI  95
            VK(IK)=C1+B*C2                                       JACI  96
            VK(JK)=C2+A*C1                                       JACI  97
            C1=VM(IK)                                            JACI  98
            C2=VM(JK)                                            JACI  99
            VM(IK)=C1+B*C2                                       JACI 100
            VM(JK)=C2+A*C1                                       JACI 101
70          IK=IK+1                                              JACI 102
80          IF(I.EQ.JM1) GO TO 100                               JACI 103
            IK=II+I                                              JACI 104
            J2=IJ+1                                              JACI 105
            IM=I                                                 JACI 106
            DO 90 JK=J2,J3                                       JACI 107
            C1=VK(IK)                                            JACI 108
            C2=VK(JK)                                            JACI 109
            VK(IK)=C1+B*C2                                       JACI 110
            VK(JK)=C2+A*C1                                       JACI 111
            C1=VM(IK)                                            JACI 112
            C2=VM(JK)                                            JACI 113
            VM(IK)=C1+B*C2                                       JACI 114
            VM(JK)=C2+A*C1                                       JACI 115
            IM=IM+1                                              JACI 116
90          IK=IK+IM                                             JACI 117
100         IF(J.EQ.N) GO TO 120                                 JACI 118
            IK=IJ+J                                              JACI 119
            JK=JJ+J                                              JACI 120
            IM=J                                                 JACI 121
            DO 110 JJK=JP1,N                                     JACI 122
            C1=VK(IK)                                            JACI 123
            C2=VK(JK)                                            JACI 124
            VK(IK)=C1+B*C2                                       JACI 125
            VK(JK)=C2+A*C1                                       JACI 126
            C1=VM(IK)                                            JACI 127
            C2=VM(JK)                                            JACI 128
            VM(IK)=C1+B*C2                                       JACI 129
            VM(JK)=C2+A*C1                                       JACI 130
            IM=IM+1                                              JACI 131
            IK=IK+IM                                             JACI 132
110         JK=JK+IM                                             JACI 133
120         C1=VK(II)                                            JACI 134
            C2=VK(IJ)                                            JACI 135
            C3=VK(JJ)                                            JACI 136
            B2=B*B                                               JACI 137
            BB=DEUX*B                                            JACI 138
            A2=A*A                                               JACI 139
            AA=DEUX*A                                            JACI 140
            VK(II)=C1+BB*C2+B2*C3                                JACI 141
            VK(IJ)=ZERO                                          JACI 142
            VK(JJ)=C3+AA*C2+A2*C1                                JACI 143
            C1=VM(II)                                            JACI 144
            C2=VM(IJ)                                            JACI 145
            C3=VM(JJ)                                            JACI 146
            VM(II)=C1+BB*C2+B2*C3                                JACI 147
            VM(IJ)=ZERO                                          JACI 148
```

Figure 5.25 (*Contd.*)

```
      VM(JJ)=C3+AA*C2+A2*C1                                          JACI 149
C------- UPDATE EIGENVECTORS                                         JACI 150
      DO 170 IJ1=1,N                                                 JACI 151
      C1=VECT(IJ1,I)                                                 JACI 152
      C2=VECT(IJ1,J)                                                 JACI 153
      VECT(IJ1,I)=C1+B*C2                                            JACI 154
170   VECT(IJ1,J)=C2+A*C1                                            JACI 155
180   IJ=IJ+J                                                        JACI 156
C------- UPDATE EIGENVALUES                                          JACI 157
      II=0                                                           JACI 158
      DO 190 I=1,N                                                   JACI 159
      II=II+I                                                        JACI 160
      IF(VK(II).GT.ZERO.AND.VM(II).GT.ZERO) GO TO 190                JACI 161
      WRITE(MP,2000) I                                               JACI 162
      STOP                                                           JACI 163
190   VALP(I)=VK(II)/VM(II)                                          JACI 164
      IF(M.GT.1) WRITE(MP,2010)IC,(VALP(I),I=1,N)                    JACI 165
2010  FORMAT(/' EIGENVALUES, SWEEP ',I4/(1X,10E12.5))                JACI 166
C------- CHECK FOR CONVERGENCE OF EIGENVALUES                        JACI 167
      DO 200 I=1,N                                                   JACI 168
      IF(ABS(VALP(I)-VALPO(I)).GT.(EPS*VALPO(I))) GO TO 230          JACI 169
200   CONTINUE                                                       JACI 170
C------- CHECK FOR CONVERGENCE ON DIAGONAL TERMS                     JACI 171
      JJ=1                                                           JACI 172
      DO 210 J=2,N                                                   JACI 173
      JJ=JJ+J                                                        JACI 174
      JM1=J-1                                                        JACI 175
      II=0                                                           JACI 176
      DO 210 I=1,JM1                                                 JACI 177
      II=II+I                                                        JACI 178
      IJ=JJ-J+I                                                      JACI 179
      FK=VK(IJ)*VK(IJ)/(VK(II)*VK(JJ))                               JACI 180
      FM=VM(IJ)*VM(IJ)/(VM(II)*VM(JJ))                               JACI 181
      IF(FK.GT.EPS2.OR.FM.GT.EPS2) GO TO 230                         JACI 182
210   CONTINUE                                                       JACI 183
C------- NORMALIZE EIGENVECTORS                                      JACI 184
      JJ=0                                                           JACI 185
      DO 220 J=1,N                                                   JACI 186
      JJ=JJ+J                                                        JACI 187
      C1=SQRT(VM(JJ))                                                JACI 188
      DO 220 I=1,N                                                   JACI 189
220   VECT(I,J)=VECT(I,J)/C1                                         JACI 190
C------- ACHIEVED CONVERGENCE                                        JACI 191
      IF(M.GT.0) WRITE(MP,2020) IC,ITR                               JACI 192
2020  FORMAT(15X,'CONVERGENCE IN ',I4,' SWEEPS AND ',I5,' TRANSFORMATIONJACI 193
     1S IN JACOBI')                                                  JACI 194
      RETURN                                                         JACI 195
C------- TRANSFER VALP INTO VALPO                                    JACI 196
230   DO 240 I=1,N                                                   JACI 197
240   VALPO(I)=VALP(I)                                               JACI 198
250   CONTINUE                                                       JACI 199
C------- FAIL TO CONVERGE                                            JACI 200
      WRITE(MP,2030) NCYM                                            JACI 201
2030  FORMAT(' ** ERROR, CONVERGENCE FAILURE IN JACOBI IN',I4,' SWEEPS')JACI 202
      STOP                                                           JACI 203
      END                                                            JACI 204
```

Figure 5.25 Subroutine JACOBI, for the computation of the eigenvalues and eigenvectors of a system, used in program MEF of Chapter 6.

5.5.3.3 *Ritz Method*

Very large eigen systems can be reduced in size with a technique inspired by the Ritz method. The reduced system is then solved by Jacobi-like methods.

We postulate that each eigenvalue of system (5.169) can be expressed as a linear combination of p independent vector q_i, called Ritz vectors:

$$\{X\} = a_1\{q_1\} + a_2\{q_2\} + \cdots + a_p\{q_p\} \tag{5.209}$$

$$\{X\} = [\{q_1\}\{q_2\}\ldots\{q_p\}]\begin{Bmatrix} a_1 \\ a_2 \\ \vdots \\ a_p \end{Bmatrix} \tag{5.210}$$

$$(n \times 1) \qquad (n \times p) \qquad (p \times 1)$$

$$\{X\} = [Q]\{a\}$$

We look for values of coefficients $\{a\}$ such that vector $\{X\}$ will be as close as possible to an eigenvector of (5.169). For this purpose, we try to render the Rayleigh quotient stationary (5.185):

$$R(\{X\}) = \frac{\langle X\rangle[K]\{X\}}{\langle X\rangle[M]\{X\}} = \frac{\langle a\rangle[\bar{K}]\{a\}}{\langle a\rangle[\bar{M}]\{a\}} \tag{5.211}$$

$$[\bar{K}] = [Q]^T[K][Q]$$
$$[\bar{M}] = [Q]^T[M][Q]$$

The stationarity condition $\delta R = 0$ for any $\langle \delta a \rangle$ is:

$$([\bar{K}] - R[\bar{M}])\{a\} = 0 \tag{5.212}$$

This last expression defines an eigenvalue problem of size p, the solution of which can be written as:

$$[\bar{K}][A] = [\bar{M}][A][\bar{\lambda}] \tag{5.213}$$

where:

$$[A] = [\{A_1\}\{A_2\}\ldots\{A_p\}]; [\bar{\lambda}] = \begin{bmatrix} \bar{\lambda}_1 & & 0 \\ & \bar{\lambda}_2 & \\ 0 & & \ddots \bar{\lambda}_p \end{bmatrix}$$

Eigenvalues $\bar{\lambda}_i$ are approximations of the original system (5.169). These approximations improve when the Ritz vectors span a subspace containing the desired vectors. Moreover, the approximate and exact eigenvalues $\bar{\lambda}_i$ and λ_i verify an expression similar to (5.193):

$$\lambda_1 \leq \bar{\lambda}_1; \lambda_2 \leq \bar{\lambda}_2; \ldots; \lambda_p \leq \bar{\lambda}_p \leq \lambda_n \tag{5.214}$$

To improve the search for the smallest eigenvalues, the Ritz vectors are obtained from:

$$[K]\{q_i\} = \{F_i\} \tag{5.215}$$

where $\{F_i\}$ are unit vectors exciting the degrees of freedom i corresponding to the smallest values of K_{ii}/M_{ii}:

$$\{F_i\} = \begin{Bmatrix} 0 \\ 0 \\ \vdots \\ 1 \\ 0 \\ \vdots \end{Bmatrix} \rightarrow \text{row } i$$

The approximate eigenvectors of (5.169) are obtained from the vectors $\{A_i\}$ with the help of (5.210).

5.5.3.4 Subspace Iteration

This is the most frequently used method to compute the first p eigenvalues of a very large system. A Ritz reduction coupled with the Jacobi method is first employed to obtain Ritz eigenvectors. Then, inverse iteration is used to improve the Ritz vectors. The process is repeated until a convergence criteria is satisfied. The method proceeds by steps as follows:

(a) choose p initial vectors:

$$[X] = [\{X_1\}\{X_2\}...\{X_p\}] \tag{5.216}$$

(b) perform an inverse iteration (Figure 5.23) to compute the p Ritz vector $\{q_i\}$ from:

$$[K]\{q_i\} = [M]\{X_i\} = \{F_i\} \quad i = 1, 2, ..., p$$
$$[K][Q] = [M][X] \tag{5.217}$$

(c) use the Jacobi method to find the eigenvectors of the problem transformed to Ritz subspace:

$$([\bar{K}] - \bar{\lambda}_i[\bar{M}])\{A_i\} = 0 \text{ (Jacobi)}$$

where:

$$[\bar{K}] = [Q]^T[K][Q]$$
$$[\bar{M}] = [Q]^T[M][Q] \tag{5.218}$$
$$\{X_i\} = [Q]\{A_i\}$$

(d) apply a convergence test for $\bar{\lambda}_i$ and repeat steps (b), (c), (d), if necessary.

The method converges towards the p smallest eigenvalues if the initial vectors $\{X_i\}$ are not M-orthogonal to one of the desired vectors. This can be checked by using a Strum sequence (5.194) as follows: matrix $[K] - (\lambda_p + \varepsilon)[M]$ is decomposed, then we verify that there are p negative pivots in the decomposition (ε is taken of the order of 10^{-2} or 10^{-3}). The method is implemented in block 'VALP' of program MEF in Chapter 6.

Remarks

— Matrices $[\bar{K}]$ and $[\bar{M}]$ tend to become diagonal; this improves the Jacobi method.

— If p eigenvalues are desired, it is more efficient to use a subspace q bigger than p, then apply the convergence test to the smallest p eigenvalues only. In reference [11], it is suggested that $q = \text{Min}\,(p+8, 2p)$ be used.

— Initial vectors are often chosen as follows:

$\{X_i\}$ is chosen randomly; the other vectors are:

$$\{X_2\} = \begin{Bmatrix} 0 \\ \vdots \\ 0 \\ 1 \\ 0 \\ 0 \\ \vdots \\ 0 \end{Bmatrix} \ldots \text{row } i_1 \qquad \{X_3\} = \begin{Bmatrix} 0 \\ \vdots \\ 0 \\ 0 \\ 0 \\ 1 \\ \vdots \\ 0 \end{Bmatrix} \ldots \text{row } i_2 \qquad \text{etc.}$$

where $i_1, i_2, \ldots$ are the indices corresponding to the smallest successive values of K_{ii}/M_{ii}

Important Results

Numerical integration:

$$[k] = \sum_{i=1}^{r} w_i [k^*(\boldsymbol{\xi}_i)]$$

$$\{f\} = \sum_{i=1}^{r} w_i \{f_v^*(\boldsymbol{\xi})\} \tag{5.4}$$

$$\int_{-1}^{1} y(\xi)\,d\xi = \sum_{i=1}^{r} w_i y(\xi_i) \tag{5.5}$$

$$\int_{-1}^{1}\int_{-1}^{1} y(\xi,\eta)\,d\xi\,d\eta = \sum_{i=1}^{r} w_i y(\xi_i, \eta_i) \tag{5.14}$$

or
$$= \sum_{i=1}^{r_1}\sum_{j=1}^{r_2} w_i w_j y(\xi_i, \eta_j) \tag{5.15}$$

$$\int_{-1}^{1}\int_{-1}^{1}\int_{-1}^{1} y(\xi,\eta,\zeta)\,d\xi\,d\eta\,d\zeta = \sum_{i=1}^{r_1}\sum_{j=1}^{r_2}\sum_{k=1}^{r_3} w_i w_j w_k y(\xi_i, \eta_j, \zeta_k) \tag{5.20}$$

Solution of linear systems:

$$[K]\{U_n\} = \{F\} \tag{5.22}$$

— Gaussian elimination algorithm (5.26)
$$[K] = [L][S] \quad \text{(decomposition)} \tag{5.28}$$
$$[K] = [L][D][U] \tag{5.35}$$

Decomposition algorithm for skyline matrices:
— non-symmetric matrix: (5.46)
— symmetric matrix: (5.48)

Solution of non-linear systems:
$$\{R^i\} = \{F\} - [K(U^{i-1})]\{U^{i-1}\} \tag{5.53}$$
— Substitution method:
$$[K(U^{i-1})]\{\Delta U^i\} = \{R^i\} \tag{5.53}$$
— Modified Newton–Raphson method:
$$[K_1]\{\Delta U^i\} = \{R^i\} \tag{5.60}$$
— Newton–Raphson method:
$$[K_t(U^{i-1})]\{\Delta U^i\} = \{R^i\} \tag{5.65}$$
— Generalized method: (Figure 5.19)

Solution of time dependent systems:
$$[C]\{\dot{U}\} + [K]\{U\} = \{F(t)\} \tag{5.85}$$
$$[M]\{\ddot{U}\} + [C]\{\dot{U}\} + [K]\{U\} = \{F(t)\} \tag{5.86}$$

— First order system:
  - explicit Euler: (5.100)
  - implicit Euler: (5.104) and (5.108)
  - predictor–corrector: (5.123) and (5.126)
  - Runge–Kutta: (5.133)
  - mode superposition: (5.134) and (5.140)

— Second order system:
  - direct methods: (Figure 5.22)
  - mode superposition: (5.163)

Computation of eigenvalues and eigenvectors:
$$[K]\{X_i\}\lambda_i[M]\{X_i\} \tag{5.165}$$
$$[X]^T[K][X] = [\lambda] \tag{5.175a}$$
$$[X]^T[M][X] = [I] \tag{5.175b}$$

— Rayleigh quotient: (5.185)
— Inverse iteration method: (Figure 5.23)

— General Jacobi method: (Figure 5.24)
— Ritz method: (5.213)
— Subspace iteration method: (5.216) to (5.218)

Notation

Numerical integration

$a_1, a_2, \ldots, a_{2r}$	polynomial coefficients
$[B], [B_\delta]$	matrices of transformation between nodal gradient variables (and their variations) and the nodal variables
C_r	constant in the expression for errors
dV^e, dV^r	differential volume for the real and reference element
$[D]$	material property matrix
e	error
f, f_S, f_V	load vectors
$\{f\}, \{f_S\}, \{f_V\}$	element load vectors
$\{f_S^*\}, \{f_V^*\}$	vectors to be integrated to get f_S and f_V
$[J]$	Jacobian matrix
$[K]$	element matrix
$[K^*]$	matrix to be integrated to get k
$[m]$	element mass matrix
$[N], \langle N \rangle$	matrix and vector of interpolation functions
P_r	Legendre's polynomial
r, r_1, r_2, r_3	number of integration points and order of polynomials
w_i	weighting coefficients of numerical integration
$y(\xi)$	function to be integrated numerically
$\xi = \langle \xi \ \eta \ \zeta \rangle$	coordinates of a point of the reference element
ξ_i	coordinates of a point for numerical integration

Solution of Linear Systems of Equations

$[D]$	diagonal matrix during decomposition
$\{F\}$	global load vector
F_i	component i of F
$\{F'\}$	vector F after Gaussian elimination
$\{F^s\}$	vector F after the elimination of the first s variables
h_J	height of column J
i_{0j}, i_{0s}	row number of the highest term in columns j and s
$[K]$	global matrix
$[K^s]$	global matrix after elimination of the first variables

$[l]$	lower triangular matrix equal to L^{-1}
$[l^i]$	lower triangular matrix corresponding to the elimination of variable i
$[L]$	lower triangular matrix for the decomposition
$[L^i]$	inverse of l^i
$[L_c]$	decomposition matrix for the Cholesky algorithm
$[S]$	upper triangular matrix for the decomposition
$\{U_n\}$	global variables
$[U]$	upper triangular matrix of the LDU decomposition

Solution of Non-Linear Systems of Equations

$\{F\}$	global load vector
$\{F_0\}$	global reference load vector
$\{\Delta F\}$	increment of F
$[K]$	global matrix
$[K_c]$	matrix used for the computation of ΔU
$[K_l]$	global matrix, linear part
$[K_{nl}]$	global matrix, non-linear part
$[K_t]$	global tangent matrix
$\|m\|, \|n\|$	norm of R and of ΔU
$\{R\}$	residual
$\{R^i\}$	residual corresponding to U^{i-1}
$\{U\}, \mathbf{U}, \{U_n\}$	global vector of nodal variables
$\{U^i\}$	vector U at integration i
$\{\Delta U\}, \{\delta U_n\}$	increment and variation of U
$W, W^e, \delta W$	global and element integral form, first variation of W
δ	variational symbol
ε	admissible error for a norm
$\lambda, \Delta\lambda$	load vector parameter, corresponding increment
Π	functional
ω	over-relaxation factor

Solution of Unsteady Systems

a, b	coefficients for the Newmark–Wilson method
a_0, a_1, a_2	coefficients for the direct integration of second order systems
$[A], [B]$	matrices used in stability computations
b_j	coefficients for predictor corrector formulas and Runge–Kutta formulas
$[C]$	damping matrix
f	right-hand side member of the standard expression $du/dt = f(u, t)$

$\{F\}, \{F_t\}$ global load vectors at instant t
$[K], [\bar{K}], [K_{nl}], [K_t]$ global, modified, non-linear and tangent matrices
l_{max} largest eigenvalue (absolute value)
$[M]$ global mass matrix
$\{R\}, \{R_t\}, \{\bar{R}\}, \{R_{nl}\}$ residues
$t, t_0, \Delta t$ time, initial time and increment of time
$\{U\}, \{U_0\}, \{U_t^i\}$ nodal variables, initial values, value at instant t after iteration i
$\{\dot{U}\}, \{\ddot{U}\}$ first and second derivatives of U with respect to time
$\{V\}$ solution of decoupled equations
$\{X_i\}, [X]$ eigenvalues and modal matrix
$\lambda, \lambda_i, [\lambda]$ eigenvalues
ξ_i damping factor of mode i
$\rho(A)$ spectral radius of matrix A
τ, θ parameters of the Wilson method
ω_i frequency corresponding to eigenvalue λ_i

Computations of Eigenvalues and Eigenvectors

a amplitude of a shift
a, b, c_1, c_2, c_3, d coefficients of the Jacobi method
$a_1, a_2, \ldots, a_p, \{a\}$ coefficients of Ritz vectors
$\{A_i\}$ eigenvectors of the Ritz method
c_i components of a vector in the basis of eigenvectors
$[K], [\bar{K}]$ global matrix, modified matrix
$[K^i]$ global matrix modified by i iterations of Jacobi
l_i eigenvalues
$[L]$ lower triangular matrix in the decomposition of $[K]$
$[M], [\bar{M}]$ global mass matrix, modified matrix
$[M^i]$ global mass matrix after i iterations of Jacobi
p exponent of $[K]$, index
$P(\lambda)$ characteristic polynomial of a matrix
$[P_i]$ projectors of a matrix
$\{q_i\}$ Ritz vectors
$[Q]$ transformation matrix, matrix of Ritz vectors
$[Q^i]$ transformation matrix for iteration i (Jacobi)
$R(V)$ Rayleigh's quotient corresponding to vector V
V any vector
$\{X_i\}, [X][\bar{X}]$ eigenvalues
$\{\bar{Y}\}$ vectors $[M]\{X\}$
θ Jacobi rotation
λ, λ_i eigenvectors
μ shift
$\rho(K)$ spectral radius of $[K]$

References

[1] J. J. Davis and P. Rabinowitz, *Methods of Numerical Integration*, Academic Press, 1975.
[2] A. H. Stroud, *Approximate Calculation of Multiple Integrals*, Prentice-Hall, New Jersey, 1971.
[3] Z. Kopal, *Numerical Analysis*, 2nd edn, Chapman & Hall, 1961.
[4] B. M. Irons, Engineering application of numerical integration in stiffness method, *J. AIAA*, **14**, 2035–2037, 1966.
[5] P. C. Hammer, O. P. Marlow and A. H. Stroud, Numerical integration over simplexes and cones, *Math. Tables Aids Comp.*, **10**, 130–137, 1956.
[6] B. M. Irons, Quadrature rules for brick based finite elements, *J. AIAA*, **9**, 293–294, 1971.
[7] T. K. Hellen, Effective quadrature rules for quadratic solid isoparametric finite elements, *Int. J. Num. Meth. Eng.*, **4**, 597–600, 1972.
[8] O. C. Zienkiewicz, *The Finite Element Method*, McGraw-Hill, New York, 1977.
[9] G. Strang and G. FIX, *An Analysis of Finite Element Method*, Prentice-Hall, New Jersey, 1973.
[10] B. Carnahan, H. A. Luther, and J. O. Wiles, *Applied Numerical Methods*, Wiley, 1969.
[11] K. J. Bathe and E. L. Wilson, *Numerical Methods in Finite Element Analysis*, Prentice-Hall, New Jersey, 1976.
[12] J. L. Batoz and G. S. Dhatt, Incremental displacement algorithms for non-linear problems, *Int. J. Num. Meth. Eng.*, **14**, 8, 1262–1267, 1979.
[13] S. H. Crandall, *Engineering Analysis*, McGraw-Hill, New York, 1956.
[14] L. Collatz, *The Numerical Treatment of Differential Equations*, Springer-Verlag, 1966.
[15] A. Ralston, *A First Course in Numerical Analysis*, McGraw-Hill, New York 1965.
[16] E. L. Wilson, *CAL Computer Analysis Language*, Report No. UC SESM 77-2, Department of Civil Engineering, University of California, Berkeley, 1977.
[17] R. W. Clough and J. Penzien, *Dynamics of Structures*, McGraw-Hill, New York, 1975.
[18] J. C. Houbolt, A recurrence matrix solution for the dynamic response of elastic Aircraft, *J. of Aeronautical Sciences*, **17**, 540–550, 1950.
[19] M. G. Katona, R. Thompson, and J. Smith, *Efficiency Study of Implicit and Explicit Time Integration Operators for Finite Element Applications*, Report No. R 856, Naval Construction Battalion Center, Port Hueneme, Cal., 1977.
[20] N. M. Newmark, A method of computation of structural dynamics, *ASCE Journal of Eng. Mech. Division*, **85**, 67–94, 1959.
[21] J. H. Wilkinson, *The Algebraic Eigenvalue Problem*, Oxford University Press, London, 1965.

6

CODING TECHNIQUES

6.0 Introduction

In this chapter, we survey some coding techniques that can be used to implement finite element methods in computers. We first identify various activities common to all finite element codes and illustrate with a program for beginners called BBMEF. We further refine the characteristics of multipurpose codes and illustrate with a program of medium complexity called MEF.

The construction of effective computational codes requires expertise in computer science as well as an intimate knowledge of the present state of the art in finite element research. One may make certain observations in this direction:

— Programs have a tendency to become very lenghty (thousands to hundreds of thousands of statements) since they must provide the user with a variety of analyses and services such as pre- and post-processing of data, resolution of large matrices and interactive graphic facilities.

— Owing to very high development costs, one should produce general-purpose programs covering wide classes of problems.

— The amount of data generated and manipulated during the solution of problems makes it mandatory that the best methods of computer science be used to store, retrieve, and process all the information needed.

— Finally, the issue of portability between computers of various brands must be considered to ensure wider usage of programs in an economic manner.

We give a complete listing of the program MEF as well as the necessary documentation for usage. This will permit readers to appreciate various aspects of a general-purpose finite element program. Moreover, with this a user would be capable of introducing the improvements and additions necessary to adapt the present code to different types of needs without any profound modifications.

6.1 Common Features of Finite Element Programs

All finite element computer codes include functional blocks similar to the following:

(a) input, generation and verification of all the information required to define the problem to be solved, including the loadings;

(b) construction of element matrices and element vectors, followed by an assembly process giving the overall system equations and load vectors;

(c) application of boundary conditions and solution of the system equations;

(d) computations of auxiliary quantities depending on the solution of the system equations and post-processing of results.

Figure 6.1 illustrates the flow of common operations for most finite element codes.

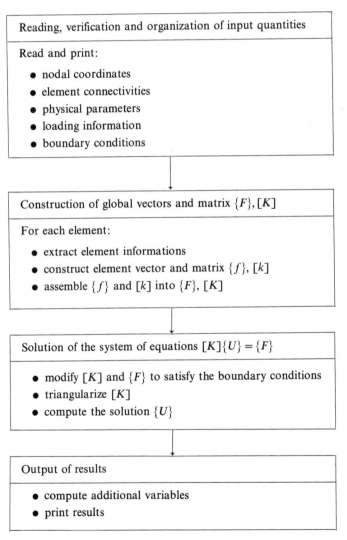

Figure 6.1 Characteristic functional blocks of a finite element code.

6.2 Beginner's Program BBMEF

We now describe a simple code designed to solve Poisson's equation:

$$d\left(\frac{\partial^2 u}{\partial x^2} + \frac{\partial^2 u}{\partial y^2}\right) + f_v = 0$$

The element matrix and element load vector of a linear triangular element for Poisson's equation have been obtained in paragraph 4.3.1. Although BBMEF was constructed to provide solutions of the Poisson's equation, it could be modified to solve other problems after modification of the basic parameters defined in the DATA statement. Other elements could also be obtained by modifying subroutine ELEM00. Figure 6.2 describes the various subroutines of BBMEF, listed in Figures 6.3 and 6.4.

Names	Functions	Figure containing listings
Comments		6.3
BBMEF	Main program	6.4
GRILLE	Reading of the mesh	1.2
LOCEF	Creation of destination table of element	4.7
ELEM00	Creation of element matrix	4.3
ASSEMB	Assembly of element	4.6
RESOL	Solution	5.12

Figure 6.2 Subroutines of program BBMEF.

```
C========================================================================BBMC   1
C                                                                        BBMC   2
C       BEBE-MEF                                                         BBMC   3
C                                                                        BBMC   4
C       INTRODUCTORY PROGRAM TO FINITE ELEMENTS                          BBMC   5
C                                                                        BBMC   6
C                                                                        BBMC   7
C       MAIN VARIABLES                                                   BBMC   8
C       --------------                                                   BBMC   9
C                                                                        BBMC  10
C           NNT     TOTAL NUMBER OF NODES (<100)                         BBMC  11
C           NELT    TOTAL NUMBER OF ELEMENTS (<60)                       BBMC  12
C           NDIM    DIMENSIONS OF THE PROBLEM (2)                        BBMC  13
C           NDLN    NUMBER OF DEGREES OF FREEDOM PER NODE (1)            BBMC  14
C           NEQ     NUMBER OF SYSTEM EQUATIONS                           BBMC  15
C           NNEL    NUMBER OF NODES PER ELEMENT                          BBMC  16
C           NDLE    NUMBER OF DEGREES OF FREEDOM PER ELEMENT             BBMC  17
C                   (NDLE=NNEL*NDLN)                                     BBMC  18
C           MR,MP   LOGICAL UNIT NUMBER FOR READING AND                  BBMC  19
C                   PRINTING DATA (MR=5,MP=6)                            BBMC  20
C           NSYM    INDEX OF SYMMETRIC MATRIX (.EQ.1 : NON SYMMETRIC)    BBMC  21
C           NEQMAX  MAXIMUM NUMBER OF PERMISSIBLE EQUATIONS              BBMC  22
C                                                                        BBMC  23
C       ARRAYS AND MINIMAL DIMENSIONS :                                  BBMC  24
C       -------------------------------                                  BBMC  25
C                                                                        BBMC  26
C           VCORG(NDIM,NNT)                                              BBMC  27
C                   NODAL COORDINATES                                    BBMC  28
```

```
C            KCONEC(NNEL,NELT)                                   BBMC  29
C                    CONNECTIVITY TABLE                          BBMC  30
C            VKG(NEQMAX*NEQMAX)                                  BBMC  31
C                    GLOBAL MATRIX                               BBMC  32
C            VFG(NEQMAX)                                         BBMC  33
C                    GLOBAL LOAD VECTOR                          BBMC  34
C            VCORE(NDIM,NNEL)                                    BBMC  35
C                    ELEMENT NODAL COORDINATES                   BBMC  36
C            KLOCE(NDLE)                                         BBMC  37
C                    ELEMENT LOCALIZATION TABLE                  BBMC  38
C            VPREE(...)                                          BBMC  39
C                    PHYSICAL PROPERTIES ARRAY REQUIRED FOR      BBMC  40
C                    ELEMENT COMPUTATIONS                        BBMC  41
C            VKE(NDLE,NDLE)                                      BBMC  42
C                    ELEMENT MATRIX                              BBMC  43
C            VFE(NDLE)                                           BBMC  44
C                    ELEMENT VECTOR                              BBMC  45
C                                                                BBMC  46
C     REQUIRED SUBPROGRAMS :                                     BBMC  47
C     --------------------                                       BBMC  48
C                                                                BBMC  49
C            GRILLE   TO READ NODAL COORDINATES AND CONNECTIVITIES BBMC  50
C            LOCEF    TO FORM ELEMENT LOCALIZATION TABLE LOCE    BBMC  51
C            ELEM00   TO FORM ELEMENT MATRIX KE AND              BBMC  52
C                     ELEMENT VECTOR FE                          BBMC  53
C            ASSEMB   TO ASSEMBLE AN ELEMENT                     BBMC  54
C            RESOL    TO SOLVE BY GAUSS METHOD                   BBMC  55
C                                                                BBMC  56
C     DATA DEFINITION :                                          BBMC  57
C     ----------------                                           BBMC  58
C                                                                BBMC  59
C            1 CARD (2I5)                                        BBMC  60
C                    NUMBER OF NODES(NNT),NUMBER OF ELEMENTS (NELT) BBMC  61
C                    MAXIMUM :    NNT < 100    NELT < 60         BBMC  62
C            NNT CARDS (2F10.0)                                  BBMC  63
C                    X,Y COORDINATES FOR EACH NODE               BBMC  64
C            NELT CARDS (3I5)                                    BBMC  65
C                    NODE NUMBERS FOR EACH ELEMENT               BBMC  66
C            1 CARD (1F10.0)                                     BBMC  67
C                    COEFFICIENT D                               BBMC  68
C            1 CARD (1F10.0)                                     BBMC  69
C                    VOLUMIC LOAD FV                             BBMC  70
C            1 CARD PER LOAD FOR A DEGREE OF FREEDOM (1I5,1F10.0) BBMC  71
C                    D.O.F. NUMBER, LOAD VALUE                   BBMC  72
C                    (THIS GROUP OF CARDS IS TERMINATED BY A BLANK CARD) BBMC  73
C                    (RESPECT D.O.F. NUMBER !)                   BBMC  74
C            1 CARD PER PRESCRIBED DEGREE OF FREEDOM (1I5,1F10.0) BBMC  75
C                    D.O.F. NUMBER, PRESCRIBED VALUE AT THIS D.O.F. BBMC  76
C                    (THIS GROUP OF CARDS IS TERMINATED BY A BLANK CARD) BBMC  77
C                    (RESPECT D.O.F. NUMBER)                     BBMC  78
C                                                                BBMC  79
C     REMARK :                                                   BBMC  80
C     --------                                                   BBMC  81
C                                                                BBMC  82
C            BY MODIFYING DIMENSIONS AND PARAMETERS DEFINED BY DATA, BBMC  83
C            IT IS POSSIBLE TO DEAL WITH PROBLEMS CORRESPONDING  BBMC  84
C            TO DIFFERENT VALUES OF NDIM,NDLN,NNE.               BBMC  85
C            THE NUMBER OF D.O.F. MUST BE CONSTANT.              BBMC  86
C                                                                BBMC  87
C                                                                BBMC  88
C*********************************************************************BBMC  89
C                                                                BBMC  90
C                                                                BBMC  91
C                                                                BBMC  92
C                                                                BBMC  93
C================================================================BBMC  94
```

Figure 6.3 Comments on program BBMEF.

```
      IMPLICIT REAL*8(A-H,O-Z)                                BBME  1
      COMMON/PARAM/NNT,NELT,NDIM,NDLN,NDLT,NEQ,NNEL,NDLE      BBME  2
C....... CARDS TO BE MODIFIED IF GENERAL PARAMETERS ARE CHANGED BBME 3
      DIMENSION VCORG(2,100),KCONEC(3,60),VKG(10000),VFG(100), BBME  4
     1          VCORE(2,3),KLOCE(3),VPREE(2),VKE(3,3),VFE(3)   BBME  5
      DATA VKG/10000*0.D0/,VFG/100*0.D0/,GRAND/1.D12/,         BBME  6
     1     MR/5/,MP/6/,NDLE/3/,NDLN/1/,NNEL/3/,NDIM/2/,NSYM/0/ BBME  7
C.......                                                      BBME  8
C                                                             BBME  9
C------- BLOCK TO READ THE DATA                               BBME 10
C                                                             BBME 11
C------- INPUT COORDINATES AND CONNECTIVITIES                 BBME 12
      CALL GRILLE(NDIM,NNEL,MR,MP,NNT,NELT,VCORG,KCONEC)       BBME 13
      NEQ=NNT*NDLN                                            BBME 14
C------- READ PHYSICAL PARAMETER D                            BBME 15
      READ(MR,1000) VPREE(1)                                  BBME 16
1000  FORMAT(1F10.0)                                          BBME 17
      WRITE(MP,2000) VPREE(1)                                 BBME 18
2000  FORMAT(/' CONDUCTIVITY COEFFICIENT(D)=',E12.5/)          BBME 19
C------- READ VOLUMIC LOAD FV                                 BBME 20
      READ(MR,1000) VPREE(2)                                  BBME 21
      WRITE(MP,2010) VPREE(2)                                 BBME 22
2010  FORMAT(/' VOLUMIC LOAD (FV)=',E12.5/)                    BBME 23
C------- READ NODAL LOADS (SOURCES)                           BBME 24
      DO 10 I=1,NEQ                                           BBME 25
      READ(MR,1010) IN,F                                      BBME 26
1010  FORMAT(I5,F10.0)                                        BBME 27
      IF(IN.LE.0) GO TO 20                                    BBME 28
      VFG(IN)=F                                               BBME 29
10    WRITE(MP,2020)IN,F                                      BBME 30
2020  FORMAT(' LOAD AT NODE ',I5,' =',E12.5)                   BBME 31
C                                                             BBME 32
C------- FORM GLOBAL MATRIX AND GLOBAL VECTOR                 BBME 33
C                                                             BBME 34
20    DO 40 IE=1,NELT                                         BBME 35
C------- FIND ELEMENT COORDINATES                             BBME 36
      DO 30 IN=1,NNEL                                         BBME 37
      J=KCONEC(IN,IE)                                         BBME 38
      VCORE(1,IN)=VCORG(1,J)                                  BBME 39
30    VCORE(2,IN)=VCORG(2,J)                                  BBME 40
C------- FORM THE LOCALIZATION TABLE                          BBME 41
      CALL LOCEF(KCONEC(1,IE),NNEL,NDLN,KLOCE)                 BBME 42
C------- EVALUATE ELEMENT MATRIX AND VECTOR                   BBME 43
      CALL ELEM00(VCORE,VPREE,VKE,VFE,NDIM,NNEL,NDLE)          BBME 44
C------- ASSEMBLE                                             BBME 45
      CALL ASSEMB(VKE,VFE,KLOCE,NDLE,NEQ,VKG,VFG)              BBME 46
40    CONTINUE                                                BBME 47
C                                                             BBME 48
C------- SOLVE RESULTING SYSTEM                               BBME 49
C                                                             BBME 50
C------- BOUNDARY CONDITIONS      BBME 51
      DO 50 I=1,NEQ                                           BBME 52
      READ(MR,1010) IN,F                                      BBME 53
      IF(IN.LE.0) GO TO 60                                    BBME 54
      WRITE(MP,2030) IN,F                                     BBME 55
2030  FORMAT(' PRESCRIBED VALUE AT D.O.F. :',I5,' =',E12.5)    BBME 56
      INI=(IN-1)*NEQ+IN                                       BBME 57
      VKG(INI)=VKG(INI)+GRAND                                 BBME 58
50    VFG(IN)=F*GRAND                                         BBME 59
C------- SOLVE                                                BBME 60
60    CALL RESOL(NSYM,NEQ,VKG,VFG)                             BBME 61
C                                                             BBME 62
C------- PRINT THE RESULTS                                    BBME 63
C                                                             BBME 64
```

Figure 6.4 (*Contd.*)

```
        WRITE(MP,2040)                                              BBME  65
 2040   FORMAT(/' NODE         X          Y         D.O.F.'/)       BBME  66
        DO 80 IN=1,NNT                                              BBME  67
        I1=(IN-1)*NDLN+1                                            BBME  68
        I2=IN*NDLN                                                  BBME  69
 80     WRITE(MP,2050)IN,VCORG(1,IN),VCORG(2,IN),(VFG(I),I=I1,I2)   BBME  70
 2050   FORMAT(I5,3F10.5)                                           BBME  71
        STOP                                                        BBME  72
        END                                                         BBME  73
```

Figure 6.4 Main program of BBMEF.

Example 6.1 *Solution of Poisson's equation over a square region using BBMEF*

Reconsidering Example 3.17; taking advantage of the double symmetry, we discretize only one-eight of the domain. We use the mesh shown having 16 elements:

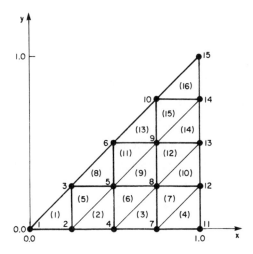

As in Example 3.17, we take:

$$d = 1.0$$
$$f_v = 1.0,$$

Input quantities required by BBMEF and described in paragraph 6.3, are:

15	16
0.0	0.0
0.1	0.0
0.1	0.1
0.25	0.0
0.25	0.10
0.25	0.25
0.5	0.0
0.5	0.10
0.5	0.25
0.5	0.5
1.0	0.0
1.0	0.10
1.0	0.25
1.0	0.5

```
        1.      1.
    1   2   3
    2   4   5
    4   7   8
    7  11  12
    2   5   3
    4   8   5
    7  12   8
    3   5   6
    5   8   9
    8  12  13
    5   9   6
    8  13   9
    6   9  10
    9  13  14
    9  14  10
   10  14  15
       1.0
       1.0
        0
   11      0.
   12      0.
   13      0.
   14      0.
   15      0.
    0
```

The results obtained by **BBMEF** are:

NUMBER OF NODES = 15 NUMBER OF ELEMENTS= 16

```
    NODES       COORDINATES

      1     0.00000   0.00000
      2     0.10000   0.00000
      3     0.10000   0.10000
      4     0.25000   0.00000
      5     0.25000   0.10000
      6     0.25000   0.25000
      7     0.50000   0.00000
      8     0.50000   0.10000
      9     0.50000   0.25000
     10     0.50000   0.50000
     11     1.00000   0.00000
     12     1.00000   0.10000
     13     1.00000   0.25000
     14     1.00000   0.50000
     15     1.00000   1.00000

   ELEMENT        CONNECTIVITY

      1         1    2    3
      2         2    4    5
      3         4    7    8
      4         7   11   12
      5         2    5    3
      6         4    8    5
      7         7   12    8
      8         3    5    6
      9         5    8    9
     10         8   12   13
     11         5    9    6
     12         8   13    9
     13         6    9   10
     14         9   13   14
     15         9   14   10
     16        10   14   15
```

CONDUCTIVITY COEFFICIENT(D)= 0.10000E+01

```
VOLUMIC LOAD (FV)= 0.10000E+01

PRESCRIBED VALUE AT D.O.F. :   11 = 0.00000E+00
PRESCRIBED VALUE AT D.O.F. :   12 = 0.00000E+00
PRESCRIBED VALUE AT D.O.F. :   13 = 0.00000E+00
PRESCRIBED VALUE AT D.O.F. :   14 = 0.00000E+00
PRESCRIBED VALUE AT D.O.F. :   15 = 0.00000E+00

NODE      X          Y         D.O.F.
  1    0.00000    0.00000    0.29131
  2    0.10000    0.00000    0.28798
  3    0.10000    0.10000    0.28506
  4    0.25000    0.00000    0.27391
  5    0.25000    0.10000    0.27119
  6    0.25000    0.25000    0.25811
  7    0.50000    0.00000    0.22352
  8    0.50000    0.10000    0.22124
  9    0.50000    0.25000    0.21078
 10    0.50000    0.50000    0.17293
 11    1.00000    0.00000    0.00000
 12    1.00000    0.10000    0.00000
 13    1.00000    0.25000    0.00000
 14    1.00000    0.50000    0.00000
 15    1.00000    1.00000    0.00000
```

The value of u at the centre (node 1) is:

$$u_c = 0.2913$$

whereas the exact value given in Example 3.24 is:

$$u = 0.2947$$

6.3 Multipurpose Programs

6.3.1 *Capabilities of General Codes*

In contrast to the simple program described in the previous section, a general program should be able to:

— solve a variety of problems from a number of disciplines: linear and non-linear, static and dynamic problems of elasticity, fluid mechanics, heat transfer, etc.;
— solve problems of large size involving a variety of elements.

6.3.1.1 *Problem types*

Depending on the nature of the problems to be solved, a large number of elements and load vectors must be available. The elements should be able to accommodate complex geometries with straight and curved edges in one, two, and three dimensions. The numerical algorithms should be able to cope with linear and non-linear, stationary and transient problems. In summary, a general program should have facilities to cope with:

— problems in one, two, and three dimensions;
— different numbers of degrees of freedom at each node;
— new elements to be added;
— symmetric and non-symmetric element matrices;
— linear and non-linear problems;

- stationary and non-stationary problems;
- eigenvalue problems.

6.3.1.2 *Problem size*

Many problems require a large number of elements. The number of unknowns vary from several hundreds to tens of thousands. The size of a problem depends upon one or more of the following factors:

- dimensionality (one, two, or three);
- number of unknown variables per node;
- complexity of the geometry;
- number of elements required to achieve the desired degree of accuracy.

(a) *Description of the problem*

The complete definition of a problem requires the preparation of nodal coordinates and element connectivities, the specification of material properties and boundary conditions. This part of the solution is generally voluminous and subject to human error. A general program is often accompanied by a variety of utility programs to assist in the preparation and verification of a problem definition.

(b) *Storage of data*

Because of the volume of data needed to define and solve even moderately sized problems, it is generally more convenient to construct a data base on auxiliary mass storage devices for each problem. Techniques of in-core and out-core data base management must be incorporated in efficient codes.

(c) *Volume of computations*

Computation time can be very high for large non-linear and non-stationary problems. Thus high cost of computing may play an important role in making a finite element procedure economical or uneconomical. One should therefore construct efficient codes for repetitive operations such as: calculation of elementary matrices, assemblage and solution algorithm, etc.

Furthermore, various parameters which influence the computer cost must be properly chosen:

- type of element and mesh configuration;
- numerical integration schemes for evaluating elementary matrices;
- solution strategy for non-linear problems;
- direct time integration schemes for non-stationary problems;
- methods for calculating eigenvalues and eigenvectors.

(d) *Presentation of results*

Final results are in the form of voluminous numberical tables that are very difficult to digest. Auxiliary utility programs must generally be prepared to generate final reports in easily readable numerical tables and graphs.

6.3.2 *Modularity*

A general program having attributes described in paragraph 6.3.1 is going to be voluminous and complex. It is, however, desirable that:

— its logic be easily understood;
— one or many of its parts be easily modifiable;
— many workers participate in its development;
— it offers possibilities to tailor its facilities for the solution of particular classes of problems.

For all these reasons, the program should have a modular structure, with the modules made as independent from one another as is practicable. The following modular operations are recognized:

(a) problem definition (data base);

— node coordinates and element connectivities;
— nodal and element properties;
— boundary conditions;

(b) element computations:

— integration points and associated weights;
— interpolation functions and their derivatives;
— Jacobian matrix, inverse, and determinant;
— element matrices and vectors: $[k]$, $[m]$, $\{f\}$, $[k_t]$, $\{r\}$, etc.;

(c) assembly operations:

— assemblage of master matrices and vectors, $[K]$, $[M]$, $\{F\}$, etc.;

(d) solution:

— factorization of master matrices and solution of equations;

(e) results:

— output of nodal variables and other calculated quantities: gradients, reactions, etc.

Subroutines implementing the various operations described above are contained in all finite element codes. The flow of information between these operations is problem-dependent; linear, non-linear, static, and dynamic problems all require logics of their own. The flow diagram for the solution of a problem can become very complex. The program described below is divided into functional blocks through which a user can steer the flow of computations in a logical order of his own choice. A functional block can perform one or more operations, for instance: definition of a problem, creation and organization of the data base, construction of element matrices and vectors, etc. When the functional blocks are very simple, they can be used in various orders; the user must then steer the flow of information correctly.

6.4 Program MEF

6.4.1 *Introduction*

A program of medium complexity, called MEF, implementing the techniques of paragraph 6.3 is now going to be described. It is written in FORTRAN IV and should be easily adaptable to various computers. It can be used to solve a large variety of boundary value problems of mathematical physics.

6.4.2 *Overall Organization*

6.4.2.1 *Flow chart of functional blocks*

The main program controls the logical flow of information through the various functional blocks. Subroutine BLnnnn executes the following preliminary operations of block 'nnnn':

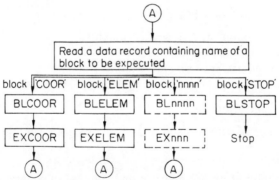

— sets logical identification of the files used in the block, providing adequate default values;

— reads various control parameters required for the creation of various files; appropriate default values are provided to resolve internal conflicts;

— allocates memory to various arrays using pseudo-dynamic in-core memory management techniques described in the next paragraph;

— calls subroutine EXnnnn.

Subroutine Exnnnn executes all the operations of the functional block, calling general utility subroutines from a library.

6.4.2.2 Pseudo-dynamic Memory Management

FORTRAN does not allow of directly changing an array size during the execution of a code. To circumvent this language limitation, the following pseudo-dynamic allocation of memory is used to vary the size of arrays during execution:

— all arrays are defined as one-dimensional arrays, and stored sequentially in one overall array (workspace) called VA;

— each array 'tttt' is located by its pointer 'Ltttt' containing the address of its first term in array VA;

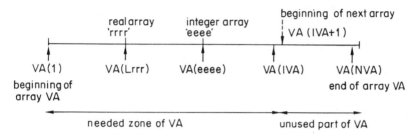

— the total space occupied by all the tables is limited only by the size of array VA defined in the main program:

$$\begin{cases} \text{COMMON} \quad \text{VA (20000)} \\ \ldots \ldots \\ \text{NVA} = 20000 \end{cases}$$

— the pointers 'Ltttt' of each table are contained in COMMON block:

COMMON/LOC/LCORG, LDLNG, ...

— the management of the memory used by the working space is done in a subroutine called ESPACE; the deletion of arrays from VA is done in a subroutine called VIDE (empty); these two subroutines are listed in Figure 6.5.

6.4.2.3 Programming Norms (programmer's reference guide)

When a program must be constructed and maintained by several programmers, it is absolutely essential to define certain rules to provide a degree of homogeneity in the various arbitrary decisions needed during the creation of the code. For program MEF, the following rules were used.

(a) *Functional Blocks*

The basic name of a functional block is restricted to four characters 'nnnn'. To each block there is a control subroutine named BLnnnn, an execution subroutine EXnnnn and a named common block COMMON/nnnn/.

```
      SUBROUTINE ESPACE(ILONG,IREEL,TBL,IDEB)                   ESPA   1
C=========================================================================ESPA   2
C     TO ALLOCATE A REAL OR INTEGER TABLE IN ARRAY VA            ESPA   3
C        INPUT                                                   ESPA   4
C           ILONG            LENGTH OF THE TABLE TO BE ALLOCATED ESPA   5
C                            (IN REAL OR INTEGER WORDS)          ESPA   6
C           IREEL            TABLE TYPE :                        ESPA   7
C                              .EQ.0      INTEGER                ESPA   8
C                              .EQ.1      REAL                   ESPA   9
C           TBL              NAME OF THE TABLE (A4)              ESPA  10
C        OUTPUT                                                  ESPA  11
C           IDEB             TABLE TO BE ALLOCATED STARTS IN VA(IDEB) ESPA 12
C=========================================================================ESPA  13
      IMPLICIT REAL*8(A-H,O-Z)                                   ESPA  14
      REAL*4 TBL                                                 ESPA  15
      COMMON/ES/M,MR,MP                                          ESPA  16
      COMMON/ALLOC/NVA,IVA,IVAMAX,NREEL                          ESPA  17
      COMMON VA(1)                                               ESPA  18
      DIMENSION KA(1)                                            ESPA  19
      EQUIVALENCE (VA(1),KA(1))                                  ESPA  20
      DATA ZERO/0.D0/                                            ESPA  21
C-------------------------------------------------------------------ESPA  22
C------- CALCULATE THE TABLE LENGTH IN REAL WORDS                ESPA  23
      ILGR=ILONG                                                 ESPA  24
      IF(IREEL.EQ.0) ILGR=(ILONG+NREEL-1)/NREEL                  ESPA  25
      IVA1=IVA+ILGR                                              ESPA  26
C------- CHECK IF ENOUGH SPACE IS AVAILABLE                      ESPA  27
      IF(IVA1.LE.NVA) GO TO 20                                   ESPA  28
C....... AUTOMATIC EXTENSION OF THE BLANK COMMON IF CORRESPONDING ESPA  29
C        SYSTEM COMMAND EXIST ON THE COMPUTER USED               ESPA  30
C     CALL EXTEND(IVA1,IERR)                                     ESPA  31
C     IF(IERR.EQ.1) GO TO 10                                     ESPA  32
C     NVA=IVA1                                                   ESPA  33
C     GO TO 20                                                   ESPA  34
C------- ALLOCATION ERROR (NOT ENOUGH SPACE)                     ESPA  35
   10 WRITE(MP,2000) TBL,IVA1,NVA                                ESPA  36
 2000 FORMAT(' **** ALLOCATION ERROR, TABLE ',A4/' REQUIRED SPACE:',I9,'ESPA 37
     1 REAL WORDS, AVAILABLE SPACE:',I9,' REAL WORDS')           ESPA  38
      STOP                                                       ESPA  39
C------- ALLOCATE TABLE                                          ESPA  40
   20 IDEB=IVA+1                                                 ESPA  41
      IVA=IVA1                                                   ESPA  42
      IF(IVA.GT.IVAMAX) IVAMAX=IVA                               ESPA  43
      IF(M.GT.0) WRITE(MP,2010) TBL,IDEB,IVA1                    ESPA  44
 2010 FORMAT(60X,'TABLE ',A4,' GOES FROM VA(',I7,') TO VA(',I7,')') ESPA 45
C------- INITIALIZE THE ALLOCATED TABLE TO ZERO                  ESPA  46
      I1=IDEB                                                    ESPA  47
      IF(IREEL.EQ.0) I1=(I1-1)*NREEL+1                           ESPA  48
      I2=I1+ILONG-1                                              ESPA  49
      IF(IREEL.EQ.0) GO TO 40                                    ESPA  50
      DO 30 I=I1,I2                                              ESPA  51
   30 VA(I)=ZERO                                                 ESPA  52
      RETURN                                                     ESPA  53
   40 DO 50 I=I1,I2                                              ESPA  54
   50 KA(I)=0                                                    ESPA  55
      RETURN                                                     ESPA  56
      END                                                        ESPA  57

      SUBROUTINE VIDE(IDEB,IREEL,TBL)                            VIDE   1
C=========================================================================VIDE   2
C     TO DELETE A TABLE FROM VA, FOLLOWED BY COMPACTING          VIDE   3
C        INPUT                                                   VIDE   4
C           IDEB             FIRST POSITION OF TABLE TO BE DELETED VIDE  5
C           IREEL            TYPE OF TABLE (SEE ESPACE)          VIDE   6
C           TBL              NAME OF THE TABLE (A4)              VIDE   7
C=========================================================================VIDE   8
```

Figure 6.4 (*Contd.*)

```
      IMPLICIT REAL*8(A-H,O-Z)                                  VIDE  9
      REAL*4 TBL                                                VIDE 10
      COMMON/ES/M,MR,MP                                         VIDE 11
      COMMON/ALLOC/NVA,IVA,IVAMAX,NREEL,NTBL                    VIDE 12
      COMMON/LOC/LXX(25)                                        VIDE 13
      COMMON VA(1)                                              VIDE 14
C-----------------------------------------------------------------VIDE 15
C------ SEARCH FOR THE FIRST POSITION OF NEXT TABLE             VIDE 16
      I1=IVA+1                                                  VIDE 17
      DO 10 I=1,NTBL                                            VIDE 18
      IF(LXX(I).LE.IDEB) GO TO 10                               VIDE 19
      IF(LXX(I).LT.I1) I1=LXX(I)                                VIDE 20
10    CONTINUE                                                  VIDE 21
C------ SHIFT ALL TABLES AFTER THIS                             VIDE 22
      ID=I1-IDEB                                                VIDE 23
      IF(I1.EQ.IVA+1) GO TO 40                                  VIDE 24
      DO 20 I=1,NTBL                                            VIDE 25
      IF(LXX(I).GT.IDEB) LXX(I)=LXX(I)-ID                       VIDE 26
20    CONTINUE                                                  VIDE 27
      DO 30 I=I1,IVA                                            VIDE 28
      J=I-ID                                                    VIDE 29
30    VA(J)=VA(I)                                               VIDE 30
C------ PRINT                                                   VIDE 31
40    IVA=IVA-ID                                                VIDE 32
      IF(M.GT.0) WRITE(MP,2000) TBL,ID,IDEB                     VIDE 33
2000  FORMAT(60X,'DELETED TABLE ',A4,' COMPACTING ',I7,' REAL WORDS AFTEVIDE 34
     1R VA(',I7,')')                                            VIDE 35
      RETURN                                                    VIDE 36
      END                                                       VIDE 37
```

Figure 6.5 Subroutines ESPACE and VIDE for pseudo-dynamic memory allocation in the arrays of MEF.

(b) *Arrays*

The basic identification of an array is a four-character word 'tttt'. Its first term is in VA (Ltttt). Its name in execution subroutines is:

— Vtttt for an array of real words (for example, VCORG);
— Ktttt for an array of integers (for example, KLOCE).

(c) *Variables*

The first character of the name of a variable is chosen to identify the nature of the variable according to the following convention:

V..... real variables
K..... integer variables
L..... pointer for the first term of an array in VA
M..... integer variables associated with input and output.
       Examples: M controls the amount of output
               MP printer logical unit
               ME storage device logical unit for element files.
N..... integer parameters of a problem.
       Examples: NNT total number of nodes
               NNEL maximum number of nodes per element.
I..... J..... integer variables for FORTRAN do loops.

Examples: INEL number of nodes of a particular element
ID, JD do loop indices of the number of degrees of freedom
IN, JN do loop indices of the number of nodes.

6.4.3 *Organization of the Problem Data Base*

6.4.3.1 *Entry and Execution functional blocks*

MEF has specialized functional blocks for the entry, verification and organization of the data required to define a problem. For example:

— block 'COOR' reads the nodal coordinates and the number of degrees of freedom for each node; after verification, it creates arrays: VCORG (node coordinates) and KDLNC (number of degrees of freedom of each node, in a cumulative sequence);

— block 'COND' reads the boundary conditions and creates arrays KNEQ (equation identification number for each degree of freedom) and VDIMP (specified values of degrees of freedom variable at a boundary);

— block 'ELEM' reads the connectivities and other element properties and creates a file containing that information.

Each record contains the complete description of an element. This block also creates array KLD (containing the location or the beginning of each column in the global skyline matrix).

Other functional blocks of MEF for the execution of particular finite element computations use the data base constructed by entry blocks and augment it by their results. For example:

— block LINM assembles and solves a linear system of equations residing in core;

— block LIND is similar to block LINM but the system of equations resides out of core in a mass storage device;

— block KLIN assembles and solves a non-linear problem.

6.4.3.2 *Core and out of core storage*

In the version of MEF described here, arrays needed for the solution of a problem normally reside in core. However, for the very large arrays, out of core storage is essential. The following arrays are stored out of core:

— the description of each element—connectivities, material properties, nodal coordinates—is contained in an 'element' file; two subroutines called WRELEM and RDELEM are used to create and retrieve the information in the file;

— global matrix VKG in execution block LIND is subdivided and stored by partition in a 'global matrix' file; during solution, only two partitions are needed in core for the computations; the matrix VKG after factorization is written on a file named 'Global triangularized matrix'.

FORTRAN name of array	Dimension	Block in which array is created	Description
VA	NVA	Main	General work space
Arrays describing the physical problem			
VCORG	NNT × NDIM	COOR	Coordinates of all nodes in global system of reference: for one dimension $x\ x\ x\ \ldots$ two dimension $x\ y\ x\ y\ \ldots$ three dim. $x\ y\ z\ x\ y\ z\ \ldots$
KDLNC	NNT + 1	COOR	Number of degrees of freedom per node, (cumulative) $KDLNC(I+1)$ contains the total number of DOF of nodes $1, 2, 3, \ldots, I-2, I-1, I$. $KDLNC(NNT+1) = NDLT$; $KDLNC(1) = 0$
KNEQ	NDLT	ELEM	Equation number for each DOF $$J = KNEQ(I)$$ (a) $J.GT.0$: DOF I is unknown and corresponds to equation J (b) $J.LT.0$: DOF I is specified and its value is $VDIMP(-J)$
VDIMP	NCLT	COND	Values of all DOF specified in boundary conditions
VPRNG	NNT × NPRN	PRND	Ensemble of nodal properties: $$\frac{P_1 P_2 \ldots P_{NPRN}}{\text{node 1}}$$ $$\frac{P_1 P_2 \ldots P_{NPRN}}{\text{node 2}} \quad \ldots \quad \text{node NNT}$$
VPREG	NGPE × NPRE	PREL	Ensemble of element properties: $$\frac{P_1 P_2 \ldots P_{NPRE}}{\text{group 1}}$$ $$\frac{P_1 P_2 \ldots P_{NPRE}}{\text{group 2}} \quad \ldots \quad \text{group NGPE}$$
KDLD	NEQ + 1	ELEM	Location of start of each column of global matrix KG $KLD(NEQ+1) - 1$ is the number of terms of the upper or lower triangular matrix KG excluding the diagonal terms. $KLD(1) = KLD(2) = 1$.

Figure 6.6 (*Contd.*)

FORTRAN name of array	Dimension	Block in which array is created	Description
Global matrices and vectors			
VKGS	NKG = KLD(NEQ + 1) − 1	Execution blocks	Terms of KG, upper triangle, off diagonal, descending column, stored by skyline method
VKGD	NEQ	Execution blocks	Diagonal terms of KG
VKGI	NKG	Execution blocks	Terms of KG, lower triangle, off diagonal, by rows left to right for non-symmetric matrices
VFG	NEQ	Execution blocks	Global load vectors (forces)
VDLG	NDLT	NLIN, TEMP	Global vector of variables (solution)
VRES	NDLT	LINM, LIND	Residues and reaction vector

Figure 6.6 Global arrays of MEF (the variables defining the dimensions are described in Figure 6.9).

FORTRAN name of array	Dimension	Description
Element description		
KNE	NNEL	Element connectivity (node numbers of an element)
KLOCE	NDLE	Element destination table, obtained by extracting **KNEQ** from the nodal information of an element
VCORE	NNEL × NDIM	Element coordinates extracted from VCORG
VPRNE	NNEL × NPRN	Nodal properties of an element extracted from VPRNG
VPREE	NPRE	Element properties extracted from VPREG
Element matrices and vectors		
VKE	If NSYM.EQ.0: NKE = NDLE × (NDLE + 1)/2 If NSYM.EQ.1: NKE = NDLE × NDLE	Element matrix k, descending columns upper triangle only for symmetrical matrix
VME	NDLE × (NDLE + 1)/2	Element matrix m, descending columns upper triangle only
VFE	NDLE	Element load vector
VDLE	NDLE	Nodal values of an element (element solution vector)

Figure 6.7 Element arrays of MEF (the variables defining the dimensions are described in Figure 6.9).

FORTRAN name of array	Dimension	Description
Numerical integration		
VKPG	IPG × NDIM	Coordinates of integration points in the element of reference organized in a manner similar to (see VCORG)
VCPG	IPG	Numerical weights of integrating points
Storage of functions N and Jacobian matrix		
VNI	IPG × NNEL × (NDIM + 1)	Values of interpolation functions at all integration points as well as their derivatives with respect to $\xi\ \eta\ \zeta$. In two dimensions: $$N_1\ N_2 \ldots \underbrace{\frac{\partial N_1}{\partial \xi}\ \frac{\partial N_2}{\partial \xi}}_{\text{point 1}} \ldots$$ $$\frac{\partial N_1}{\partial \eta}\ \frac{\partial N_2}{\partial \eta} \ldots \underbrace{\vdots \ldots \vdots}_{\text{point 2}} \ldots ;$$
VJ, VJI	NDIM × NDIM	Jacobian matrix and its inverse
VNIX	NNEL × NDIM	Values at a given point of the derivatives of N with respect to $x\ y\ z$ $$\frac{\partial N_1}{\partial x}\ \frac{\partial N_2}{\partial x} \ldots ; \frac{\partial N_1}{\partial y}\ \frac{\partial N_2}{\partial y} \ldots ;$$ $$\frac{\partial N_1}{\partial z}\ \frac{\partial N_2}{\partial z} \ldots ;$$
Automatic computation of functions N		
VKSI	INEL × NDIM	Coordinates of the nodes of the element of reference (similar to VCORG)
KEXP	INEL × NDIM	Monomial exponents of the polynomial basis. For example: $1\ \xi\ \eta\ \xi\eta \rightarrow$ KEXP = 0 0; 1 0; 0 1; 1 1
KDER	—	Index defining the derivative order for the interpolation functions for 2 dimensions: $$\frac{\partial^2 N}{\partial \xi \partial \eta} \rightarrow \text{KDER} = 1\quad 1$$ for 3 dimensions: $$\frac{\partial^3 N}{\partial \xi^2 \partial \zeta} \rightarrow \text{KDER} = 2\quad 0\quad 1$$
VPN	INEL × INEL	Matrix PN or its inverse
VP	INEL	Value of the polynomial basis at a point

Figure 6.8 (*Contd.*)

FORTRAN name of array	Dimension	Description
Block TEMP		
VDLEO	NDLE	Temporary vector for the DOF of an element
VDLGO	NEQ	Global vector of degrees of freedom (reference values)
VFGO	NEQ	Global load vector (reference value)
Block VALP		
VMGS	KNG	Terms of global mass matrix MG, upper triangle, by descending columns, skyline method of storage
VMGD	NEQ	Diagonal terms of MG
VEC	NEQ × NSS	Desired eigenvectors
VLAMB	NSS	Desired eigenvalues
VLAM1	NSS	Temporary storage of eigenvalues
VKSS	NSS × (NSS + 1)/2	Projection of KG in the subspace upper triangle, by descending columns
VMSS	NSS × (NSS + 1)/2	Projection of MG in the subspace upper triangle, by descending columns
VX	NSS × NSS	Eigenvectors of the subspace
V1	NEQ	Work space
Block LIND		
KEB	NBLM + 1	Column number at the start of each block of KG; KEB(NBLM + 1) = NEQ + 1
KPB	NBLM	Number of the first block connected to each block of KG

Figure 6.8 Local arrays of MEF (the variables defining the dimensions are described in Figure 6.9).

/COOR/	
NDIM	Number of dimensions in the problem (1, 2 or 3)
NNT	Total number of nodes
NDLN	Maximum number of degrees of freedom per node
NDLT	Total number of degrees of freedom
FAC(3)	Scaling factors in direction x, y and z
/COND/	
NCLT	Total number of boundary conditions
NCLZ	Number of zero valued specified boundary conditions
NCLNZ	Number of non-zero valued specified B. C.
/PRND/	
NPRN	Number of nodal properties per node
/PERL/	
NGPE	Number of groups of element properties
NPRE	Number of properties per element
/ELEM/	
NELT	Total number of elements
NNEL	Maximum number of nodes per element
NTPE	Element type (default)
NGRE	Number of element groups
ME	Logical number of element file
NIDENT	.EQ.1 if all matrices k are identical
NPG	Number of integration points

Figure 6.9 (*Contd.*)

/ASSE/
  NSYM   .EQ.0 if matrix KG is symmetrical
                         .EQ.1 if matrix KG is not symmetrical
  NKG   Number of terms of upper (or lower) triangle of KG excluding diagonal terms
  NKE   Maximum number of terms of array VKE
  NDLE   Maximum number of degrees of freedom per element
/RESO/
  NEQ   Number of equations of a problem
  NRES   .EQ.1 if the residues are desired
  MRES   Logical number of the residues file
/RGDT/
  IEL   Element number
  ITPE   Element type
  ITPE1   Type of the element preceding element IEL
  IGRE   Group number of an element
  IDLE   Number of degrees of freedom of an element
  ICE   Number of nodal coordinates for an element (INEL × NDIM)
  IPRNE   Number of nodal properties of an element (NPRN × INEL)
  IPREE   Number of properties of an element
  INEL   Number of nodes of an element
  IDEG   Maximum degree of the polynomial basis
  IPG   Number of integration points (Gauss points)
  ICODE   Index defining the type of element functions to be executed by the subroutines ELEMxx
  IDLEO   Check variable for IDLE
  INELO   Check variable for INEL
  IPGO   Check variable for IPG
/LIND/
  NLBL   Length of the blocks of KG
  NBLM   Maximum number of blocks
  MKG1   Logical number of the file containing matrix KG
  MKG2   Logical number of the file containing triangularized matrix KG
/NLIN/
  EPSLD   Admissible error for the norm of the DOF
  XNORM   Norm of U
  OMEGA   Relaxation factor for Euler's method
  XPAS   Load level reached
  DPAS   Load increment
  DPASO   Preceding load increment
  NPAS   Maximum number of load increments
  IPAS   Number of the current load increment
  NITER   Maximum number of iterations per increment
  ITER   Iteration number
  IMETH   Type of method used
/VALP/
  NITER1   Maximum number of iterations
  NMDIAG   (not used)
  EPSLB   Admissible error for the eigenvalues
  SHIFT   (not used)
  NSS   Subspace dimension
  NSWM   Maximum number of cycles in the JACOBI subroutine
  TOLJAC   Tolerance in JACOBI
  NVALP   Number of eigenvalues desired

Figure 6.9   (*Contd.*)

```
/ES/
  M              Printing index: 0 production run
                                1 reduced printing
                                2 verification
                                3 maximum
  MR             Logical unit number of input device
  MP             Logical unit number for the output device
  MLUN(10)       Logical number of the various files used by a block
/ALLOC/
  NVA            Number of words in the general work vector VA
  IVA            Position of the last real word used in VA
  IVAMAX         Maximum number of floating point values stored in VA for a
                 problem
  NREEL          Number of integer variables that can be stored in a floating point
                 variable. Example:
                        IBM-370 single precision NREEL = 1
                        IBM-370 double precision          2
                        CDC-660 single precision          1
  NTBL           Number of arrays for which the pointers are stored in
                 COMMON/LOC/
/LOC/
  LCORG, etc... Position in VA of the first term of arrays VCORG,...
  Blank COMMON
  VA             Work space for use in many subroutines
```

Figure 6.9 Variables of the COMMON blocks.

— for the computation of residues and reactions, the global matrix VKG and the load vector VF are saved temporarily in a file named 'residue'.

6.4.3.3 *Description of the most important arrays and variables in COMMOM*

Three different groups of arrays are constructed by MEF. The global arrays, used in a majority of functional blocks, define the problem to be solved and its system of equations. The *element* arrays define the elements; they are used to construct the element matrices and vectors. *Local* arrays are used in a limited number of blocks. Figure 6.6, 6.7, and 6.8 describe all the arrays of MEF. Figure 6.9, describes the variables contained in the various COMMON statements.

6.5 Description of Functional Blocks

6.5.1 *Principal Program*

The principal or main program is divided into two parts:

(a) a control section for the flow of operations through the functional blocks; it reads an instruction containing:

— the name of a functional block to be executed; BLOC;
— the quantity of output desired: M;

— the integer identification numbers of up to ten files to be used in the block: MLUN (10).

This part of the program transfers logical control to one of the subroutines called in the second section.
(b) the execution section calls the various subroutines that execute the operations and returns control to the beginning of control section after execution.

Flow chart

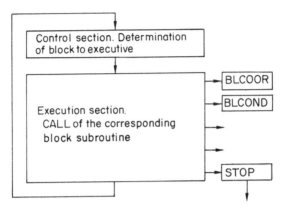

Subroutines to be called

BLCOORD
BLCOND   } control section of the various functional blocks to be used
...

Creation of a new block

To add a new functional block in the code, for example, a block named 'PLUS', one must do the following:

— add the name PLUS in array BLOCS defined in the DATA statement for example, in the 16th position.
— identify the corresponding tag where a CALL BLPLUS statement will be found; in the following computed GO TO statement
— 30 GO TO (110, 120,..., 260,...), I
— (tag 260 occupies 16th position for I = 16)
— replace instruction 260 CONTINUE by

        C------- BLOC'PLUS'
        260 CALL BLPLUS
        GO TO 10

Furthermore, the user adds the subroutine BLPLUS and other necessary programs required by BLPLUS and not existing in the MEF library.

Listing

Figure 6.10 contains the listing of the MAIN PROGRAM and BLOCK DATA subroutine of MEF, as well as the error subroutine ERREUR used by all functional blocks.

```
C                                                                          FEM   1
C==========================================================================FEM   2
C                                                                          FEM   3
C        F . E . M .  - 3 -  PROGRAM, 'BOOK' VERSION   OCTOBER 1979        FEM   4
C      ( G.TOUZOT , G.DHATT , COMPIEGNE UNIVERSITY OF TECHNOLOGY, FRANCE)  FEM   5
C        MAIN PROGRAM                                                      FEM   6
C                                                                          FEM   7
C==========================================================================FEM   8
       IMPLICIT REAL*8(A-H,O-Z)                                            FEM   9
       REAL*4 BLOC,BLOCS                                                   FEM  10
       COMMON/ALLOC/NVA,IVA,IVAMAX,NREEL,NTBL                              FEM  11
       COMMON/ES/M,MR,MP,MLUN(10)                                          FEM  12
       COMMON VA(20000)                                                    FEM  13
       DIMENSION BLOCS(21)                                                 FEM  14
       DATA BLOCS/4HIMAG,4HCOMT,4HCOOR,4HDLPN,4HCOND,4HPRND,4HPREL,        FEM  15
      1           4HELEM,4HSOLC,4HSOLR,4HLINM,4HLIND,4HNLIN,4HTEMP,        FEM  16
      2           4HVALP,4H....,4H....,4H....,4H....,4H....,4HSTOP/        FEM  17
       DATA NB/21/                                                         FEM  18
C--------------------------------------------------------------------------FEM  19
C....... LENGTH OF BLANK COMMON IN REAL WORDS (TABLE VA)                   FEM  20
       NVA=20000                                                           FEM  21
C------- HEADING                                                           FEM  22
       WRITE(MP,2000)                                                      FEM  23
 2000  FORMAT(1H1,30X,'F.E.M.3.'/23X,' G.TOUZOT , G.DHATT'/23X,22('-')//)  FEM  24
C------- READ BLOCK TITLE                                                  FEM  25
  10   READ(MR,1000) BLOC,M,MLUN                                           FEM  26
 1000  FORMAT(A4,I6,10I5)                                                  FEM  27
C------- SEARCH FOR THE BLOCK TO BE EXECUTED                               FEM  28
       DO 20 I=1,NB                                                        FEM  29
       IF(BLOC.EQ.BLOCS(I)) GO TO 30                                       FEM  30
  20     CONTINUE                                                          FEM  31
       WRITE(MP,2010)                                                      FEM  32
 2010  FORMAT(' ** ERROR, MISSING BLOCK CALLING CARD')                     FEM  33
       GO TO 10                                                            FEM  34
  30   GO TO (110,120,130,140,150,160,170,                                 FEM  35
      1       180,190,200,210,220,230,240,                                 FEM  36
      2       250,260,270,280,290,300,999),I                               FEM  37
C------- BLOCK TO PRINT IMAGES OF DATA CARDS              'IMAG'           FEM  38
 110   CALL BLIMAG                                                         FEM  39
       GO TO 10                                                            FEM  40
C------- BLOCK TO READ AND PRINT COMMENTS                 'COMT'           FEM  41
 120   CALL BLCOMT                                                         FEM  42
       GO TO 10                                                            FEM  43
C------- BLOCK TO READ NODAL POINTS COORDINATES           'COOR'           FEM  44
 130   CALL BLCOOR                                                         FEM  45
       GO TO 10                                                            FEM  46
C------- BLOCK TO READ DEGREES OF FREEDOM PER NODE        'DLPN'           FEM  47
 140   CALL BLDLPN                                                         FEM  48
       GO TO 10                                                            FEM  49
C------- BLOCK TO READ BOUNDARY CONDITIONS                'COND'           FEM  50
 150   CALL BLCOND                                                         FEM  51
       GO TO 10                                                            FEM  52
C------- BLOCK TO READ NODAL PROPERTIES                   'PRND'           FEM  53
 160   CALL BLPRND                                                         FEM  54
       GO TO 10                                                            FEM  55
C------- BLOCK TO READ ELEMENT PROPERTIES                 'PREL'           FEM  56
```

Figure 6.10 (*Contd.*)

```
      170 CALL BLPREL                                                    FEM  57
          GO TO 10                                                       FEM  58
C------- BLOCK TO READ ELEMENT DATA                          'ELEM'      FEM  59
      180 CALL BLELEM                                                    FEM  60
          GO TO 10                                                       FEM  61
C------- BLOCK TO READ CONCENTRATED LOADS                    'SOLC'      FEM  62
                                                                         FEM  63
      190 CALL BLSOLC                                                    FEM  64
          GO TO 10
C------- BLOCK TO READ DISTRIBUTED LOADS                     'SOLR'      FEM  65
                                                                         FEM  66
      200 CALL BLSOLR                                                    FEM  67
          GO TO 10
C------- BLOCK FOR IN CORE ASSEMBLING AND LINEAR SOLUTION 'LINM'         FEM  68
                                                                         FEM  69
      210 CALL BLLINM                                                    FEM  70
          GO TO 10
C------- BLOCK FOR ON DISK ASSEMBLING AND LINEAR SOLUTION 'LIND'         FEM  71
                                                                         FEM  72
      220 CALL BLLIND                                                    FEM  73
          GO TO 10
C------- BLOCK FOR NON LINEAR PROBLEM SOLUTION               'NLIN'      FEM  74
                                                                         FEM  75
      230 CALL BLNLIN                                                    FEM  76
          GO TO 10
C------- BLOCK FOR UNSTEADY PROBLEM                          'TEMP'      FEM  77
                                                                         FEM  78
      240 CALL BLTEMP                                                    FEM  79
          GO TO 10
C------- BLOCK TO COMPUTE EIGENVALUES (SUBSPACE)             'VALP'      FEM  80
                                                                         FEM  81
      250 CALL BLVALP                                                    FEM  82
          GO TO 10
C------- UNDEFINED BLOCKS                                                FEM  83
      260 CONTINUE                                                       FEM  84
      270 CONTINUE                                                       FEM  85
      280 CONTINUE                                                       FEM  86
      290 CONTINUE                                                       FEM  87
      300 CONTINUE                                                       FEM  88
          GO TO 10                                                       FEM  89
C------- END OF PROBLEM                                      'STOP'      FEM  90
      999 WRITE(MP,2020) IVAMAX,NVA                                      FEM  91
     2020 FORMAT(//' END OF PROBLEM, ',I10,' UTILIZED REAL WORDS OVER ',I10) FEM 92
          STOP                                                           FEM  93
          END                                                            FEM  94

          BLOCK DATA                                                     BLOC  1
C===============================================================================BLOC  2
C        INITIALIZE LABELLED COMMONS                                     BLOC  3
C===============================================================================BLOC  4
          IMPLICIT REAL*8(A-H,O-Z)                                       BLOC  5
          COMMON/COOR/NDIM,NNT,NDLN,NDLT,FAC(3)                          BLOC  6
          COMMON/COND/NCLT,NCLZ,NCLNZ                                    BLOC  7
          COMMON/PRND/NPRN                                               BLOC  8
          COMMON/PREL/NGPE,NPRE                                          BLOC  9
          COMMON/ELEM/NELT,NNEL,NTPE,NGRE,ME,NIDENT,NPG                  BLOC 10
          COMMON/ASSE/NSYM,NKG,NKE,NDLE                                  BLOC 11
          COMMON/RESO/NEQ,NRES,MRES                                      BLOC 12
          COMMON/RGDT/IEL,ITPE,ITPE1,IGRE,IDLE,ICE,IPRNE,IPREE,INEL,IDEG, BLOC 13
         1 IPG,ICOD,IDLEO,INELO,IPGO                                     BLOC 14
          COMMON/LIND/NLBL,NBLM,MKG1,MKG2                                BLOC 15
          COMMON/NLIN/EPSDL,XNORM,OMEGA,XPAS,DPAS,DPASO,NPAS,IPAS,NITER, BLOC 16
         1 ITER,IMETH                                                    BLOC 17
          COMMON/VALP/NITER1,NMDIAG,EPSLB,SHIFT,NSS,NSWM,TOLJAC,NVALP    BLOC 18
          COMMON/ES/M,MR,MP,MLUN(10)                                     BLOC 19
          COMMON/ALLOC/NVA,IVA,IVAMAX,NREEL,NTBL                         BLOC 20
          COMMON/LOC/LCORG,LDLNC,LNEQ,LDIMP,LPRNG,LPREG,LLD,LLOCE,LCORE,LNE,BLOC 21
         1 LPRNE,LPREE,LDLE,LKE,LFE,LKGS,LKGD,LKGI,LFG,LRES,LDLG,LME,    BLOC 22
         2 LDLEO,LDLGO,LFGO                                              BLOC 23
          DIMENSION LXX(25)                                              BLOC 24
          EQUIVALENCE (LXX(1),LCORG)                                     BLOC 25
```

Figure 6.10 (*Contd.*)

```
C------ COMMON /COOR/                                              BLOC 26
       DATA NNT/20/,NDLN/2/,NDIM/2/,FAC/3*1.D0/                    BLOC 27
C------ COMMON /PRND/                                              BLOC 28
       DATA NPRN/0/                                                BLOC 29
C------ COMMON /PREL/                                              BLOC 30
       DATA NGPE/0/,NPRE/0/                                        BLOC 31
C------ COMMON /ELEM/                                              BLOC 32
       DATA NELT/20/,NNEL/8/,NTPE/1/,NGRE/1/,ME/1/,NIDENT/0/       BLOC 33
C------ COMMON/ASSE/                                               BLOC 34
       DATA NSYM/0/                                                BLOC 35
C------ COMMON /RESO/                                              BLOC 36
       DATA NRES/0/,MRES/2/                                        BLOC 37
C------ COMMON /RGDT/                                              BLOC 38
       DATA ITPE1/0/                                               BLOC 39
C------ COMMON /LIND/                                              BLOC 40
       DATA MKG1/4/,MKG2/7/                                        BLOC 41
C------ COMMON /NLIN/                                              BLOC 42
       DATA EPSDL/1.D-2/,OMEGA/1.D0/,DPAS/.2D0/,NPAS/1/,NITER/5/,IMETH/1/BLOC 43
C------ COMMON /VALP/                                              BLOC 44
       DATA NITER1/10/,NMDIAG/0/,EPSLB/1.D-3/,SHIFT/0.D0/,NSS/5/,  BLOC 45
      1 NSWM/12/,TOLJAC/1.D-12/,NVALP/3/                           BLOC 46
C------ COMMON /ES/                                                BLOC 47
       DATA MR/5/,MP/6/                                            BLOC 48
C------ COMMON /ALLOC/                                             BLOC 49
       DATA IVA/1/,IVAMAX/1/,NTBL/25/                              BLOC 50
C....... DEFINE HERE THE NUMBER OF INTEGERS CONTAINED IN A REAL    BLOC 51
C           FOR THE COMPUTER EMPLOYED                              BLOC 52
C           EXAMPLES:   IBM    SIMPLE PRECISION    NREEL.EQ.1      BLOC 53
C                       IBM    DOUBLE PRECISION    NREEL.EQ.2      BLOC 54
C                       CDC                        NREEL.EQ.1      BLOC 55
       DATA NREEL/2/                                               BLOC 56
C.......                                                           BLOC 57
C------ COMMON /LOC/                                               BLOC 58
       DATA LXX/25*1/                                              BLOC 59
       END                                                         BLOC 60

       SUBROUTINE ERREUR(IERR,I1,I2,INIV)                          ERRE  1
C===============================================================   ERRE  2
C      PRINT ERROR MESSAGES FOR BLOCKS READING DATA                ERRE  3
C===============================================================   ERRE  4
       COMMON/ES/M,MR,MP                                           ERRE  5
C---------------------------------------------------------------   ERRE  6
C------ BLOCK 'COOR'                                               ERRE  7
       IF(IERR.GT.19) GO TO 200                                    ERRE  8
       IE=IERR-10                                                  ERRE  9
       GO TO (110,120,130,140,150,160,160,180),IE                  ERRE 10
  110  WRITE(MP,2110)I1,I2                                         ERRE 11
 2110  FORMAT(' *** ERROR, FIRST NODE NUMBER(',I4,') IS GREATER THAN NNT=ERRE 12
      1',I4)                                                       ERRE 13
       GO TO 900                                                   ERRE 14
  120  WRITE(MP,2120)I1,I2                                         ERRE 15
 2120  FORMAT(' ** ERROR, SECOND NODE NUMBER(',I4,') IS GREATER THAN NNT=ERRE 16
      1',I4)                                                       ERRE 17
       GO TO 900                                                   ERRE 18
  130  WRITE(MP,2130)I1,I2                                         ERRE 19
 2130  FORMAT(' ** ERROR, NODAL NUMBER OF D.O.F.(',I4,') IS GREATER THAN ERRE 20
      1NDLN=',I4)                                                  ERRE 21
       GO TO 900                                                   ERRE 22
  140  WRITE(MP,2140)                                              ERRE 23
 2140  FORMAT(' ** ERROR, FIRST AND SECOND NODE NUMBERS ARE INCOMPATIBLE ERRE 24
      1WITH THE GENERATION PARAMETER')                             ERRE 25
       GO TO 900                                                   ERRE 26
  150  WRITE(MP,2150)I1                                            ERRE 27
 2150  FORMAT(' ** ERROR, NODE ',I4,' IS DEFINED MORE THAN ONCE')  ERRE 28
```

Figure 6.10 (*Contd.*)

```
              GO TO 900                                            ERRE  29
        160   WRITE(MP,2160)I1                                     ERRE  30
        2160  FORMAT(' ** ERROR, NODE ',I4,' IS NOT DEFINED')      ERRE  31
              GO TO 900                                            ERRE  32
        180   WRITE(MP,2180)I2,I1                                  ERRE  33
        2180  FORMAT(' ** ERROR, GENERATED NODES NUMBER(',I4,') IS LESS THAN NNTERRE 34
             1=',I4)                                               ERRE  35
              GO TO 900                                            ERRE  36
        C------- BLOCK 'DLPN'                                      ERRE  37
        200   IF(IERR.GT.29) GO TO 300                             ERRE  38
              IE=IERR-20                                           ERRE  39
              GO TO (210,220),IE                                   ERRE  40
        210   WRITE(MP,2210)I1,I2                                  ERRE  41
        2210  FORMAT(' ** ERROR, NUMBER OF D.O.F. (',I2,') IS GREATER THAN NDLN=ERRE 42
             1',I2)                                                ERRE  43
              GO TO 900                                            ERRE  44
        220   WRITE(MP,2220)I1,I2                                  ERRE  45
        2220  FORMAT(' ** ERROR, NODE NUMBER(',I4,') IS GREATER THAN ERRE 46
             1NNT=',I4)                                            ERRE  47
              GO TO 900                                            ERRE  48
        C------- BLOCK 'COND'                                      ERRE  49
        300   IF(IERR.GT.39)GO TO 400                              ERRE  50
              IE=IERR-30                                           ERRE  51
              GO TO (900,320,900),IE                               ERRE  52
        320   GO TO 220                                            ERRE  53
        C------- BLOCK 'PREL'                                      ERRE  54
        400   IF(IERR.GT.49) GO TO 500                             ERRE  55
              IE=IERR-40                                           ERRE  56
              GO TO (410,900),IE                                   ERRE  57
        410   WRITE(MP,2410)I1,I2                                  ERRE  58
        2410  FORMAT(' ** ERROR, GROUP NUMBER (',I3,') IS GREATER THAN NGPE=',I3ERRE 59
             1)                                                    ERRE  60
              GO TO 900                                            ERRE  61
        C------- BLOCK 'ELEM'                                      ERRE  62
        500   IF(IERR.GT.59) GO TO 900                             ERRE  63
              IE=IERR-50                                           ERRE  64
              GO TO (510,900,530,540,550,560,570),IE               ERRE  65
        510   WRITE(MP,2510)I1,I2                                  ERRE  66
        2510  FORMAT(' ** ERROR, NUMBER OF NODES (',I3,') IS GREATER THAN NNEL='ERRE 67
             1,I3)                                                 ERRE  68
              GO TO 900                                            ERRE  69
        530   WRITE(MP,2530)I1,I2                                  ERRE  70
        2530  FORMAT(' ** ERROR, PROPERTY NUMBER (',I3,') IS GREATER THAN NGPE='ERRE 71
             ,1I3)                                                 ERRE  72
              GO TO 900                                            ERRE  73
        540   WRITE(MP,2540)I1,I2                                  ERRE  74
        2540  FORMAT(' ** ERROR, GROUP NUMBER (',I3,') IS GREATER THAN NGRE=',I3ERRE 75
             1)                                                    ERRE  76
              GO TO 900                                            ERRE  77
        550   WRITE(MP,2550)I1,I2                                  ERRE  78
        2550  FORMAT(' ** ERROR, ELEMENT NUMBER (',I4,') IS GREATER THAN NELT=',ERRE 79
             1I4)                                                  ERRE  80
              GO TO 900                                            ERRE  81
        560   GO TO 220                                            ERRE  82
        570   WRITE(MP,2570)I1,I2                                  ERRE  83
        2570  FORMAT(' ** ERROR, NUMBER OF ELEMENTS (',I4,') IS GREATER THAN NELERRE 84
             1T=',I4)                                              ERRE  85
        C------- END                                               ERRE  86
        900   I1=I2                                                ERRE  87
              IF(INIV.GE.2) STOP                                   ERRE  88
              RETURN                                               ERRE  89
              END                                                  ERRE  90
```

Figure 6.10 Main program of MEF.

6.5.2 *Functional Blocks Reading Data*

The functional blocks used for reading all the information needed to define a problem are:

Name	Function
'IMAG'	printing of all input quantities
'COMT'	reading and printing of comments
'COOR'	reading of nodal coordinates
'DLPN'	reading of the number of DOF per node
'COND'	reading of boundary conditions
'PRND'	reading of nodal properties
'PREL'	reading of element properties
'ELEM'	reading of connectivities
'SOLC'	reading of concentrated loads

6.5.2.1 *Block 'IMAG'*

Function

This block echoes back all the quantities read into the computer for verification.

Flow chart

Subroutines

   BLIMAG: executes the functional block 'IMAG' (Figure 6.11).

Remark

Block 'IMAG' must be the first one to be executed.

6.5.2.2 *Block 'COMT'*

Function

   This block reads and outputs the comments.

Flow chart

```
      SUBROUTINE BLIMAG                                              BLIM   1
C================================================================BLIM   2
C     TO CALL AND EXECUTE BLOCK 'IMAG'                               BLIM   3
C     TO PRINT OUT THE IMAGE OF DATA CARDS                           BLIM   4
C================================================================BLIM   5
      IMPLICIT REAL*8(A-H,O-Z)                                       BLIM   6
      COMMON/ES/M,MR,MP,M1                                           BLIM   7
      COMMON/TRVL/CART(20)                                           BLIM   8
      DATA ICARTM/40/                                                BLIM   9
C----------------------------------------------------------------BLIM  10
      IF(M1.EQ.0) M1=MR                                              BLIM  11
      WRITE(MP,2000)                                                 BLIM  12
 2000 FORMAT(///,1X,'IMAGE OF DATA CARDS'/1X,28('='),/)              BLIM  13
      WRITE(MP,2005)                                                 BLIM  14
 2005 FORMAT(/                                                       BLIM  15
     1 50X,'C O L U M N    N U M B E R',/,13X,'CARD',6X,             BLIM  16
     2 10X,'1',9X,'2',9X,'3',9X,'4',9X,'5',9X,'6',9X,'7',9X,'8',/,   BLIM  17
     3 12X,'NUMBER',6X,8('1234567890'),/,12X,8('-'),6X,80('-'))      BLIM  18
      ICART=0                                                        BLIM  19
      ICART1=0                                                       BLIM  20
 10   READ(M1,1000,END=30) CART                                      BLIM  21
 1000 FORMAT(20A4)                                                   BLIM  22
      ICART=ICART+1                                                  BLIM  23
      ICART1=ICART1+1                                                BLIM  24
      IF(ICART1.LE.ICARTM) GO TO 20                                  BLIM  25
      WRITE(MP,2010)                                                 BLIM  26
 2010 FORMAT(12X,8(1H-),6X,80(1H-),/,13X,'CARD',7X,8('1234567890'),/,BLIM  27
     1 12X,'NUMBER',6X,9X,'1',9X,'2',9X,'3',9X,'4',9X,'5',9X,'6',    BLIM  28
     2 9X,'7',9X,'8',/,50X,'C O L U M N    N U M B E R')             BLIM  29
      WRITE(MP,2015)                                                 BLIM  30
 2015 FORMAT(1H1,//)                                                 BLIM  31
      WRITE(MP,2005)                                                 BLIM  32
      ICART1=0                                                       BLIM  33
 20   WRITE(MP,2020) ICART,CART                                      BLIM  34
 2020 FORMAT(10X,I10,6X,20A4)                                        BLIM  35
      GO TO 10                                                       BLIM  36
 30   WRITE(MP,2010)                                                 BLIM  37
      WRITE(MP,2030)                                                 BLIM  38
 2030 FORMAT(///,51X,'E N D    O F    D A T A',/,1H1)                BLIM  39
      REWIND M1                                                      BLIM  40
      READ(M1,1000) CART                                             BLIM  41
      RETURN                                                         BLIM  42
      END                                                            BLIM  43
```

Figure 6.11 Block 'IMAG'.

Subroutines

BLCOMT: executes block 'COMT' (Figure 6.12).

Remark

The block can be executed as many times as comments are to be introduced in the output.

6.5.2.3 *Block 'COOR'*

Function

This block reads the nodal coordinates and the number of degrees of freedom per node, generates, some nodes, and creates arrays VCORG and KDLNC.

```
      SUBROUTINE BLCOMT                                        BLCM  1
C================================================================BLCM  2
C     TO CALL AND EXECUTE BLOCK 'COMT'                         BLCM  3
C================================================================BLCM  4
      IMPLICIT REAL*8(A-H,O-Z)                                 BLCM  5
      REAL*4 BLANC,CART                                        BLCM  6
      COMMON/ES/M,MR,MP                                        BLCM  7
      COMMON/TRVL/CART(20)                                     BLCM  8
      DATA BLANC/4H    /                                       BLCM  9
C---------------------------------------------------------------BLCM 10
      WRITE(MP,2000)                                           BLCM 11
 2000 FORMAT(//' COMMENTS'/' ',10('=')/)                       BLCM 12
C------ READ A COMMENT CARD                                    BLCM 13
   10 READ(MR,1000) CART                                       BLCM 14
 1000 FORMAT(20A4)                                             BLCM 15
C------ SEARCH FOR A WHOLLY BLANK CARD                         BLCM 16
      DO 20 I=1,20                                             BLCM 17
      IF(CART(I).NE.BLANC) GO TO 30                            BLCM 18
   20 CONTINUE                                                 BLCM 19
      RETURN                                                   BLCM 20
   30 WRITE(MP,2010) CART                                      BLCM 21
 2010 FORMAT(1X,20A4)                                          BLCM 22
      GO TO 10                                                 BLCM 23
      END                                                      BLCM 24
```

Figure 6.12 Block 'COMT'.

Flow chart

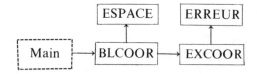

Subroutines

BLCOOR: reads and outputs an entry containing variables NNT, NDLN, NDIM, FAC defined in Figure 6.9 (Figure 6.13).
ESPACE: allocates space for arrays VCORG and KDLNC (Figure 6.5).
EXCOOR: reads nodal records containing the nodes number and coordinates, generates missing nodes by interpolation, and constructs arrays VCORG and KDLNC (Figure 6.13).
ERREUR: outputs error diagnostics (Figure 6.10).

6.5.2.4 *Block 'DLPN'*

Function

This block reads the number of degrees of freedom per node and modifies array KDLNC. It is used only when the number of degrees of freedom varies from node to node.

```
      SUBROUTINE BLCOOR                                          BLCR   1
C================================================================BLCR   2
C     TO CALL BLOCK COOR                                         BLCR   3
C     TO READ NODAL COORDINATES                                  BLCR   4
C================================================================BLCR   5
      IMPLICIT REAL*8(A-H,O-Z)                                   BLCR   6
      REAL*4 TBL                                                 BLCR   7
      COMMON/COOR/NDIM,NNT,NDLN,NDLT,FAC(3)                      BLCR   8
      COMMON/ES/M,MR,MP,M1                                       BLCR   9
      COMMON/ALLOC/NVA                                           BLCR  10
      COMMON/LOC/LCORG,LDLNC                                     BLCR  11
      COMMON/TRVL/FAC1(3),IN(3)                                  BLCR  12
      COMMON VA(1)                                               BLCR  13
      DIMENSION TBL(2)                                           BLCR  14
      DATA ZERO/0.D0/,TBL/4HCORG,4HDLNC/                         BLCR  15
C----------------------------------------------------------------BLCR  16
C------- BLOCK HEADING                                           BLCR  17
      IF(M1.EQ.0) M1=MR                                          BLCR  18
      READ(M1,1000) IN,FAC1                                      BLCR  19
1000  FORMAT(3I5,3F10.0)                                         BLCR  20
C------- DEFAULT OPTIONS                                         BLCR  21
      IF(IN(1).GT.0) NNT=IN(1)                                   BLCR  22
      IF(IN(2).GT.0) NDLN=IN(2)                                  BLCR  23
      IF(IN(3).GT.0) NDIM=IN(3)                                  BLCR  24
      DO 10 I=1,3                                                BLCR  25
      IF(FAC1(I).NE.ZERO) FAC(I)=FAC1(I)                         BLCR  26
10    CONTINUE                                                   BLCR  27
C------- PRINT BLOCK PARAMETERS                                  BLCR  28
      WRITE(MP,2000) M,NNT,NDLN,NDIM,FAC,NVA                     BLCR  29
2000  FORMAT(//' INPUT OF NODES (M=',I2,')'/' ',18('=')/         BLCR  30
     1    15X,'MAX. NUMBER OF NODES            (NNT)=',I5/       BLCR  31
     2    15X,'MAX. NUMBER OF D.O.F. PER NODE  (NDLN)=',I5/      BLCR  32
     3    15X,'DIMENSIONS OF THE PROBLEM       (NDIM)=',I5/      BLCR  33
     4    15X,'COORDINATE SCALE FACTORS        (FAC)=',3E12.5/   BLCR  34
     5    15X,'WORKSPACE IN REAL WORDS         (NVA)=',I10)      BLCR  35
C------- ALLOCATE SPACE                                          BLCR  36
      IF(LCORG.EQ.1) CALL ESPACE(NNT*NDIM,1,TBL(1),LCORG)        BLCR  37
      IF(LDLNC.EQ.1) CALL ESPACE(NNT+1,0,TBL(2),LDLNC)           BLCR  38
C------- EXECUTE THE BLOCK                                       BLCR  39
      CALL EXCOOR(VA(LCORG),VA(LDLNC))                           BLCR  40
      RETURN                                                     BLCR  41
      END                                                        BLCR  42

      SUBROUTINE EXCOOR(VCORG,KDLNC)                             EXCR   1
C================================================================EXCR   2
C     TO EXECUTE BLOCK 'COOR'                                    EXCR   3
C     READ NODAL COORDINATES                                     EXCR   4
C================================================================EXCR   5
      IMPLICIT REAL*8(A-H,O-Z)                                   EXCR   6
      COMMON/COOR/NDIM,NNT,NDLN,NDLT,FAC(3)                      EXCR   7
      COMMON/ES/M,MR,MP,M1                                       EXCR   8
      COMMON/TRVL/X1(3),X2(3)                                    EXCR   9
      DIMENSION VCORG(1),KDLNC(1)                                EXCR  10
      DATA SPECL/1.23456789D31/                                  EXCR  11
C----------------------------------------------------------------EXCR  12
C------- INITIALIZE COORDINATES                                  EXCR  13
      I1=(NNT-1)*NDIM+1                                          EXCR  14
      DO 10 I=1,I1,NDIM                                          EXCR  15
10    VCORG(I)=SPECL                                             EXCR  16
C------- READ NODAL DATA CARDS                                   EXCR  17
      IF(M.GT.0) WRITE(MP,2000)                                  EXCR  18
2000  FORMAT(//' NODAL DATA CARDS'/)                             EXCR  19
20    READ(M1,1000) IN1,X1,IN2,X2,INCR,IDLN                      EXCR  20
```

Figure 6.13 (*Contd.*)

```
1000    FORMAT(2(I5,3F10.0),2I5)                                EXCR 21
        IF(M.GT.0) WRITE(MP,2010) IN1,X1,IN2,X2,INCR,IDLN       EXCR 22
2010    FORMAT(' >>>>>',2(I5,3E12.5),2I5)                       EXCR 23
        IF(IN1.LE.0) GO TO 60                                   EXCR 24
C------- DECODE THE CARD                                        EXCR 25
        IF(IN1.GT.NNT) CALL ERREUR(11,IN1,NNT,0)                EXCR 26
        IF(IN2.GT.NNT) CALL ERREUR(12,IN2,NNT,0)                EXCR 27
        IF(IN2.LE.0) IN2=IN1                                    EXCR 28
        IF(IDLN.GT.NDLN) CALL ERREUR(13,IDLN,NDLN,0)            EXCR 29
        IF(IDLN.LE.0) IDLN=NDLN                                 EXCR 30
        IF(INCR.EQ.0) INCR=1                                    EXCR 31
        I1=(IN2-IN1)/INCR                                       EXCR 32
        I2=IN1+I1*INCR                                          EXCR 33
        IF(I1.EQ.0)I1=1                                         EXCR 34
        IF(IN2.NE.I2) CALL ERREUR(14,IN2,IN2,0)                 EXCR 35
C------- GENERATE NODES BY INTERPOLATION                        EXCR 36
        DO 30 I=1,NDIM                                          EXCR 37
        X1(I)=X1(I)*FAC(I)                                      EXCR 38
        X2(I)=X2(I)*FAC(I)                                      EXCR 39
30      X2(I)=(X2(I)-X1(I))/I1                                  EXCR 40
        I1=0                                                    EXCR 41
        I2=(IN1-1)*NDIM+1                                       EXCR 42
        I3=(INCR-1)*NDIM                                        EXCR 43
        DO 50 IN=IN1,IN2,INCR                                   EXCR 44
        KDLNC(IN+1)=IDLN                                        EXCR 45
        IF(VCORG(I2).NE.SPECL) CALL ERREUR(15,IN,IN,0)          EXCR 46
        DO 40 I=1,NDIM                                          EXCR 47
        VCORG(I2)=X1(I)+X2(I)*I1                                EXCR 48
40      I2=I2+1                                                 EXCR 49
        I1=I1+1                                                 EXCR 50
50      I2=I2+I3                                                EXCR 51
        GO TO 20                                                EXCR 52
C------- CHECK FOR MISSING NODES                                EXCR 53
60      I1=NNT*NDIM+1                                           EXCR 54
        I2=0                                                    EXCR 55
        I3=NNT+1                                                EXCR 56
        DO 90 I=1,NNT                                           EXCR 57
        I1=I1-NDIM                                              EXCR 58
        I3=I3-1                                                 EXCR 59
        IF(VCORG(I1)-SPECL) 70,80,70                            EXCR 60
70      IF(I2.EQ.0) I2=I3                                       EXCR 61
        GO TO 90                                                EXCR 62
80      IF(I2.EQ.0) CALL ERREUR(16,I3,I3,0)                     EXCR 63
        IF(I2.NE.0) CALL ERREUR(17,I3,I3,1)                     EXCR 64
90      CONTINUE                                                EXCR 65
        IF(I2.NE.NNT) CALL ERREUR(18,NNT,I2,0)                  EXCR 66
C------- TOTAL NUMBER OF D.O.F.                                 EXCR 67
        NDLT=0                                                  EXCR 68
        I1=NNT+1                                                EXCR 69
        DO 100 I=2,I1                                           EXCR 70
100     NDLT=NDLT+KDLNC(I)                                      EXCR 71
C------- OUTPUT                                                 EXCR 72
        IF(M.LT.2) GO TO 120                                    EXCR 73
        WRITE(MP,2020)                                          EXCR 74
2020    FORMAT(/10X,'NODE D.O.F.',5X,'X',11X,'Y',11X,'Z'/)      EXCR 75
        I1=1                                                    EXCR 76
        I2=NDIM                                                 EXCR 77
        DO 110 IN=1,NNT                                         EXCR 78
        WRITE(MP,2030) IN,KDLNC(IN+1),(VCORG(I),I=I1,I2)        EXCR 79
2030    FORMAT(10X,2I5,3E12.5)                                  EXCR 80
        I1=I1+NDIM                                              EXCR 81
110     I2=I2+NDIM                                              EXCR 82
120     RETURN                                                  EXCR 83
        END                                                     EXCR 84
```

Figure 6.13 Block 'COOR'.

Flow chart

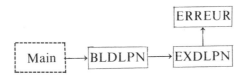

```
      SUBROUTINE BLDLPN                                           BLDL   1
C===============================================================BLDL    2
C     TO CALL BLOCK 'DLPN'                                        BLDL   3
C     TO READ NUMBER OF D.O.F. PER NODE                           BLDL   4
C===============================================================BLDL    5
      IMPLICIT REAL*8(A-H,O-Z)                                    BLDL   6
      COMMON/ES/M,MR,MP,M1                                        BLDL   7
      COMMON/LOC/LCORG,LDLNC                                      BLDL   8
      COMMON VA(1)                                                BLDL   9
C---------------------------------------------------------------BLDL   10
      IF(M1.EQ.0) M1=MR                                           BLDL  11
      WRITE(MP,2000) M                                            BLDL  12
 2000 FORMAT(//' INPUT OF D.O.F. (M=',I2,')'/' ',17('='))         BLDL  13
      CALL EXDLPN(VA(LDLNC))                                      BLDL  14
      RETURN                                                      BLDL  15
      END                                                         BLDL  16

      SUBROUTINE EXDLPN(KDLNC)                                    EXDL   1
C===============================================================EXDL    2
C     TO EXECUTE BLOCK 'DLPN'                                     EXDL   3
C     TO READ THE NUMBER OF D.O.F. PER NODE                       EXDL   4
C===============================================================EXDL    5
      IMPLICIT REAL*8(A-H,O-Z)                                    EXDL   6
      COMMON/COOR/NDIM,NNT,NDLN,NDLT                              EXDL   7
      COMMON/ES/M,MR,MP,M1                                        EXDL   8
      COMMON/TRVL/K1(15)                                          EXDL   9
      DIMENSION KDLNC(1)                                          EXDL  10
C---------------------------------------------------------------EXDL   11
      IF(M.GT.0) WRITE(MP,2000)                                   EXDL  12
 2000 FORMAT(//'GROUP OF D.O.F.'/)                                EXDL  13
C------ READ A GROUP CARD                                         EXDL  14
 10   READ(M1,1000) IDLN,K1                                       EXDL  15
 1000 FORMAT(16I5)                                                EXDL  16
      IF(M.GT.0) WRITE(MP,2010)IDLN,K1                            EXDL  17
 2010 FORMAT(' >>>>>',16I5)                                       EXDL  18
      IF(IDLN.LE.0) GO TO 40                                      EXDL  19
      IF(IDLN.GT.NDLN) CALL ERREUR(21,IDLN,NDLN,1)                EXDL  20
C------ STORE D.O.F. NUMBERS                                      EXDL  21
 20   DO 30 I=1,15                                                EXDL  22
      J=K1(I)                                                     EXDL  23
      IF(J.LE.0) GO TO 10                                         EXDL  24
      IF(J.GT.NNT) CALL ERREUR(22,J,NNT,1)                        EXDL  25
 30   KDLNC(J+1)=IDLN                                             EXDL  26
      READ(M1,1010) K1                                            EXDL  27
 1010 FORMAT(5X,15I5)                                             EXDL  28
      IF(M.GT.0) WRITE(MP,2020) K1                                EXDL  29
 2020 FORMAT(' >>>>>',5X,15I5)                                    EXDL  30
      GO TO 20                                                    EXDL  31
C------ TOTAL NUMBER OF D.O.F.                                    EXDL  32
 40   NDLT=0                                                      EXDL  33
      J=NNT+1                                                     EXDL  34
      DO 50 I=2,J                                                 EXDL  35
 50   NDLT=NDLT+KDLNC(I)                                          EXDL  36
      RETURN                                                      EXDL  37
      END                                                         EXDL  38
```

Figure 6.14 Block 'DLPN'.

Subroutines

    BLDLPN: (Figure 6.14).
    EXDLPN: reads records containing the number of degrees of liberty and corresponding node numerical identification (Figure 6.14).
    ERREUR: (Figure 6.10).

6.5.2.5 Block 'COND'

Function

Reads and interprets records containing boundary conditions and creates arrays VDIMP and KNEQ.

Flow chart

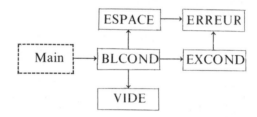

Subroutines

    BLCOND: creates arrays KNEQ and VDIMP (Figure 6.15).
    ESPACE: (Figure 6.5).
    VIDE: releases the unused part of array VDIMP (Figure 6.5).
    EXCOND: transforms array KDLNC defined in Figure 6.6 in a sequence of cumulative numbers; then, it reads records defining boundary conditions and corresponding node identification; finally, EXCOND constructs and outputs for verification arrays VDIMP and KNEQ containing the specified variable values and global equation number of each DOF (Figure 6.15).
    ERREUR: (Figure 6.10).

Remark

This block is always required to construct array KNEQ.

6.5.2.6 Block 'PRND'

Function

This block reads nodal properties and creates array VPRNG.

```
      SUBROUTINE BLCOND                                          BLCN   1
C================================================================BLCN   2
C     TO CALL BLOCK 'COND'                                       BLCN   3
C     TO READ BOUNDARY CONDITIONS AND GENERATE TABLE (NEQ)       BLCN   4
C================================================================BLCN   5
      IMPLICIT REAL*8(A-H,O-Z)                                   BLCN   6
      REAL*4 TBL                                                 BLCN   7
      COMMON/COOR/NDIM,NNT,NDLN,NDLT                             BLCN   8
      COMMON/COND/NCLT,NCLZ,NCLNZ                                BLCN   9
      COMMON/ALLOC/NVA,IVA                                       BLCN  10
      COMMON/ES/M,MR,MP,M1                                       BLCN  11
      COMMON/LOC/LCORG,LDLNC,LNEQ,LDIMP                          BLCN  12
      COMMON VA(1)                                               BLCN  13
      DIMENSION TBL(2)                                           BLCN  14
      DATA TBL/4HNEQ ,4HDIMP/                                    BLCN  15
C----------------------------------------------------------------BLCN  16
      IF(M1.EQ.0) M1=MR                                          BLCN  17
      WRITE(MP,2000) M                                           BLCN  18
 2000 FORMAT(//' INPUT OF BOUNDARY CONDITIONS (M=',I2,')'/' ',   BLCN  19
     1 33('=')/)                                                 BLCN  20
      IF(LNEQ.EQ.1) CALL ESPACE(NDLT,0,TBL(1),LNEQ)              BLCN  21
      IF(LDIMP.EQ.1) CALL ESPACE(NDLT,1,TBL(2),LDIMP)            BLCN  22
      CALL EXCOND(VA(LCORG),VA(LDLNC),VA(LNEQ),VA(LDIMP))        BLCN  23
      CALL VIDE(LDIMP+NCLT,1,TBL(2))                             BLCN  24
      RETURN                                                     BLCN  25
      END                                                        BLCN  26

      SUBROUTINE EXCOND(VCORG,KDLNC,KNEQ,VDIMP)                  EXCN   1
C================================================================EXCN   2
C     TO EXECUTE BLOCK 'COND'                                    EXCN   3
C     READ BOUNDARY CONDITIONS AND GENERALE TABLE (NEQ)          EXCN   4
C================================================================EXCN   5
      IMPLICIT REAL*8 (A-H,O-Z)                                  EXCN   6
      COMMON/COOR/NDIM,NNT,NDLN,NDLT                             EXCN   7
      COMMON/COND/NCLT,NCLZ,NCLNZ                                EXCN   8
      COMMON/RESO/NEQ                                            EXCN   9
      COMMON/ES/M,MR,MP,M1                                       EXCN  10
      COMMON/TRVL/ KV(16),V(10),H(20),ICOD(10)                   EXCN  11
      DIMENSION VCORG(1),KDLNC(1),KNEQ(1),VDIMP(1)               EXCN  12
      DATA L7/7/,L8/8/,L16/16/ ,X1/0.0D0/,X2/0.0D0/,X3/0.0D0/,ZERO/0.D0/EXCN 13
C----------------------------------------------------------------EXCN  14
C------- CUMULATIVE TABLE KDLNC                                  EXCN  15
      DO 10 IN=1,NNT                                             EXCN  16
   10 KDLNC(IN+1)=KDLNC(IN)+KDLNC(IN+1)                          EXCN  17
      I1=NNT+1                                                   EXCN  18
      IF(M.GE.2) WRITE(MP,2000) (KDLNC(IN),IN=1,I1)              EXCN  19
 2000 FORMAT(//' NUMBER OF D.O.F. PRECEDING EACH NODE   (DLNC)'/ EXCN  20
     1 (1X,10I10))                                               EXCN  21
C------- INITIALIZE                                              EXCN  22
      NCLT=0                                                     EXCN  23
      NCLNZ=0                                                    EXCN  24
      NCLZ=0                                                     EXCN  25
      IF(M.GE.0) WRITE(MP,2010)                                  EXCN  26
 2010 FORMAT(//' BOUNDARY CONDITIONS CARDS'/)                    EXCN  27
C------- READ A B.C. GROUP CARD : 10 CODES + PRESCRIBED VAL.     EXCN  28
   20 READ(M1,1000) ICOD,(V(I),I=1,L7)                           EXCN  29
 1000 FORMAT(10I1,7F10.0)                                        EXCN  30
      IF(M.GE.0) WRITE(MP,2020) ICOD,(V(I),I=1,L7)               EXCN  31
 2020 FORMAT(' >>>>>',10I1,7E12.5)                               EXCN  32
C------- CHECK FOR A BLANK CARD                                  EXCN  33
      J=0                                                        EXCN  34
      DO 30 I=1,10                                               EXCN  35
   30 J=J+ICOD(I)                                                EXCN  36
      IF(J.EQ.0) GO TO 110                                       EXCN  37
```

Figure 6.15 (*Contd.*)

```
C------- READ ADDITIONAL CARD IF REQUIRED                         EXCN  38
      I2=0                                                        EXCN  39
      DO 40 ID=1,NDLN                                             EXCN  40
      IF(ICOD(ID).LT.2) GO TO 40                                  EXCN  41
      I2=I2+1                                                     EXCN  42
      IF(I2.NE.L8) GO TO 40                                       EXCN  43
      READ(M1,1010) (V(I),I=L8,NDLN)                              EXCN  44
1010  FORMAT(10X,7F10.0)                                          EXCN  45
      IF(M.GE.0) WRITE(MP,2030) (V(I),I=L8,NDLN)                  EXCN  46
2030  FORMAT(' >>>>>',10X,7E12.5)                                 EXCN  47
40    CONTINUE                                                    EXCN  48
C------- READ NODE CARDS                                          EXCN  49
50    READ(M1,1020) (KV(IN),IN=1,L16)                             EXCN  50
1020  FORMAT(16I5)                                                EXCN  51
      IF(M.GE.0) WRITE(MP,2040) (KV(IN),IN=1,L16)                 EXCN  52
2040  FORMAT(' >>>>>',10X,16I5)                                   EXCN  53
C------- FORM NEQ                                                 EXCN  54
      DO 100 IN=1,L16                                             EXCN  55
      I2=KV(IN)                                                   EXCN  56
C------- END OF GROUP OF B.C. OR END OF NODES OR ANALYSIS OF A NODE  EXCN 57
      IF(I2) 20,20,60                                             EXCN  58
60    IF(I2.GT.NNT) CALL ERREUR(32,I2,NNT,1)                      EXCN  59
      I1=KDLNC(I2)                                                EXCN  60
      IDN=KDLNC(I2+1)-I1                                          EXCN  61
C-------   GENERATE VDIMP, PUT IT IN KNEQ (THE PRESCRIBED D.O.F. ADDRESS)EXCN 62
      IV=0                                                        EXCN  63
      DO 90 ID=1,IDN                                              EXCN  64
      I1=I1+1                                                     EXCN  65
      IC=ICOD(ID)-1                                               EXCN  66
      IF(IC) 90,70,80                                             EXCN  67
70    NCLT=NCLT+1                                                 EXCN  68
      VDIMP(NCLT)=ZERO                                            EXCN  69
      NCLZ=NCLZ+1                                                 EXCN  70
      KNEQ(I1)=-NCLT                                              EXCN  71
      GO TO 90                                                    EXCN  72
80    NCLT=NCLT+1                                                 EXCN  73
      IV=IV+1                                                     EXCN  74
      VDIMP(NCLT)=V(IV)                                           EXCN  75
      NCLNZ=NCLNZ+1                                               EXCN  76
      KNEQ(I1)=-NCLT                                              EXCN  77
90    CONTINUE                                                    EXCN  78
100   CONTINUE                                                    EXCN  79
C------- ADDITIONAL CARD OF NODE NUMBERS                          EXCN  80
      GO TO 50                                                    EXCN  81
C------- GENERATE EQUATION NUMBERS IN NEQ                         EXCN  82
110   I1=0                                                        EXCN  83
      DO 150 IN=1,NNT                                             EXCN  84
      ID=KDLNC(IN)                                                EXCN  85
120   ID=ID+1                                                     EXCN  86
      IF(ID.GT.KDLNC(IN+1)) GO TO 150                             EXCN  87
      IF(KNEQ(ID)) 120,130,120                                    EXCN  88
130   I1=I1+1                                                     EXCN  89
      KNEQ(ID)=I1                                                 EXCN  90
      GO TO 120                                                   EXCN  91
150   CONTINUE                                                    EXCN  92
      NEQ=I1                                                      EXCN  93
C------- OUTPUT                                                   EXCN  94
      IF(M.LT.0) GO TO 170                                        EXCN  95
      WRITE(MP,2050) NNT,NDLT,NEQ,NCLNZ,NCLZ,NCLT                 EXCN  96
2050  FORMAT(//                                                   EXCN  97
     1  15X,'TOTAL NUMBER OF NODES                 (NNT)=',I5/    EXCN  98
     2  15X,'TOTAL NUMBER OF D.O.F.               (NDLT)=',I5/    EXCN  99
     3  15X,'NUMBER OF EQUATIONS TO BE SOLVED      (NEQ)=',I5/    EXCN 100
     4  15X,'NUMBER OF PRESCRIBED NON ZERO D.O.F.(NCLNZ)=',I5/    EXCN 101
```

Figure 6.15 (*Contd.*)

```
        5 15X,'MUMBER OF PRESCRIBED ZERO D.O.F.          (NCLZ)=',I5/     EXCN 102
        6 15X,'TOTAL NUMBER OF PRESCRIBED D.O.F.         (NCLT)=',I5/)    EXCN 103
        IF(M.GE.2.AND.NCLT.GT.0) WRITE(MP,2060)(VDIMP(I),I=1,NCLT)        EXCN 104
 2060   FORMAT(//' PRESCRIBED VALUES   (VDIMP)'//(10X,10E12.5))           EXCN 105
        WRITE(MP,2070)                                                    EXCN 106
 2070   FORMAT(//' NODAL COORDINATES ARRAY'//                             EXCN 107
       1 '   NO  D.L.',5X,'X',12X,'Y',12X,'Z',10X,'EQUATION NUMBER        EXCN 108
       2(NEQ)'/)                                                          EXCN 109
        I2=0                                                              EXCN 110
        DO 160 IN=1,NNT                                                   EXCN 111
        I1=I2+1                                                           EXCN 112
        I2=I2+NDIM                                                        EXCN 113
        ID1=KDLNC(IN)+1                                                   EXCN 114
        ID2=KDLNC(IN+1)                                                   EXCN 115
        ID=ID2-ID1+1                                                      EXCN 116
        IF(ID2.LT.ID1) ID2=ID1                                            EXCN 117
        X1=VCORG(I1)                                                      EXCN 118
        IF(NDIM.GE.2) X2=VCORG(I1+1)                                      EXCN 119
        IF(NDIM.GE.3) X3=VCORG(I1+2)                                      EXCN 120
  160   WRITE(MP,2080) IN,ID,X1,X2,X3,(KNEQ(I),I=ID1,ID2)                 EXCN 121
 2080   FORMAT(1X,2I5,3E12.5,10X,10I6)                                    EXCN 122
  170   RETURN                                                            EXCN 123
        END                                                               EXCN 124
```

Figure 6.15 Block 'COND'.

Flow chart

Subroutines

BLPRND: creates array VPRNG (Figure 6.16).
EXPRND: reads all the nodal properties (Figure 6.16).

6.5.2.7 Block 'PREL'

Function

This block reads and writes element properties and records and constructs array VPREG containing all the element properties.

Flow chart

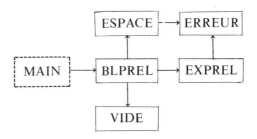

```
      SUBROUTINE BLPRND                                              BLPN   1
C================================================================BLPN   2
C     TO CALL BLOCK 'PRND'                                           BLPN   3
C     TO READ NODAL PROPERTIES                                       BLPN   4
C================================================================BLPN   5
      IMPLICIT REAL*8(A-H,O-Z)                                       BLPN   6
      REAL*4 TBL                                                     BLPN   7
      COMMON/COOR/NDIM,NNT                                           BLPN   8
      COMMON/PRND/NPRN                                               BLPN   9
      COMMON/ES/M,MR,MP,M1                                           BLPN  10
      COMMON/LOC/LXX(4),LPRNG                                        BLPN  11
      COMMON VA(1)                                                   BLPN  12
      DATA TBL/4HPRNG/                                               BLPN  13
C----------------------------------------------------------------BLPN  14
      IF(M1.EQ.0) M1=MR                                              BLPN  15
      READ(M1,1000) NPRN                                             BLPN  16
 1000 FORMAT(I5)                                                     BLPN  17
      WRITE(MP,2000) M,NPRN                                          BLPN  18
 2000 FORMAT(//' INPUT OF NODAL PROPERTIES (M=',I2,')'/' ',30('=')/  BLPN  19
     1 15X,'NUMBER OF PROPERTIES PER NODE   (NPRN)=',I5)             BLPN  20
      IF(LPRNG.EQ.1) CALL ESPACE(NNT*NPRN,1,TBL,LPRNG)               BLPN  21
      CALL EXPRND(VA(LPRNG))                                         BLPN  22
      RETURN                                                         BLPN  23
      END                                                            BLPN  24


      SUBROUTINE EXPRND(VPRNG)                                       EXPN   1
C================================================================EXPN   2
C     TO EXECUTE BLOCK 'PRND'                                        EXPN   3
C     READ NODAL PROPERTIES                                          EXPN   4
C================================================================EXPN   5
      IMPLICIT REAL*8(A-H,O-Z)                                       EXPN   6
      COMMON/COOR/NDIM,NNT                                           EXPN   7
      COMMON/PRND/NPRN                                               EXPN   8
      COMMON/ES/M,MR,MP,M1                                           EXPN   9
      DIMENSION VPRNG(1)                                             EXPN  10
C----------------------------------------------------------------EXPN  11
C------ READ PROPERTIES NODEWISE                                     EXPN  12
      I1=NNT*NPRN                                                    EXPN  13
      READ(M1,1000)(VPRNG(I),I=1,I1)                                 EXPN  14
 1000 FORMAT(8F10.0)                                                 EXPN  15
      IF(M.GE.0) WRITE(MP,2000) (VPRNG(I),I=1,I1)                    EXPN  16
 2000 FORMAT(//' CARDS OF NODAL PROPERTIES'/ (' >>>>>',8E12.5))      EXPN  17
      RETURN                                                         EXPN  18
      END                                                            EXPN  19
```

Figure 6.16 Block 'PRND'.

Subroutines

BLPREL: reads the number of groups of element properties NGPE and the number of properties per group NPRE, then constructs array VPREG as well as temporary workspace V (Figure 6.17).

ESPACE: (Figure 6.5).

VIDE: (Figure 6.5).

EXPREL: for each group, reads group number and the corresponding NPRE property values, then constructs array VPREG (Figure 6.17).

ERREUR: (Figure 6.10).

```
      SUBROUTINE BLPREL                                          BLPE  1
C================================================================BLPE  2
C     TO CALL BLOCK 'PREL'                                       BLPE  3
C     TO READ ELEMENT PROPERTIES                                 BLPE  4
C================================================================BLPE  5
      IMPLICIT REAL*8(A-H,O-Z)                                   BLPE  6
      REAL*4 TBL                                                 BLPE  7
      COMMON/PREL/NGPE,NPRE                                      BLPE  8
      COMMON/ES/M,MR,MP,M1                                       BLPE  9
      COMMON/LOC/LXX(5),LPREG                                    BLPE 10
      COMMON/TRVL/IN(2)                                          BLPE 11
      COMMON VA(1)                                               BLPE 12
      DIMENSION TBL(2)                                           BLPE 13
      DATA TBL/4HPREG,4HV    /                                   BLPE 14
C----------------------------------------------------------------BLPE 15
      IF(M1.EQ.0) M1=MR                                          BLPE 16
C------ READ NUMBER OF GROUPS AND PROPERTIES PER GROUP           BLPE 17
      READ(M1,1000) IN                                           BLPE 18
1000  FORMAT(2I5)                                                BLPE 19
      IF(IN(1).GT.0) NGPE=IN(1)                                  BLPE 20
      IF(IN(2).GT.0) NPRE=IN(2)                                  BLPE 21
      WRITE(MP,2000) M,NGPE,NPRE                                 BLPE 22
2000  FORMAT(//' INPUT OF ELEMENT PROPERTIES (M=',I2,')'/' ',    BLPE 23
     1   35('=')/15X,'NUMBER OF GROUPS OF PROPERTIES   (NGPE)=',I5/ BLPE 24
     2   15X,'NUMBER OF PROPERTIES PER GROUP    (NPRE)=',I5)     BLPE 25
      IF(LPREG.EQ.1) CALL ESPACE(NGPE*NPRE,1,TBL(1),LPREG)       BLPE 26
      CALL ESPACE(NPRE,1,TBL(2),L1)                              BLPE 27
      CALL EXPREL(VA(LPREG),VA(L1))                              BLPE 28
      CALL VIDE(L1,1,TBL(2))                                     BLPE 29
      RETURN                                                     BLPE 30
      END                                                        BLPE 31

      SUBROUTINE EXPREL(VPREG,V1)                                EXPE  1
C================================================================EXPE  2
C     TO EXECUTE BLOCK 'PREL'                                    EXPE  3
C     READ ELEMENT PROPERTIES                                    EXPE  4
C================================================================EXPE  5
      IMPLICIT REAL*8(A-H,O-Z)                                   EXPE  6
      COMMON/PREL/NGPE,NPRE                                      EXPE  7
      COMMON/ES/M,MR,MP,M1                                       EXPE  8
      DIMENSION VPREG(1),V1(1)                                   EXPE  9
C----------------------------------------------------------------EXPE 10
      IF(M.GE.0) WRITE(MP,2000)                                  EXPE 11
2000  FORMAT(//' CARDS OF ELEMENT PROPERTIES'/)                  EXPE 12
C------ READ A GROUP                                             EXPE 13
      I1=MINO(7,NPRE)                                            EXPE 14
      J=1                                                        EXPE 15
10    READ(M1,1000) IGPE,(V1(I),I=1,I1)                          EXPE 16
1000  FORMAT(I5,7F10.0)                                          EXPE 17
      IF(M.GE.0) WRITE(MP,2010) IGPE,(V1(I),I=1,I1)              EXPE 18
2010  FORMAT(' >>>>>',I5,7E12.5)                                 EXPE 19
      IF(IGPE.LE.0) GO TO 40                                     EXPE 20
      IF(IGPE.GT.NGPE) CALL ERREUR(41,IGPE,NGPE,1)               EXPE 21
      IF(NPRE.LE.7) GO TO 20                                     EXPE 22
C------ READ THE PROPERTIES                                      EXPE 23
      READ(M1,1010) (V1(I),I=8,NPRE)                             EXPE 24
1010  FORMAT(5X,7F10.0)                                          EXPE 25
      IF(M.GE.0) WRITE(MP,2020) (V1(I),I=8,NPRE)                 EXPE 26
2020  FORMAT(' >>>>>',5X,7E12.5)                                 EXPE 27
20    DO 30 I=1,NPRE                                             EXPE 28
      VPREG(J)=V1(I)                                             EXPE 29
30    J=J+1                                                      EXPE 30
      GO TO 10                                                   EXPE 31
40    RETURN                                                     EXPE 32
      END                                                        EXPE 33
```

Figure 6.17 Block 'PREL'.

394

Remark

The number and values of element properties must be coherent with the element subroutines ELEM01 which use them. It is this block 'PREL' that reads physical properties like Young's modulus, thermal conductivity, etc.

6.5.2.8 Block 'ELEM'

Function

This block reads records containing element connectivities and constructs the element file.

Flow chart

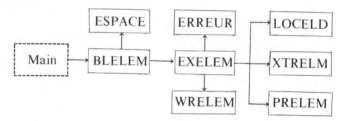

Subroutines

BLELEM: reads a record containing variables NELT, NNEL, NTPE, NGRE, NSYM, NIDENT defined in Figure 6.9 and creates arrays KLD, KLOCE, VCORE, KNE, VPRNE, VPREE (Figure 6.18).

ESPACE: (Figure 6.5).

EXELEM: reads a record containing IEL, IGEN, INCR, ITPE, IGPE, IGRE defined in Figure 6.9 as well as connectivity KNE for each element group. Then, EXELEM executes the following operations for each element:

— verifies the validity of the element information;
— transfers the equation number corresponding to each degree of freedom of an element from KNEQ to KLOCE;
— transfers the nodal coordinates of an element from VCORG to VCORE;
— updates the array containing column heights KLD;
— transfers the nodal properties of an element from VPRNG to VPRNE;
— transfers the element properties from VPREG to VPREE;
— calls WRELEM.

Finally, EXELEM transforms KLD into a location table for the beginnings of the columns of the global matrix (Figure 6.18).

LOCELD: constructs array KLOCE then updates array KLD for each element (Figure 6.18).
XTRELM: extracts nodal coordinates and nodal properties from the global arrays for a given element (Figure 6.18).
PRELEM: outputs all the information concerning one element (Figure 6.18).
WRELEM: writes all the results computed by EXELEM in the element file (Figure 6.18).
RDELEM: reads subroutines corresponding to WRELEM needed in the execution blocks (Figure 6.18).
ERREUR: (Figure 6.10).

```
      SUBROUTINE BLELEM                                             BLEL   1
C================================================================== BLEL   2
C     TO CALL BLOCK 'ELEM'                                          BLEL   3
C     TO READ ELEMENT DATA                                          BLEL   4
C================================================================== BLEL   5
      IMPLICIT REAL*8(A-H,O-Z)                                      BLEL   6
      REAL*4 TBL                                                    BLEL   7
      COMMON/COOR/NDIM,NNT,NDLN                                     BLEL   8
      COMMON/PRND/NPRN                                              BLEL   9
      COMMON/PREL/NGPE,NPRE                                         BLEL  10
      COMMON/ELEM/NELT,NNEL,NTPE,NGRE,ME,NIDENT,NPG                 BLEL  11
      COMMON/ASSE/NSYM,NKG                                          BLEL  12
      COMMON/RESO/NEQ                                               BLEL  13
      COMMON/ES/M,MR,MP,M1,M2                                       BLEL  14
      COMMON/LOC/LCORG,LDLNC,LNEQ,LDIMP,LPRNG,LPREG,LLD,LLOCE,LCORE,LNE,BLEL  15
     1 LPRNE,LPREE,LDLE,LKE,LFE,LKGS,LKGD,LKGI,LFG,LRES,LDLG        BLEL  16
      COMMON VA(1)                                                  BLEL  17
      DIMENSION TBL(6),IN(6)                                        BLEL  18
      DATA TBL/4HLD   ,4HLOCE,4HCORE,4HNE  ,4HPRNE,4HPREE/          BLEL  19
C----------------------------------------------------------------  BLEL  20
      IF(M1.EQ.0) M1=MR                                             BLEL  21
      IF(M2.EQ.0) M2=ME                                             BLEL  22
      READ(M1,1000)IN                                               BLEL  23
 1000 FORMAT(6I5)                                                   BLEL  24
      IF(IN(1).GT.0) NELT=IN(1)                                     BLEL  25
      IF(IN(2).GT.0) NNEL=IN(2)                                     BLEL  26
      IF(IN(3).GT.0) NTPE=IN(3)                                     BLEL  27
      IF(IN(4).GT.0) NGRE=IN(4)                                     BLEL  28
      IF(IN(5).NE.0) NSYM=1                                         BLEL  29
      IF(IN(6).NE.0) NIDENT=1                                       BLEL  30
      WRITE(MP,2000) M,NELT,NNEL,NTPE,NGRE,NSYM,NIDENT              BLEL  31
 2000 FORMAT(//' INPUT OF ELEMENTS (M=',I2,')'/' ',20('=')/         BLEL  32
     1  15X,'MAX. NUMBER OF ELEMENTS      (NELT)=',I5/              BLEL  33
     2  15X,'MAX. NUMBER OF NODES PER ELEMENT (NNEL)=',I5/          BLEL  34
     3  15X,'DEFAULT ELEMENT TYPE         (NTPE)=',I5/              BLEL  35
     4  15X,'NUMBER OF GROUPS OF ELEMENTS (NGRE)=',I5/              BLEL  36
     5  15X,'INDEX FOR NON SYMMETRIC PROBLEM  (NSYM)=',I5/          BLEL  37
     6  15X,'INDEX FOR IDENTICAL ELEMENTS (NIDENT)=',I5/)           BLEL  38
      IF(LLD.EQ.1) CALL ESPACE(NEQ+1,0,TBL(1),LLD)                  BLEL  39
      IF(LLOCE.EQ.1) CALL ESPACE(NNEL*NDLN,0,TBL(2),LLOCE)          BLEL  40
      IF(LCORE.EQ.1) CALL ESPACE(NNEL*NDIM,1,TBL(3),LCORE)          BLEL  41
      IF(LNE.EQ.1) CALL ESPACE(NNEL,0,TBL(4),LNE)                   BLEL  42
      IF(NPRN.GT.0.AND.LPRNE.EQ.1) CALL ESPACE(NNEL*NPRN,1,TBL(5),LPRNE)BLEL  43
      IF(NPRE.GT.0.AND.LPREE.EQ.1) CALL ESPACE(NPRE,1,TBL(6),LPREE) BLEL  44
      CALL EXELEM(VA(LCORG),VA(LDLNC),VA(LPRNG),VA(LPREG),VA(LLOCE),BLEL  45
     1            VA(LCORE),VA(LNE),VA(LPRNE),VA(LPREE),VA(LNEQ),VA(LLD)) BLEL 46
      WRITE(MP,2010) NKG,NPG                                        BLEL  47
```

Figure 6.18 (*Contd.*)

```
2010  FORMAT(15X,'LENGTH OF A TRIANGLE IN KG          (NKG)=',I10/      BLEL 48
     1      15X,'NUMBER OF INTEGRATION POINTS         (NPG)=',I10/)     BLEL 49
      RETURN                                                            BLEL 50
      END                                                               BLEL 51

      SUBROUTINE EXELEM(VCORG,KDLNC,VPRNG,VPREG,KLOCE,VCORE,KNE,VPRNE,  EXEL  1
     1                  VPREE,KNEQ,KLD)                                 EXEL  2
C======================================================================EXEL  3
C            TO EXECUTE BLOCK 'ELEM'                                    EXEL  4
C            READ ELEMENTS DATA                                         EXEL  5
C======================================================================EXEL  6
      IMPLICIT REAL*8(A-H,O-Z)                                          EXEL  7
      COMMON/COOR/NDIM,NNT                                              EXEL  8
      COMMON/PRND/NPRN                                                  EXEL  9
      COMMON/PREL/NGPE,NPRE                                             EXEL 10
      COMMON/ELEM/NELT,NNEL,NTPE,NGRE,ME,NIDENT,NPG                     EXEL 11
      COMMON/ASSE/NSYM,NKG,NKE,NDLE                                     EXEL 12
      COMMON/RGDT/IEL,ITPE,ITPE1,IGRE,IDLE,ICE,IPRNE,IPREE,INEL,IDEG,IPGEXEL 13
     1 ,ICODE,IDLEO,INELO,IPGO                                          EXEL 14
      COMMON/RESO/NEQ                                                   EXEL 15
      COMMON/ES/M,MR,MP,M1,M2                                           EXEL 16
      DIMENSION VCORG(1),KDLNC(1),VPRNG(1),VPREG(1),KLOCE(1),VCORE(1)   EXEL 17
     1          KNE(1),VPRNE(1),VPREE(1),KNEQ(1),KLD(1)                 EXEL 18
      DATA I10/10/,I15/15/                                              EXEL 19
C----------------------------------------------------------------------EXEL 20
C------ INITIALIZE                                                      EXEL 21
      NDLE=0                                                            EXEL 22
      IELT=0                                                            EXEL 23
      NPG=0                                                             EXEL 24
      REWIND M2                                                         EXEL 25
      IF(M.GT.0) WRITE(MP,2000)                                         EXEL 26
2000  FORMAT(//' ELEMENTS CARDS'/)                                      EXEL 27
C------ READ AN ELEMENT CARD                                            EXEL 28
10    READ(M1,1000) IEL,IGEN,INCR,ITPE,IGPE,IGRE,(KNE(IN),IN=1,I10)     EXEL 29
1000  FORMAT(16I5)                                                      EXEL 30
      IF(M.GT.0) WRITE(MP,2010) IEL,IGEN,INCR,ITPE,IGPE,IGRE,           EXEL 31
     1                          (KNE(IN),IN=1,I10)                      EXEL 32
2010  FORMAT(' >>>>>',16I5)                                             EXEL 33
      IF(IEL) 80,80,20                                                  EXEL 34
C------ NUMBER OF NODES AND ADDITIONNAL CARDS AS REQUIRED               EXEL 35
20    INEL=0                                                            EXEL 36
      I1=1                                                              EXEL 37
      I2=I10                                                            EXEL 38
30    DO 40 IN=I1,I2                                                    EXEL 39
      IF(KNE(IN).EQ.0) GO TO 50                                         EXEL 40
      INEL=INEL+1                                                       EXEL 41
40    CONTINUE                                                          EXEL 42
      I1=I2+1                                                           EXEL 43
      I2=I1+I15                                                         EXEL 44
      READ(M1,1000) (KNE(IN),IN=I1,I2)                                  EXEL 45
      IF(M.GT.0) WRITE(MP,2010) (KNE(IN),IN=I1,I2)                      EXEL 46
      GO TO 30                                                          EXEL 47
C------ CHECKING                                                        EXEL 48
50    IF(INEL.GT.NNEL) CALL ERREUR(51,INEL,NNEL,1)                      EXEL 49
      IF(INCR.EQ.0) INCR=1                                              EXEL 50
      IF(ITPE.EQ.0) ITPE=NTPE                                           EXEL 51
      IF(IGPE.GT.NGPE) CALL ERREUR(53,IGPE,NGPE,1)                      EXEL 52
      IF(IGPE.EQ.0) IGPE=1                                              EXEL 53
      IF(IGRE.GT.NGRE) CALL ERREUR(54,IGRE,NGRE,1)                      EXEL 54
C------ ELEMENT GENERATION                                              EXEL 55
      IF(IGEN.EQ.0) IGEN=1                                              EXEL 56
      DO 70 IE=1,IGEN                                                   EXEL 57
      IF(IEL.GT.NELT) CALL ERREUR(55,IEL,NELT,1)                        EXEL 58
C------ GENERATE KLOCE AND UPDATE KLD                                   EXEL 59
```

Figure 6.18 *(Contd.)*

```
      CALL LOCELD(KDLNC,KNE,KNEQ,KLOCE,KLD)                      EXEL  60
C------ GENERATE ELEMENT COORDINATES AND PROPERTIES              EXEL  61
      CALL XTRELM(IGPE,VCORG,VPRNG,VPREG,KNE,VCORE,VPRNE,VPREE)  EXEL  62
C------ CHECK ELEMENT NODE NUMBERS AND D.O.F.                    EXEL  63
      IPGO=0                                                     EXEL  64
      ICODE=1                                                    EXEL  65
      CALL ELEMLB(VCORE,VPRNE,VPREE,VDLE,VKE,VFE)                EXEL  66
      IF(INEL.EQ.INELO.AND.IDLE.EQ.IDLEO) GO TO 55               EXEL  67
      WRITE(MP,2020) IEL,INEL,INELO,IDLE,IDLEO                   EXEL  68
2020  FORMAT(' ** ELEMENT',I5,' INCONSISTENT'/5X,'INEL=',I4,' INELO=',I5EXEL 69
     1/ 5X,'IDLE=',I5,' IDLEO=',I5)                              EXEL  70
C------ UPDATE TOTAL NUMBER OF INTEGRATION POINTS                EXEL  71
55    NPG=NPG+IPGO                                               EXEL  72
C------ STORE ON ELEMENT FILE                                    EXEL  73
      CALL WRELEM(M2,KLOCE,VCORE,VPRNE,VPREE,KNE)                EXEL  74
      IELT=IELT+1                                                EXEL  75
C------ PRINT ELEMENT CHARACTERISTICS                            EXEL  76
      CALL PRELEM(KLOCE,VCORE,VPRNE,VPREE,KNE)                   EXEL  77
C------ NEXT ELEMENT TO BE GENERATED OR READ                     EXEL  78
      DO 60 IN=1,INEL                                            EXEL  79
60    KNE(IN)=KNE(IN)+INCR                                       EXEL  80
      IF(IDLE.GT.NDLE) NDLE=IDLE                                 EXEL  81
70    IEL=IEL+1                                                  EXEL  82
      GO TO 10                                                   EXEL  83
C------ CHECK IF TOTAL NUMBER OF ELEMENT IS EXCEEDED             EXEL  84
80    IF(IELT.NE.NELT) CALL ERREUR(57,IELT,NELT,1)               EXEL  85
C------ PRINT BAND HEIGHTS                                       EXEL  86
      IMA=0                                                      EXEL  87
      IMO=0                                                      EXEL  88
      I1=NEQ+1                                                   EXEL  89
      DO 90 I=2,I1                                               EXEL  90
      J=KLD(I)                                                   EXEL  91
      IF(J.GT.IMA)IMA=J                                          EXEL  92
90    IMO=IMO+J                                                  EXEL  93
      C=IMO                                                      EXEL  94
      C=C/NEQ                                                    EXEL  95
      WRITE(MP,2030) C,IMA                                       EXEL  96
2030  FORMAT(/15X,'MEAN BAND HEIGHT=',F8.1,' MAXIMUM=',I5)       EXEL  97
      IF(M.GE.2) WRITE(MP,2040) (KLD(I),I=1,I1)                  EXEL  98
2040  FORMAT(//' TABLE OF BAND HEIGHTS'/(10X,20I5))              EXEL  99
C------ TRANSFORM KLD INTO ADDRESSES OF COLUMN TOP TERM          EXEL 100
      IF(NSYM.EQ.0) NKE=(NDLE*(NDLE+1))/2                        EXEL 101
      IF(NSYM.EQ.1) NKE=NDLE*NDLE                                EXEL 102
      KLD(1)=1                                                   EXEL 103
      DO 100 ID=2,I1                                             EXEL 104
100   KLD(ID)=KLD(ID-1)+KLD(ID)                                  EXEL 105
      NKG=KLD(I1)-1                                              EXEL 106
      IF(M.GE.2) WRITE(MP,2050) (KLD(ID),ID=1,I1)                EXEL 107
2050  FORMAT(//' TABLE OF ADDRESSES OF COLUMN TOP TERMS   (LD)'/ EXEL 108
     1        (10X,20I6))                                        EXEL 109
      RETURN                                                     EXEL 110
      END                                                        EXEL 111

      SUBROUTINE LOCELD(KDLNC,KNE,KNEQ,KLOCE,KLD)                LOCL   1
C================================================================LOCL   2
C     TO FORM THE ELEMENT LOCALIZATION TABLE (LOCE)              LOCL   3
C     AND UPDATE COLUMN HEIGHTS FOR A GIVEN ELEMENT              LOCL   4
C================================================================LOCL   5
      COMMON/COOR/NDIM,NNT                                       LOCL   6
      COMMON/RGDT/NUL(4),IDLE,NUL1(3),INEL                       LOCL   7
      DIMENSION KDLNC(1),KNE(1),KNEQ(1),KLOCE(1),KLD(1)          LOCL   8
      DATA NDLMAX/32000/                                         LOCL   9
C----------------------------------------------------------------LOCL  10
C------ GENERATE KLOCE FROM KNEQ                                 LOCL  11
```

Figure 6.18 (*Contd.*)

```
      IDLE=0                                                      LOCL  12
      LOCMIN=NDLMAX                                               LOCL  13
      DO 20 IN=1,INEL                                             LOCL  14
      INN=KNE(IN)                                                 LOCL  15
      IF(INN.GT.NNT) CALL ERREUR(56,INN,NNT,1)                    LOCL  16
      IEQ=KDLNC(INN)                                              LOCL  17
      IEQ1=KDLNC(INN+1)                                           LOCL  18
   10 IF(IEQ.GE.IEQ1) GO TO 20                                    LOCL  19
      IEQ=IEQ+1                                                   LOCL  20
      IDLE=IDLE+1                                                 LOCL  21
      J=KNEQ(IEQ)                                                 LOCL  22
      KLOCE(IDLE)=J                                               LOCL  23
      IF(J.LT.LOCMIN.AND.J.GT.0) LOCMIN=J                         LOCL  24
      GO TO 10                                                    LOCL  25
   20 CONTINUE                                                    LOCL  26
C------- UPDATE TABLE OF COLUMN HEIGHTS (KLD)                     LOCL  27
      DO 30 ID=1,IDLE                                             LOCL  28
      J=KLOCE(ID)                                                 LOCL  29
      IF(J.LE.0) GO TO 30                                         LOCL  30
      IH=J-LOCMIN                                                 LOCL  31
      IF(IH.GT.KLD(J+1))KLD(J+1)=IH                               LOCL  32
   30 CONTINUE                                                    LOCL  33
      RETURN                                                      LOCL  34
      END                                                         LOCL  35

      SUBROUTINE XTRELM(IGPE,VCORG,VPRNG,VPREG,KNE,VCORE,VPRNE,VPREE)  XTRE   1
C===========================================================================XTRE   2
C     TO GENERATE ELEMENT COORDINATES AND PROPERTIES FROM        XTRE   3
C     GLOBAL ARRAYS                                              XTRE   4
C     (IGPE: GROUP NUMBER FOR ELEMENT PROPERTIES)                XTRE   5
C===========================================================================XTRE   6
      IMPLICIT REAL*8(A-H,O-Z)                                   XTRE   7
      COMMON/COOR/NDIM                                           XTRE   8
      COMMON/PRND/NPRN                                           XTRE   9
      COMMON/PREL/NGPE,NPRE                                      XTRE  10
      COMMON/RGDT/NUL(5),ICE,IPRNE,IPREE,INEL                    XTRE  11
      DIMENSION VCORG(1),VPRNG(1),VPREG(1),KNE(1),VCORE(1),      XTRE  12
     1 VPRNE(1),VPREE(1)                                         XTRE  13
C---------------------------------------------------------------XTRE  14
C------- GENERATE ELEMENT COORDINATES                           XTRE  15
      IPRNE=0                                                    XTRE  16
      ICE=0                                                      XTRE  17
      DO 30 IN=1,INEL                                            XTRE  18
      IC=(KNE(IN)-1)*NDIM                                        XTRE  19
      DO 10 I=1,NDIM                                             XTRE  20
      ICE=ICE+1                                                  XTRE  21
      IC=IC+1                                                    XTRE  22
   10 VCORE(ICE)=VCORG(IC)                                       XTRE  23
C------- GENERATE ELEMENT NODAL PROPERTIES                      XTRE  24
      IF(NPRN.EQ.0) GO TO 30                                     XTRE  25
      IC=(KNE(IN)-1)*NPRN                                        XTRE  26
      DO 20 I=1,NPRN                                             XTRE  27
      IPRNE=IPRNE+1                                              XTRE  28
      IC=IC+1                                                    XTRE  29
   20 VPRNE(IPRNE)=VPRNG(IC)                                     XTRE  30
   30 CONTINUE                                                   XTRE  31
C------- GENERATE ELEMENT PROPERTIES                            XTRE  32
      IPREE=0                                                    XTRE  33
      IF(NPRE.EQ.0) GO TO 50                                     XTRE  34
      IC=(IGPE-1)*NPRE                                           XTRE  35
      DO 40 I=1,NPRE                                             XTRE  36
      IPREE=IPREE+1                                              XTRE  37
      IC=IC+1                                                    XTRE  38
   40 VPREE(IPREE)=VPREG(IC)                                     XTRE  39
```

Figure 6.18 (*Contd.*)

```
 50     RETURN                                                         XTRE  40
        END                                                            XTRE  41

        SUBROUTINE PRELEM(KLOCE,VCORE,VPRNE,VPREE,KNE)                 PREL   1
C==================================================================PREL   2
C       PRINT DATA DEFINING AN ELEMENT                                 PREL   3
C==================================================================PREL   4
        IMPLICIT REAL*8(A-H,O-Z)                                       PREL   5
        COMMON/PRND/NPRN                                               PREL   6
        COMMON/PREL/NGPE,NPRE                                          PREL   7
        COMMON/RGDT/IEL,ITPE,ITPE1,IGRE,IDLE,ICE,IPRNE,IPREE,INEL      PREL   8
        COMMON/ES/M,MR,MP                                              PREL   9
        DIMENSION KLOCE(1),VCORE(1),VPRNE(1),VPREE(1),KNE(1)           PREL  10
C------------------------------------------------------------------PREL  11
        IF(M.GE.0) WRITE(MP,2000) IEL,ITPE,INEL,IDLE,IPRNE,IPREE,IGRE  PREL  12
 2000   FORMAT(10X,'ELEMENT:',I5,' TYPE:',I2,' N.P.:',I2,' D.O.F.:',   PREL  13
       1  I3,' N. PROP:',I3,' EL. PROP:',I3,' GROUP:',I3)              PREL  14
        IF(M.GE.0) WRITE(MP,2010) (KNE(I),I=1,INEL)                    PREL  15
 2010   FORMAT(15X,'CONNECTIVITY (NE)',20I5/(32X,20I5))                PREL  16
        IF(M.LT.1) GO TO 10                                            PREL  17
        WRITE(MP,2020) (KLOCE(I),I=1,IDLE)                             PREL  18
 2020   FORMAT(15X,'LOCALIZATN (LOCE)',20I5/(32X,20I5))                PREL  19
        WRITE(MP,2030) (VCORE(I),I=1,ICE)                              PREL  20
 2030   FORMAT(15X,'COORDINATES(CORE)',8E12.5/(32X,8E12.5))            PREL  21
        IF(NPRN.GT.0) WRITE(MP,2040) (VPRNE(I),I=1,IPRNE)              PREL  22
 2040   FORMAT(15X,'NOD.PROP.  (PRNE)',8E12.5/(32X,8E12.5))            PREL  23
        IF(IPREE.GT.0) WRITE(MP,2050) (VPREE(I),I=1,IPREE)             PREL  24
 2050   FORMAT(15X,'ELEM. PROP.(PREE)',8E12.5/(32X,8E12.5))            PREL  25
 10     RETURN                                                         PREL  26
        END                                                            PREL  27

        SUBROUTINE WRELEM(ME,KLOCE,VCORE,VPRNE,VPREE,KNE)              WREL   1
C==================================================================WREL   2
C       WRITE ELEMENT PROPERTIES ON FILE ME                            WREL   3
C==================================================================WREL   4
        IMPLICIT REAL*8(A-H,O-Z)                                       WREL   5
        COMMON/RGDT/IEL,ITPE,ITPE1,IGRE,IDLE,ICE,IPRNE,IPREE,INEL      WREL   6
        DIMENSION KLOCE(1),VCORE(1),VPRNE(1),VPREE(1),KNE(1)           WREL   7
C------------------------------------------------------------------WREL   8
        IPRNE1=IPRNE                                                   WREL   9
        IF(IPRNE1.EQ.0) IPRNE1=1                                       WREL  10
        IPREE1=IPREE                                                   WREL  11
        IF(IPREE1.EQ.0) IPREE1=1                                       WREL  12
        WRITE(ME)IEL,ITPE,IGRE,IDLE,ICE,IPRNE1,IPREE1,INEL,            WREL  13
       1         (KLOCE(I),I=1,IDLE),(VCORE(I),I=1,ICE),               WREL  14
       2         (VPRNE(I),I=1,IPRNE1),(VPREE(I),I=1,IPREE1),          WREL  15
       3         (KNE(I),I=1,INEL)                                     WREL  16
        RETURN                                                         WREL  17
        END                                                            WREL  18

        SUBROUTINE RDELEM(ME,KLOCE,VCORE,VPRNE,VPREE,KNE)              RDEL   1
C==================================================================RDEL   2
C       READ ELEMENT PROPERTIES FROM FILE ME                           RDEL   3
C==================================================================RDEL   4
        IMPLICIT REAL*8(A-H,O-Z)                                       RDEL   5
        COMMON/RGDT/IEL,ITPE,ITPE1,IGRE,IDLE,ICE,IPRNE,IPREE,INEL      RDEL   6
        DIMENSION KLOCE(1),VCORE(1),VPRNE(1),VPREE(1),KNE(1)           RDEL   7
C------------------------------------------------------------------RDEL   8
        READ(ME) IEL,ITPE,IGRE,IDLE,ICE,IPRNE,IPREE,INEL,              RDEL   9
       1         (KLOCE(I),I=1,IDLE),(VCORE(I),I=1,ICE),               RDEL  10
       2         (VPRNE(I),I=1,IPRNE),(VPREE(I),I=1,IPREE),             RDEL  11
       3         (KNE(I),I=1,INEL)                                     RDEL  12
        RETURN                                                         RDEL  13
        END                                                            RDEL  14
```

Figure 6.18 Block 'ELEM'.

6.5.2.9 *Block 'SOLC'*

Function

This block reads the loads and superposes them in array VFG.

Flow chart

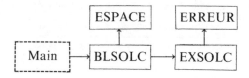

Subroutines

    BLSOLC: constructs array VFG (Figure 6.19).
    ESPACE: (Figure 6.5).
    EXSLOC: for each group of concentrated nodal loads, the routine reads the group number, the load value associated with each degree of freedom and the nodal numbers where the concentrated load is applied. Then, EXSOL superposes all the loads in array VFG (Figure 6.19).
    ERREUR: (Figure 6.10).

```
      SUBROUTINE BLSOLC                                            BLSC  1
C=================================================================BLSC  2
C     TO CALL BLOCK "SOLC'                                         BLSC  3
C     TO READ CONCENTRATED LOADS                                   BLSC  4
C=================================================================BLSC  5
      IMPLICIT REAL*8(A-H,O-Z)                                     BLSC  6
      REAL*4 TBL                                                   BLSC  7
      COMMON/RESO/NEQ                                              BLSC  8
      COMMON/ES/M,MR,MP,M1                                         BLSC  9
      COMMON/LOC/LCORG,LDLNC,LNEQ,LXX(15),LFG                      BLSC 10
      COMMON VA(1)                                                 BLSC 11
      DATA TBL/4HFG  /                                             BLSC 12
C-----------------------------------------------------------------BLSC 13
      IF(M1.EQ.0) M1=MR                                            BLSC 14
      WRITE(MP,2000) M                                             BLSC 15
 2000 FORMAT(//' INPUT OF CONCENTRADED LOADS (M=',I2,')'/' ',      BLSC 16
     1 39('='))                                                    BLSC 17
      IF(LFG.EQ.1) CALL ESPACE(NEQ,1,TBL,LFG)                      BLSC 18
      CALL EXSOLC(VA(LFG),VA(LDLNC),VA(LNEQ))                      BLSC 19
      RETURN                                                       BLSC 20
      END                                                          BLSC 21

      SUBROUTINE EXSOLC(VFG,KDLNC,KNEQ)                            EXSC  1
C=================================================================EXSC  2
C     TO EXECUTE BLOCK 'SOLC'                                      EXSC  3
C     READ CONCENTRATED LOADS                                      EXSC  4
C================================================================= EXSC  5
      IMPLICIT REAL*8 (A-H,O-Z)                                    EXSC  6
      COMMON/COOR/NDIM,NNT,NDLN                                    EXSC  7
      COMMON/RESO/NEQ                                              EXSC  8
```

Figure 6.19   (*Contd.*)

```
       COMMON/ES/M,MR,MP,M1                                    EXSC   9
       COMMON/TRVL/KV(16),V(14)                                EXSC  10
       DIMENSION VFG(1),KDLNC(1),KNEQ(1)                       EXSC  11
       DATA L16/16/                                            EXSC  12
C------------------------------------------------------------- EXSC  13
C------ READ DATA                                              EXSC  14
       IF(M.GE.0)WRITE(MP,2000)                                EXSC  15
 2000  FORMAT(//' CARDS OF NODAL LOADS'//)                     EXSC  16
       IO=MIN0(7,NDLN)                                         EXSC  17
 10    READ(M1,1000) IG,(V(I),I=1,IO)                          EXSC  18
 1000  FORMAT(I5,7F10.0)                                       EXSC  19
       IF(NDLN.GT.7) READ(M1,1005) (V(I),I=8,NDLN)             EXSC  20
 1005  FORMAT(5X,7F10.0)                                       EXSC  21
       IF(M.GE.0)WRITE(MP,2010)IG,(V(I),I=1,NDLN)              EXSC  22
 2010  FORMAT(' >>>>>',I5,7E12.5/(' >>>>>',5X,7E12.5))         EXSC  23
       IF(IG.LE.0) GO TO 60                                    EXSC  24
 20    READ(M1,1010)(KV(I),I=1,L16)                            EXSC  25
 1010  FORMAT(16I5)                                            EXSC  26
       IF(M.GE.0)WRITE(MP,2020)(KV(I),I=1,L16)                 EXSC  27
 2020  FORMAT(' >>>>>',16I5)                                   EXSC  28
C----- DECODE NODAL DATA                                       EXSC  29
       DO 50 IN=1,L16                                          EXSC  30
       I1=KV(IN)                                               EXSC  31
       IF(I1.GT.NNT) CALL ERREUR(61,I1,NNT,1)                  EXSC  32
       IF(I1)10,10,30                                          EXSC  33
 30    ID1=KDLNC(I1)+1                                         EXSC  34
       ID2=KDLNC(I1+1)                                         EXSC  35
       J=0                                                     EXSC  36
       DO 50 ID=ID1,ID2                                        EXSC  37
       J=J+1                                                   EXSC  38
       IEQ=KNEQ(ID)                                            EXSC  39
       IF(IEQ)50,50,40                                         EXSC  40
 40    VFG(IEQ)=VFG(IEQ)+V(J)                                  EXSC  41
 50    CONTINUE                                                EXSC  42
       GO TO 20                                                EXSC  43
C----- OUTPUT                                                  EXSC  44
 60    IF(M.GE.1)WRITE(MP,2030)(VFG(I),I=1,NEQ)                EXSC  45
 2030  FORMAT(///' TOTAL LOAD VECTOR'/(10X,10E12.5))            EXSC  46
       RETURN                                                  EXSC  47
       END                                                     EXSC  48
```

Figure 6.19 Block 'SOLC'.

6.5.3 *Execution Functional Blocks*

The execution functional blocks contained in the present version of MEF are:

Block	Function
'SOLR'	Assemblage of distributed loads
'LINM'	Solution of linear problem with global matrix in core
'LIND'	Solution of linear problem with global matrix out of core
'NLIN'	Solution of stationary non-linear problem
'TEMP'	Solution of unsteady problem (linear or non-linear)
'VALP'	Eigenvalues and eigenvectors

6.5.3.1 *Organization of Execution Blocks*

The various blocks designed for execution of the various computations have similar structures since they all have to:

— construct element and load matrices;
— assemble global matrices and vectors;
— factorize and solve the system of equations;
— output the results.

The following general structure is common to all these blocks:

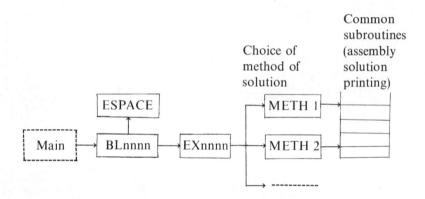

Subroutine BLnnnn allocates space for all the necessary arrays.

Subroutine EXnnnn loops over load increments, time increment, or equilibrium iterations within each step. Furthermore, it applies convergence tests to iterative processes and, according to the chosen method of solution, calls METH 1, METH 2, etc. For example, to solve a linear problem, METH 1 would simply consist in assembling global matrices, solve the system and output the results. For a transient problem, METH 2 could for example, use implicit Euler's formula. Subroutines METH 1, METH 2, etc. Contain only the operations that differ from one method to the other. Common operations are performed by other subroutines: see Figure 6.20.

6.5.3.2 *Block 'SOLR'*

Function

This block assembles element loads corresponding to body forces.

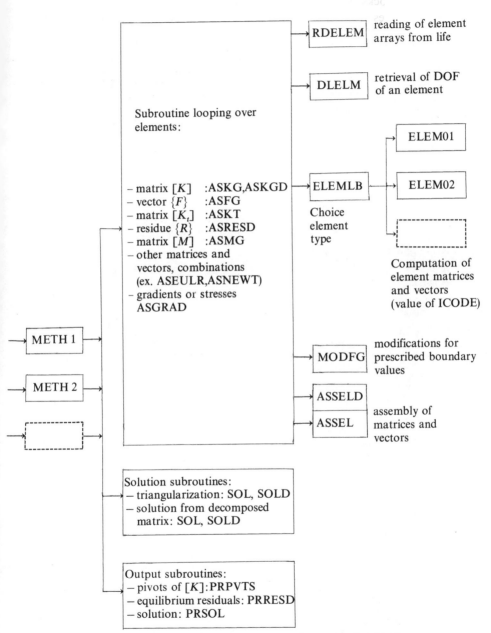

Figure 6.20 Subroutines common to various execution blocks.

Flow chart

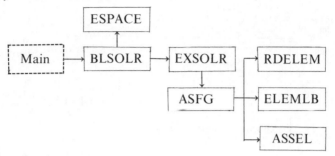

Subroutines

BLSOLR: creates array VFG if it does not yet exist (Figure 6.21).
ESPACE: (Figure 6.5).
EXSOLR: adds body forces to element vector FG (Figure 6.21).
ASFG: assembles element body forces (Figure 6.21).
RDELEM: (Figure 6.18).
ELEMLB: (Figure 6.22).
ASSEL: (Figure 6.22).

```
      SUBROUTINE BLSOLR                                             BLSR   1
C=====================================================================BLSR   2
C     TO CALL BLOCK 'SOLR'                                          BLSR   3
C     TO ASSEMBLE DISTRIBUTED LOADS (ELEMENT FUNCTION 7)            BLSR   4
C=====================================================================BLSR   5
      IMPLICIT REAL*8(A-H,O-Z)                                      BLSR   6
      REAL*4 TBL                                                    BLSR   7
      COMMON/COOR/NDIM,NNT,NDLN,NDLT                                BLSR   8
      COMMON/ELEM/NUL(4),ME                                         BLSR   9
      COMMON/ASSE/NSYM,NKG,NKE,NDLE                                 BLSR  10
      COMMON/RESO/NEQ,NRES,MRES                                     BLSR  11
      COMMON/ES/M,MR,MP,M1,M2                                       BLSR  12
      COMMON/LOC/LCORG,LDLNC,LNEQ,LDIMP,LPRNG,LPREG,LLD,LLOCE,LCORE,LNE,BLSR  13
     1 LPRNE,LPREE,LDLE,LKE,LFE,LKGS,LKGD,LKGI,LFG,LRES,LDLG        BLSR  14
      COMMON VA(1)                                                  BLSR  15
      DIMENSION TBL(8)                                              BLSR  16
      DATA TBL/4HFG   ,4HKE   ,4HFE   ,4HDLE ,4HKGS ,4HKGD ,4HKGI , BLSR  17
     1 4HRES /                                                      BLSR  18
C---------------------------------------------------------------------BLSR  19
      IF(M1.EQ.0) M1=MR                                             BLSR  20
      IF(M2.EQ.0) M2=ME                                             BLSR  21
      WRITE(MP,2000) M                                              BLSR  22
 2000 FORMAT(//' ASSEMBLING OF DISTRIBUTED LOADS (M=',I2,')'/        BLSR  23
     1 1X,40('=')/)                                                 BLSR  24
      IF(LFG.EQ.1) CALL ESPACE(NEQ,1,TBL(1),LFG)                    BLSR  25
      IF(LKE.EQ.1) CALL ESPACE(NKE,1,TBL(2),LKE)                    BLSR  26
      IF(LFE.EQ.1) CALL ESPACE(NDLE,1,TBL(3),LFE)                   BLSR  27
      IF(LDLE.EQ.1) CALL ESPACE(NDLE,1,TBL(4),LDLE)                 BLSR  28
      IF(LKGS.EQ.1) CALL ESPACE(NKG,1,TBL(5),LKGS)                  BLSR  29
      IF(LKGD.EQ.1) CALL ESPACE(NEQ,1,TBL(6),LKGD)                  BLSR  30
      IF(NSYM.EQ.1.AND.LKGI.EQ.1) CALL ESPACE(NKG,1,TBL(7),LKGI)    BLSR  31
      IF(LRES.EQ.1) CALL ESPACE(NDLT,1,TBL(8),LRES)                 BLSR  32
      CALL EXSOLR(VA(LLD),VA(LDIMP),VA(LLOCE),VA(LCORE),VA(LPRNE),  BLSR  33
     1          VA(LPREE),VA(LNE),VA(LKE),VA(LFE),VA(LKGS),VA(LKGD),BLSR  34
```

Figure 6.21 (*Contd.*)

```
      2         VA(LKGI),VA(LFG),VA(LCORG),VA(LDLNC),VA(LNEQ),      BLSR  35
      3         VA(LRES),VA(LDLE))                                   BLSR  36
            RETURN                                                   BLSR  37
            END                                                      BLSR  38

            SUBROUTINE EXSOLR(KLD,VDIMP,KLOCE,VCORE,VPRNE,VPREE,KNE,VKE,VFE, EXSR  1
      1     VKGS,VKGD,VKGI,VFG,VCORG,KDLNC,KNEQ,VRES,VDLE)           EXSR  2
C=================================================================EXSR  3
C     TO EXECUTE BLOCK 'SOLR'                                       EXSR  4
C     ASSEMBLE DISTRIBUTED LOADS (ELEMENT FUNCTION 7)                EXSR  5
C=================================================================EXSR  6
            IMPLICIT REAL*8(A-H,O-Z)                                 EXSR  7
            COMMON/ASSE/NSYM,NKG,NKE,NDLE                            EXSR  8
            COMMON/RESO/NEQ,NRES                                     EXSR  9
            COMMON/ES/M,MR,MP,M1,M2                                  EXSR 10
            DIMENSION KLD(1),VDIMP(1),KLOCE(1),VCORE(1),VPRNE(1),VPREE(1), EXSR 11
      1     KNE(1),VKE(1),VFE(1),VKGS(1),VKGD(1),VKGI(1),VFG(1),VCORG(1), EXSR 12
      2     KDLNC(1),KNEQ(1),VRES(1),VDLE(1)                         EXSR 13
C-----------------------------------------------------------------EXSR 14
C----- ASSEMBLE FG                                                  EXSR 15
            CALL ASFG(KLD,VDIMP,KLOCE,VCORE,VPRNE,VPREE,KNE,VKE,VFE,VKGS, EXSR 16
      1     VKGD,VKGI,VFG,VDLE,VRES)                                 EXSR 17
C------- OUTPUT                                                     EXSR 18
            IF(M.GE.1) WRITE(MP,2000) (VFG(I),I=1,NEQ)               EXSR 19
 2000     FORMAT(/' GLOBAL LOAD VECTOR  (FG)'/(1X,10E12.5))          EXSR 20
            RETURN                                                   EXSR 21
            END                                                      EXSR 22

            SUBROUTINE ASFG(KLD,VDIMP,KLOCE,VCORE,VPRNE,VPREE,KNE,VKE,VFE, ASFG  1
      1     VKGS,VKGD,VKGI,VFG,VDLE,VRES)                            ASFG  2
C=================================================================ASFG  3
C     ASSEMBLING DISTRIBUTED LOADS IN FG                            ASFG  4
C=================================================================ASFG  5
            IMPLICIT REAL*8(A-H,O-Z)                                 ASFG  6
            COMMON/ELEM/NELT,NNEL,NTPE,NGRE,ME,NIDENT                ASFG  7
            COMMON/ASSE/NSYM                                         ASFG  8
            COMMON/RESO/NEQ                                          ASFG  9
            COMMON/RGDT/IEL,ITPE,ITPE1,IGRE,IDLE,ICE,IPRNE,IPREE,INEL,IDEG,IPGASFG 10
      1     ,ICOD                                                    ASFG 11
            COMMON/ES/M,MR,MP,M1,M2                                  ASFG 12
            DIMENSION KLD(1),VDIMP(1),KLOCE(1),VCORE(1),VPRNE(1),VPREE(1), ASFG 13
      1     KNE(1),VKE(1),VFE(1),VKGS(1),VKGD(1),VKGI(1),VFG(1),VDLE(1), ASFG 14
      2     VRES(1)                                                  ASFG 15
C-----------------------------------------------------------------ASFG 16
C------- REWIND ELEMENT FILE M2                                     ASFG 17
            REWIND M2                                                ASFG 18
C------- LOOP OVER THE ELEMENTS                                     ASFG 19
            DO 20 IE=1,NELT                                          ASFG 20
C------- READ AN ELEMENT FROM FILE M2                               ASFG 21
            CALL RDELEM(M2,KLOCE,VCORE,VPRNE,VPREE,KNE)              ASFG 22
C------- EVALUATE INTERPOLATION FUNCTIONS IF REQUIRED                ASFG 23
            IF(ITPE.EQ.ITPE1) GO TO 10                               ASFG 24
            ICOD=2                                                   ASFG 25
            CALL ELEMLB(VCORE,VPRNE,VPREE,VDLE,VKE,VFE)              ASFG 26
C------- EVALUATE ELEMENT VECTOR                                    ASFG 27
 10       ICOD=7                                                     ASFG 28
            CALL ELEMLB(VCORE,VPRNE,VPREE,VDLE,VKE,VFE)              ASFG 29
C------- PRINT ELEMENT VECTOR VFE                                   ASFG 30
            IF(M.GE.2) WRITE(MP,2000) IEL,(VFE(I),I=1,IDLE)          ASFG 31
 2000     FORMAT(/' VECTOR (FE) , ELEMENT:',I5/(10X,10E12.5))        ASFG 32
C------- ASSEMBLE                                                   ASFG 33
            CALL ASSEL(0,1,IDLE,NSYM,KLOCE,KLD,VKE,VFE,VKGS,VKGD,VKGI,VFG) ASFG 34
 20       ITPE1=ITPE                                                 ASFG 35
            RETURN                                                   ASFG 36
            END                                                      ASFG 37
```

Figure 6.21 Block 'SOLR'.

6.5.3.3 Block 'LINM'

We now describe an execution functional block for the solution of linear systems.

The basic structure described in paragraph 6.5.3.1 is greatly simplified since only one method of solution is necessary.

Function

This block assembles the global matrices, solves the resulting system of equations and outputs the results for an in-core problem.

Flow chart

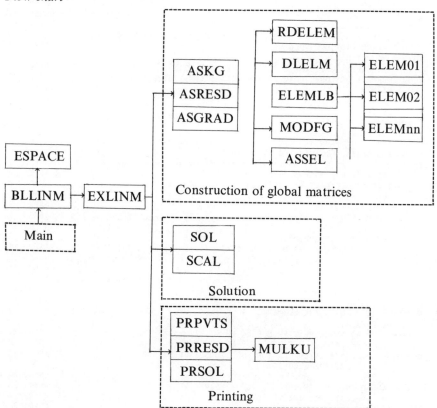

Subroutines

BLLINM: reads a record containing an index NRES of the residue computation, then creates arrays VKGS, VKGI, VFG, VKE, VFE, VRES, VDLE (Figure 6.22).

ESPACE: (Figure 6.5).

EXLINM: assembles matrix KG after calling ASKG, then solves the linear system with a call to SOL. Finally, EXLINM computes and outputs the following values:

— global matrix pivots (PRPVTS);
— solution residuals if NRES equals one (PRRESD);
— the global solution (PRSOL);
— the gradients (or stresses) (ASGRAD);
— the equilibrium residuals and reactions (ASRESD).

EXLINM is listed in Figure 6.22.

ASKG: assembles global matrix (Figure 6.22).
ASGRAD: computes gradients at numerical integration points (Figure 6.22).
ASRESD: assembles residuals and gradients (Figure 6.22).
ELEMLB: chooses routine ELEM corresponding to the type of element desired (Figure 6.22).
ASSEL: assembles an element matrix or vector (Figure 6.22).
MODFG: modifies vector FG to account for non-zero specified value of a variable at a node (Figure 6.22).
PRPVTS: outputs the pivots of global matrix (Figure 6.22).
PRRESD: outputs the solution residuals

$$[K] \cdot \{U\} - F \quad \text{(Figure 6.22)}.$$

PRSOL: outputs the solution $\{U\}$ (Figure 6.22).
DLELM: extracts the degrees of freedom of a given element (Figure 6.22).
MULKU: multiplies global matrix by a vector (Figure 6.22).
ELEM01: element subroutine for one; two-, and three-dimensional quadratic elements corresponding to the quasi-harmonic equation of mathematical physics (Figure 4.4).
ELEM02: element subroutine for an eight-node plane elasticity quadrilateral element (Figure 4.5).
SOL,
SCAL: solve linear systems (Figure 5.15).
RDELEM: (Figure 6.18).

Remarks

Block 'LINM' could be split into several independent blocks, for example:

'ASKG'   for the assembly process;
'RESO'   for the solution of linear systems;
'RESD'   for the computation of equilibrium residuals and reactions;
'GRAD'   for the computation of gradients and stresses.

The structure of 'LINM' is valid for all linear problems (elasticity, irrotational inviscid flows, heat conduction, etc.) using any element of any type (defined by their shape and interpolation functions). Subroutines ELEMnn must be added for problems other than those already included.

```
      SUBROUTINE BLLINM                                              BLLM   1
C===========================================================================BLLM   2
C     TO CALL BLOCK 'LINM'                                           BLLM   3
C     ASSEMBLE AND SOLVE A LINEAR PROBLEM IN CORE                    BLLM   4
C===========================================================================BLLM   5
      IMPLICIT REAL*8(A-H,O-Z)                                       BLLM   6
      REAL*4 TBL                                                     BLLM   7
      COMMON/COOR/NDIM,NNT,NDLN,NDLT                                 BLLM   8
      COMMON/ELEM/NUL(4),ME                                          BLLM   9
      COMMON/ASSE/NSYM,NKG,NKE,NDLE                                  BLLM  10
      COMMON/RESO/NEQ,NRES,MRES                                      BLLM  11
      COMMON/ES/M,MR,MP,M1,M2,M3                                     BLLM  12
      COMMON/LOC/LCORG,LDLNC,LNEQ,LDIMP,LPRNG,LPREG,LLD,LLOCE,LCORE,LNE,BLLM  13
     1 LPRNE,LPREE,LDLE,LKE,LFE,LKGS,LKGD,LKGI,LFG,LRES,LDLG         BLLM  14
      COMMON VA(1)                                                   BLLM  15
      DIMENSION TBL(8)                                               BLLM  16
      DATA TBL/4HKGS ,4HKGD ,4HKGI ,4HFG  ,4HKE  ,4HFE  ,4HRES ,4HDLE / BLLM  17
C---------------------------------------------------------------------------BLLM  18
      IF(M1.EQ.0) M1=MR                                              BLLM  19
      IF(M2.EQ.0) M2=ME                                              BLLM  20
      IF(M3.EQ.0) M3=MRES                                            BLLM  21
      READ(M1,1000) IN                                               BLLM  22
 1000 FORMAT(1I5)                                                    BLLM  23
      IF(IN.NE.0) NRES=1                                             BLLM  24
      WRITE(MP,2000) M,NRES                                          BLLM  25
 2000 FORMAT(//' ASSEMBLING AND LINEAR SOLUTION (M=',I2,')'/' ',30('=')/BLLM  26
     1 15X,' INDEX FOR RESIDUAL COMPUTATION      (NRES)=',I5)        BLLM  27
      IF(LKGS.EQ.1) CALL ESPACE(NKG,1,TBL(1),LKGS)                   BLLM  28
      IF(LKGD.EQ.1) CALL ESPACE(NEQ,1,TBL(2),LKGD)                   BLLM  29
      IF(NSYM.EQ.1.AND.LKGI.EQ.1) CALL ESPACE(NKG,1,TBL(3),LKGI)     BLLM  30
      IF(LFG.EQ.1) CALL ESPACE(NEQ,1,TBL(4),LFG)                     BLLM  31
      IF(LKE.EQ.1) CALL ESPACE(NKE,1,TBL(5),LKE)                     BLLM  32
      IF(LFE.EQ.1) CALL ESPACE(NDLE,1,TBL(6),LFE)                    BLLM  33
      IF(LRES.EQ.1) CALL ESPACE(NDLT,1,TBL(7),LRES)                  BLLM  34
      IF(LDLE.EQ.1) CALL ESPACE(NDLE,1,TBL(8),LDLE)                  BLLM  35
      CALL EXLINM(VA(LLD),VA(LDIMP),VA(LLOCE),VA(LCORE),VA(LPRNE),   BLLM  36
     1            VA(LPREE),VA(LNE),VA(LKE),VA(LFE),VA(LKGS),VA(LKGD),BLLM 37
     2            VA(LKGI),VA(LFG),VA(LCORG),VA(LDLNC),VA(LNEQ),     BLLM  38
     3            VA(LRES),VA(LDLE))                                 BLLM  39
      RETURN                                                         BLLM  40
      END                                                            BLLM  41

      SUBROUTINE EXLINM(KLD,VDIMP,KLOCE,VCORE,VPRNE,VPREE,KNE,VKE,VFE, EXLM  1
     1 VKGS,VKGD,VKGI,VFG,VCORG,KDLNC,KNEQ,VRES,VDLE)                EXLM   2
C===========================================================================EXLM   3
C     TO EXECUTE BLOCK 'LINM'                                        EXLM   4
C     ASSEMBLE AND SOLVE A LINEAR PROBLEM IN CORE                    EXLM   5
C===========================================================================EXLM   6
      IMPLICIT REAL*8(A-H,O-Z)                                       EXLM   7
      COMMON/ASSE/NSYM,NKG,NKE,NDLE                                  EXLM   8
      COMMON/RESO/NEQ,NRES,MRES                                      EXLM   9
      COMMON/ES/M,MR,MP,M1,M2,M3                                     EXLM  10
      DIMENSION KLD(1),VDIMP(1),KLOCE(1),VCORE(1),VPRNE(1),VPREE(1), EXLM  11
     1 KNE(1),VKE(1),VFE(1),VKGS(1),VKGD(1),VKGI(1),VFG(1),VCORG(1), EXLM  12
     2 KDLNC(1),KNEQ(1),VRES(1),VDLE(1)                              EXLM  13
C---------------------------------------------------------------------------EXLM  14
      REWIND M3                                                      EXLM  15
C                                                                    EXLM  16
C------ ASSEMBLE KG                                                   EXLM  17
C                                                                    EXLM  18
C------ SAVE UNMODIFIED VECTOR FG (BY B.C.) ON FILE M3                EXLM  19
      WRITE(M3) (VFG(I),I=1,NEQ)                                     EXLM  20
      IF(M.GE.2) WRITE(MP,2000) (VFG(I),I=1,NEQ)                     EXLM  21
```

Figure 6.22 (*Contd.*)

```
2000  FORMAT(/' GLOBAL LOAD VECTOR NON MODIFIED BY B.C. (FG)'          EXLM 22
     1/(1X,10E12.5))                                                    EXLM 23
C------- ASSEMBLE KG, MODIFY FG FOR THE B.C. AND SAVE THEM              EXLM 24
      CALL ASKG(KLD,VDIMP,KLOCE,VCORE,VPRNE,VPREE,KNE,VKE,VFE,VKGS,     EXLM 25
     1 VKGD,VKGI,VFG,VDLE,VRES)                                         EXLM 26
      WRITE(M3) (VFG(I),I=1,NEQ)                                        EXLM 27
      WRITE(M3) (VKGS(I),I=1,NKG),(VKGD(I),I=1,NEQ)                     EXLM 28
      IF(NSYM.EQ.1) WRITE(M3) (VKGI(I),I=1,NKG)                         EXLM 29
C------- PRINT KG AND FG                                                EXLM 30
      IF(M.LT.2) GO TO 20                                               EXLM 31
      WRITE(MP,2005) (VKGS(I),I=1,NKG)                                  EXLM 32
2005  FORMAT(/' GLOBAL MATRIX (KG)'/'     UPPER TRIANGLE'/              EXLM 33
     1 (1X,10E12.5))                                                    EXLM 34
      WRITE(MP,2010) (VKGD(I),I=1,NEQ)                                  EXLM 35
2010  FORMAT('    DIAGONAL'/(1X,10E12.5))                               EXLM 36
      IF(NSYM.EQ.1) WRITE(MP,2020) (VKGI(I),I=1,NKG)                    EXLM 37
2020  FORMAT('    LOWER TRIANGLE'/(1X,10E12.5))                         EXLM 38
      WRITE(MP,2030) (VFG(I),I=1,NEQ)                                   EXLM 39
2030  FORMAT(/' GLOBAL LOAD VECTOR MODIFIED BY THE B.C. (FG)'           EXLM 40
     1 /(1X,10E12.5))                                                   EXLM 41
C                                                                       EXLM 42
C------- SOLVE                                                          EXLM 43
C                                                                       EXLM 44
20    CALL SOL(VKGS,VKGD,VKGI,VFG,KLD,NEQ,MP,1,1,NSYM,ENERG)            EXLM 45
      IF(NSYM.NE.1) WRITE(MP,2035) ENERG                                EXLM 46
2035  FORMAT(15X,'ENERGY     (ENERG)=',1E12.5)                          EXLM 47
      IF(M.LT.2) GO TO 30                                               EXLM 48
      WRITE(MP,2040) (VKGS(I),I=1,NKG)                                  EXLM 49
2040  FORMAT(/' TRIANGULARIZED MATRIX (KG)'/'     UPPER TRIANGLE'/      EXLM 50
     1 (1X,10E12.5))                                                    EXLM 51
      WRITE(MP,2010) (VKGD(I),I=1,NEQ)                                  EXLM 52
      IF(NSYM.EQ.1) WRITE(MP,2020) (VKGI(I),I=1,NKG)                    EXLM 53
C------- PIVOTS OF KG AND DETERMINANT                                   EXLM 54
30    CALL PRPVTS(VKGD)                                                 EXLM 55
C------- EVALUATE AND PRINT RESIDUAL VECTOR K.U - F                     EXLM 56
      IF(NRES.EQ.1) CALL PRRESD(VKGS,VKGD,VKGI,VFG,KLD,VRES)            EXLM 57
C------- PRINT THE SOLUTION                                             EXLM 58
      WRITE(MP,2050)                                                    EXLM 59
2050  FORMAT(//' SOLUTION'//)                                           EXLM 60
      CALL PRSOL(KDLNC,VCORG,VDIMP,KNEQ,VFG)                            EXLM 61
C                                                                       EXLM 62
C------- EVALUATE AND PRINT GRADIENTS (STRESSES)                        EXLM 63
C                                                                       EXLM 64
      CALL ASGRAD(KLD,VDIMP,KLOCE,VCORE,VPRNE,VPREE,KNE,VKE,VFE,VKGS,   EXLM 65
     1 VKGD,VKGI,VFG,VDLE,VRES)                                         EXLM 66
C                                                                       EXLM 67
C------- EVALUATE AND PRINT EQUILIBRIUM RESIDUAL VECTOR                 EXLM 68
C                                                                       EXLM 69
C------- READ VECTOR FG AND CHANGE SIGN                                 EXLM 70
      REWIND M3                                                         EXLM 71
      READ(M3) (VRES(I),I=1,NEQ)                                        EXLM 72
      DO 40 I=1,NEQ                                                     EXLM 73
40    VRES(I)=-VRES(I)                                                  EXLM 74
C------- ASSEMBLE THE RESIDUALS                                         EXLM 75
      CALL ASRESD(1,1,KLD,VDIMP,KLOCE,VCORE,VPREE,KNE,VKE,VFE,          EXLM 76
     1 VKGS,VKGD,VKGI,VFG,VDLE,VRES,VRES(NEQ+1))                        EXLM 77
C------- PRINT THE RESIDUALS                                            EXLM 78
      WRITE(MP,2060)                                                    EXLM 79
2060  FORMAT(//' EQUILIBRIUM RESIDUALS AND REACTIONS'//)                EXLM 80
      CALL PRSOL(KDLNC,VCORG,VRES(NEQ+1),KNEQ,VRES)                     EXLM 81
      RETURN                                                            EXLM 82
      END                                                               EXLM 83
```

Figure 6.22 (*Contd.*)

```
      SUBROUTINE ASKG(KLD,VDIMP,KLOCE,VCORE,VPRNE,VPREE,KNE,VKE,VFE,    ASKG   1
     1 VKGS,VKGD,VKGI,VFG,VDLE,VRES)                                    ASKG   2
C=======================================================================ASKG   3
C        TO ASSEMBLE GLOBAL MATRIX KG (ELEMENT FUNCTION 3)              ASKG   4
C           TAKING INTO ACCOUNT OF NON ZERO PRESCRIBED D.O.F.           ASKG   5
C=======================================================================ASKG   6
      IMPLICIT REAL*8(A-H,O-Z)                                          ASKG   7
      COMMON/COND/NCLT,NCLZ,NCLNZ                                       ASKG   8
      COMMON/ELEM/NELT,NNEL,NTPE,NGRE,ME,NIDENT                         ASKG   9
      COMMON/ASSE/NSYM                                                  ASKG  10
      COMMON/RESO/NEQ                                                   ASKG  11
      COMMON/RGDT/IEL,ITPE,ITPE1,IGRE,IDLE,ICE,IPRNE,IPREE,INEL,IDEG,IPGASKG 12
     1 ,ICOD                                                            ASKG  13
      COMMON/ES/M,MR,MP,M1,M2                                           ASKG  14
      DIMENSION KLD(1),VDIMP(1),KLOCE(1),VCORE(1),VPRNE(1),VPREE(1),    ASKG  15
     1 KNE(1),VKE(1),VFE(1),VKGS(1),VKGD(1),VKGI(1),VFG(1),VDLE(1),     ASKG  16
     2 VRES(1),KEB(1)                                                   ASKG  17
C----------------------------------------------------------------------ASKG  18
C------- REWIND ELEMENT FILE (M2)                                       ASKG  19
      REWIND M2                                                         ASKG  20
C------- LOOP OVER THE ELEMENTS                                         ASKG  21
      DO 30 IE=1,NELT                                                   ASKG  22
C------- READ AN ELEMENT ON FILE M2                                     ASKG  23
      CALL RDELEM(M2,KLOCE,VCORE,VPRNE,VPREE,KNE)                       ASKG  24
C------- SKIP COMPUTATION IF IDENTICAL ELEMENTS ENCOUNTERED             ASKG  25
      IF(NIDENT.EQ.1.AND.IE.GT.1) GO TO 20                              ASKG  26
C------- EVALUATE INTERPOLATION FUNCTIONS IF REQUIRED                   ASKG  27
      IF(ITPE.EQ.ITPE1) GO TO 10                                        ASKG  28
      ICOD=2                                                            ASKG  29
      CALL ELEMLB(VCORE,VPRNE,VPREE,VDLE,VKE,VFE)                       ASKG  30
C------- FORM ELEMENT MATRIX                                            ASKG  31
 10   ICOD=3                                                            ASKG  32
      CALL ELEMLB(VCORE,VPRNE,VPREE,VDLE,VKE,VFE)                       ASKG  33
C------- PRINT ELEMENT MATRIX                                           ASKG  34
      IF(M.LT.2) GO TO 20                                               ASKG  35
      IF(NSYM.EQ.0) IKE=IDLE*(IDLE+1)/2                                 ASKG  36
      IF(NSYM.EQ.1) IKE=IDLE*IDLE                                       ASKG  37
      WRITE(MP,2000) IEL,(VKE(I),I=1,IKE)                               ASKG  38
 2000 FORMAT(/' MATRIX (KE) , ELEMENT:',I5/(10X,10E12.5))               ASKG  39
C------- MODIFY FG FOR NON ZERO PRESCRIBED D.O.F.                       ASKG  40
 20   IF(NCLNZ.NE.0) CALL MODFG(IDLE,NSYM,KLOCE,VDIMP,VKE,VFG)          ASKG  41
C------- ASSEMBLE                                                       ASKG  42
      CALL ASSEL(1,0,IDLE,NSYM,KLOCE,KLD,VKE,VFE,VKGS,VKGD,VKGI,VFG)    ASKG  43
 30   ITPE1=ITPE                                                        ASKG  44
      RETURN                                                            ASKG  45
      END                                                               ASKG  46

      SUBROUTINE ASGRAD(KLD,VDIMP,KLOCE,VCORE,VPRNE,VPREE,KNE,VKE,VFE,  ASGR   1
     1 VKGS,VKGD,VKGI,VFG,VDLE,VRES)                                    ASGR   2
C=======================================================================ASGR   3
C        TO EVALUATE AND PRINT GRADIENTS (STRESSES) AT ELEMENT G.P.     ASGR   4
C          (ELEMENT FUNCTION 8)                                         ASGR   5
C=======================================================================ASGR   6
      IMPLICIT REAL*8(A-H,O-Z)                                          ASGR   7
      COMMON/ELEM/NELT,NNEL,NTPE,NGRE,ME,NIDENT                         ASGR   8
      COMMON/ASSE/NSYM                                                  ASGR   9
      COMMON/RESO/NEQ                                                   ASGR  10
      COMMON/RGDT/IEL,ITPE,ITPE1,IGRE,IDLE,ICE,IPRNE,IPREE,INEL,IDEG,IPGASGR 11
     1 ,ICOD                                                            ASGR  12
      COMMON/ES/M,MR,MP,M1,M2                                           ASGR  13
      DIMENSION KLD(1),VDIMP(1),KLOCE(1),VCORE(1),VPRNE(1),VPREE(1),    ASGR  14
     1 KNE(1),VKE(1),VFE(1),VKGS(1),VKGD(1),VKGI(1),VFG(1),VDLE(1),     ASGR  15
     2 VRES(1)                                                          ASGR  16
```

Figure 6.22 (*Contd.*)

```
C-----------------------------------------------------------------------ASGR  17
C------- REWIND ELEMENTS FILE (M2)                                      ASGR  18
      REWIND M2                                                         ASGR  19
C------- LOOP OVER THE ELEMENTS                                         ASGR  20
      DO 20 IE=1,NELT                                                   ASGR  21
C------- READ THE ELEMENT                                               ASGR  22
      CALL RDELEM(M2,KLOCE,VCORE,VPRNE,VPREE,KNE)                       ASGR  23
C------- EVALUATE INTERPOLATION FUNCTION IF REQUIRED                    ASGR  24
      IF(ITPE.EQ.ITPE1) GO TO 10                                        ASGR  25
      ICOD=2                                                            ASGR  26
      CALL ELEMLB(VCORE,VPRNE,VPREE,VDLE,VKE,VFE)                       ASGR  27
C------- FIND ELEMENT D.O.F.                                            ASGR  28
10    CALL DLELM(KLOCE,VFG,VDIMP,VDLE)                                  ASGR  29
C------- COMPUTE AND PRINT STRESSES OR GRADIENTS                        ASGR  30
      ICOD=8                                                            ASGR  31
      CALL ELEMLB(VCORE,VPRNE,VPREE,VDLE,VKE,VFE)                       ASGR  32
20    ITPE1=ITPE                                                        ASGR  33
      RETURN                                                            ASGR  34
      END                                                               ASGR  35

      SUBROUTINE ASRESD(IRESD,IREAC,KLD,VDIMP,KLOCE,VCORE,VPRNE,VPREE,  ASRE   1
     1 KNE,VKE,VFE,VKGS,VKGD,VKGI,VFG,VDLE,VRES,VREAC)                  ASRE   2
C======================================================================ASRE   3
C     TO ASSEMBLE INTERNAL RESIDUALS IN VRES (IF IRESD .EQ.1)           ASRE   4
C     AND EXTERNAL REACTIONS IN VREAC (IF IREAC.EQ.1)                   ASRE   5
C======================================================================ASRE   6
      IMPLICIT REAL*8(A-H,O-Z)                                          ASRE   7
      COMMON/ELEM/NELT,NNEL,NTPE,NGRE,ME,NIDENT                         ASRE   8
      COMMON/ASSE/NSYM                                                  ASRE   9
      COMMON/RESO/NEQ                                                   ASRE  10
      COMMON/RGDT/IEL,ITPE,ITPE1,IGRE,IDLE,ICE,IPRNE,IPREE,INEL,IDEG,IPGASRE  11
     1 ,ICOD                                                            ASRE  12
      COMMON/ES/M,MR,MP,M1,M2                                           ASRE  13
      DIMENSION KLD(1),VDIMP(1),KLOCE(1),VCORE(1),VPRNE(1),VPREE(1),    ASRE  14
     1 KNE(1),VKE(1),VFE(1),VKGS(1),VKGD(1),VKGI(1),VFG(1),VDLE(1),     ASRE  15
     2 VRES(1),VREAC(1)                                                 ASRE  16
C-----------------------------------------------------------------------ASRE  17
C------- REWIND ELEMENT FILE (M2)                                       ASRE  18
      REWIND M2                                                         ASRE  19
C------- LOOP OVER THE ELEMENTS                                         ASRE  20
      DO 60 IE=1,NELT                                                   ASRE  21
C------- READ AN ELEMENT ON FILE M2                                     ASRE  22
      CALL RDELEM(M2,KLOCE,VCORE,VPRNE,VPREE,KNE)                       ASRE  23
C------- EVALUATE INTERPOLATION FUNCTION IF REQUIRED                    ASRE  24
      IF(ITPE.EQ.ITPE1) GO TO 10                                        ASRE  25
      ICOD=2                                                            ASRE  26
      CALL ELEMLB(VCORE,VPRNE,VPREE,VDLE,VKE,VFE)                       ASRE  27
C------- FIND ELEMENT D.O.F.                                            ASRE  28
10    CALL DLELM(KLOCE,VFG,VDIMP,VDLE)                                  ASRE  29
C------- EVALUATE ELEMENT REACTIONS                                     ASRE  30
      ICOD=6                                                            ASRE  31
      CALL ELEMLB(VCORE,VPRNE,VPREE,VDLE,VKE,VFE)                       ASRE  32
C------- PRINT ELEMENT REACTIONS                                        ASRE  33
      IF(M.GE.2) WRITE(MP,2000) IEL,(VFE(I),I=1,IDLE)                   ASRE  34
2000  FORMAT(/' REACTIONS (FE)  , ELEMENT:',I5/(10X,10E12.5))           ASRE  35
      IF(IRESD.NE.1) GO TO 20                                           ASRE  36
C------- ASSEMBLE INTERNAL RESIDUALS                                    ASRE  37
      CALL ASSEL(0,1,IDLE,NSYM,KLOCE,KLD,VKE,VFE,VKGS,VKGD,VKGI,VRES)   ASRE  38
20    IF(IREAC.NE.1) GO TO 60                                           ASRE  39
C------- ASSEMBLE EXTERNAL REACTIONS                                    ASRE  40
C          MODIFY TERMS IN KLOCE SUCH THAT PRESCRIBED D.O.F. ARE THE ONLYASRE  41
C          ASSEMBLED ONES                                               ASRE  42
      DO 50 ID=1,IDLE                                                   ASRE  43
      IF(KLOCE(ID)) 30,50,40                                            ASRE  44
```

Figure 6.22 (*Contd.*)

```
30    KLOCE(ID)=-KLOCE(ID)                                              ASRE 45
      GO TO 50                                                          ASRE 46
40    KLOCE(ID)=0                                                       ASRE 47
50    CONTINUE                                                          ASRE 48
      CALL ASSEL(0,1,IDLE,NSYM,KLOCE,KLD,VKE,VFE,VKGS,VKGD,VKGI,VREAC)   ASRE 49
60    ITPE1=ITPE                                                        ASRE 50
      RETURN                                                            ASRE 51
      END                                                               ASRE 52

      SUBROUTINE ELEMLB(VCORE,VPRNE,VPREE,VDLE,VKE,VFE)                  ELLB  1
C=======================================================================ELLB  2
C     TO COMPUTE ELEMENT INFORMATIONS FOR ALL TYPES OF ELEMENTS          ELLB  3
C=======================================================================ELLB  4
      IMPLICIT REAL*8(A-H,O-Z)                                           ELLB  5
      COMMON/RGDT/IEL,ITPE                                               ELLB  6
      DIMENSION VCORE(1),VPRNE(1),VPREE(1),VDLE(1),VKE(1),VFE(1)         ELLB  7
C----------------------------------------------------------------------- ELLB  8
      GO TO ( 10, 20, 30, 40, 50, 60, 70, 80, 90,100),ITPE               ELLB  9
C------    ELEMENT OF TYPE 1                                             ELLB 10
10    CALL ELEM01(VCORE,VPRNE,VPREE,VDLE,VKE,VFE)                        ELLB 11
      GO TO 900                                                          ELLB 12
C------    ELEMENT OF TYPE 2                                             ELLB 13
20    CALL ELEM02(VCORE,VPRNE,VPREE,VDLE,VKE,VFE)                        ELLB 14
      GO TO 900                                                          ELLB 15
C------    ELEMENT OF TYPE 3                                             ELLB 16
30    CALL ELEM03(VCORE,VPRNE,VPREE,VDLE,VKE,VFE)                        ELLB 17
      GO TO 900                                                          ELLB 18
C------    ELEMENT OF TYPE 4                                             ELLB 19
40    CALL ELEM04(VCORE,VPRNE,VPREE,VDLE,VKE,VFE)                        ELLB 20
      GO TO 900                                                          ELLB 21
C------    ELEMENT OF TYPE 5                                             ELLB 22
50    CALL ELEM05(VCORE,VPRNE,VPREE,VDLE,VKE,VFE)                        ELLB 23
      GO TO 900                                                          ELLB 24
C------    ELEMENT OF TYPE 6                                             ELLB 25
60    CALL ELEM06(VCORE,VPRNE,VPREE,VDLE,VKE,VFE)                        ELLB 26
      GO TO 900                                                          ELLB 27
C------    ELEMENT OF TYPE 7                                             ELLB 28
70    CALL ELEM07(VCORE,VPRNE,VPREE,VDLE,VKE,VFE)                        ELLB 29
      GO TO 900                                                          ELLB 30
C------    ELEMENT OF TYPE 8                                             ELLB 31
80    CALL ELEM08(VCORE,VPRNE,VPREE,VDLE,VKE,VFE)                        ELLB 32
      GO TO 900                                                          ELLB 33
C------    ELEMENT OF TYPE 9                                             ELLB 34
90    CALL ELEM09(VCORE,VPRNE,VPREE,VDLE,VKE,VFE)                        ELLB 35
      GO TO 900                                                          ELLB 36
C------    ELEMENT OF TYPE 10                                            ELLB 37
100   CALL ELEM10(VCORE,VPRNE,VPREE,VDLE,VKE,VFE)                        ELLB 38
      GO TO 900                                                          ELLB 39
C------    OTHER ELEMENTS                                                ELLB 40
C         ........                                                       ELLB 41
900   RETURN                                                             ELLB 42
      END                                                                ELLB 43
```

Figure 6.22 (*Contd.*)

```
      SUBROUTINE ASSEL(IKG,IFG,IDLE,NSYM,KLOCE,KLD,VKE,VFE,VKGS,        ASSE   1
     1 VKGD,VKGI,VFG)                                                   ASSE   2
C=====================================================================ASSE   3
C        TO ASSEMBLE AN ELEMENT MATRIX AND/OR VECTOR                    ASSE   4
C        (MATRIX SYMMETRICAL OR NOT)                                    ASSE   5
C          INPUT                                                        ASSE   6
C             IKG      IF IKG.EQ.1 ASSEMBLE ELEMENT MATRIX KE            ASSE   7
C             IFG      IF IFG.EQ.1 ASSEMBLE ELEMENT VECTOR FE            ASSE   8
C             IDLE     ELEMENT NUMBER OF D.O.F.                         ASSE   9
C             NSYM     0=SYMMETRIC PROBLEM, 1=UNSYMMETRIC PROBLEM        ASSE  10
C             KLOCE    ELEMENT LOCALIZATION VECTOR                      ASSE  11
C             KLD      CUMULATIVE COLUMN HEIGHTS OF KG                  ASSE  12
C             VKE      ELEMENT MATRIX KE (FULL OR UPPER TRIANGLE BY     ASSE  13
C                      DESCENDING COLUMNS)                              ASSE  14
C             VFE      ELEMENT VECTOR FE                                ASSE  15
C          OUTPUT                                                       ASSE  16
C             VKGS,VKGD,VKGI    GLOBAL MATRIX (SKYLINES)                ASSE  17
C                               (SYMMETRIC OR NOT)                      ASSE  18
C             VFG      GLOBAL LOAD VECTOR                               ASSE  19
C=====================================================================ASSE  20
      IMPLICIT REAL*8(A-H,O-Z)                                          ASSE  21
      DIMENSION KLOCE(1),KLD(1),VKE(1),VFE(1),VKGS(1),VKGD(1),          ASSE  22
     1 VKGI(1),VFG(1)                                                   ASSE  23
C---------------------------------------------------------------------ASSE  24
C                                                                       ASSE  25
C------- ASSEMBLE ELEMENT MATRIX                                        ASSE  26
C                                                                       ASSE  27
      IF(IKG.NE.1) GO TO 100                                            ASSE  28
      IEQ0=IDLE                                                         ASSE  29
      IEQ1=1                                                            ASSE  30
C------- FOR EACH COLUMN OF KE                                          ASSE  31
      DO 90 JD=1,IDLE                                                   ASSE  32
      IF(NSYM.NE.1) IEQ0=JD                                             ASSE  33
      JL=KLOCE(JD)                                                      ASSE  34
      IF(JL) 90,90,10                                                   ASSE  35
10    I0=KLD(JL+1)                                                      ASSE  36
      IEQ=IEQ1                                                          ASSE  37
      IQ=1                                                              ASSE  38
C------- FOR EACH ROW OF KE                                             ASSE  39
      DO 80 ID=1,IDLE                                                   ASSE  40
      IL=KLOCE(ID)                                                      ASSE  41
      IF(NSYM.EQ.1) GO TO 30                                            ASSE  42
      IF(ID-JD) 30,20,20                                                ASSE  43
20    IQ=ID                                                             ASSE  44
30    IF(IL) 80,80,40                                                   ASSE  45
40    IJ=JL-IL                                                          ASSE  46
      IF(IJ) 70,50,60                                                   ASSE  47
C------- DIAGONAL TERMS OF KG                                           ASSE  48
50    VKGD(IL)=VKGD(IL)+VKE(IEQ)                                        ASSE  49
      GO TO 80                                                          ASSE  50
C------- UPPER TRIANGLE TERMS OF KG                                     ASSE  51
60    I=I0-IJ                                                           ASSE  52
      VKGS(I)=VKGS(I)+VKE(IEQ)                                          ASSE  53
      GO TO 80                                                          ASSE  54
C------- LOWER TRIANGLE TERMS OF KG                                     ASSE  55
70    IF(NSYM.NE.1) GO TO 80                                            ASSE  56
      I=KLD(IL+1)+IJ                                                    ASSE  57
      VKGI(I)=VKGI(I)+VKE(IEQ)                                          ASSE  58
80    IEQ=IEQ+IQ                                                        ASSE  59
90    IEQ1=IEQ1+IEQ0                                                    ASSE  60
C                                                                       ASSE  61
C------- ASSEMBLE ELEMENT LOAD VECTOR                                   ASSE  62
C                                                                       ASSE  63
100   IF(IFG.NE.1) GO TO 130                                            ASSE  64
```

Figure 6.22 (*Contd.*)

```
      DO 120 ID=1,IDLE                                            ASSE 65
      IL=KLOCE(ID)                                                ASSE 66
      IF(IL) 120,120,110                                          ASSE 67
110   VFG(IL)=VFG(IL)+VFE(ID)                                     ASSE 68
120   CONTINUE                                                    ASSE 69
130   RETURN                                                      ASSE 70
      END                                                         ASSE 71

      SUBROUTINE MODFG(IDLE,NSYM,KLOCE,VDIMP,VKE,VFG)              MODF  1
C===============================================================MODF  2
C     TO MODIFY VECTOR FG TO TAKE INTO ACCOUNT OF PRESCRIBED NON ZERO MODF 3
C     D.O.F. FOR A GIVEN ELEMENT                                  MODF  4
C        INPUT                                                    MODF  5
C           IDLE    ELEMENT NUMBER OF D.O.F.                      MODF  6
C           NSYM    0=SYMMETRIC PROBLEM, 1=NON SYMMETRIC PROBLEM   MODF  7
C           KLOCE   ELEMENT LOCALIZATION VECTOR                   MODF  8
C           VDIMP   VALUES OF PRESCRIBED D.O.F.                   MODF  9
C           VKE     ELEMENT MATRIX (FULL OR UPPER TRIANGLE        MODF 10
C                   BY DESCENDING COLUMNS)                        MODF 11
C        OUTPUT                                                   MODF 12
C           VFG     GLOBAL LOAD VECTOR                            MODF 13
C===============================================================MODF 14
      IMPLICIT REAL*8(A-H,O-Z)                                    MODF 15
      DIMENSION KLOCE(1),VDIMP(1),VKE(1),VFG(1)                   MODF 16
      DATA ZERO/0.D0/                                             MODF 17
C---------------------------------------------------------------MODF 18
      IEQ0=IDLE                                                   MODF 19
      IEQ1=1                                                      MODF 20
C------- FOR EACH ROW OF ELEMENT MATRIX                           MODF 21
      DO 50 JD=1,IDLE                                             MODF 22
      IF(NSYM.NE.1) IEQ0=JD                                       MODF 23
      IEQ=IEQ1                                                    MODF 24
      JL=KLOCE(JD)                                                MODF 25
      IQ=1                                                        MODF 26
      IF(JL) 10,50,50                                             MODF 27
10    JL=-JL                                                      MODF 28
      DIMP=VDIMP(JL)                                              MODF 29
      IF(DIMP.EQ.ZERO) GO TO 50                                   MODF 30
C------- FOR EACH COLUMN OF ELEMENT MATRIX                        MODF 31
      DO 40 ID=1,IDLE                                             MODF 32
      IL=KLOCE(ID)                                                MODF 33
      IF(NSYM.EQ.1) GO TO 30                                      MODF 34
      IF(ID-JD) 30,20,20                                          MODF 35
20    IQ=ID                                                       MODF 36
30    IF(IL.GT.0) VFG(IL)=VFG(IL)-VKE(IEQ)*DIMP                   MODF 37
40    IEQ=IEQ+IQ                                                  MODF 38
50    IEQ1=IEQ1+IEQ0                                              MODF 39
      RETURN                                                      MODF 40
      END                                                         MODF 41

      SUBROUTINE PRPVTS(VKGD)                                     PRPV  1
C===============================================================PRPV  2
C     TO EVALUATE AND TO PRINT THE PIVOTS AND THE DETERMINANT OF MATRIX KGPRPV 3
C===============================================================PRPV  4
      IMPLICIT REAL*8(A-H,O-Z)                                    PRPV  5
      COMMON/RESO/NEQ                                             PRPV  6
      COMMON/ES/M,MR,MP                                           PRPV  7
      DIMENSION VKGD(1)                                           PRPV  8
      DATA UN/1.D0/,GROS/1.D38/                                   PRPV  9
      ABS(X)=DABS(X)                                              PRPV 10
C---------------------------------------------------------------PRPV 11
      X1=GROS                                                     PRPV 12
      X2=GROS                                                     PRPV 13
      DET=UN                                                      PRPV 14
```

Figure 6.22 (*Contd*)

```
            IDET=0                                              PRPV  15
C------- PRINT PIVOTS OF MATRIX KG                              PRPV  16
            IF(M.GE.2) WRITE(MP,2000)(VKGD(I),I=1,NEQ)          PRPV  17
 2000   FORMAT(/' GLOBAL MATRIX PIVOTS'/(1X,10E12.5))           PRPV  18
            DO 50 I=1,NEQ                                       PRPV  19
C------- ABSOLUTE VALUE OF MINIMUM PIVOT                        PRPV  20
            X=ABS(VKGD(I))                                      PRPV  21
            IF(X.GT.X1) GO TO 10                                PRPV  22
            X1=X                                                PRPV  23
            I1=I                                                PRPV  24
C------- ALGEBRAIC VALUE OF MINIMUM PIVOT                       PRPV  25
            X=VKGD(I)                                           PRPV  26
 10         IF(X.GT.X2) GO TO 20                                PRPV  27
            X2=X                                                PRPV  28
            I2=I                                                PRPV  29
C------- DETERMINANT (BOUNDS : 10 EXPONENT + OR - 10)           PRPV  30
 20         DET=DET*VKGD(I)                                     PRPV  31
 30         DET1=ABS(DET)                                       PRPV  32
            IF(DET1.LT.1.D10) GO TO 40                          PRPV  33
            DET=DET*1.D-10                                      PRPV  34
            IDET=IDET+10                                        PRPV  35
 40         IF(DET1.GT.1.D-10) GO TO 50                         PRPV  36
            DET=DET*1.D10                                       PRPV  37
            IDET=IDET-10                                        PRPV  38
            GO TO 30                                            PRPV  39
 50         CONTINUE                                            PRPV  40
C------- OUTPUT                                                 PRPV  41
            WRITE(MP,2010) X1,I1,X2,I2,DET,IDET                 PRPV  42
 2010   FORMAT(/15X,'ABSOLUTE VALUE OF MINIMUM PIVOT    =',E12.5,' EQUATIONPRPV 43
       1:',I5   /29X,               'ALGEBRAIC VALUE=',E12.5,' EQUATION:', PRPV 44
       2    I5  /29X,               'DETERMINANT     =',E12.5,' * 10 ** ', PRPV 45
       3  I5/)                                                  PRPV  46
            RETURN                                              PRPV  47
            END                                                 PRPV  48

            SUBROUTINE PRRESD(VKGS,VKGD,VKGI,VFG,KLD,VRES)      PRRE   1
C=====================================================================PRRE 2
C     TO COMPUTE AND PRINT THE RESIDUAL VECTOR K.U - F          PRRE   3
C=====================================================================PRRE 4
            IMPLICIT REAL*8(A-H,O-Z)                            PRRE   5
            COMMON/ASSE/NSYM,NKG                                PRRE   6
            COMMON/RESO/NEQ,NRES,MRES                           PRRE   7
            COMMON/ES/M,MR,MP,M1,M2,M3                          PRRE   8
            DIMENSION VKGS(1),VKGD(1),VKGI(1),VFG(1),KLD(1),VRES(1)  PRRE 9
            DATA ZERO/0.D0/                                     PRRE  10
            ABS(X)=DABS(X)                                      PRRE  11
C-------------------------------------------------------------------PRRE 12
            REWIND M3                                           PRRE  13
C------- SKIP VECTOR FG NON MODIFIED BY B.C. ON FILE M3         PRRE  14
            READ(M3) (VRES(I),I=1,NEQ)                          PRRE  15
C------- READ VECTOR FG MODIFIED BY B.C. AND MATRIX KG          PRRE  16
            READ(M3) (VRES(I),I=1,NEQ)                          PRRE  17
            READ(M3) (VKGS(I),I=1,NKG),(VKGD(I),I=1,NEQ)        PRRE  18
            IF(NSYM.EQ.1) READ(M3) (VKGI(I),I=1,NKG)            PRRE  19
C------- EVALUATE THE RESIDUAL VECTOR                           PRRE  20
            DO 10 I=1,NEQ                                       PRRE  21
 10         VRES(I)=-VRES(I)                                    PRRE  22
            CALL MULKU(VKGS,VKGD,VKGI,KLD,VFG,NEQ,NSYM,VRES)    PRRE  23
            DO 20 I=1,NEQ                                       PRRE  24
 20         VRES(I)=-VRES(I)                                    PRRE  25
            X1=ZERO                                             PRRE  26
            DO 30 I=1,NEQ                                       PRRE  27
            X=ABS(VRES(I))                                      PRRE  28
            IF(X1.GE.X) GO TO 30                                PRRE  29
```

Figure 6.22 (*Contd.*)

```
      X1=X                                                      PRRE  30
      I1=I                                                      PRRE  31
   30 CONTINUE                                                  PRRE  32
      IF(M.GE.2) WRITE(MP,2000) (VRES(I),I=1,NEQ)               PRRE  33
 2000 FORMAT(/' RESIDUALS VECTOR'/(1X,10E12.5))                 PRRE  34
      WRITE(MP,2010) X1,I1                                      PRRE  35
 2010 FORMAT(/' MAX. RESIDUAL VALUE=',E12.5,' EQUATION',I5)     PRRE  36
      RETURN                                                    PRRE  37
      END                                                       PRRE  38

      SUBROUTINE PRSOL(KDLNC,VCORG,VDIMP,KNEQ,VFG)              PRSO   1
C=================================================================PRSO   2
C     TO PRINT THE SOLUTION                                     PRSO   3
C=================================================================PRSO   4
      IMPLICIT REAL*8(A-H,O-Z)                                  PRSO   5
      COMMON/COOR/NDIM,NNT                                      PRSO   6
      COMMON/ES/M,MR,MP                                         PRSO   7
      COMMON/TRVL/V(10),FX(10)                                  PRSO   8
      DIMENSION VDIMP(1),KDLNC(1),VCORG(1),KNEQ(1),VFG(1)       PRSO   9
      DATA RF/4H * /,RL/4H    /,ZERO/0.D0/                      PRSO  10
C----------------------------------------------------------------PRSO  11
      X2=ZERO                                                   PRSO  12
      X3=ZERO                                                   PRSO  13
      WRITE(MP,2000)                                            PRSO  14
 2000 FORMAT(/' NODES',4X,'X',11X,'Y',11X,'Z',10X,'DEGREES OF FREEDOM (*PRSO  15
     1 = PRESCRIBED)'/)                                         PRSO  16
      I2=0                                                      PRSO  17
      DO 50 IN=1,NNT                                            PRSO  18
      I1=I2+1                                                   PRSO  19
      I2=I2+NDIM                                                PRSO  20
      ID1=KDLNC(IN)+1                                           PRSO  21
      ID2=KDLNC(IN+1)                                           PRSO  22
      ID=ID2-ID1+1                                              PRSO  23
      IF(ID2.LT.ID1) GO TO 50                                   PRSO  24
      X1=VCORG(I1)                                              PRSO  25
      IF(NDIM.GE.2) X2=VCORG(I1+1)                              PRSO  26
      IF(NDIM.GE.3) X3=VCORG(I1+2)                              PRSO  27
      J=ID1                                                     PRSO  28
      DO 40 I=1,ID                                              PRSO  29
      JJ=KNEQ(J)                                                PRSO  30
      IF(JJ) 10,20,30                                           PRSO  31
   10 V(I)=VDIMP(-JJ)                                           PRSO  32
      FX(I)=RF                                                  PRSO  33
      GO TO 40                                                  PRSO  34
   20 V(I)=ZERO                                                 PRSO  35
      FX(I)=RF                                                  PRSO  36
      GO TO 40                                                  PRSO  37
   30 V(I)=VFG(JJ)                                              PRSO  38
      FX(I)=RL                                                  PRSO  39
   40 J=J+1                                                     PRSO  40
      WRITE(MP,2010)IN,X1,X2,X3,(V(II),FX(II),II=1,ID)          PRSO  41
 2010 FORMAT(1X,I5,3E12.5,5X,5(E12.5,A4)/47X,5(E12.5,A4))       PRSO  42
   50 CONTINUE                                                  PRSO  43
      RETURN                                                    PRSO  44
      END                                                       PRSO  45

      SUBROUTINE DLELM(KLOCE,VDLG,VDIMP,VDLE)                   DLEL   1
C=================================================================DLEL   2
C     TO GENERATE ELEMENT D.O.F.                                DLEL   3
C=================================================================DLEL   4
      IMPLICIT REAL*8(A-H,O-Z)                                  DLEL   5
      COMMON/RGDT/IEL,INUL(3),IDLE                              DLEL   6
      COMMON/ES/M,MR,MP                                         DLEL   7
      DIMENSION KLOCE(1),VDLG(1),VDIMP(1),VDLE(1)               DLEL   8
```

Figure 6.22 (*Contd.*)

```
      DATA ZERO/0.D0/                                              DLEL   9
C-----------------------------------------------------------------DLEL  10
      DO 40 ID=1,IDLE                                              DLEL  11
      IL=KLOCE(ID)                                                 DLEL  12
      IF(IL) 10,20,30                                              DLEL  13
   10 VDLE(ID)=VDIMP(-IL)                                          DLEL  14
      GO TO 40                                                     DLEL  15
   20 VDLE(ID)=ZERO                                                DLEL  16
      GO TO 40                                                     DLEL  17
   30 VDLE(ID)=VDLG(IL)                                            DLEL  18
   40 CONTINUE                                                     DLEL  19
      IF(M.GE.2) WRITE(MP,2000) IEL,(VDLE(ID),ID=1,IDLE)           DLEL  20
 2000 FORMAT(' DEGREES OF FREEDOM OF ELEMENT ',I5/(1X,10E12.5))    DLEL  21
      RETURN                                                       DLEL  22
      END                                                          DLEL  23

      SUBROUTINE MULKU(VKGS,VKGD,VKGI,KLD,VFG,NEQ,NSYM,VRES)       MULK   1
C=============================================================MULK    2
C     SUBPROGRAM :    MULK   3
C     TO ADD VECTOR RES TO THE PRODUCT OF MATRIX KG AND THE VECTOR FG   MULK   4
C     INPUT                                                        MULK   5
C        VKGS,VKGD,VKGI  MATRIX KG STORED BY SKYLINE    MULK   6
C                        (SYM. OR NON SYM.)                        MULK   7
C        KLD      ARRAY OF ADDRESS OF COLUMN TOP TERMS IN KG       MULK   8
C        VFG      VECTOR FG                                        MULK   9
C        NEQ      ORDER OF VECTORS FG AND RES                      MULK  10
C        NSYM     .EQ.1 IF NON SYMMETRIC PROBLEM                   MULK  11
C        VRES     VECTOR RES                                       MULK  12
C     OUTPUT                                                       MULK  13
C        VRES     VECTOR RES                                       MULK  14
C=============================================================MULK   15
      IMPLICIT REAL*8(A-H,O-Z)                                     MULK  16
      DIMENSION VKGS(1),VKGD(1),VKGI(1),KLD(1),VFG(1),VRES(1)      MULK  17
C-----------------------------------------------------------------MULK  18
C------ FOR EACH COLUMN OF MATRIX KG                               MULK  19
      DO 20 IK=1,NEQ                                               MULK  20
      JHK=KLD(IK)                                                  MULK  21
      JHK1=KLD(IK+1)                                               MULK  22
      LHK=JHK1-JHK                                                 MULK  23
C------ DIAGONAL TERMS                                             MULK  24
      C=VKGD(IK)*VFG(IK)                                           MULK  25
      IF(LHK.LE.0) GO TO 20                                        MULK  26
      I0=IK-LHK                                                    MULK  27
C------ ROW TERMS                                                  MULK  28
      IF(NSYM.NE.1) C=C+SCAL(VKGS(JHK),VFG(I0),LHK)                MULK  29
      IF(NSYM.EQ.1) C=C+SCAL(VKGI(JHK),VFG(I0),LHK)                MULK  30
C------ COLUMN TERMS                                               MULK  31
      J=JHK                                                        MULK  32
      I1=IK-1                                                      MULK  33
      DO 10 IJ=I0,I1                                               MULK  34
      VRES(IJ)=VRES(IJ)+VKGS(J)*VFG(IK)                            MULK  35
   10 J=J+1                                                        MULK  36
   20 VRES(IK)=VRES(IK)+C                                          MULK  37
      RETURN                                                       MULK  38
      END                                                          MULK  39
```

Figure 6.22 Block 'LINM'.

6.5.3.4 Block 'LIND'

Function

This block is similar to block 'LINM'; however, the global matrix is partitioned and stored out of core using the method described in paragraph 4.6.3g. The matrix is contained in file MKG1 and, after being factored by the method of paragraph 5.2.4.2, is written out in file MKG2.

Flow chart

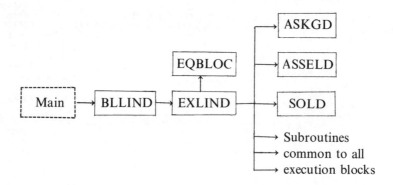

Subroutines

BLLIND: is similar to BLLINM; it reads variables NRES, NLBL and NBLM; creates arrays VKGD, VFG, VKE, VFE, VRES, VDLE, KEB, KPB, VKGS, VKGI (Figure 6.23).
EXLIND: is similar to EXLINM (Figure 6.23).
EQBLOC: constructs arrays KEB and KPB describing the partitions of the global matrix (Figure 6.23).
ASKGD: assembles a global matrix partitioned out of core (Figure 6.23).
SOLD: solves an out of core system (Figure 5.16).
ASSELD: is similar to ASSEL of block 'LINM' but it assembles only for the degrees of freedom included between IE1 and IE2.

Remark

Subroutines ASKGD and SOLD would be more efficient with direct access files instead of sequential files; on the other hand, they would no longer be independent of the type of computer used.

6.5.3.5 Block 'NLIN'

The structure of block 'NLIN' conforms to the one described in paragraph 6.5.3.1. The subroutines used for the construction of global matrices, solution of

```
      SUBROUTINE BLLIND                                              BLLD   1
C===============================================================BLLD   2
C     TO CALL BLOCK 'LIND'                                           BLLD   3
C     TO ASSEMBLE AND TO SOLVE A LINEAR PROBLEM WHEN MATRIX KG IS    BLLD   4
C     STORED BLOCKWISE ON DISK                                       BLLD   5
C===============================================================BLLD   6
      IMPLICIT REAL*8(A-H,O-Z)                                       BLLD   7
      REAL*4 TBL                                                     BLLD   8
      COMMON/COOR/NDIM,NNT,NDLN,NDLT                                 BLLD   9
      COMMON/ELEM/NUL(4),ME                                          BLLD  10
      COMMON/ASSE/NSYM,NKG,NKE,NDLE                                  BLLD  11
      COMMON/RESO/NEQ,NRES,MRES                                      BLLD  12
      COMMON/LIND/NLBL,NBLM,MKG1,MKG2                                BLLD  13
      COMMON/ES/M,MR,MP,M1,M2,M3,M4,M5                               BLLD  14
      COMMON/ALLOC/NVA,IVA,IVAMAX,NREEL                              BLLD  15
      COMMON/LOC/LCORG,LDLNC,LNEQ,LDIMP,LPRNG,LPREG,LLD,LLOCE,LCORE,LNE,BLLD  16
     1 LPRNE,LPREE,LDLE,LKE,LFE,LKGS,LKGD,LKGI,LFG,LRES,LDLG         BLLD  17
      COMMON VA(1)                                                   BLLD  18
      DIMENSION TBL(10),IN(3)                                        BLLD  19
      DATA TBL/4HKGS ,4HKGD ,4HKGI ,4HFG  ,4HKE  ,4HFE  ,4HRES ,4HDLE , BLLD  20
     1 4HEB  ,4HPB  /,DEUX/2.D0/,NBLMAX/100/                         BLLD  21
C---------------------------------------------------------------BLLD  22
C------- FILE NUMBERS                                                BLLD  23
      IF(M1.EQ.0) M1=MR                                              BLLD  24
      IF(M2.EQ.0) M2=ME                                              BLLD  25
      IF(M3.EQ.0) M3=MRES                                            BLLD  26
      IF(M4.EQ.0) M4=MKG1                                            BLLD  27
      IF(M5.EQ.0) M5=MKG2                                            BLLD  28
C------- READ BLOCK PARAMETERS                                       BLLD  29
      READ(M1,1000) IN                                               BLLD  30
 1000 FORMAT(3I5)                                                    BLLD  31
      IF(IN(1).NE.0) NRES=1                                          BLLD  32
      NLBL=IN(2)                                                     BLLD  33
      NBLM=IN(3)                                                     BLLD  34
      WRITE(MP,2000) M,NRES                                          BLLD  35
 2000 FORMAT(//' ON DISK ASSEMBLAGE AND LINEAR SOLUTION (M=',I2,')'/ BLLD  36
     1 ' ',42('=')/15X,'INDEX FOR RESIDUAL COMPUTATION   (NRES)=',I5) BLLD  37
      IF(LKGD.EQ.1) CALL ESPACE(NEQ,1,TBL(2),LKGD)                   BLLD  38
      IF(LFG.EQ.1) CALL ESPACE(NEQ,1,TBL(4),LFG)                     BLLD  39
      IF(LKE.EQ.1) CALL ESPACE(NKE,1,TBL(5),LKE)                     BLLD  40
      IF(LFE.EQ.1) CALL ESPACE(NDLE,1,TBL(6),LFE)                    BLLD  41
      IF(LRES.EQ.1) CALL ESPACE(NDLT,1,TBL(7),LRES)                  BLLD  42
      IF(LDLE.EQ.1) CALL ESPACE(NDLE,1,TBL(8),LDLE)                  BLLD  43
C------- FIND BLOCK LENGTH                                 BLLD  44
      I3=2                                                           BLLD  45
      I2=1+NSYM                                                      BLLD  46
      IF(NLBL.EQ.0) GO TO 10                                         BLLD  47
      IF(NBLM.EQ.0) NBLM=NKG/NLBL+2                                  BLLD  48
      GO TO 30                                                       BLLD  49
   10 I1=NVA-IVA-(2*NBLMAX+2)/NREEL-1                                BLLD  50
      IF(I1.GE.(NKG*I2+2)) GO TO 20                                  BLLD  51
C------- CASE WHERE MATRIX IS TO BE SEGMENTED                        BLLD  52
      NLBL=I1/(DEUX*I2)                                              BLLD  53
      NBLM=NKG/NLBL+2                                                BLLD  54
      GO TO 30                                                       BLLD  55
C------- CASE WHERE MATRIX IS IN CORE                                BLLD  56
   20 NLBL=NKG                                                       BLLD  57
      NBLM=1                                                         BLLD  58
      I3=1                                                           BLLD  59
   30 WRITE(MP,2010) NLBL,NBLM                                       BLLD  60
 2010 FORMAT(                                                        BLLD  61
     1 15X,'BLOCKS LENGTH IN KG              (NLBL)=',I5/            BLLD  62
     2 15X,'MAX. NUMBER OF BLOCKS IN KG           =',I5)             BLLD  63
      CALL ESPACE(NBLM+1,0,TBL(9),LEB)                               BLLD  64
```

Figure 6.23 (*Contd.*)

```
      CALL ESPACE(NBLM,0,TBL(10),LPB)                              BLLD  65
      IF(LKGS.EQ.1) CALL ESPACE(NLBL*I3,1,TBL(1),LKGS)             BLLD  66
      IF(NSYM.EQ.1.AND.LKGI.EQ.1) CALL ESPACE(NLBL*I3,1,TBL(3),LKGI)  BLLD 67
      CALL EXLIND(VA(LLD),VA(LDIMP),VA(LLOCE),VA(LCORE),VA(LPRNE), BLLD  68
     1            VA(LPREE),VA(LNE),VA(LKE),VA(LFE),VA(LKGS),VA(LKGD), BLLD 69
     2            VA(LKGI),VA(LFG),VA(LCORG),VA(LDLNC),VA(LNEQ),   BLLD  70
     3            VA(LRES),VA(LDLE),VA(LEB),VA(LPB))               BLLD  71
      RETURN                                                       BLLD  72
      END                                                          BLLD  73

      SUBROUTINE EXLIND(KLD,VDIMP,KLOCE,VCORE,VPRNE,VPREE,KNE,VKE,VFE, EXLD 1
     1     VKGS,VKGD,VKGI,VFG,VCORG,KDLNC,KNEQ,VRES,VDLE,KEB,KPB)  EXLD   2
C================================================================EXLD   3
C        TO EXECUTE BLOCK 'LIND'                                   EXLD   4
C        ASSEMBLE AND SOLVE A LINEAR PROBLEM WHEN MATRIX KG IS STORED EXLD 5
C        BLOCKWISE ON DISK                                         EXLD   6
C================================================================EXLD   7
      IMPLICIT REAL*8(A-H,O-Z)                                     EXLD   8
      COMMON/ASSE/NSYM,NKG,NKE,NDLE                                EXLD   9
      COMMON/RESO/NEQ,NRES,MRES                                    EXLD  10
      COMMON/LIND/NLBL,NBLM,MKG1,MKG2                              EXLD  11
      COMMON/ES/M,MR,MP,M1,M2,M3                                   EXLD  12
      DIMENSION KLD(1),VDIMP(1),KLOCE(1),VCORE(1),VPRNE(1),VPREE(1), EXLD 13
     1 KNE(1),VKE(1),VFE(1),VKGS(1),VKGD(1),VKGI(1),VFG(1),VCORG(1), EXLD 14
     2 KDLNC(1),KNEQ(1),VRES(1),VDLE(1),KEB(1),KPB(1)              EXLD  15
C----------------------------------------------------------------EXLD  16
      REWIND M3                                                    EXLD  17
C------- FORM TABLES EB AND PB DEFINING EQUATION BLOCKS            EXLD  18
      CALL EQBLOC(KLD,NLBL,NBLM,NEQ,KEB,KPB)                       EXLD  19
      WRITE(MP,2000) NBLM                                          EXLD  20
 2000 FORMAT(15X,'NUMBER OF BLOCKS IN KG (NBLM)=',I5)              EXLD  21
      IF(M.LT.2) GO TO 10                                          EXLD  22
      I1=NBLM+1                                                    EXLD  23
      WRITE(MP,2010) (KEB(I),I=1,I1)                               EXLD  24
 2010 FORMAT(/' FIRST EQUATION IN EACH BLOCK (EB)'/(5X,20I5))      EXLD  25
      WRITE(MP,2020) (KPB(I),I=1,NBLM)                             EXLD  26
 2020 FORMAT(/' FIRST BLOCK CONNECTED TO EACH BLOCK: (PB)'/(5X,20I5)) EXLD 27
C------ SAVE FG UNMODIFIED FOR PRESCRIBED B.C.                     EXLD  28
 10   WRITE (M3) (VFG(I),I=1,NEQ)                                  EXLD  29
      IF(M.GE.2) WRITE(MP,2030) (VFG(I),I=1,NEQ)                   EXLD  30
 2030 FORMAT(/' GLOBAL LOAD VECTOR UNMODIFIED FOR THE B.C. (FG)'   EXLD  31
     1/(1X,10E12.5))                                               EXLD  32
C------ ASSEMBLE KG, MODIFY FG FOR B.C. AND SAVE MODIFIED FG       EXLD  33
      CALL ASKGD(KLD,VDIMP,KLOCE,VCORE,VPRNE,VPREE,KNE,VKE,VFE,VKGS, EXLD 34
     1 VKGD,VKGI,VFG,VDLE,VRES,KEB)                                EXLD  35
      WRITE(M3) (VFG(I),I=1,NEQ)                                   EXLD  36
C------ PRINT FG                                                   EXLD  37
      IF(M.GE.2) WRITE(MP,2040) (VFG(I),I=1,NEQ)                   EXLD  38
 2040 FORMAT(/' GLOBAL LOAD VECTOR MODIFIED FOR THE B.C. (FG)'     EXLD  39
     1 /(1X,10E12.5))                                              EXLD  40
C                                                                  EXLD  41
C------ SOLVE                                                      EXLD  42
C                                                                  EXLD  43
 20   CALL SOLD(VKGS,VKGD,VKGI,VFG,KLD,NEQ,MP,1,1,NSYM,ENERG,KEB,KPB) EXLD 44
      IF(NSYM.NE.1) WRITE(MP,2050) ENERG                           EXLD  45
 2050 FORMAT(15X,'ENERGY     (ENERG)=',1E12.5)                     EXLD  46
C------ KG PIVOTS AND DETERMINANT                                  EXLD  47
 30   CALL PRPVTS(VKGD)                                            EXLD  48
C------ PRINT OUT THE SOLUTION                                     EXLD  49
      WRITE(MP,2060)                                               EXLD  50
 2060 FORMAT(//' SOLUTION'//)                                      EXLD  51
      CALL PRSOL(KDLNC,VCORG,VDIMP,KNEQ,VFG)                       EXLD  52
C                                                                  EXLD  53
C------ EVALUATE AND PRINT GRADIENTS                               EXLD  54
```

Figure 6.23 (*Contd.*)

```
C                                                                       EXLD 55
      CALL ASGRAD(KLD,VDIMP,KLOCE,VCORE,VPRNE,VPREE,KNE,VKE,VFE,VKGS,   EXLD 56
     1 VKGD,VKGI,VFG,VDLE,VRES)                                         EXLD 57
C                                                                       EXLD 58
C------ EVALUATE AND PRINT EQUILIBRIUM RESIDUALS AND REACTIONS          EXLD 59
C                                                                       EXLD 60
C------ READ VECTOR FG AND CHANGE ITS SIGN                              EXLD 61
      REWIND M3                                                         EXLD 62
      READ(M3) (VRES(I),I=1,NEQ)                                        EXLD 63
      DO 40 I=1,NEQ                                                     EXLD 64
   40 VRES(I)=-VRES(I)                                                  EXLD 65
C------ ASSEMBLE RESIDUALS AND REACTIONS                                EXLD 66
      CALL ASRESD(1,1,KLD,VDIMP,KLOCE,VCORE,VPRNE,VPREE,KNE,VKE,VFE,    EXLD 67
     1 VKGS,VKGD,VKGI,VFG,VDLE,VRES,VRES(NEQ+1))                        EXLD 68
C------ OUTPUT                                                          EXLD 69
      WRITE(MP,2070)                                                    EXLD 70
 2070 FORMAT(//' EQUILIBRIUM RESIDUALS AND REACTIONS'//)                EXLD 71
      CALL PRSOL(KDLNC,VCORG,VRES(NEQ+1),KNEQ,VRES)                     EXLD 72
      RETURN                                                            EXLD 73
      END                                                               EXLD 74

      SUBROUTINE EQBLOC(KLD,NLBL,NBLMAX,NEQ,KEB,KPB)                    EQBL  1
C=====================================================================EQBL  2
C   TO FORM TABLES KEB AND KPB DEFINING EQUATION BLOCKS                 EQBL  3
C     INPUT                                                             EQBL  4
C       KLD     ARRAY OF A ADDRESS OF COLUMN TOP TERMS IN KG            EQBL  5
C       NLBL    BLOCKS LENGTH                                           EQBL  6
C       NBLMAX  MAX. NUMBER OF BLOCKS ALLOWED                           EQBL  7
C       NEQ     NUMBER OF EQUATIONS                                     EQBL  8
C     OUTPUT                                                            EQBL  9
C       KEB     ARRAY CONTAINING THE NUMBERS OF FIRST EQUATIONS IN      EQBL 10
C               EACH BLOCK (DIMENSION  NEQ+1)                           EQBL 11
C       KPB     ARRAY CONTAINING THE NUMBER OF FIRST BLOCKS CONNECTED   EQBL 12
C               TO EACH BLOCK (DIMENSION NEQ)                           EQBL 13
C       NBLMAX  NUMBER OF BLOCKS                                        EQBL 14
C=====================================================================EQBL 15
      COMMON/ES/M,MR,MP                                                 EQBL 16
      DIMENSION KLD(1),KEB(1),KPB(1)                                    EQBL 17
C---------------------------------------------------------------------EQBL 18
C------ FIRST BLOCK                                                     EQBL 19
      ILBL=0                                                            EQBL 20
      NBL=1                                                             EQBL 21
      KEB(1)=1                                                          EQBL 22
      KPB(1)=1                                                          EQBL 23
      IMIN=1                                                            EQBL 24
C------ FOR EACH EQUATION                                               EQBL 25
      DO 70 IK=1,NEQ                                                    EQBL 26
C------ ADDRESSES FOR COLUMN IK                                         EQBL 27
      JHK=KLD(IK)                                                       EQBL 28
      JHK1=KLD(IK+1)                                                    EQBL 29
      LBK1=JHK1-JHK                                                     EQBL 30
      IF(LBK1.LE.NLBL) GO TO 10                                         EQBL 31
      WRITE(MP,2000) IK,LBK1,NLBL                                       EQBL 32
 2000 FORMAT(' *** ERROR,COLUMN',I5,' GREATER(',I5,')THAN BLOCK(' ,I5,'EQBL 33
     1)')                                                               EQBL 34
      STOP                                                              EQBL 35
C------ CHECK FOR NEW BLOCK                                             EQBL 36
   10 ILBL=ILBL+LBK1                                                    EQBL 37
      IF(ILBL.LE.NLBL) GO TO 60                                         EQBL 38
      NBL=NBL+1                                                         EQBL 39
      IF(NBL.LE.NBLMAX) GO TO 20                                        EQBL 40
      WRITE(MP,2010) IK                                                 EQBL 41
 2010 FORMAT(' *** ERROR, EXCESSIVE NUMBER OF BLOCKS, EQUATION',I5)     EQBL 42
      STOP                                                              EQBL 43
```

Figure 6.23 (*Contd.*)

```
20      KEB(NBL)=IK                                                 EQBL  44
        ILBL=LBK1                                                   EQBL  45
C------- SEARCH FOR FIRST BLOCK CONNECTED TO COMPLETED BLOCK        EQBL  46
        IB=NBL                                                      EQBL  47
40      IF(IMIN.GE.KEB(IB)) GO TO 50                                EQBL  48
        IB=IB-1                                                     EQBL  49
        GO TO 40                                                    EQBL  50
50      KPB(NBL-1)=IB                                               EQBL  51
        IMIN=IK                                                     EQBL  52
C------- SEARCH FOR MINIMUM ROW NUMBER FOR COLUMN TOP TERMS         EQBL  53
60      I=IK-LBK1+1                                                 EQBL  54
        IF(I.LT.IMIN)IMIN=I                                         EQBL  55
70      CONTINUE                                                    EQBL  56
C------- FIRST BLOCK CONNECTED TO LAST BLOCK                        EQBL  57
        IB=NBL                                                      EQBL  58
80      IF(IMIN.GE.KEB(IB)) GO TO 90                                EQBL  59
        IB=IB-1                                                     EQBL  60
        GO TO 80                                                    EQBL  61
90      KPB(NBL)=IB                                                 EQBL  62
        KEB(NBL+1)=NEQ+1                                            EQBL  63
        NBLMAX=NBL                                                  EQBL  64
        RETURN                                                      EQBL  65
        END                                                         EQBL  66

        SUBROUTINE ASKGD(KLD,VDIMP,KLOCE,VCORE,VPRNE,VPREE,KNE,VKE,VFE, ASKD 1
      1 VKGS,VKGD,VKGI,VFG,VDLE,VRES,KEB)                           ASKD   2
C========================================================================ASKD 3
C       TO ASSEMBLE GLOBAL MATRIX KG (ELEMENT FUNCTION TYPE 3)      ASKD   4
C       TAKING INTO ACCOUNT OF PRESCRIBED NON ZERO D.O.F.           ASKD   5
C       VERSION : MATRIX KG STORED BLOCKWISE ON FILE M4             ASKD   6
C========================================================================ASKD 7
        IMPLICIT REAL*8(A-H,O-Z)                                    ASKD   8
        COMMON/COND/NCLT,NCLZ,NCLNZ                                 ASKD   9
        COMMON/ELEM/NELT,NNEL,NTPE,NGRE,ME,NIDENT                   ASKD  10
        COMMON/ASSE/NSYM                                            ASKD  11
        COMMON/RESO/NEQ                                             ASKD  12
        COMMON/RGDT/IEL,ITPE,ITPE1,IGRE,IDLE,ICE,IPRNE,IPREE,INEL,IDEG,IPGASKD 13
      1 ,ICOD                                                       ASKD  14
        COMMON/LIND/NLBL,NBLM,MKG1,MKG2                             ASKD  15
        COMMON/ES/M,MR,MP,M1,M2,M3,M4,M5                            ASKD  16
        DIMENSION KLD(1),VDIMP(1),KLOCE(1),VCORE(1),VPRNE(1),VPREE(1), ASKD 17
      1 KNE(1),VKE(1),VFE(1),VKGS(1),VKGD(1),VKGI(1),VFG(1),VDLE(1), ASKD 18
      1 VRES(1),KEB(1)                                              ASKD  19
        DATA ZERO/0.D0/                                             ASKD  20
C------------------------------------------------------------------ASKD    21
C------- REWIND FILE M4                                             ASKD  22
        REWIND M4                                                   ASKD  23
C------- LOOP OVER THE BLOCKS                                       ASKD  24
        DO 80 IB=1,NBLM                                             ASKD  25
C------- INITIALIZE THE BLOCK                                       ASKD  26
        DO 10 I=1,NLBL                                              ASKD  27
        IF(NSYM.EQ.1) VKGI(I)=ZERO                                  ASKD  28
10      VKGS(I)=ZERO                                                ASKD  29
        IE1=KEB(IB)                                                 ASKD  30
        IE2=KEB(IB+1)-1                                             ASKD  31
C------- REWIND ELEMENT FILE (M2)                                   ASKD  32
        REWIND M2                                                   ASKD  33
C------- LOOP OVER THE ELEMENTS      ASKD   34
        DO 70 IE=1,NELT                                             ASKD  35
C------- READ AN ELEMENT                                            ASKD  36
        CALL RDELEM(M2,KLOCE,VCORE,VPRNE,VPREE,KNE)                 ASKD  37
C------- CHECK IF BLOCK IS AFFECTED BY THIS ELEMENT                 ASKD  38
        DO 20 ID=1,IDLE                                             ASKD  39
        J=KLOCE(ID)                                                 ASKD  40
```

Figure 6.23 (*Contd.*)

```
            IF(J.LT.IE1.OR.J.GT.IE2) GO TO 20                         ASKD  41
            GO TO 40                                                  ASKD  42
     20     CONTINUE                                                  ASKD  43
     30     IF(IB.NE.1.OR.(NCLNZ.EQ.0.AND.IB.EQ.1)) GO TO 70          ASKD  44
     C------ EVALUATE INTERPOLATION FUNCTIONS IF REQUIRED             ASKD  45
            IF(ITPE.EQ.ITPE1) GO TO 50                                ASKD  46
     40     ICOD=2                                                    ASKD  47
            CALL ELEMLB(VCORE,VPRNE,VPREE,VDLE,VKE,VFE)               ASKD  48
     C------ FORM ELEMENT MATRIX                                      ASKD  49
     50     ICOD=3                                                    ASKD  50
            CALL ELEMLB(VCORE,VPRNE,VPREE,VDLE,VKE,VFE)               ASKD  51
     C------ PRINT ELEMENT MATRIX                                     ASKD  52
            IF(M.LT.2) GO TO 60                                       ASKD  53
            IF(NSYM.EQ.0) IKE=IDLE*(IDLE+1)/2                         ASKD  54
            IF(NSYM.EQ.1) IKE=IDLE*IDLE                               ASKD  55
            WRITE(MP,2000) IEL,(VKE(I),I=1,IKE)                       ASKD  56
     2000   FORMAT(/' MATRIX (KE) , ELEMENT:',I5/(10X,10E12.5))       ASKD  57
     C------ MODIFY FG FOR THE PRESCRIBED NON ZERO D.O.F.             ASKD  58
     60     IF(NCLNZ.NE.0.AND.IB.EQ.1) CALL MODFG(IDLE,NSYM,KLOCE,VDIMP,VKE, ASKD 59
          1 VFG)                                                      ASKD  60
     C------ ASSEMBLE           ASKD  61
            CALL ASSELD(1,0,IDLE,NSYM,IE1,IE2,KLOCE,KLD,VKE,VFE,VKGS,VKGD, ASKD 62
          1 VKGI,VFG)                                                 ASKD  63
            ITPE1=ITPE                                                ASKD  64
     70     CONTINUE                                                  ASKD  65
     C------ END OF A BLOCK                                           ASKD  66
            WRITE(M4) (VKGS(I),I=1,NLBL)                              ASKD  67
            IF(NSYM.EQ.1) WRITE(M4) (VKGI(I),I=1,NLBL)                ASKD  68
            IF(M.LT.2) GO TO 80                                       ASKD  69
            WRITE(MP,2010) IB,(VKGS(I),I=1,NLBL)                      ASKD  70
     2010   FORMAT(' UPPER TRIANGLE BLOCK OF (KG) NO:',I5/(1X,10E12.5)) ASKD 71
            IF(NSYM.EQ.1) WRITE(MP,2020) IB,(VKGI(I),I=1,NLBL)        ASKD  72
     2020   FORMAT(' LOWER TRIANGLE BLOCK OF (KG) NO:',I5/(1X,10E12.5)) ASKD 73
     80     CONTINUE                                                  ASKD  74
            IF(M.GE.2) WRITE(MP,2030) (VKGD(I),I=1,NEQ)               ASKD  75
     2030   FORMAT(' DIAGONAL OF (KG)'/(1X,10E12.5))                  ASKD  76
            RETURN                                                    ASKD  77
            END                                                       ASKD  78

            SUBROUTINE ASSELD(IKG,IFG,IDLE,NSYM,IE1,IE2,KLOCE,KLD,VKE,VFE, ASSD  1
          1 VKGS,VKGD,VKGI,VFG)                                       ASSD   2
     C===============================================================ASSD   3
     C      TO ASSEMBLE ELEMENT MATRIX (SYMMETRIC OR NOT) AND/OR VECTOR. ASSD 4
     C      THE MATRIX IS STORED BLOCKWISE ON DISK                    ASSD   5
     C         INPUT         ASSD   6
     C            IKG        IF IKG.EQ.1 ASSEMBLE ELEMENT MATRIX KE   ASSD   7
     C            IFG        IF IFG.EQ.1 ASSEMBLE ELEMENT VECTOR FE   ASSD   8
     C            IDLE       NUMBER OF D.O.F. OF THE ELEMENT          ASSD   9
     C            NSYM       0=SYMMETRIC PROBLEM, 1=NON SYMMETRIC PROBLEM ASSD 10
     C            IE1,IE2    FIRST AND LAST COLUMN OF KG TO BE ASSEMBLED ASSD 11
     C            KLOCE      ELEMENT LOCALIZATION VECTOR              ASSD  12
     C            KLD        CUMULATIVE COLUMN HEIGHTS IN KG          ASSD  13
     C            VKE        ELEMENT MATRIX KE (FULL OR UPPER TRIANGLE BY ASSD 14
     C                       DESCENDING COLUMNS)                      ASSD  15
     C            VFE        ELEMENT VECTOR FE                        ASSD  16
     C         OUTPUT                                                 ASSD  17
     C            VKGS,VKGD,VKGI   GLOBAL MATRIX (SKYLINE)            ASSD  18
     C                       (SYMMETRIC OR NOT)                       ASSD  19
     C            VFG        GLOBAL LOAD VECTOR                       ASSD  20
     C===============================================================ASSD  21
            IMPLICIT REAL*8(A-H,O-Z)                                  ASSD  22
            DIMENSION KLOCE(1),KLD(1),VKE(1),VFE(1),VKGS(1),VKGD(1),  ASSD  23
          1 VKGI(1),VFG(1)                                            ASSD  24
     C----------------------------------------------------------------ASSD  25
```

Figure 5.23 (*Contd.*)

```
      C                                                       ASSD  26
      C------   ASSEMBLE ELEMENT MATRIX                       ASSD  27
      C                                                       ASSD  28
             IF(IKG.NE.1) GO TO 100                           ASSD  29
             IOBLOC=KLD(IE1)-1                                ASSD  30
             IEQ0=IDLE                                        ASSD  31
             IEQ1=1                                           ASSD  32
      C------   FOR EACH COLUMN OF KE                         ASSD  33
             DO 90 JD=1,IDLE                                  ASSD  34
             IF(NSYM.NE.1) IEQ0=JD                            ASSD  35
             JL=KLOCE(JD)                                     ASSD  36
             IF(JL) 90,90,10                                  ASSD  37
      10     IO=KLD(JL+1)-IOBLOC                              ASSD  38
             IEQ=IEQ1                                         ASSD  39
             IQ=1                                             ASSD  40
             IF(JL.LT.IE1.OR.JL.GT.IE2) GO TO 90              ASSD  41
      C------   FOR EACH ROW OF KE                            ASSD  42
             DO 80 ID=1,IDLE                                  ASSD  43
             IL=KLOCE(ID)                                     ASSD  44
             IF(NSYM.EQ.1) GO TO 30                           ASSD  45
             IF(ID-JD) 30,20,20                               ASSD  46
      20     IQ=ID                                            ASSD  47
      30     IF(IL) 80,80,40                                  ASSD  48
      40     IJ=JL-IL                                         ASSD  49
             IF(IJ) 80,50,60                                  ASSD  50
      C------   DIAGONAL TERMS IN KG                          ASSD  51
      50     VKGD(IL)=VKGD(IL)+VKE(IEQ)                       ASSD  52
             GO TO 80                                         ASSD  53
      C------   UPPER TRIANGLE TERMS IN KG                    ASSD  54
      60     I=IO-IJ                                          ASSD  55
             VKGS(I)=VKGS(I)+VKE(IEQ)                         ASSD  56
             IF(NSYM.NE.1) GO TO 80                           ASSD  57
      C------   LOWER TRIANGLE TERMS IN KG                    ASSD  58
             IEQI=(ID-1)*IDLE+JD                              ASSD  59
             VKGI(I)=VKGI(I)+VKE(IEQI)                        ASSD  60
      80     IEQ=IEQ+IQ                                       ASSD  61
      90     IEQ1=IEQ1+IEQ0                                   ASSD  62
      C                                                       ASSD  63
      C------   ASSEMBLE ELEMENT VECTOR                       ASSD  64
      C                                                       ASSD  65
      100    IF(IFG.NE.1) GO TO 130                           ASSD  66
             DO 120 ID=1,IDLE                                 ASSD  67
             IL=KLOCE(ID)                                     ASSD  68
             IF(IL) 120,120,110                               ASSD  69
      110    VFG(IL)=VFG(IL)+VFE(ID)                          ASSD  70
      120    CONTINUE                                         ASSD  71
      130    RETURN                                           ASSD  72
             END                                              ASSD  73
```

Figure 6.23   Block 'LIND'.

equations, and output of results are identical to those of block 'LINM'; they are described in paragraph 6.5.3.3.

Function

This block solves a non-linear problem using one of the Newton–Raphson techniques described in paragraph 5.3.

Flow chart

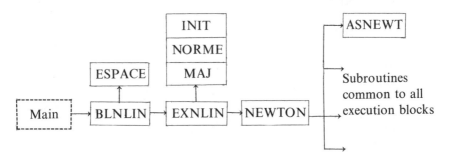

Subroutines

BLNLIN: creates arrays VKGS, VKGD, VKGI, VFG, VKE, VFE, VRES, VDLE, VDLG, VME (Figure 6.24).

ESPACE: (Figure 6.5).

EXNLIN: implements algorithm of Figure 5.19. It reads variables DPAS, NPAS, NITER, EPSDL, IMETH, and OMEGA defined in Figure 6.9. Then, it executes Newton–Raphson method by looping over the load steps and equilibrium iterations (Figure 6.24).

NEWTON: executes all the operations required by one iteration of the Newton–Raphson method (Figure 6.24).

ASNEWT: assembles the residual and the global matrix $[K]$ or $[K_t]$ (Figure 6.24).

INIT: initialize a vector (Figure 6.24).

MAJ: executes the following vector operation:

$$\{V_2\} = x_1\{V_1\} + x_2\{V_2\} \quad \text{(Figure 6.24)}$$

NORME: computes the following vector norm (Figure 6.24)

$$\left(\frac{\langle \Delta U \rangle \{\Delta U\}}{\langle U \rangle \{U\}}\right)^{1/2}$$

Remarks

Depending on the type of non-linearities of the problem under study, the method of resolution (NEWTON) and the chosen norm (NORME) could be modified.

6.5.3.6 Block 'TEMP'

Block 'TEMP' is similar to block 'NLIN' in which subroutines NEWTON and ASNEWT are replaced by EULER and ASEULR.

```
      SUBROUTINE BLNLIN                                              BLNL    1
C================================================================BLNL    2
C     TO CALL BLOCK 'NLIN'                                           BLNL    3
C     TO SOLVE A STEADY NON LINEAR PROBLEM                           BLNL    4
C================================================================BLNL    5
      IMPLICIT REAL*8(A-H,O-Z)                                       BLNL    6
      REAL*4 TBL                                                     BLNL    7
      COMMON/ELEM/NUL(4),ME                                          BLNL    8
      COMMON/ASSE/NSYM,NKG,NKE,NDLE                                  BLNL    9
      COMMON/RESO/NEQ                                                BLNL   10
      COMMON/NLIN/EPSDL,XNORM,OMEGA,XPAS,DPAS,DPASO,NPAS,IPAS,NITER, BLNL   11
     1 ITER,IMETH                                                    BLNL   12
      COMMON/ES/M,MR,MP,M1,M2,M3,M4                                  BLNL   13
      COMMON/LOC/LCORG,LDLNC,LNEQ,LDIMP,LPRNG,LPREG,LLD,LLOCE,LCORE,LNE,BLNL 14
     1 LPRNE,LPREE,LDLE,LKE,LFE,LKGS,LKGD,LKGI,LFG,LRES,LDLG,LME     BLNL   15
      COMMON VA(1)                                                   BLNL   16
      DIMENSION TBL(10),IN(2),XIN(3)                                 BLNL   17
      DATA TBL/4HKGS ,4HKGD ,4HKGI ,4HFG  ,4HKE  ,                   BLNL   18
     1   4HFE  ,4HRES ,4HDLE ,4HDLG ,4HME  /                         BLNL   19
C----------------------------------------------------------------BLNL   20
      IF(M1.EQ.0) M1=MR                                              BLNL   21
      IF(M2.EQ.0) M2=ME                                              BLNL   22
      WRITE(MP,2000) M                                               BLNL   23
 2000 FORMAT(//' NON LINEAR SOLUTION (M=',I2,')'/1X,23('='))         BLNL   24
C------ TO ALLOCATE SPACE                                            BLNL   25
      IF(LKGS.EQ.1) CALL ESPACE(NKG,1,TBL(1),LKGS)                   BLNL   26
      IF(LKGD.EQ.1) CALL ESPACE(NEQ,1,TBL(2),LKGD)                   BLNL   27
      IF(NSYM.EQ.1.AND.LKGI.EQ.1) CALL ESPACE(NKG,1,TBL(3),LKGI)     BLNL   28
      IF(LFG.EQ.1) CALL ESPACE (NEQ,1,TBL(4),LFG)                    BLNL   29
      IF(LKE.EQ.1) CALL ESPACE(NKE,1,TBL(5),LKE)                     BLNL   30
      IF(LFE.EQ.1) CALL ESPACE(NDLE,1,TBL(6),LFE)                    BLNL   31
      IF(LRES.EQ.1) CALL ESPACE(NEQ,1,TBL(7),LRES)                   BLNL   32
      IF(LDLE.EQ.1) CALL ESPACE(NDLE,1,TBL(8),LDLE)                  BLNL   33
      IF(LDLG.EQ.1) CALL ESPACE(NEQ,1,TBL(9),LDLG)                   BLNL   34
      IF(LME.EQ.1) CALL ESPACE(NKE,1,TBL(10),LME)                    BLNL   35
C------ TO EXECUTE THE BLOCK                                         BLNL   36
      CALL EXNLIN(VA(LCORG),VA(LDLNC),VA(LDIMP),VA(LNEQ),VA(LLD),    BLNL   37
     1 VA(LLOCE),VA(LCORE),VA(LPRNE),VA(LPREE),VA(LNE),VA(LKE),VA(LME),BLNL 38
     2 VA(LFE),VA(LDLE),VA(LKGS),VA(LKGD),VA(LKGI),VA(LFG),VA(LRES), BLNL   39
     3 VA(LDLG))                                                     BLNL   40
      RETURN                                                         BLNL   41
      END                                                            BLNL   42

      SUBROUTINE EXNLIN(VCORG,KDLNC,VDIMP,KNEQ,KLD,KLOCE,VCORE,VPRNE, EXNL   1
     1 VPREE,KNE,VKE,VME,VFE,VDLE,VKGS,VKGD,VKGI,VFG,VRES,VDLG)      EXNL    2
C================================================================EXNL    3
C     TO EXECUTE BLOCK 'NLIN'                                        EXNL    4
C     TO SOLVE A STEADY NON LINEAR PROBLEM                           EXNL    5
C================================================================EXNL    6
      IMPLICIT REAL*8(A-H,O-Z)                                       EXNL    7
      COMMON/RESO/NEQ                                                EXNL    8
      COMMON/COND/NCLT,NCLZ,NCLNZ                                    EXNL    9
      COMMON/ASSE/NSYM                                               EXNL   10
      COMMON/NLIN/EPSDL,XNORM,OMEGA,XPAS,DPAS,DPASO,NPAS,IPAS,NITER, EXNL   11
     1 ITER,IMETH                                                    EXNL   12
      COMMON/ES/M,MR,MP,M1,M2,M3,M4                                  EXNL   13
      DIMENSION VCORG(1),KDLNC(1),VDIMP(1),KNEQ(1),KLD(1),KLOCE(1),  EXNL   14
     1 VCORE(1),VPRNE(1),VPREE(1),KNE(1),VKE(1),VME(1),VFE(1),VDLE(1),EXNL  15
     2 VKGS(1),VKGD(1),VKGI(1),VFG(1),VRES(1),VDLG(1)                EXNL   16
      DATA ZERO/0.D0/                                                EXNL   17
C----------------------------------------------------------------EXNL   18
      DPASO=ZERO                                                     EXNL   19
      XPAS=ZERO                                                      EXNL   20
      IPAS=0                                                         EXNL   21
```

Figure 6.24 (*Contd.*)

```
C------- READY INITIAL D.O.F. ON FILE M3                        EXNL  22
        IF(M3.EQ.0) GO TO 10                                    EXNL  23
        REWIND M3                                               EXNL  24
        READ(M3) (VDLG(I),I=1,NEQ)                              EXNL  25
C------- READ A CARD DEFINING A SET OF IDENTICAL STEPS          EXNL  26
10      READ(M1,1000) DPAS,I1,I2,I3,X1,X2                       EXNL  27
1000    FORMAT(F10.0,3I5,2F10.0)                                EXNL  28
        IF(DPAS.EQ.ZERO) GO TO 140                              EXNL  29
        IF(I1.GT.0) NPAS=I1                                     EXNL  30
        IF(I2.GT.0) NITER=I2                                    EXNL  31
        IF(I3.GT.0) IMETH=I3                                    EXNL  32
        IF(X1.GT.ZERO) EPSDL=X1                                 EXNL  33
        IF(X2.GT.ZERO) OMEGA=X2                                 EXNL  34
C                                                               EXNL  35
C------- LOOP OVER ALL STEPS                                    EXNL  36
C                                                               EXNL  37
        DO 130 IP=1,NPAS                                        EXNL  38
        IPAS=IPAS+1                                             EXNL  39
        XPAS=XPAS+DPAS                                          EXNL  40
        WRITE(MP,2000) IPAS,DPAS,XPAS,NITER,IMETH,EPSDL,OMEGA   EXNL  41
2000    FORMAT(/1X,13('-'),'STEP NUMBER (IPAS):',I5//           EXNL  42
     1              14X,'INCREMENT              (DPAS)=',E12.5/ EXNL  43
     2              14X,'TOTAL LEVEL            (XPAS)=',E12.5/ EXNL  44
     3              14X,'NUMBER OF ITERATIONS  (NITER)=',I12/   EXNL  45
     4              14X,'METHOD NUMBER         (IMETH):',I12/   EXNL  46
     5              14X,'TOLERANCE             (EPSDL)=',E12.5/ EXNL  47
     6              14X,'OVER RELAXATION FACTOR (OMEGA)=',E12.5/) EXNL 48
C                                                               EXNL  49
C------- LOOP OVER EQUILIBRIUM ITERATIONS                       EXNL  50
C                                                               EXNL  51
        DO 110 ITER=1,NITER                                     EXNL  52
C------- CHOOSE THE METHOD                                      EXNL  53
        IF(IMETH.GT.3) GO TO 20                                 EXNL  54
C------- NEWTON TYPE METHODS                                    EXNL  55
        CALL NEWTON(VCORG,KDLNC,VDIMP,KNEQ,KLD,KLOCE,VCORE,VPRNE, EXNL 56
     1   KNE,VKE,VME,VFE,VDLE,VKGS,VKGD,VKGI,VFG,VRES,VDLG)     EXNL  57
        GO TO 100                                               EXNL  58
C------- OTHER METHODS ......                                   EXNL  59
20      CONTINUE                                                EXNL  60
        WRITE(MP,2010) IMETH                                    EXNL  61
2010    FORMAT(' ** ERROR, METHOD:',I3,' UNKNOWN')              EXNL  62
        STOP                                                    EXNL  63
C------- COMPUTE THE NORM                                       EXNL  64
100     CALL NORME(NEQ,VRES,VDLG,XNORM)                         EXNL  65
        IF(M.GT.0) WRITE(MP,2020) ITER,XNORM                    EXNL  66
2020    FORMAT(5X,'ITERATION (ITER):',I3,' NORM (XNORM)=',E12.5) EXNL 67
        IF(M.GE.2) CALL PRSOL(KDLNC,VCORG,VDIMP,KNEQ,VDLG)      EXNL  68
        IF(XNORM.LE.EPSDL) GO TO 120                            EXNL  69
110     CONTINUE                                                EXNL  70
        ITER=NITER                                              EXNL  71
C------- END OF STEP                                            EXNL  72
120     DPAS0=DPAS                                              EXNL  73
        WRITE(MP,2030) ITER,NITER                               EXNL  74
2030    FORMAT(/10X,I4,' PERFORMED ITERATIONS OVER',I4/)        EXNL  75
        IF(M.LT.2) CALL PRSOL(KDLNC,VCORG,VDIMP,KNEQ,VDLG)      EXNL  76
130     CONTINUE                                                EXNL  77
        GO TO 10                                                EXNL  78
C------- SAVE THE SOLUTION ON FILE M4                           EXNL  79
140     IF(M4.NE.0) WRITE(M4) (VDLG(I),I=1,NEQ)                 EXNL  80
        RETURN                                                  EXNL  81
        END                                                     EXNL  82

        SUBROUTINE NEWTON(VCORG,KDLNC,VDIMP,KNEQ,KLD,KLOCE,VCORE,VPRNE, NEWT 1
     1   VPREE,KNE,VKE,VME,VFE,VDLE,VKGS,VKGD,VKGI,VFG,VRES,VDLG)       NEWT 2
```

Figure 6.24 (*Contd.*)

```
C==============================================================================NEWT   3
C      ALGORITHM FOR NEWTON-RAPHSON TYPE METHODS                        NEWT   4
C         IMETH.EQ.1   COMPUTE K AT EACH ITERATION                      NEWT   5
C         IMETH.EQ.2   K IS CONSTANT                                    NEWT   6
C         IMETH.EQ.3   RECOMPUTE K AT THE BEGINNING OF EACH STEP        NEWT   7
C==============================================================================NEWT   8
       IMPLICIT REAL*8(A-H,O-Z)                                         NEWT   9
       COMMON/ASSE/NSYM,NKG                                             NEWT  10
       COMMON/RESO/NEQ                                                  NEWT  11
       COMMON/NLIN/EPSDL,XNORM,OMEGA,XPAS,DPAS,DPASO,NPAS,IPAS,NITER,   NEWT  12
      1 ITER,IMETH                                                      NEWT  13
       COMMON/ES/M,MR,MP                                                NEWT  14
       DIMENSION VCORG(1),KDLNC(1),VDIMP(1),KNEQ(1),KLD(1),KLOCE(1),    NEWT  15
      1 VCORE(1),VPRNE(1),VPREE(1),KNE(1),VKE(1),VME(1),VFE(1),VDLE(1), NEWT  16
      2 VKGS(1),VKGD(1),VKGI(1),VFG(1),VRES(1),VDLG(1)                  NEWT  17
       DATA ZERO/0.D0/,UN/1.D0/                                         NEWT  18
C------------------------------------------------------------------------NEWT  19
C------- DECIDE IF GLOBAL MATRIX IS TO REASSEMBLED                      NEWT  20
       IKT=0                                                            NEWT  21
       IF(IMETH.EQ.1) GO TO 10                                          NEWT  22
       IF(IPAS.EQ.1.AND.ITER.EQ.1) GO TO 10                             NEWT  23
       IF(IMETH.EQ.3.AND.ITER.EQ.1) GO TO 10                            NEWT  24
       GO TO 20                                                         NEWT  25
10     IKT=1                                                            NEWT  26
C------- INITIALIZE GLOBAL MATRIX TO ZERO IF IT IS TO BE ASSEMBLED      NEWT  27
20     IF(IKT.EQ.0)GO TO 30                                             NEWT  28
       CALL INIT(ZERO,NKG,VKGS)                                         NEWT  29
       CALL INIT(ZERO,NEQ,VKGD)                                         NEWT  30
       IF(NSYM.EQ.1) CALL INIT(ZERO,NKG,VKGI)                           NEWT  31
C------- STORE LOADS IN THE RESIDUAL VECTOR                             NEWT  32
30     CALL MAJ(XPAS,ZERO,NEQ,VFG,VRES)                                 NEWT  33
C------- ASSEMBLE RESIDUAL VECTOR, AND EVENTUALLY THE GLOBAL MATRIX     NEWT  34
       CALL ASNEWT(IKT,KLD,VDIMP,KLOCE,VCORE,VPRNE,VPREE,KNE,VKE,VFE,   NEWT  35
      1 VKGS,VKGD,VKGI,VDLG,VDLE,VRES)                                  NEWT  36
C------- SOLVE                                                          NEWT  37
       CALL SOL(VKGS,VKGD,VKGI,VRES,KLD,NEQ,MP,IKT,1,NSYM,ENERG)        NEWT  38
       IF(IKT.EQ.1.AND.M.GT.1) CALL PRPVTS(VKGD)                        NEWT  39
C------- UPDATE THE SOLUTION       NEWT  40
       CALL MAJ(OMEGA,UN,NEQ,VRES,VDLG)                                 NEWT  41
       RETURN                                                           NEWT  42
       END                                                              NEWT  43

       SUBROUTINE ASNEWT(IKT,KLD,VDIMP,KLOCE,VCORE,VPRNE,VPREE,         ASNE   1
      1 KNE,VKE,VFE,VKGS,VKGD,VKGI,VFG,VDLE,VRES)                       ASNE   2
C==============================================================================ASNE   3
C      TO ASSEMBLE THE RESIDUALS AND THE GLOBAL MATRIX (IF IKT.EQ.1)    ASNE   4
C      WHILE LOOPING OVER THE ELEMENTS        ASNE   5
C      (FOR THE NEWTON-RAPHSON METHOD)                                  ASNE   6
C==============================================================================ASNE   7
       IMPLICIT REAL*8(A-H,O-Z)                                         ASNE   8
       COMMON/ELEM/NELT,NNEL,NTPE,NGRE,ME,NIDENT                        ASNE   9
       COMMON/ASSE/NSYM                                                 ASNE  10
       COMMON/RESO/NEQ                                                  ASNE  11
       COMMON/RGDT/IEL,ITPE,ITPE1,IGRE,IDLE,ICE,IPRNE,IPREE,INEL,IDEG,IPGASNE  12
      1 ,ICOD                                                           ASNE  13
       COMMON/ES/M,MR,MP,M1,M2                                          ASNE  14
       DIMENSION KLD(1),VDIMP(1),KLOCE(1),VCORE(1),VPRNE(1),VPREE(1),   ASNE  15
      1 KNE(1),VKE(1),VFE(1),VKGS(1),VKGD(1),VKGI(1),VFG(1),VDLE(1),    ASNE  16
      2 VRES(1)                                                         ASNE  17
C------------------------------------------------------------------------ASNE  18
C------- REWIND ELEMENT FILE M2           ASNE  19
       REWIND M2                                                        ASNE  20
C------- LOOP OVER THE ELEMENTS                                         ASNE  21
       DO 40 IE=1,NELT                                                  ASNE  22
C------- READ AN ELEMENT                                                ASNE  23
```

Figure 6.24 (*Contd.*)

```
      CALL RDELEM(M2,KLOCE,VCORE,VPRNE,VPREE,KNE)              ASNE 24
C------- EVALUATE INTERPOLATION FUNCTIONS IF REQUIRED          ASNE 25
      IF(ITPE.EQ.ITPE1) GO TO 10                               ASNE 26
      ICOD=2                                                   ASNE 27
      CALL ELEMLB(VCORE,VPRNE,VPREE,VDLE,VKE,VFE)              ASNE 28
C------- FIND THE D.O.F. OF THE ELEMENT FROM VFG               ASNE 29
10    CALL DLELM(KLOCE,VFG,VDIMP,VDLE)                         ASNE 30
C------- CALCULATE ELEMENT RESIDUALS AND CHANGE THEIR SIGN     ASNE 31
      ICOD=6                                                   ASNE 32
      CALL ELEMLB(VCORE,VPRNE,VPREE,VDLE,VKE,VFE)              ASNE 33
      DO 20 I=1,IDLE                                           ASNE 34
20    VFE(I)=-VFE(I)                                           ASNE 35
C------- EVALUATE GLOBAL MATRIX                                ASNE 36
      IF(IKT.EQ.0) GO TO 30                                    ASNE 37
      ICOD=4                                                   ASNE 38
      CALL ELEMLB(VCORE,VPRNE,VPREE,VDLE,VKE,VFE)              ASNE 39
C------- ASSEMBLE THE RESIDUALS AND THE GLOBAL MATRIX          ASNE 40
30    CALL ASSEL(IKT,1,IDLE,NSYM,KLOCE,KLD,VKE,VFE,VKGS,VKGD,VKGI,VRES) ASNE 41
40    ITPE1=ITPE                                               ASNE 42
      RETURN                                                   ASNE 43
      END                                                      ASNE 44

      SUBROUTINE INIT(X,N,V)                                   INIT  1
C==================================================================INIT  2
C     INITIALIZE VECTOR V TO VALUE X                           INIT  3
C==================================================================INIT  4
      IMPLICIT REAL*8(A-H,O-Z)                                 INIT  5
      DIMENSION V(1)                                           INIT  6
C------------------------------------------------------------------INIT  7
      DO 10 I=1,N                                              INIT  8
10    V(I)=X                                                   INIT  9
      RETURN                                                   INIT 10
      END                                                      INIT 11

      SUBROUTINE MAJ(X1,X2,N,V1,V2)                            MAJ   1
C==================================================================MAJ   2
C     EXECUTE THE VECTOR OPERATION: V2=X1*V1 + X2*V2            MAJ   3
C        X1,X2:SCALARS    V1,V2:VECTORS                         MAJ   4
C==================================================================MAJ   5
      IMPLICIT REAL*8(A-H,O-Z)                                 MAJ   6
      DIMENSION V1(1),V2(1)                                    MAJ   7
C------------------------------------------------------------------MAJ   8
      DO 10 I=1,N                                              MAJ   9
10    V2(I)=X1*V1(I)+X2*V2(I)                                  MAJ  10
      RETURN                                                   MAJ  11
      END                                                      MAJ  12

      SUBROUTINE NORME(N,VDEL,V,XNORM)                         NORM  1
C==================================================================NORM  2
C     COMPUTE THE LENGTHS RATIO OF VECTORS VDEL AND V          NORM  3
C==================================================================NORM  4
      IMPLICIT REAL*8(A-H,O-Z)                                 NORM  5
      DIMENSION VDEL(1),V(1)                                   NORM  6
      DATA ZERO/0.D0/,UN/1.D0/,FAC/1.D-3/                      NORM  7
      SQRT(X)=DSQRT(X)                                         NORM  8
C------------------------------------------------------------------NORM  9
      C1=ZERO                                                  NORM 10
      C2=ZERO                                                  NORM 11
      DO 10 I=1,N                                              NORM 12
      C1=C1+VDEL(I)*VDEL(I)                                    NORM 13
10    C2=C2+V(I)*V(I)                                          NORM 14
      C=C1*FAC                                                 NORM 15
      IF(C2.LE.C) C2=UN                                        NORM 16
      XNORM=SQRT(C1/C2)                                        NORM 17
      RETURN                                                   NORM 18
      END                                                      NORM 19
```

Figure 6.24 Block 'NLIN'.

Function

This block solves a linear or non-linear time-dependent problem using one of the EULER formulas described in paragraph 5.4 (algorithm of Figure 5.21).

Flow chart

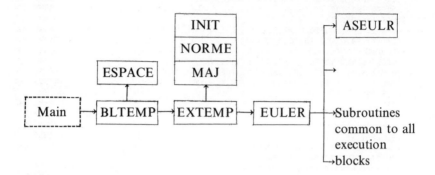

Subroutines

BLTEMP⎫
EXTEMP⎭ are almost identical to subroutines BLNLIN and EXNLIN (Figure 6.25).

EULER: executes all the operations required by one iteration of the EULER algorithm (Figure 6.25).

ASEULR: assembles the residual and eventually the global matrix of the EULER algorithm (Figure 6.25).

INIT ⎫
MAJ ⎬ (Figure 6.24).
NORME ⎭

Remarks

All other methods of solution for time-dependent problems (Runge–Kutta, predictor–corrector, Wilson) can easily be included by adding other execution subroutines besides EULER. Some storage might be needed for methods requiring several past step values.

6.5.3.7 Block 'VALP'

Function

This block computes eigenvalues and eigenvectors by the subspace iteration method described in paragraph 5.5.3.4 when $[K]$ and $[M]$ are symmetric and positive definite. When the user seeks only one eigenvalue, this block performs

```
      SUBROUTINE BLTEMP                                                BLTE  1
C======================================================================BLTE  2
C     TO CALL BLOCK 'TEMP'                                             BLTE  3
C     TO SOLVE AN UNSTEADY PROBLEM (LINEAR OR NOT)                     BLTE  4
C======================================================================BLTE  5
      IMPLICIT REAL*8(A-H,O-Z)                                         BLTE  6
      REAL*4 TBL                                                       BLTE  7
      COMMON/ELEM/NUL(4),ME                                            BLTE  8
      COMMON/ASSE/NSYM,NKG,NKE,NDLE                                    BLTE  9
      COMMON/RESO/NEQ                                                  BLTE 10
      COMMON/NLIN/EPSDL,XNORM,OMEGA,XPAS,DPAS,DPASO,NPAS,IPAS,NITER,   BLTE 11
     1 ITER,IMETH                                                      BLTE 12
      COMMON/ES/M,MR,MP,M1,M2,M3,M4                                    BLTE 13
      COMMON/LOC/LCORG,LDLNC,LNEQ,LDIMP,LPRNG,LPREG,LLD,LLOCE,LCORE,LNE,BLTE 14
     1 LPRNE,LPREE,LDLE,LKE,LFE,LKGS,LKGD,LKGI,LFG,LRES,LDLG,LME,      BLTE 15
     1 LDLEO,LDLGO,LFGO                                                BLTE 16
      COMMON VA(1)                                                     BLTE 17
      DIMENSION TBL(13),IN(2),XIN(3)                                   BLTE 18
      DATA TBL/4HKGS ,4HKGD ,4HKGI ,4HFG  ,4HKE  ,                     BLTE 19
     1 4HFE  ,4HRES ,4HDLE ,4HDLG ,4HME  ,4HDLEO,4HDLGO,4HFGO /        BLTE 20
C----------------------------------------------------------------------BLTE 21
      IF(M1.EQ.0) M1=MR                                                BLTE 22
      IF(M2.EQ.0) M2=ME                                                BLTE 23
      WRITE(MP,2000) M                                                 BLTE 24
 2000 FORMAT(//' UNSTEADY SOLUTION (M=',I2,')'/1X,23('='))              BLTE 25
C------ TO ALLOCATE SPACE                                              BLTE 26
      IF(LKGS.EQ.1) CALL ESPACE(NKG,1,TBL(1),LKGS)                     BLTE 27
      IF(LKGD.EQ.1) CALL ESPACE(NEQ,1,TBL(2),LKGD)                     BLTE 28
      IF(NSYM.EQ.1.AND.LKGI.EQ.1) CALL ESPACE(NKG,1,TBL(3),LKGI)       BLTE 29
      IF(LFG.EQ.1)  CALL ESPACE (NEQ,1,TBL(4),LFG)                     BLTE 30
      IF(LKE.EQ.1)  CALL ESPACE(NKE,1,TBL(5),LKE)                      BLTE 31
      IF(LFE.EQ.1)  CALL ESPACE(NDLE,1,TBL(6),LFE)                     BLTE 32
      IF(LRES.EQ.1) CALL ESPACE(NEQ,1,TBL(7),LRES)                     BLTE 33
      IF(LDLE.EQ.1) CALL ESPACE(NDLE,1,TBL(8),LDLE)                    BLTE 34
      IF(LDLG.EQ.1) CALL ESPACE(NEQ,1,TBL(9),LDLG)                     BLTE 35
      IF(LME.EQ.1)  CALL ESPACE(NKE,1,TBL(10),LME)                     BLTE 36
      IF(LDLEO.EQ.1) CALL ESPACE(NDLE,1,TBL(11),LDLEO)                 BLTE 37
      IF(LDLGO.EQ.1) CALL ESPACE(NEQ,1,TBL(12),LDLGO)                  BLTE 38
      IF(LFGO.EQ.1)  CALL ESPACE(NEQ,1,TBL(13),LFGO)                   BLTE 39
C------ TO EXECUTE THE BLOCK                                           BLTE 40
      CALL EXTEMP(VA(LCORG),VA(LDLNC),VA(LDIMP),VA(LNEQ),VA(LLD),      BLTE 41
     1 VA(LLOCE),VA(LCORE),VA(LPRNE),VA(LPREE),VA(LNE),VA(LKE),VA(LME),BLTE 42
     2 VA(LFE),VA(LDLE),VA(LKGS),VA(LKGD),VA(LKGI),VA(LFG),VA(LRES),   BLTE 43
     3 VA(LDLG),VA(LDLEO),VA(LDLGO),VA(LFGO))                          BLTE 44
      RETURN                                                           BLTE 45
      END                                                              BLTE 46

      SUBROUTINE EXTEMP(VCORG,KDLNC,VDIMP,KNEQ,KLD,KLOCE,VCORE,VPRNE,  EXTE  1
     1 VPREE,KNE,VKE,VME,VFE,VDLE,VKGS,VKGD,VKGI,VFG,VRES,VDLG,        EXTE  2
     2 VDLEO,VDLGO,VFGO)                                               EXTE  3
C======================================================================EXTE  4
C     TO EXECUTE BLOCK 'TEMP'                                          EXTE  5
C     TO SOLVE AN UNSTEADY PROBLEM (LINEAR OR NOT)                     EXTE  6
C======================================================================EXTE  7
      IMPLICIT REAL*8(A-H,O-Z)                                         EXTE  8
      COMMON/RESO/NEQ                                                  EXTE  9
      COMMON/COND/NCLT,NCLZ,NCLNZ                                      EXTE 10
      COMMON/ASSE/NSYM                                                 EXTE 11
      COMMON/NLIN/EPSDL,XNORM,OMEGA,XPAS,DPAS,DPASO,NPAS,IPAS,NITER,   EXTE 12
     1 ITER,IMETH                                                      EXTE 13
      COMMON/ES/M,MR,MP,M1,M2,M3,M4                                    EXTE 14
      DIMENSION VCORG(1),KDLNC(1),VDIMP(1),KNEQ(1),KLD(1),KLOCE(1),    EXTE 15
     1 VCORE(1),VPRNE(1),VPREE(1),KNE(1),VKE(1),VME(1),VFE(1),VDLE(1), EXTE 16
     2 VKGS(1),VKGD(1),VKGI(1),VFG(1),VRES(1),VDLG(1),VDLEO(1),        EXTE 17
     3 VDLGO(1),VFGO(1)                                                EXTE 18
      DATA ZERO/0.D0/,UN/1.D0/                                         EXTE 19
C----------------------------------------------------------------------EXTE 20
```

Figure 6.25 (*Contd.*)

```
              DPAS0=ZERO                                                  EXTE 21
              XPAS=ZERO                                                   EXTE 22
              IPAS=0                                                      EXTE 23
       C------- READ INITIAL D.O.F. ON FILE M3                            EXTE 24
              IF(M3.EQ.0) GO TO 5                                         EXTE 25
              REWIND M3                                                   EXTE 26
              READ(M3) (VDLG(I),I=1,NEQ)                                  EXTE 27
              CALL MAJ(UN,ZERO,NEQ,VDLG,VDLG0)                            EXTE 28
       C------- SAVE THE REFERENCE LOAD CONDITIONS                        EXTE 29
       5      CALL MAJ(UN,ZERO,NEQ,VFG,VFG0)                              EXTE 30
       C------- READ A CARD DEFINING A SET OF IDENTICAL STEPS             EXTE 31
       10     READ(M1,1000) DPAS,I1,I2,I3,X1,X2                           EXTE 32
       1000   FORMAT(F10.0,3I5,2F10.0)                                    EXTE 33
              IF(DPAS.EQ.ZERO) GO TO 140                                  EXTE 34
              IF(I1.GT.0) NPAS=I1                                         EXTE 35
              IF(I2.GT.0) NITER=I2                                        EXTE 36
              IF(I3.GT.0) IMETH=I3                                        EXTE 37
              IF(X1.GT.ZERO) EPSDL=X1                                     EXTE 38
              IF(X2.NE.ZERO) OMEGA=X2                                     EXTE 39
       C                                                                  EXTE 40
       C------- LOOP OVER THE STEPS                                       EXTE 41
       C                                                                  EXTE 42
              DO 130 IP=1,NPAS                                            EXTE 43
              CALL INIT(ZERO,NEQ,VFG)                                     EXTE 44
              IPAS=IPAS+1                                                 EXTE 45
              XPAS=XPAS+DPAS                                              EXTE 46
              WRITE(MP,2000) IPAS,DPAS,XPAS,NITER,IMETH,EPSDL,OMEGA       EXTE 47
       2000   FORMAT(/1X,13('-'),'STEP NUMBER (IPAS):',I5//                EXTE 48
            1              14X,'INCREMENT           (DPAS)=',E12.5/       EXTE 49
            2              14X,'TOTAL LEVEL         (XPAS)=',E12.5/       EXTE 50
            3              14X,'NUMBER OF ITERATIONS (NITER)=',I12/       EXTE 51
            4              14X,'METHOD NUMBER       (IMETH):',I12/        EXTE 52
            5              14X,'TOLERANCE           (EPSDL)=',E12.5/      EXTE 53
            6              14X,'COEFFICIENT ALPHA   (OMEGA)=',E12.5/)     EXTE 54
       C                                                                  EXTE 55
       C------- LOOP OVER EQUILIBRIUM ITERATIONS                          EXTE 56
       C                                                                  EXTE 57
              DO 110 ITER=1,NITER                                         EXTE 58
       C------- CHOOSE THE METHOD                                         EXTE 59
              IF(IMETH.GT.3) GO TO 20                                     EXTE 60
       C------- EULER TYPE METHODS                                        EXTE 61
              CALL EULER(VCORG,KDLNC,VDIMP,KNEQ,KLD,KLOCE,VCORE,VPRNE,VPREE, EXTE 62
            1 KNE,VKE,VME,VFE,VDLE,VKGS,VKGD,VKGI,VFG,VRES,VDLG,          EXTE 63
            2 VDLE0,VDLG0,VFG0)                                           EXTE 64
              GO TO 100                                                   EXTE 65
       C------- OTHER METHODS ......                                      EXTE 66
       20     CONTINUE                                                    EXTE 67
              WRITE(MP,2010) IMETH                                        EXTE 68
       2010   FORMAT(' ** ERROR, METHOD:',I3,' UNKNOWN')                  EXTE 69
              STOP                                                        EXTE 70
       C------- COMPUTE THE NORM                                          EXTE 71
       100    CALL NORME(NEQ,VRES,VDLG,XNORM)                             EXTE 72
              IF(M.GT.0) WRITE(MP,2020) ITER,XNORM                        EXTE 73
       2020   FORMAT(5X,'ITERATION (ITER):',I3,' NORM (XNORM)=',E12.5)    EXTE 74
              IF(M.GE.2) CALL PRSOL(KDLNC,VCORG,VDIMP,KNEQ,VDLG)          EXTE 75
              IF(XNORM.LE.EPSDL) GO TO 120                                EXTE 76
       110    CONTINUE                                                    EXTE 77
       C------- END OF STEP                                               EXTE 78
       120    DPAS0=DPAS                                                  EXTE 79
              CALL MAJ(UN,ZERO,NEQ,VDLG,VDLG0)                            EXTE 80
              CALL PRSOL(KDLNC,VCORG,VDIMP,KNEQ,VDLG)                     EXTE 81
       130    CONTINUE                                                    EXTE 82
              GO TO 10                                                    EXTE 83
       C------- SAVE THE SOLUTION ON FILE M4                              EXTE 84
       140    IF(M4.NE.0) WRITE(M4) (VDLG(I),I=1,NEQ)                     EXTE 85
              RETURN                                                      EXTE 86
              END                                                         EXTE 87
```

Figure 6.25 (*Contd.*)

```
      SUBROUTINE EULER(VCORG,KDLNC,VDIMP,KNEQ,KLD,KLOCE,VCORE,VPRNE,     EULE   1
     1 VPREE,KNE,VKE,VME,VFE,VDLE,VKGS,VKGD,VKGI,VFG,VRES,VDLG,          EULE   2
     2 VDLEO,VDLGO,VFGO)                                                 EULE   3
C=====================================================================EULE   4
C     ALGORITHM FOR EULER TYPE METHODS (IMPLICIT, EXPLICIT OR BOTH       EULE   5
C     ACCORDING TO OMEGA) FOR LINEAR OR NON LINEAR PROBLEMS.             EULE   6
C     THE NON LINEAR PROBLEM IS SOLVED BY A NEWTON-RAPHSON               EULE   7
C     METHOD          EULE   8
C       IMETH.EQ.1   STANDARD NEWTON-RAPHSON                             EULE   9
C       IMETH.EQ.2   K IS CONSTANT                                       EULE  10
C       IMETH.EQ.3   K IS RECOMPUTED AT THE BEGINNING OF EACH STEP       EULE  11
C=====================================================================EULE  12
      IMPLICIT REAL*8(A-H,O-Z)                                           EULE  13
      COMMON/ASSE/NSYM,NKG                                               EULE  14
      COMMON/RESO/NEQ                                                    EULE  15
      COMMON/NLIN/EPSDL,XNORM,OMEGA,XPAS,DPAS,DPASO,NPAS,IPAS,NITER,     EULE  16
     1 ITER,IMETH                                                        EULE  17
      COMMON/ES/M,MR,MP                                                  EULE  18
      DIMENSION VCORG(1),KDLNC(1),VDIMP(1),KNEQ(1),KLD(1),KLOCE(1),      EULE  19
     1 VCORE(1),VPRNE(1),VPREE(1),KNE(1),VKE(1),VME(1),VFE(1),           EULE  20
     2 VDLE(1),VKGS(1),VKGD(1),VKGI(1),VFG(1),VRES(1),VDLG(1),           EULE  21
     3 VDLEO(1),VDLGO(1),VFGO(1)                                         EULE  22
      DATA ZERO/0.D0/,UN/1.D0/                                           EULE  23
C---------------------------------------------------------------------EULE  24
C------- DECIDE IF GLOBAL MATRIX IS TO BE REASSEMBLED                     EULE  25
      IKT=0                                                              EULE  26
      IF(IMETH.EQ.1) GO TO 10                                            EULE  27
      IF(DPAS.NE.DPASO.AND.ITER.EQ.1) GO TO 10                           EULE  28
      IF(IMETH.EQ.3.AND.ITER.EQ.1) GO TO 10                              EULE  29
      GO TO 20                                                           EULE  30
10    IKT=1                                                              EULE  31
C------- INITIALIZE GLOBAL MATRIX TO ZERO IF NECESSARY                    EULE  32
20    IF(IKT.EQ.0) GO TO 30                                              EULE  33
      CALL INIT(ZERO,NKG,VKGS)                                           EULE  34
      CALL INIT(ZERO,NEQ,VKGD)                                           EULE  35
      IF(NSYM.EQ.1) CALL INIT(ZERO,NKG,VKGI)                             EULE  36
C------- ASSEMBLE RESIDUALS AND GLOBAL MATRIX IF REQUIRED         EULE  37
30    CALL MAJ(UN,ZERO,NEQ,VFGO,VRES)                                    EULE  38
      CALL ASEULR(IKT,VCORG,KDLNC,VDIMP,KNEQ,KLD,KLOCE,VCORE,VPRNE,      EULE  39
     1 VPREE,KNE,VKE,VME,VFE,VDLE,VKGS,VKGD,VKGI,VFG,VRES,VDLG,          EULE  40
     2 VDLEO,VDLGO,VFGO)                                                 EULE  41
      C1=UN                                                              EULE  42
      IF(ITER.GT.1) C1=C1-OMEGA                                          EULE  43
      DO 40 I=1,NEQ                                                      EULE  44
40    VRES(I)=DPAS*(VRES(I)-C1*VFG(I))                                   EULE  45
C------- SOLVE                                                            EULE  46
      CALL SOL(VKGS,VKGD,VKGI,VRES,KLD,NEQ,MP,IKT,1,NSYM,ENERG)          EULE  47
C------- UPDATE THE SOLUTION                                              EULE  48
      CALL MAJ(UN,UN,NEQ,VRES,VDLG)                                      EULE  49
      RETURN                                                             EULE  50
      END                                                                EULE  51

      SUBROUTINE ASEULR(IKT,VCORG,KDLNC,VDIMP,KNEQ,KLD,KLOCE,VCORE,      ASEU   1
     1 VPRNE,VPREE,KNE,VKE,VME,VFE,VDLE,VKGS,VKGD,VKGI,VFG,VRES,         ASEU   2
     2 VDLG,VDLEO,VDLGO,VFGO)                                            ASEU   3
C=====================================================================ASEU   4
C     TO ASSEMBLE THE RESIDUALS AND THE GLOBAL MATRIX (IF IKT.EQ.1)      ASEU   5
C     WHILE LOOPING OVER THE ELEMENTS (FOR EULER METHOD)                 ASEU   6
C=====================================================================ASEU   7
      IMPLICIT REAL*8(A-H,O-Z)                                           ASEU   8
      COMMON/ELEM/NELT,NNEL,NTPE,NGRE,ME,NIDENT                          ASEU   9
      COMMON/ASSE/NSYM                                                   ASEU  10
      COMMON/RESO/NEQ                                                    ASEU  11
      COMMON/RGDT/IEL,ITPE,ITPE1,IGRE,IDLE,ICE,IPRNE,IPREE,INEL,IDEG,IPGASEU  12
     1 ,ICOD                                                             ASEU  13
      COMMON/NLIN/EPSDL,XNORM,OMEGA,XPAS,DPAS,DPASO,NPAS,IPAS,NITER,     ASEU  14
     1 ITER,IMETH                                                        ASEU  15
```

Figure 6.25 (*Contd.*)

```
      COMMON/ES/M,MR,MP,M1,M2                                     ASEU 16
      DIMENSION VCORG(1),KDLNC(1),VDIMP(1),KNEQ(1),KLD(1),KLOCE(1), ASEU 17
     1 VCORE(1),VPRNE(1),VPREE(1),KNE(1),VKE(1),VME(1),VFE(1),VDLE(1), ASEU 18
     2 VKGS(1),VKGD(1),VKGI(1),VFG(1),VRES(1),VDLG(1),VDLE0(1),    ASEU 19
     3 VDLG0(1),VFG0(1)                                            ASEU 20
      DATA UN/1.D0/                                                ASEU 21
C-----------------------------------------------------------------ASEU 22
      CC=DPAS*OMEGA                                                ASEU 23
      IFE=0                                                        ASEU 24
      IF(ITER.GT.1) IFE=1                                          ASEU 25
C------- REWIND ELEMENT FILE (ME)                                  ASEU 26
      REWIND M2                                                    ASEU 27
C------- LOOP OVER THE ELEMENTS                                    ASEU 28
      DO 90 IE=1,NELT                                              ASEU 29
C------- READ AN ELEMENT                                           ASEU 30
      CALL RDELEM(M2,KLOCE,VCORE,VPRNE,VPREE,KNE)                  ASEU 31
C------- EVALUATE INTERPOLATION FUNCTIONS IF REQUIRED              ASEU 32
      IF(ITPE.EQ.ITPE1) GO TO 10                                   ASEU 33
      ICOD=2                                                       ASEU 34
      CALL ELEMLB(VCORE,VPRNE,VPREE,VDLE,VKE,VFE)                  ASEU 35
C------- FIND ELEMENT D.O.F. FROM VFG                              ASEU 36
10    CALL DLELM(KLOCE,VDLG,VDIMP,VDLE)                            ASEU 37
C------- COMPUTE THE RESIDUAL K.U.                                 ASEU 38
      ICOD=6                                                       ASEU 39
      CALL ELEMLB(VCORE,VPRNE,VPREE,VDLE,VKE,VFE)                  ASEU 40
C------- COMPUTE MATRIX M                                          ASEU 41
      ICOD=5                                                       ASEU 42
      CALL ELEMLB(VCORE,VPRNE,VPREE,VDLE,VME,VFE)                  ASEU 43
C------- COMPUTE MATRIX K IF REQUIRED                              ASEU 44
      IF(IKT.EQ.0) GO TO 15                                        ASEU 45
      ICOD=3                                                       ASEU 46
      CALL ELEMLB(VCORE,VPRNE,VPREE,VDLE,VKE,VFE)                  ASEU 47
C------- RESIDUALS OF THE FIRST ITERATION IN EACH STEP (LINEAR)    ASEU 48
15    IF(ITER.GT.1) GO TO 20                                       ASEU 49
      CALL ASSEL(0,1,IDLE,NSYM,KLOCE,KLD,VKE,VFE,VKGS,VKGD,VKGI,VFG) ASEU 50
      GO TO 60                                                     ASEU 51
C------- RESIDUALS AFTER FIRST ITERATION                           ASEU 52
20    CALL DLELM(KLOCE,VDLG0,VDIMP,VDLE0)                          ASEU 53
      DO 30 I=1,IDLE                                               ASEU 54
      VDLE(I)=(VDLE0(I)-VDLE(I))/DPAS                              ASEU 55
30    VFE(I)=-OMEGA*VFE(I)                                         ASEU 56
C------- PRODUCT M . U                                             ASEU 57
      VFE(1)=VFE(1)+VME(1)*VDLE(1)                                 ASEU 58
      II=1                                                         ASEU 59
      DO 50 J=2,IDLE                                               ASEU 60
      J1=J-1                                                       ASEU 61
      DO 40 I=1,J1                                                 ASEU 62
      II=II+1                                                      ASEU 63
      VFE(I)=VFE(I)+VME(II)*VDLE(J)                                ASEU 64
40    VFE(J)=VFE(J)+VME(II)*VDLE(I)                                ASEU 65
      II=II+1                                                      ASEU 66
50    VFE(J)=VFE(J)+VME(II)*VDLE(J)                                ASEU 67
C------- MATRIX   M + DPAS.OMEGA. K                                ASEU 68
60    IF(IKT.EQ.0) GO TO 80                                        ASEU 69
      II=0                                                         ASEU 70
      DO 70 I=1,IDLE                                               ASEU 71
      DO 70 J=I,IDLE                                               ASEU 72
      II=II+1                                                      ASEU 73
70    VKE(II)=VKE(II)*CC+VME(II)                                   ASEU 74
C------- ASSEMBLE THE RESIDUAL AND THE GLOBAL MATRIX               ASEU 75
80    CALL ASSEL(IKT,IFE,IDLE,NSYM,KLOCE,KLD,VKE,VFE,VKGS,VKGD,VKGI, ASEU 76
     1 VRES)                                                       ASEU 77
90    ITPE1=ITPE                                                   ASEU 78
      RETURN                                                       ASEU 79
      END                                                          ASEU 80
```

Figure 6.25 Block 'TEMP'.

simply inverse iteration requiring that $[M]$ be symmetrical and only $[K]$ be positive-definite. Thus this block may be employed for stability problems where $[M]$ representing the geometrical matrix may not be positive definite.

Flow chart

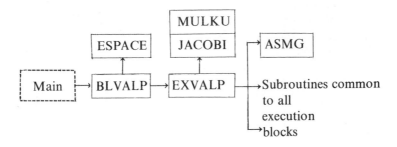

Subroutines

BLVALP: reads a record containing variables NVALP, NITER, EPSDL, SHIFT, NSS, NMDIAG, NSWM, and TOLJAC described in Figure 6.9. Then, it creates all the required arrays (Figure 6.26).

EXVALP: executes the operations required by the subspace algorithm (Figure 6.26).

ASMG: assembles global matrix (Figure 6.26).

MULKU: executes the product $[K] \cdot \{U\}$ (Figure 6.26).

JACOBI: executes the general JACOBI algorithm presented in Figure 5.24 (Figure 5.25).

```
      SUBROUTINE BLVALP                                              BLVA   1
C======================================================================BLVA   2
C     TO CALL BLOCK 'VALP'                                            BLVA   3
C     TO COMPUTE EIGENVALUES AND EIGENVECTORS BY THE SUBSPACE         BLVA   4
C     ITERATION TECHNIQUE                                             BLVA   5
C======================================================================BLVA   6
      IMPLICIT REAL*8(A-H,O-Z)                                        BLVA   7
      REAL*4 TBL                                                      BLVA   8
      COMMON/ELEM/NUL(4),ME                                           BLVA   9
      COMMON/ASSE/NSYM,NKG,NKE,NDLE                                   BLVA  10
      COMMON/RESO/NEQ                                                 BLVA  11
      COMMON/VALP/NITER,NMDIAG,EPSLB,SHIFT,NSS,NSWM,TOLJAC,NVALP      BLVA  12
      COMMON/ES/M,MR,MP,M1,M2                                         BLVA  13
      COMMON/LOC/LCORG,LDLNC,LNEQ,LDIMP,LPRNG,LPREG,LLD,LLOCE,LCORE,LNE,BLVA 14
     1 LPRNE,LPREE,LDLE,LKE,LFE,LKGS,LKGD,LKGI,LFG,LRES,LDLG          BLVA  15
      COMMON/TRVL/X1,X2,X3,I1,I2,I3,I4,I5                             BLVA  16
      COMMON VA(1)                                                    BLVA  17
      DIMENSION TBL(20)                                               BLVA  18
      DATA TBL/4HKGS ,4HKGD ,4HMGS ,4HMGD ,4HFG  ,4HKE  ,4HFE  ,4HDLE , BLVA 19
     1 4HRES ,4HDLG ,4HP   ,4HLAMB,4HLAM1,4HR   ,4HPHI ,4HKSS ,4HMSS , BLVA 20
     1 4HV1  ,4HVX  ,4HV2  /,ZERO/0.D0/                               BLVA  21
C----------------------------------------------------------------------BLVA  22
      IF(M1.EQ.0) M1=MR                                               BLVA  23
```

Figure 6.26 (*Contd.*)

```
      IF(M2.EQ.0) M2=ME                                         BLVA  24
      READ(M1,1000) I1,I2,X1,X2,I3,I4,I5,X3                     BLVA  25
 1000 FORMAT(2I5,2F10.0,3I5,1F10.0)                             BLVA  26
      IF(I1.NE.0) NVALP=I1                                      BLVA  27
      IF(I2.NE.0) NITER=I2                                      BLVA  28
      NSS=I3                                                    BLVA  29
      IF(I4.NE.0) NMDIAG=I4                                     BLVA  30
      IF(I5.NE.0) NSWM=I5                                       BLVA  31
      IF(X1.NE.ZERO) EPSLB=X1                                   BLVA  32
      IF(X2.NE.ZERO) SHIFT=X2                                   BLVA  33
      IF(X3.NE.ZERO) TOLJAC=X3                                  BLVA  34
      IF(NSS.NE.0) GO TO 10                                     BLVA  35
      NSS=MIN0(NVALP+8,2*NVALP)                                 BLVA  36
      NSS=MIN0(NSS,NEQ)                                         BLVA  37
   10 CONTINUE                                                  BLVA  38
      WRITE(MP,2000) M,NVALP,NITER,NMDIAG,EPSLB,SHIFT,NSS,NSWM,TOLJAC BLVA 39
 2000 FORMAT(//' SUBSPACE ITERATION (M=',I2,')'/' ',26('=')/    BLVA  40
     1 15X,'NUMBER OF DESIRED EIGENVALUES        (NVALP)=',I12/ BLVA  41
     2 15X,'MAX. NUMBER OF ITERATIONS PERMITTED  (NITER)=',I12/ BLVA  42
     3 15X,'INDEX FOR DIAGONAL MATRIX           (NMDIAG)=',I12/ BLVA  43
     4 15X,'CONVERGENCE TOLERANCE ON EIGENVALUES (EPSLB)=',E12.5/ BLVA 44
     5 15X,'SHIFT                                (SHIFT)=',E12.5/ BLVA 45
     6 15X,'SUBSPACE DIMENSION                     (NSS)=',I12/ BLVA  46
     7 15X,'MAX. NUMBER OF ITERATION IN JACOBI    (NSWM)=',I12/ BLVA  47
     8 15X,'CONVERGENCE TOLERANCE IN JACOBI     (TOLJAC)=',1E12.5/)BLVA 48
      IF(NVALP.LE.NEQ.AND.NSS.LE.NEQ) GO TO 20                  BLVA  49
      WRITE(MP,2010)                                            BLVA  50
 2010 FORMAT(//'---ERROR--- NVALP OR NSS GREATER THAN NEQ',/,   BLVA  51
     1        '---STOP EXECUTION---')                           BLVA  52
      GO TO 30                                                  BLVA  53
   20 IF(LKGS.EQ.1) CALL ESPACE(NKG,1,TBL(1),LKGS)              BLVA  54
      IF(LKGD.EQ.1) CALL ESPACE(NEQ,1,TBL(2),LKGD)              BLVA  55
      CALL ESPACE(NKG,1,TBL(3),LMGS)                            BLVA  56
      CALL ESPACE(NEQ,1,TBL(4),LMGD)                            BLVA  57
      IF(LFG.EQ.1) CALL ESPACE(NEQ,1,TBL(5),LFG)                BLVA  58
      IF(LKE.EQ.1) CALL ESPACE(NKE,1,TBL(6),LKE)                BLVA  59
      IF(LFE.EQ.1) CALL ESPACE(NDLE,1,TBL(7),LFE)               BLVA  60
      IF(LDLE.EQ.1) CALL ESPACE(NDLE,1,TBL(8),LDLE)             BLVA  61
      IF(LRES.EQ.1) CALL ESPACE(NEQ,1,TBL(9),LRES)              BLVA  62
      IF(LDLG.EQ.1) CALL ESPACE(NEQ,1,TBL(10),LDLG)             BLVA  63
      CALL ESPACE(NEQ*NSS,1,TBL(11),LVEC)                       BLVA  64
      CALL ESPACE(NSS,1,TBL(12),LLAMB)                          BLVA  65
      CALL ESPACE(NSS,1,TBL(13),LLAM1)                          BLVA  66
      CALL ESPACE(NSS*(NSS+1)/2,1,TBL(16),LKSS)                 BLVA  67
      CALL ESPACE(NSS*(NSS+1)/2,1,TBL(17),LMSS)                 BLVA  68
      CALL ESPACE(NEQ,1,TBL(18),LV1)                            BLVA  69
      CALL ESPACE(NSS*NSS,1,TBL(19),LX)                         BLVA  70
      CALL EXVALP(VA(LLD),VA(LDIMP),VA(LLOCE),VA(LCORE),VA(LPRNE), BLVA 71
     1 VA(LPREE),VA(LNE),VA(LFE),VA(LKE),VA(LKGS),VA(LKGD),VA(LFG), BLVA 72
     2 VA(LCORG),VA(LDLNC),VA(LNEQ),VA(LRES),VA(LDLE),VA(LDLG), BLVA 73
     3 VA(LMGS),VA(LMGD),VA(LVEC),VA(LLAMB),VA(LLAM1),VA(LKSS),VA(LMSS)BLVA 74
     4 ,VA(LV1),VA(LX),NEQ,NSS)                                 BLVA  75
   30 RETURN                                                    BLVA  76
      END                                                       BLVA  77

      SUBROUTINE EXVALP(KLD,VDIMP,KLOCE,VCORE,VPRNE,VPREE,KNE,VFE,VKE, EXVA  1
     1 VKGS,VKGD,VFG,VCORG,KDLNC,KNEQ,VRES,VDLE,VDLG,VMGS,VMGD, EXVA  2
     2 VEC,VLAMB,VLAM1,VKSS,VMSS,V1,VX,NEQ,NSS)                 EXVA  3
C===============================================================EXVA  4
C     TO EXECUTE BLOCK 'VALP'                                   EXVA  5
C     TO COMPUTE EIGENVALUES AND EIGENVECTORS BY SUBSPACE       EXVA  6
C     ITERATION         EXVA   7
C     (IF  NVALP.EQ.1  INVERSE ITERATION METHOD)                EXVA  8
C===============================================================EXVA  9
```

Figure 6.26 (*Contd.*)

```
      IMPLICIT REAL*8(A-H,O-Z)                                    EXVA  10
      COMMON/ASSE/NSYM,NKG,NKE,NDLE                               EXVA  11
      COMMON/VALP/NITER,NMDIAG,EPSLB,SHIFT,NSS1,NSWM,TOLJAC,NVALP EXVA  12
      COMMON/ES/M,MR,MP                                           EXVA  13
      DIMENSION KLD(1),VDIMP(1),KLOCE(1),VCORE(1),VPRNE(1),VPREE(1), EXVA 14
     1 KNE(1),VFE(1),VKE(1),VKGS(1),VKGD(1),VFG(1),VCORG(1),KDLNC(1), EXVA 15
     2 KNEQ(1),VRES(1),VDLE(1),VDLG(1),VMGS(1),VMGD(1),VEC(NEQ,1), EXVA 16
     3 VLAMB(1),VLAM1(1),VKSS(1),VMSS(1),V1(1),VX(NSS,1)           EXVA 17
      DATA ZERO/0.D0/,UN/1.0D0/,GRAND/1.0D32/                     EXVA  18
      ABS(X)=DABS(X)                                              EXVA  19
C-----------------------------------------------------------------EXVA  20
C                                                                 EXVA  21
C------- PRELIMINARY COMPUTATIONS                                 EXVA  22
C                                                                 EXVA  23
C------- ASSEMBLE KG AND MG                                       EXVA  24
      CALL ASKG(KLD,VDIMP,KLOCE,VCORE,VPRNE,VPREE,KNE,VKE,VFE,VKGS,VKGD,EXVA 25
     1 VKGI,VFG,VDLE,VRES)                                        EXVA  26
      CALL ASMG(KLD,VDIMP,KLOCE,VCORE,VPRNE,VPREE,KNE,VKE,VFE,VMGS, EXVA 27
     1 VMGD,VMGS,VFG,VDLE,VRES)                                   EXVA  28
C------- TRIANGULARIZE KG                                         EXVA  29
      CALL SOL(VKGS,VKGD,VKGI,VFG,KLD,NEQ,MP,1,0,0,ENERG)         EXVA  30
C------- LOAD VECTOR EQUAL TO DIAGONAL OF M                       EXVA  31
      CMAX=ZERO                                                   EXVA  32
      ICONT=0                                                     EXVA  33
      DO 10 ID=1,NEQ                                              EXVA  34
      C=GRAND                                                     EXVA  35
C------- CHECK FOR ZERO DIAGONAL TERM IN VMGD                     EXVA  36
      IF(VMGD(ID).EQ.ZERO) GO TO 5                                EXVA  37
      ICONT=ICONT+1                                               EXVA  38
      C=VKGD(ID)/VMGD(ID)                                         EXVA  39
    5 V1(ID)=C                                                    EXVA  40
      IF(C.GT.CMAX) CMAX=C                                        EXVA  41
      VEC(ID,1)=VMGD(ID)                                          EXVA  42
      DO 10 JS=2,NSS                                              EXVA  43
   10 VEC(ID,JS)=ZERO                                             EXVA  44
C------- CHECK IF SUBSPACE DIMENSION IS EQUAL TO MASS D.O.F.      EXVA  45
      IF(ICONT.LT.NSS) GO TO 250                                  EXVA  46
C------- UNIT LOAD VECTORS CORRESPONDING TO MIN. OF               EXVA  47
C            K(I,I)/M(I,I)                                        EXVA  48
      DO 30 JS=2,NSS                                              EXVA  49
      C=CMAX                                                      EXVA  50
      DO 20 ID=1,NEQ                                              EXVA  51
      IF(V1(ID).GT.C) GO TO 20                                    EXVA  52
      C=V1(ID)                                                    EXVA  53
      II=ID                                                       EXVA  54
   20 CONTINUE                                                    EXVA  55
      V1(II)=CMAX                                                 EXVA  56
      VEC(II,JS)=UN                                               EXVA  57
   30 VLAMB(JS)=UN                                                EXVA  58
      VLAMB(1)=UN                                                 EXVA  59
      IF(NVALP.EQ.1) NSS=1                                        EXVA  60
C------- INVERSE ITERATION IF NVALP=1                             EXVA  61
C------- START  ITERATIONS LOOP                                   EXVA  62
C                                                                 EXVA  63
      ITERM=0                                                     EXVA  64
      ITMAX=NITER+1                                               EXVA  65
      DO 200 ITER=1,ITMAX                                         EXVA  66
C------- COMPUTE RITZ VECTORS                                     EXVA  67
      II0=0                                                       EXVA  68
      DO 80 JS=1,NSS                                              EXVA  69
      II0=II0+JS                                                  EXVA  70
      DO 40 ID=1,NEQ                                              EXVA  71
   40 V1(ID)=VEC(ID,JS)                                           EXVA  72
      CALL SOL(VKGS,VKGD,VKGI,V1,KLD,NEQ,MP,0,1,0,ENERG)          EXVA  73
```

Figure 6.26 (*Contd.*)

```
C------- CALCULATE THE PROJECTION OF K                          EXVA  74
       II=II0                                                   EXVA  75
       DO 60 IS=JS,NSS                                          EXVA  76
       C=ZERO                                                   EXVA  77
       DO 50 ID=1,NEQ                                           EXVA  78
 50    C=C+V1(ID)*VEC(ID,IS)                                    EXVA  79
       VKSS(II)=C                                               EXVA  80
 60    II=II+IS                                                 EXVA  81
       DO 70 ID=1,NEQ                                           EXVA  82
 70    VEC(ID,JS)=V1(ID)                                        EXVA  83
 80    CONTINUE                                                 EXVA  84
C------- CALCULATE THE PROJECTION OF M                          EXVA  85
       II0=0                                                    EXVA  86
       DO 120 JS=1,NSS                                          EXVA  87
       II0=II0+JS                                               EXVA  88
       DO 85 ID=1,NEQ                                           EXVA  89
 85    V1(ID)=ZERO                                              EXVA  90
       CALL MULKU(VMGS,VMGD,VMGS,KLD,VEC(1,JS),NEQ,0,V1)        EXVA  91
       II=II0                                                   EXVA  92
       DO 100 IS=JS,NSS                                         EXVA  93
       C=ZERO                                                   EXVA  94
       DO 90 ID=1,NEQ                                           EXVA  95
 90    C=C+V1(ID)*VEC(ID,IS)                                    EXVA  96
       IF(ITERM.GT.0) GO TO 120                                 EXVA  97
       VMSS(II)=C                                               EXVA  98
 100   II=II+IS                                                 EXVA  99
       DO 110 ID=1,NEQ                                          EXVA 100
 110   VEC(ID,JS)=V1(ID)                                        EXVA 101
 120   CONTINUE                                                 EXVA 102
       IF (NSS.GT.1) GO TO 125                                  EXVA 103
       VLAM1(1)=VKSS(1)/VMSS(1)                                 EXVA 104
       GO TO 165                                                EXVA 105
C------- CALCULATE EIGENVALUES IN THE SUBSPACE                  EXVA 106
 125   CALL JACOBI(VKSS,VMSS,NSS,NSWM,TOLJAC,V1,VLAM1,VX)       EXVA 107
C------- NEW LOAD VECTOR       EXVA 108
       DO 160 ID=1,NEQ                                          EXVA 109
       DO 130 JS=1,NSS                                          EXVA 110
 130   V1(JS)=VEC(ID,JS)                                        EXVA 111
       DO 150 JS=1,NSS                                          EXVA 112
       C=ZERO                                                   EXVA 113
       DO 140 IS=1,NSS                                          EXVA 114
 140   C=C+V1(IS)*VX(IS,JS)                                     EXVA 115
 150   VEC(ID,JS)=C                                             EXVA 116
 160   CONTINUE                                                 EXVA 117
 165   CONTINUE                                                 EXVA 118
C------- PRINT THE ITERATION VALUES                             EXVA 119
       IF(M.LT.1) GO TO 180                                     EXVA 120
       WRITE(MP,2000) ITER                                      EXVA 121
 2000  FORMAT(//' . . . . . . ITERATION ',I5/)                  EXVA 122
       DO 170 IS=1,NSS                                          EXVA 123
       WRITE(MP,2010) IS,VLAM1(IS)                              EXVA 124
 2010  FORMAT(/' EIGENVALUE NO. ',I5,' =',E12.5//' EIGENVECTOR:') EXVA 125
 170   CALL PRSOL(KDLNC,VCORG,VDIMP,KNEQ,VEC(1,IS))             EXVA 126
C------- CHECK FOR CONVERGENCE                                  EXVA 127
 180   IF(ITERM.GT.0) GO TO 210                                 EXVA 128
       C=ZERO                                                   EXVA 129
       IEX=0                                                    EXVA 130
       DO 190 IS=1,NSS                                          EXVA 131
       C1=ABS((VLAM1(IS)-VLAMB(IS))/VLAMB(IS))                  EXVA 132
       IF(C1.GT.C) C=C1                                         EXVA 133
       IF(C1.LE.EPSLB) IEX=IEX+1                                EXVA 134
 190   CONTINUE                                                 EXVA 135
       WRITE(MP,2015) ITER,C,IEX                                EXVA 136
 2015  FORMAT(' ITERATION ',I4,' MAX. ERROR=',E9.1,' EXACT EIGENVALUES:'EXVA 137
```

Figure 6.26 (*Contd.*)

```
              1,I4)                                                EXVA 138
              IF(IEX.GE.NVALP) ITERM=1                             EXVA 139
C------- NON CONVERGENCE                                           EXVA 140
              IF(ITER.LT.NITER.OR.ITERM.EQ.1) GO TO 195            EXVA 141
              WRITE(MP,2020) NITER                                 EXVA 142
 2020   FORMAT(' ** NON CONVERGENCE AFTER ',I5,' ITERATIONS')      EXVA 143
              ITERM=1                                              EXVA 144
C------- SAVE THE EIGENVALUES                                      EXVA 145
 195    DO 200 IS=1,NSS                                            EXVA 146
 200    VLAMB(IS)=VLAM1(IS)                                        EXVA 147
C                                                                  EXVA 148
C------- RESULT                                                    EXVA 149
C                                                                  EXVA 150
C------- ARRANGE EIGENVALUES IN ASCENDING ORDER                    EXVA 151
 210    IS1=NSS-1                                                  EXVA 152
              IF (IS1.EQ.0) GO TO 235                              EXVA 153
              DO 230 IS=1,IS1                                      EXVA 154
              I1=IS+1                                              EXVA 155
              C=VLAMB(IS)                                          EXVA 156
              II=IS                                                EXVA 157
              DO 220 JS=I1,NSS                                     EXVA 158
              IF(C.LT.VLAMB(JS)) GO TO 220                         EXVA 159
              C=VLAMB(JS)                                          EXVA 160
              II=JS                                                EXVA 161
 220    CONTINUE                                                   EXVA 162
              VLAMB(II)=VLAMB(IS)                                  EXVA 163
              VLAMB(IS)=C                                          EXVA 164
              DO 230 ID=1,NEQ                                      EXVA 165
              C=VEC(ID,IS)                                         EXVA 166
              VEC(ID,IS)=VEC(ID,II)                                EXVA 167
 230    VEC(ID,II)=C                                               EXVA 168
C------- PRINT RESULT                                              EXVA 169
              WRITE(MP,2030) ITER                                  EXVA 170
 2030   FORMAT(/' . . . . CONVERGENCE IN',I4,' ITERATIONS'/)       EXVA 171
 235    CONTINUE                                                   EXVA 172
              DO 240 IS=1,NVALP                                    EXVA 173
              WRITE(MP,2010) IS,VLAMB(IS)                          EXVA 174
 240    CALL PRSOL(KDLNC,VCORG,VDIMP,KNEQ,VEC(1,IS))               EXVA 175
              GO TO 260                                            EXVA 176
 250    CONTINUE                                                   EXVA 177
              WRITE(MP,2040)                                       EXVA 178
 2040   FORMAT(' ** NSS IS LARGER THAN MASS D.O.F.')               EXVA 179
 260    RETURN                                                     EXVA 180
              END                                                  EXVA 181

              SUBROUTINE ASMG(KLD,VDIMP,KLOCE,VCORE,VPRNE,VPREE,KNE,VKE,VFE,  ASMG  1
            1 VKGS,VKGD,VKGI,VFG,VDLE,VRES)                        ASMG  2
C==================================================================ASMG  3
C       TO ASSEMBLE THE GLOBAL MASS MATRIX (ELEMENT FUNCTION 5)    ASMG  4
C==================================================================ASMG  5
              IMPLICIT REAL*8(A-H,O-Z)                             ASMG  6
              COMMON/ELEM/NELT,NNEL,NTPE,NGRE,ME,NIDENT            ASMG  7
              COMMON/ASSE/NSYM                                     ASMG  8
              COMMON/RESO/NEQ                                      ASMG  9
              COMMON/RGDT/IEL,ITPE,ITPE1,IGRE,IDLE,ICE,IPRNE,IPREE,INEL,IDEG,IPG ASMG 10
            1 ,ICOD                                                ASMG 11
              COMMON/ES/M,MR,MP,M1,M2                              ASMG 12
              DIMENSION KLD(1),VDIMP(1),KLOCE(1),VCORE(1),VPRNE(1),VPREE(1), ASMG 13
            1 KNE(1),VKE(1),VFE(1),VKGS(1),VKGD(1),VKGI(1),VFG(1),VDLE(1), ASMG 14
            2 VRES(1),KEB(1)                                       ASMG 15
C------------------------------------------------------------------ASMG 16
C------- REWIND ELEMENT FILE (M2)                                  ASMG 17
              REWIND M2                                            ASMG 18
C------- LOOP OVER THE ELEMENTS                                    ASMG 19
              DO 30 IE=1,NELT                                      ASMG 20
```

Figure 6.26 (*Contd.*)

```
C------- SKIP COMPUTATIONS IF IDENTICAL ELEMENTS              ASMG 21
        IF(NIDENT.EQ.1.AND.IE.GT.1) GO TO 20                  ASMG 22
C------- READ AN ELEMENT                                      ASMG 23
        CALL RDELEM(M2,KLOCE,VCORE,VPRNE,VPREE,KNE)           ASMG 24
C------- EVALUATE INTERPOLATION FUNCTIONS IF REQUIRED         ASMG 25
        IF(ITPE.EQ.ITPE1) GO TO 10                            ASMG 26
        ICOD=2                                                ASMG 27
        CALL ELEMLB(VCORE,VPRNE,VPREE,VDLE,VKE,VFE)           ASMG 28
 10     ICOD=5                                                ASMG 30
        CALL ELEMLB(VCORE,VPRNE,VPREE,VDLE,VKE,VFE)           ASMG 31
C------- PRINT ELEMENT MATRIX                                 ASMG 32
        IF(M.LT.2) GO TO 20                                   ASMG 33
        IF(NSYM.EQ.0) IKE=IDLE*(IDLE+1)/2                     ASMG 34
        IF(NSYM.EQ.1) IKE=IDLE*IDLE                           ASMG 35
        WRITE(MP,2000) IEL,(VKE(I),I=1,IKE)                   ASMG 36
 2000   FORMAT(/' MATRIX (ME) , ELEMENT:',I5/(10X,10E12.5))   ASMG 37
C------- ASSEMBLE                                             ASMG 38
 20     CALL ASSEL(1,0,IDLE,NSYM,KLOCE,KLD,VKE,VFE,VKGS,VKGD,VKGI,VFG)  ASMG 39
 30     ITPE1=ITPE                                            ASMG 40
        RETURN                                                ASMG 41
        END                                                   ASMG 42
```

Figure 6.26 Block 'VALP'.

6.6 Description of Input Values for Program MEF

6.6.1 *Conventions*

To each functional block corresponds a number of records to be read:

— a title;
— a record with parameters, if needed;
— additional records containing other quantities, if necessary.

All the header records use the same format:

Variable	Columns	Default	Format	Description
BLOC	1–4	—	A4	Name of block to execute
M	5–10	0	I6	Parameter controlling amount of printing $0 \leq M \leq 4$
M1	11–15	5	I5	Logical unit number of input device
M2	16–20	—	I5 ⎫	Logical unit number of the
⋮	⋮		⋮ ⎬	files used by this
M10	56–60	—	I5 ⎭	block

In general, all integer variables are read under I5 format, and real variables, under F10.0 format.

For all input records, the same method of description as the one used above is employed.

6.6.2 *Input Corresponding to Each Block*

| IMAG | Output of all inputted quantities (optional, but must be the first block if used).

— one record with the word 'IMAG'.

| COMT | Output of comments (optional, but can be used at any time).
— one record with the word 'COMT';
— comment records (maximum length of 80 characters), must be terminated with a blank record.

| COOR | Reading of coordinates and number of degrees of freedom of all the nodes.
— one record containing the word 'COOR';
— one record of parameters;

Variable	Columns	Default	Format	Description
NNT	1–5	20	I5	Maximum number of nodes
NDLN	6–10	2	I5	Maximum number of DOF per node
NDIM	11–15	2	I5	Number of dimensions (1, 2, or 3)
FAC(1)	16–25	1.0	F10.0	Scaling factor in direction x
FAC(2)	26–35	1.0	F10.0	Scaling factor in direction y
FAC(3)	36–45	1.0	F10.0	Scaling factor in direction z

— series of records for nodes; terminated by a record on which $IN1 \leq 0$; (each record may generate many nodes).

Variable	Columns	Default	Format	Description
IN1	1–5	—	I5	Number of first node to be generated
X1(1)	6–15	—	F10.0	x coordinate
X1(2)	16–25	—	F10.0	y coordinate
X1(3)	26–35	—	F10.0	z coordinate
IN2	36–40	IN1	I5	Number of the last node to be generated
X2(1)	41–50	x1(1)	F10.0	x coordinate
X2(2)	51–60	x1(2)	F10.0	y coordinate
X2(3)	61–70	x1(3)	F10.0	z coordinate
INCR	71–75	1	I5	Pitch of the node numbers
IDLN	76–80	NDLN	I5	Number of DOF generated if different from default (NDLN)

Remarks

— The number of degrees of freedom of a node must be coherent with the number of freedoms at a node for a given element.

— If the number of degrees of freedom varies from one node to the other, use the default value, then modify the data using block DLPN.

— If the nodes are all entered with no generation parameters, IN2 to IDLN (columns 36 to 80) are left null.

| DLPN | Reading of the number of degrees of freedom per node (optional).
— one record containing word 'DLPN';
— records of groups of degrees of freedom terminated by a record in which IDLN ≤ 0. |
|---|---|

Variable	Columns	Default	Format	Description
IDLN	1–5	—	I5	Number of degrees of freedom
K1	6–80	—	15I5	Node numbers with IDLN degrees of freedom, terminated with a 0

Remark

If necessary, the list K1 can continue with records of format (5X, 15I5).

| COND | Reading of boundary conditions (required).
— one record with the word 'COND';
— groups of two records terminated by a blank record:
  • one header record for each group of boundary conditions; |
|---|---|

Variable	Columns	Default	Format	Description
ICOD	1–10	—	10I1	For each degree of freedom (maximum 10) 1 if prescribed 0 if free
V	11–80	—	7F10.0	Values for prescribed DOF in the same order as ICOD, the number of values read in V equal to maximum degrees of freedom per node (NDLN)

  • one record for nodal numbers.

KV	1–80		16I5	node numbers which must terminate by zero. If all 16 values are non-zero, reading will continue on the following record.

Remarks

The list of values V can be continued with additional records of format 10X, 7E12.5. The list of nodal numbers KV can also be continued with additional records of format 16I5.

| PRND | Reading of nodal properties (optional).
— one record with the word 'PRND';
— one record of parameters;

Variable	Columns	Default	Format	Description
NPRN	1–5	0	I5	Number of properties per node

— records of properties ((NNT × NPRN) eight records)

| VPRNG | 1–80 | 0 | 8F10.0 | Values of nodal properties (NPRN values) |

| PREL | Reading of element properties (required if subroutine ELEMnn needs element properties).
— one record with word 'PREL';
— one record of parameters;

Variable	Columns	Default	Format	Description
NGPE	1–5	0	I5	Number of groups of element properties
NPRE	6–10	0	I5	Number of properties per group

— series of records of group properties terminated by a record in which IGPE ≤ 0.

| IGPE | 1–5 | — | I5 | Group number |
| V1 | 6–75 | — | 7F10.0 | Successive values of properties |

Remarks

— If NPRE > 7, the list of properties V1 is continued with additional records of format 5X, 7F10.0.

— The number of properties per group must be equal to the maximum number of properties required by the element used.

The element of type 1 corresponding to ELEM01 (harmonic equation) uses four properties: three physical parameters d_x, d_y, d_z and a specific capacity.

The element of type 2 corresponding to ELEM02 (plane elasticity) requires four properties: Young's modulus E, Poisson's ratio v, a numerical index (0 for plane stress, 1 for plane strain), the specific mass.

| ELEM | Reading of elements (connectivities) (mandatory).
— one record with the word 'ELEM';
M2: logical unit number for the element file (M2 = 1 by default).
— one record for parameters;

Variable	Columns	Default	Format	Description
NELT	1–5	20	I5	Maximum number of elements
NNEL	6–10	8	I5	Maximum number of nodes per element
NTPE	11–15	1	I5	Number of element type by default
NGRE	16–20	1	I5	Number of element group by default
NSYM	21–25	0	I5	.EQ.0 if matrix K symmetrical .EQ.1 if matrix K non-symmetrical
NIDENT	26–30	0	I5	.EQ.1 if all matrices k are identical

— element records terminated by a record in which IEL ≤ 0 (each record may generate more than one element).

Variable	Columns	Default	Format	Description
IEL	1–5	—	I5	Number of the first element
IGEN	6–10	1	I5	Total number of elements to generate
INCR	11–15	1	I5	Nodal number pitch for automatic generation
ITPE	16–20	NTPE	I5	Element type number if different from NTPE
IGPE	21–25	1	I5	Element properties group number
IGRE	26–30	1	I5	Element group number
KNE	31–80	—	10I5	Node numbers of the element, terminated by a zero node number (order must follow convention)

|SOLC| Reading of concentrated loads (optional).
— one record with the word SOLC;
— groups of two records terminated by a blank record:
  • one header record per group of forces;

Variable	Columns	Default	Format	Description
IG	1–5	—	I5	Group number
V	6–75	—	7F10.0	Load values for each degree of freedom

  • one record with node numbers;

KV	1–80	—	16I5	Node numbers so loaded, terminated with a zero node number

Remark

List V can be continued on additional records of format (5X, 7F10.0). List KV can be continued on additional records of format (16I5).

|SOLR| Computation and assemblage of distributed loads (optional).
— one record with the word 'SOLR'.

|LINM| Assemblage and solution of an in-core linear problem (optional).
— one record with the word 'LINM':
M2: logical unit number of element file ($M2 = 1$ by default);
M3: logical unit number for the storage of $[K]$ and $\{F\}$ to compute residuals ($M3 = 2$ by default);
— one record of parameters.

Variable	Columns	Default	Format	Description
NRES	1–5	0	I5	Computation of residues $[K]\{U\} - \{F\}$ if NRES.EQ.1

|LIND| Assembly and solution of an out of core problem with a partitioned matrix stored in a mass storage device (optional).
— one record with the word 'LIND':
M2, M3: see block LINM;
M4: logical unit number of the file containing $[K]$ ($M4 = 4$ by default);
M5: logical unit number of the file containing the decomposed $[K]$ matrix ($M5 = 7$ by default);
— one record of parameters.

Variable	Columns	Default	Format	Description
NRES	1–5	0	I5	(Name, see block 'LINM')
NLBL	6–10	*	I5	Block length for matrix K
NBLM	11–15	*	I5	Maximum number of blocks for K

* Computed to minimize the memory used in the workspace.

|NLIN| Solution of a non-linear problem (optional).
— one record with the word 'NLIN'
M2: logical unit number of the element file ($M2 = 1$ by default);
M3: logical unit number of the file containing specified initial values, if needed;
— records of parameters terminated by a record on which $DPAS = 0.0$.

Variable	Columns	Default	Format	Description
DPAS	1–10	0.2	F10.0	Load increment
NPAS	11–15	1	I5	Number of identical load increments
NITER	16–20	5	I5	Number of iterations per step
IMETH	21–25	1	I5	Method: 1 computation of K for each iteration 2 K constant 3 computation of K at start of each step
EPSDL	26–35	0.01	F10.0	Admissible error of norm
OMEGA	36–45	1.0	F10.0	Over-relaxation factor

TEMP Solution of a time dependent problem (optional).
— one record with the word 'TEMP':
    M2: logical unit number of the element file (M2 = 1 by default);
    M4: logical unit number of the file to save the degrees of freedom, if needed;
— records of parameters terminated by a record with DPAS = 0.0.

Variables	Columns	Default	Format	Description
DPAS	1–10	0.2	F10.0	Time step
NPAS	11–15	1	I5	Number of identical time steps
NITER	16–20	5	I5	Number of iterations per step
IMETH	21–25	1	I5	Method: 1 computation of K at each iteration 2 K constant 3 computation of K at start of each step
EPSDL	26–35	0.01	F10.0	Admissible error of the norm
OMEGA	36–45	1.0	F10.0	Coefficient of Euler method (α) ($\alpha = 0$: explicit $\alpha = 1$: implicit)

VALP Computation of eigenvalues and eigenvectors (optional).
— one record with the word 'VALP':
    M2: logical unit number of the element file (M2 = 1 by default);
— one record of parameters.

Variable	Columns	Default	Format	Description
NVALP	1–5	1	I5	Number of eigenvalues desired
NITER	6–10	10	I5	Maximum number of iterations
EPSLB	11–20	0.001	F10.0	Admissible error on the eigenvalues
SHIFT	21–30	0.0	F10.0	(Not used)
NSS	31–35	*	I5	Subspace dimension
NMDIAG	36–40	0	I5	(Not used)
NSWM	41–45	12	I5	Maximum number of cycles for the Jacobi method
TOLJAC	46–55	1.E–12	F10.0	Tolerance for the Jacobi method

* NSS = Min(NVALP + 8.2 * NVALP).

|STOP| End (required).
— one record with the word 'STOP'.

6.7 Applications of MEF

6.7.1 *Heat Conduction*

Consider the heat transfer problem for the plate shown below:

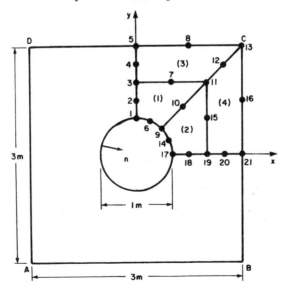

For a thermal isotropic material

$$d = d_x = d_y = 1.4 \, \text{w/(m)} \, ^\circ\text{C} \quad \text{(concrete)}$$

the temperature distribution u acrosses the plate is governed by Laplace equation

$$\frac{\partial^2 u}{\partial x^2} + \frac{\partial^2 u}{\partial y^2} = 0$$

The boundary conditions are:

$u = 0$ on AD and BC (specified temperature)

$\dfrac{\partial u}{\partial n} = \dfrac{\partial u}{\partial y} = 0$ on AB and CD (insulated edges)

$d\dfrac{\partial u}{\partial n} = 1$ specified heat flux on the inside circle

Only one quarter of the plate needs to be analysed because of the double symmetry. Four eight-node isoparametric elements of type 1 are used to model the region under study (paragraph 4.3.4). The distributed boundary condition on the inner circle is replaced by consistent concentrated nodal values of amplitude:

$$\frac{\pi}{24} \quad \text{at nodes 1 and 17}$$

$$\frac{\pi}{12} \quad \text{at node 9}$$

$$\frac{\pi}{6} \quad \text{at nodes 6 and 14}$$

Input records and results of MEF produced by block 'LINM' are shown in Figure 6.27. The corresponding transient heat transfer problem is governed by:

$$\frac{\partial u}{\partial t} = \frac{d}{\rho c}\left(\frac{\partial^2 u}{\partial x^2} + \frac{\partial^2 u}{\partial y^2}\right) \quad t > 0$$

$$u = 0 \quad \text{for} \quad t = 0$$

For concrete, the thermal capacity pc is

$$(2.03)10^6 \, \text{J}/(\text{m}^3)(°\text{C})$$

Input records and results produced by MEF using block 'TEMP' (implicit Euler's method) are shown in Figure 6.28.

F.E.M.3.
G.TOUZOT , G.DHATT

IMAGE OF DATA CARDS
=====================

```
                                    C O L U M N   N U M B E R
CARD          1         2         3         4         5         6         7         8
NUMBER  12345678901234567890123456789012345678901234567890123456789012345678901234567890
------  --------------------------------------------------------------------------------
  1     COMT
  2        HEAT TRANSFER IN A PERFORATED SQUARE PLATE
  3     COOR
  4     21      1       2       0.5
  5      1 0.0        1.0       0.0        5 0.0      3.0     0.0     1
  6      6 0.3827     0.9239    0.0        8 1.5      3.0     0.0     1
  7      9 0.707      0.707     0.0       13 3.0      3.0     0.0     1
  8     14 0.9239     0.3827    0.0       16 3.0      1.5     0.0     1
  9     17 1.0        0.0       0.0       21 3.0      0.0     0.0     1
 10      0
 11     COND
 12      1     13      16      21
 13      0
 14     PREL
 15      1      4
 16      1 1.4        1.4       1.4                   2.03E6
 17      0
 18     ELEM
 19      4      8       1       1          1   6   9  10  11   7   3   2
 20      1      2       8       1          1   7  11  12  13   8   5   4
 21      3      2       8       1          3
 22      0
 23     SOLC
 24      1 0.06545
 25      1     17
 26      2 0.1309
 27      9
 28      3 0.2618
 29      6     14
 30      0
 31     LINM
 32      1
 33     STOP
```

Figure 6.27 (*Contd.*)

450

COMMENTS
==========

HEAT TRANSFER IN A PERFORATED SQUARE PLATE

INPUT OF NODES (M= 0)
==================

```
        MAX. NUMBER OF NODES                    (NNT)=  21
        MAX. NUMBER OF D.O.F. PER NODE          (NDLN)=  1
        DIMENSIONS OF THE PROBLEM               (NDIM)=  2
        COORDINATE SCALE FACTORS                (FAC)= 0.50000E+00 0.50000E+00 0.10000E+01
        WORKSPACE IN REAL WORDS                 (NVA)=  20000
```

INPUT OF BOUNDARY CONDITIONS (M= 0)
====================================

BOUNDARY CONDITIONS CARDS

```
>>>>>1000000000 0.00000000000 0.00000E+00 0.00000E+00 0.00000E+00 0.00000E+00 0.00000E+00 0.00000E+00
>>>>>        13    16    21       0       0       0       0       0       0       0       0       0
>>>>>0000000000 0.00000000000 0.00000E+00 0.00000E+00 0.00000E+00 0.00000E+00 0.00000E+00 0.00000E+00
```

```
        TOTAL NUMBER OF NODES                    (NNT)=  21
        TOTAL NUMBER OF D.O.F.                   (NDLT)=  21
        NUMBER OF EQUATIONS TO BE SOLVED         (NEQ)=   18
        NUMBER OF PRESCRIBED NON ZERO D.O.F.     (NCLNZ)=  0
        NUMBER OF PRESCRIBED ZERO D.O.F.         (NCLZ)=   3
        TOTAL NUMBER OF PRESCRIBED D.O.F.        (NCLT)=   3
```

NODAL COORDINATES ARRAY

NO	D.L.	X	Y	Z	EQUATION NUMBER	(NEQ)
1	1	0.00000E+00	0.50000E+00	0.00000E+00	1	
2	1	0.00000E+00	0.75000E+00	0.00000E+00	2	
3	1	0.00000E+00	0.10000E+01	0.00000E+00	3	
4	1	0.00000E+00	0.12500E+01	0.00000E+00	4	
5	1	0.00000E+00	0.15000E+01	0.00000E+00	5	
6	1	0.19135E+00	0.46195E+00	0.00000E+00	6	
7	1	0.47068E+00	0.98097E+00	0.00000E+00	7	
8	1	0.75000E+00	0.15000E+01	0.00000E+00	8	
9	1	0.35350E+00	0.35350E+00	0.00000E+00	9	
10	1	0.64012E+00	0.64012E+00	0.00000E+00	10	
11	1	0.92675E+00	0.92675E+00	0.00000E+00	11	
12	1	0.12134E+01	0.12134E+01	0.00000E+00	12	
13	1	0.15000E+01	0.15000E+01	0.00000E+00	-1	
14	1	0.46195E+00	0.19135E+00	0.00000E+00	13	
15	1	0.98097E+00	0.47068E+00	0.00000E+00	14	
16	1	0.15000E+01	0.75000E+00	0.00000E+00	-2	
17	1	0.50000E+00	0.00000E+00	0.00000E+00	15	
18	1	0.75000E+00	0.00000E+00	0.00000E+00	16	
19	1	0.10000E+01	0.00000E+00	0.00000E+00	17	
20	1	0.12500E+01	0.00000E+00	0.00000E+00	18	
21	1	0.15000E+01	0.00000E+00	0.00000E+00	-3	

INPUT OF ELEMENT PROPERTIES (M= 0)
==================================
NUMBER OF GROUPS OF PROPERTIES   (NGPE)= 1
NUMBER OF PROPERTIES PER GROUP   (NPRE)= 4

CARDS OF ELEMENT PROPERTIES

>>>>> 1 0.14000E+01 0.14000E+01 0.20300E+07
>>>>> 0 0.00000E+00 0.00000E+00 0.00000E+00

Figure 6.27 (*Contd.*)

INPUT OF ELEMENTS (M= 0)
========================

```
MAX. NUMBER OF ELEMENTS              (NELT)=   4
MAX. NUMBER OF NODES PER ELEMENT     (NNEL)=   8
DEFAULT ELEMENT TYPE                 (NTPE)=   1
NUMBER OF GROUPS OF ELEMENTS         (NGRE)=   1
INDEX FOR NON SYMMETRIC PROBLEM      (NSYM)=   0
INDEX FOR IDENTICAL ELEMENTS         (NIDENT)= 0

ELEMENT:  1 TYPE: 1 N.P.: 8 D.O.F.: 8 N. PROP: 0 EL. PROP: 4 GROUP: 1
  CONNECTIVITY (NE)    1   6   9  10  11   7   3   2
ELEMENT:  2 TYPE: 1 N.P.: 8 D.O.F.: 8 N. PROP: 0 EL. PROP: 4 GROUP: 1
  CONNECTIVITY (NE)    9  14  17  18  19  15  11  10
ELEMENT:  3 TYPE: 1 N.P.: 8 D.O.F.: 8 N. PROP: 0 EL. PROP: 4 GROUP: 1
  CONNECTIVITY (NE)    3   7  11  12  13   8   5   4
ELEMENT:  4 TYPE: 1 N.P.: 8 D.O.F.: 8 N. PROP: 0 EL. PROP: 4 GROUP: 1
  CONNECTIVITY (NE)   11  15  19  20  21  16  13  12

MEAN BAND HEIGHT=    5.3 MAXIMUM=   10
LENGTH OF A TRIANGLE IN KG   (NKG)=   95
NUMBER OF INTEGRATION POINTS (NPG)=   36
```

INPUT OF CONCENTRADED LOADS (M= 0)
==================================

CARDS OF NODAL LOADS

```
>>>>>   1 0.65450E-01   0   0   0   0   0   0   0   0
>>>>>   1  17       0   0   0   0   0   0   0   0   0
>>>>>   2 0.13090E+00   0   0   0   0   0   0   0   0
>>>>>   9   0          0   0   0   0   0   0   0   0
>>>>>   3 0.26180E+00   0   0   0   0   0   0   0   0
>>>>>   6  14       0   0   0   0   0   0   0   0   0
>>>>>   0 0.00000E+00   0   0   0   0   0   0   0   0
```

```
ASSEMBLING AND LINEAR SOLUTION (M= 0)
=====================================
        INDEX FOR RESIDUAL COMPUTATION        (NRES)=    1
        ENERGY    (ENERG)= 0.42653E+00

     ABSOLUTE VALUE OF MINIMUM PIVOT   = 0.97469E+00 EQUATION:    5
                      ALGEBRAIC VALUE= 0.97469E+00 EQUATION:    5
                      DETERMINANT     = 0.14103E+09 * 10 **    0

MAX. RESIDUAL VALUE= 0.76328E-16 EQUATION    9

SOLUTION

NODES      X              Y              Z           DEGREES OF FREEDOM (* = PRESCRIBED)

  1  0.00000E+00   0.50000E+00   0.00000E+00        0.57520E+00
  2  0.00000E+00   0.75000E+00   0.00000E+00        0.44681E+00
  3  0.00000E+00   0.10000E+01   0.00000E+00        0.37137E+00
  4  0.00000E+00   0.12500E+01   0.00000E+00        0.33326E+00
  5  0.00000E+00   0.15000E+01   0.00000E+00        0.32317E+00
  6  0.19135E+00   0.46195E+00   0.00000E+00        0.56684E+00
  7  0.47068E+00   0.98097E+00   0.00000E+00        0.31756E+00
  8  0.75000E+00   0.15000E+01   0.00000E+00        0.21866E+00
  9  0.35350E+00   0.35350E+00   0.00000E+00        0.53887E+00
 10  0.64012E+00   0.64012E+00   0.00000E+00        0.33300E+00
 11  0.92675E+00   0.92675E+00   0.00000E+00        0.19818E+00
 12  0.12134E+01   0.12134E+01   0.00000E+00        0.93103E-01
 13  0.15000E+01   0.15000E+01   0.00000E+00        0.00000E+00 *
 14  0.46195E+00   0.19135E+00   0.00000E+00        0.52269E+00
 15  0.98097E+00   0.47068E+00   0.00000E+00        0.21421E+00
 16  0.15000E+01   0.75000E+00   0.00000E+00        0.00000E+00 *
 17  0.50000E+00   0.00000E+00   0.00000E+00        0.50587E+00
 18  0.75000E+00   0.00000E+00   0.00000E+00        0.35749E+00
 19  0.10000E+01   0.00000E+00   0.00000E+00        0.22597E+00
 20  0.12500E+01   0.00000E+00   0.00000E+00        0.10981E+00
 21  0.15000E+01   0.00000E+00   0.00000E+00        0.00000E+00 *
```

Figure 6.27 (*Contd.*)

GRADIENTS IN ELEMENT : 1

P.G. : 1 COORDINATES : 0.52632E-01 0.55406E+00
       GRADIENTS :-0.11372E+00-0.81069E+00
P.G. : 2 COORDINATES : 0.76523E-01 0.74818E+00
       GRADIENTS :-0.12315E+00-0.57158E+00
P.G. : 3 COORDINATES : 0.10041E+00 0.94230E+00
       GRADIENTS :-0.11487E+00-0.33465E+00
P.G. : 4 COORDINATES : 0.22283E+00 0.52044E+00
       GRADIENTS :-0.41886E+00-0.70605E+00
P.G. : 5 COORDINATES : 0.33101E+00 0.72146E+00
       GRADIENTS :-0.33351E+00-0.49289E+00
P.G. : 6 COORDINATES : 0.43919E+00 0.92248E+00
       GRADIENTS :-0.28656E+00-0.25907E+00
P.G. : 7 COORDINATES : 0.37650E+00 0.44697E+00
       GRADIENTS :-0.65190E+00-0.47047E+00
P.G. : 8 COORDINATES : 0.57236E+00 0.66307E+00
       GRADIENTS :-0.51763E+00-0.32390E+00
P.G. : 9 COORDINATES : 0.76823E+00 0.87916E+00
       GRADIENTS :-0.44680E+00-0.11983E+00

GRADIENTS IN ELEMENT : 2

P.G. : 1 COORDINATES : 0.44697E+00 0.37650E+00
       GRADIENTS :-0.71704E+00-0.40960E+00
P.G. : 2 COORDINATES : 0.66307E+00 0.57236E+00
       GRADIENTS :-0.60362E+00-0.25620E+00
P.G. : 3 COORDINATES : 0.87916E+00 0.76823E+00
       GRADIENTS :-0.45595E+00-0.14059E+00
P.G. : 4 COORDINATES : 0.52044E+00 0.22283E+00
       GRADIENTS :-0.88575E+00-0.24160E+00
P.G. : 5 COORDINATES : 0.72146E+00 0.33101E+00
       GRADIENTS :-0.73534E+00-0.17979E+00
P.G. : 6 COORDINATES : 0.92248E+00 0.43919E+00
       GRADIENTS :-0.59060E+00-0.10744E+00
P.G. : 7 COORDINATES : 0.55406E+00 0.52632E-01
       GRADIENTS :-0.89617E+00-0.17241E-01
P.G. : 8 COORDINATES : 0.74818E+00 0.76523E-01
       GRADIENTS :-0.78959E+00-0.75879E-01
P.G. : 9 COORDINATES : 0.94230E+00 0.10041E+00
       GRADIENTS :-0.69255E+00-0.56986E-01

GRADIENTS IN ELEMENT : 3

P.G. : 1 COORDINATES : 0.11432E+00 0.10553E+01
        GRADIENTS   :-0.11051E+00-0.25454E+00
P.G. : 2 COORDINATES : 0.13821E+00 0.12494E+01
        GRADIENTS   :-0.11648E+00-0.14037E+00
P.G. : 3 COORDINATES : 0.16210E+00 0.14435E+01
        GRADIENTS   :-0.13120E+00-0.25123E-01
P.G. : 4 COORDINATES : 0.50216E+00 0.10395E+01
        GRADIENTS   :-0.27711E+00-0.20122E+00
P.G. : 5 COORDINATES : 0.61034E+00 0.12405E+01
        GRADIENTS   :-0.28059E+00-0.11578E+00
P.G. : 6 COORDINATES : 0.71852E+00 0.14415E+01
        GRADIENTS   :-0.29573E+00-0.24060E-01
P.G. : 7 COORDINATES : 0.88222E+00 0.10049E+01
        GRADIENTS   :-0.43817E+00-0.86922E-01
P.G. : 8 COORDINATES : 0.10781E+01 0.12210E+01
        GRADIENTS   :-0.44335E+00-0.28634E-01
P.G. : 9 COORDINATES : 0.12740E+01 0.14371E+01
        GRADIENTS   :-0.46020E+00 0.40218E-01

GRADIENTS IN ELEMENT : 4

P.G. : 1 COORDINATES : 0.10049E+01 0.88222E+00
        GRADIENTS   :-0.46122E+00-0.99442E-01
P.G. : 2 COORDINATES : 0.12210E+01 0.10781E+01
        GRADIENTS   :-0.46380E+00-0.47956E-01
P.G. : 3 COORDINATES : 0.14371E+01 0.12740E+01
        GRADIENTS   :-0.45461E+00-0.94565E-02
P.G. : 4 COORDINATES : 0.10395E+01 0.50216E+00
        GRADIENTS   :-0.57615E+00-0.74187E-01
P.G. : 5 COORDINATES : 0.12405E+01 0.61034E+00
        GRADIENTS   :-0.55835E+00-0.36136E-01
P.G. : 6 COORDINATES : 0.14415E+01 0.71852E+00
        GRADIENTS   :-0.53562E+00-0.72341E-02
P.G. : 7 COORDINATES : 0.10553E+01 0.11432E+00
        GRADIENTS   :-0.65031E+00-0.38270E-01
P.G. : 8 COORDINATES : 0.12494E+01 0.13821E+00
        GRADIENTS   :-0.62235E+00-0.20090E-01
P.G. : 9 COORDINATES : 0.14435E+01 0.16210E+00
        GRADIENTS   :-0.59409E+00-0.42941E-02

Figure 6.27 (*Contd.*)

EQUILIBRIUM RESIDUALS AND REACTIONS

NODES	X	Y	Z	DEGREES OF FREEDOM (* = PRESCRIBED)
1	0.00000E+00	0.50000E+00	0.00000E+00	-0.17347E-17
2	0.00000E+00	0.75000E+00	0.00000E+00	-0.15870E-16
3	0.00000E+00	0.10000E+01	0.00000E+00	-0.19082E-16
4	0.00000E+00	0.12500E+01	0.00000E+00	-0.80976E-16
5	0.00000E+00	0.15000E+01	0.00000E+00	0.10910E-16
6	0.19135E+00	0.46195E+00	0.00000E+00	0.41633E-16
7	0.47068E+00	0.98097E+00	0.00000E+00	-0.41633E-16
8	0.75000E+00	0.15000E+01	0.00000E+00	-0.10734E-16
9	0.35350E+00	0.35350E+00	0.00000E+00	-0.69389E-17
10	0.64012E+00	0.64012E+00	0.00000E+00	-0.79797E-16
11	0.92675E+00	0.92675E+00	0.00000E+00	0.55511E-16
12	0.12134E+01	0.12134E+01	0.00000E+00	-0.38164E-16
13	0.15000E+01	0.15000E+01	0.00000E+00	-0.10288E+00 *
14	0.46195E+00	0.19135E+00	0.00000E+00	0.34694E-16
15	0.98097E+00	0.47068E+00	0.00000E+00	0.55511E-16
16	0.15000E+01	0.75000E+00	0.00000E+00	-0.52531E+00 *
17	0.50000E+00	0.00000E+00	0.00000E+00	0.34694E-16
18	0.75000E+00	0.00000E+00	0.00000E+00	-0.30846E-16
19	0.10000E+01	0.00000E+00	0.00000E+00	0.19082E-16
20	0.12500E+01	0.00000E+00	0.00000E+00	-0.80333E-17
21	0.15000E+01	0.00000E+00	0.00000E+00	-0.15722E+00 *

END OF PROBLEM,   314 UTILIZED REAL WORDS OVER   20000

Figure 6.27 Example of stationary heat flow.

```
                    F.E.M.3.
                G.TOUZOT , G.DHATT
                ──────────────────

IMAGE OF DATA CARDS
===================

                                 C O L U M N   N U M B E R
                   1         2         3         4         5         6         7         8
    CARD  12345678901234567890123456789012345678901234567890123456789012345678901234567890
   NUMBER ────────────────────────────────────────────────────────────────────────────────
     1    COMT
     2    HEAT TRANSFER IN A PERFORATED SQUARE PLATE
     3    COOR
     4    21       2
     5     1 0.0       0.0        0.5
     6     6 0.3827    1.0        0.0       5 0.0       3.0
     7     9 0.707     0.9239     0.0       8 1.5       3.0       1
     8    14 0.9239    0.707      0.0      13 3.0       3.0       1
     9    17 1.0       0.3827     0.0      16 3.0       1.5       1
    10     0                                21 3.0       0.0       1
    11    COND
    12     1    13    16    21
    13     0
    14    PREL
    15     1     4
    16     1 1.4       1.4       2.03E6
    17     0
    18    ELEM
    19     4     8     1
    20     1     2     8     1     1     1     6     9    10    11     7     3
    21     3     2     8     1     1     3     7    11    12    13     8     5     2
    22     0                                                                         4
    23    SOLC
    24     1 0.06545
    25     1    17
    26     2 0.1309
    27     9    14
    28     3 0.2618
    29     6    14
    30     0
    31    TEMP     1
    32    3.0E5    20
    33    0.0
    34    STOP
          ────────────────────────────────────────────────────────────────────────────────
    CARD  12345678901234567890123456789012345678901234567890123456789012345678901234567890
   NUMBER    1         2         3         4         5         6         7         8
                                 C O L U M N   N U M B E R

                                 E N D   O F   D A T A
```

Figure 6.28 (*Contd.*)

COMMENTS
=========

HEAT TRANSFER IN A PERFORATED SQUARE PLATE

INPUT OF NODES (M= 0)
=====================

```
MAX. NUMBER OF NODES              (NNT)=  21
MAX. NUMBER OF D.O.F. PER NODE    (NDLN)=  1
DIMENSIONS OF THE PROBLEM         (NDIM)=  2
COORDINATE SCALE FACTORS          (FAC)= 0.50000E+00 0.50000E+00 0.10000E+01
WORKSPACE IN REAL WORDS           (NVA)=  20000
```

INPUT OF BOUNDARY CONDITIONS (M= 0)
===================================

BOUNDARY CONDITIONS CARDS

```
>>>>>1000000000 0.00000E+00 0.00000E+00 0.00000E+00 0.00000E+00 0.00000E+00 0.00000E+00
>>>>>      13  16  21      0    0    0    0    0    0    0
>>>>>0000000000 0.00000E+00 0.00000E+00 0.00000E+00 0.00000E+00 0.00000E+00 0.00000E+00
```

```
TOTAL NUMBER OF NODES                        (NNT)=   21
TOTAL NUMBER OF D.O.F.                       (NDLT)=  21
NUMBER OF EQUATIONS TO BE SOLVED             (NEQ)=   18
NUMBER OF PRESCRIBED NON ZERO D.O.F.         (NCLNZ)=  0
NUMBER OF PRESCRIBED ZERO D.O.F.             (NCLZ)=   3
TOTAL NUMBER OF PRESCRIBED D.O.F.            (NCLT)=   3
```

NODAL COORDINATES ARRAY

NO	D.L.	X	Y	Z	EQUATION NUMBER	(NEQ)
1	1	0.00000E+00	0.50000E+00	0.00000E+00	1	
2	1	0.00000E+00	0.75000E+00	0.00000E+00	2	
3	1	0.00000E+00	0.10000E+01	0.00000E+00	3	
4	1	0.00000E+00	0.12500E+01	0.00000E+00	4	
5	1	0.00000E+00	0.15000E+01	0.00000E+00	5	
6	1	0.19135E+00	0.46195E+00	0.00000E+00	6	
7	1	0.47068E+00	0.98097E+00	0.00000E+00	7	
8	1	0.75000E+00	0.15000E+01	0.00000E+00	8	
9	1	0.35350E+00	0.35350E+00	0.00000E+00	9	
10	1	0.64012E+00	0.64012E+00	0.00000E+00	10	
11	1	0.92675E+00	0.92675E+00	0.00000E+00	11	
12	1	0.12134E+01	0.12134E+01	0.00000E+00	12	
13	1	0.15000E+01	0.15000E+01	0.00000E+00	-1	
14	1	0.46195E+00	0.19135E+00	0.00000E+00	13	
15	1	0.98097E+00	0.47068E+00	0.00000E+00	14	
16	1	0.15000E+01	0.75000E+00	0.00000E+00	-2	
17	1	0.50000E+00	0.00000E+00	0.00000E+00	15	
18	1	0.75000E+00	0.00000E+00	0.00000E+00	16	
19	1	0.10000E+01	0.00000E+00	0.00000E+00	17	
20	1	0.12500E+01	0.00000E+00	0.00000E+00	18	
21	1	0.15000E+01	0.00000E+00	0.00000E+00	-3	

INPUT OF ELEMENT PROPERTIES (M= 0)
================================
        NUMBER OF GROUPS OF PROPERTIES   (NGPE)= 1
        NUMBER OF PROPERTIES PER GROUP   (NPRE)= 4

CARDS OF ELEMENT PROPERTIES

```
>>>>>   1 0.14000E+01 0.14000E+01 0.14000E+01 0.20300E+07
>>>>>   0 0.00000E+00 0.00000E+00 0.00000E+00 0.00000E+00
```

Figure 6.28 (*Contd.*)

INPUT OF ELEMENTS (M= 0)
========================

```
            MAX. NUMBER OF ELEMENTS           (NELT)=   4
            MAX. NUMBER OF NODES PER ELEMENT  (NNEL)=   8
            DEFAULT ELEMENT TYPE              (NTPE)=   1
            NUMBER OF GROUPS OF ELEMENTS      (NGRE)=   1
            INDEX FOR NON SYMMETRIC PROBLEM   (NSYM)=   0
            INDEX FOR IDENTICAL ELEMENTS      (NIDENT)= 0

ELEMENT:  1 TYPE: 1 N.P.: 8 D.O.F.:  8 N. PROP:  0  EL. PROP:  4  GROUP:  1
        CONNECTIVITY (NE)    1   6   9  10  11   7   3   2
ELEMENT:  2 TYPE: 1 N.P.: 8 D.O.F.:  8 N. PROP:  0  EL. PROP:  4  GROUP:  1
        CONNECTIVITY (NE)    9  14  17  18  19  15  11  10
ELEMENT:  3 TYPE: 1 N.P.: 8 D.O.F.:  8 N. PROP:  0  EL. PROP:  4  GROUP:  1
        CONNECTIVII/ (NE)    3   7  11  12  13   8   5   4
ELEMENT:  4 TYPE: 1 N.P.: 8 D.O.F.:  8 N. PROP:  0  EL. PROP:  4  GROUP:  1
        CONNECTIVITY (NE)   11  15  19  20  21  16  13  12

         MEAN BAND HEIGHT=     5.3  MAXIMUM=  10
         LENGTH OF A TRIANGLE IN KG        (NKG)=  95
         NUMBER OF INTEGRATION POINTS      (NPG)=  36
```

INPUT OF CONCENTRADED LOADS (M= 0)
==================================

CARDS OF NODAL LOADS

```
>>>>>   1 0.65450E-01  0  0  0  0  0  0  0  0  0  0  0  0  0
>>>>>   1  17  0        0  0  0  0  0  0  0  0  0  0  0  0  0
>>>>>   2 0.13090E+00  0  0  0  0  0  0  0  0  0  0  0  0  0
>>>>>   9   0          0  0  0  0  0  0  0  0  0  0  0  0  0
>>>>>   3 0.26180E+00  0  0  0  0  0  0  0  0  0  0  0  0  0
>>>>>   6  14  0        0  0  0  0  0  0  0  0  0  0  0  0  0
>>>>>   0 0.00000E+00  0  0  0  0  0  0  0  0  0  0  0  0  0
```

```
UNSTEADY SOLUTION (M= 1)
========================

                                          TABLE KGS  GOES FROM VA(  129) TO VA(  223)
                                          TABLE KGD  GOES FROM VA(  224) TO VA(  241)
                                          TABLE KE   GOES FROM VA(  242) TO VA(  277)
                                          TABLE FE   GOES FROM VA(  278) TO VA(  285)
                                          TABLE RES  GOES FROM VA(  286) TO VA(  303)
                                          TABLE DLE  GOES FROM VA(  304) TO VA(  311)
                                          TABLE DLG  GOES FROM VA(  312) TO VA(  329)
                                          TABLE ME   GOES FROM VA(  330) TO VA(  365)
                                          TABLE DLEO GOES FROM VA(  366) TO VA(  373)
                                          TABLE DLGO GOES FROM VA(  374) TO VA(  391)
                                          TABLE FGO  GOES FROM VA(  392) TO VA(  409)

----------STEP NUMBER (IPAS):   1

         INCREMENT                (DPAS)= 0.30000E+06
         TOTAL LEVEL              (XPAS)= 0.30000E+06
         NUMBER OF ITERATIONS     (NITER)=     5
         METHOD NUMBER            (IMETH):     1
         TOLERANCE                (EPSDL)= 0.10000E-01
         COEFFICIENT ALPHA        (OMEGA)= 0.10000E+01

ITERATION (ITER):   1 NORM (XNORM)= 0.10000E+01
ITERATION (ITER):   2 NORM (XNORM)= 0.61492E-16

NODES    X            Y            Z          DEGREES OF FREEDOM (* = PRESCRIBED)

  1  0.00000E+00  0.50000E+00  0.00000E+00      0.22930E+00
  2  0.00000E+00  0.75000E+00  0.00000E+00      0.11715E+00
  3  0.00000E+00  0.10000E+01  0.00000E+00      0.60596E-01
  4  0.00000E+00  0.12500E+01  0.00000E+00      0.37763E-01
  5  0.00000E+00  0.15000E+01  0.00000E+00      0.31511E-01
  6  0.19135E+00  0.46195E+00  0.00000E+00      0.23278E+00
  7  0.47068E+00  0.98097E+00  0.00000E+00      0.47586E-01
  8  0.75000E+00  0.15000E+01  0.00000E+00      0.18603E-01
  9  0.35350E+00  0.35350E+00  0.00000E+00      0.22360E+00
 10  0.64012E+00  0.64012E+00  0.00000E+00      0.73699E-01
 11  0.92675E+00  0.92675E+00  0.00000E+00      0.25578E-01
 12  0.12134E+01  0.12134E+01  0.00000E+00      0.74505E-02
 13  0.15000E+01  0.15000E+01  0.00000E+00      0.00000E+00 *
 14  0.46195E+00  0.19135E+00  0.00000E+00      0.23105E+00
 15  0.98097E+00  0.47068E+00  0.00000E+00      0.41595E-01
 16  0.15000E+01  0.75000E+00  0.00000E+00      0.00000E+00 *
 17  0.50000E+00  0.00000E+00  0.00000E+00      0.22627E+00
 18  0.75000E+00  0.00000E+00  0.00000E+00      0.11319E+00
 19  0.10000E+01  0.00000E+00  0.00000E+00      0.52550E-01
 20  0.12500E+01  0.00000E+00  0.00000E+00      0.21331E-01
 21  0.15000E+01  0.00000E+00  0.00000E+00      0.00000E+00 *
```

Figure 6.28 (*Contd.*)

```
--------------STEP NUMBER (IPAS):   2

           INCREMENT                    (DPAS)= 0.30000E+06
           TOTAL LEVEL                  (XPAS)= 0.60000E+06
           NUMBER OF ITERATIONS         (NITER)=     5
           METHOD NUMBER                (IMETH):     1
           TOLERANCE                    (EPSDL)= 0.10000E-01
           COEFFICIENT ALPHA            (OMEGA)= 0.10000E+01

ITERATION (ITER):  1 NORM (XNORM)= 0.32663E+00
ITERATION (ITER):  2 NORM (XNORM)= 0.37413E-16

NODES     X            Y            Z         DEGREES OF FREEDOM (* = PRESCRIBED)

 1 0.00000E+00  0.50000E+00  0.00000E+00      0.31920E+00
 2 0.00000E+00  0.75000E+00  0.00000E+00      0.19451E+00
 3 0.00000E+00  0.10000E+01  0.00000E+00      0.12080E+00
 4 0.00000E+00  0.12500E+01  0.00000E+00      0.86570E-01
 5 0.00000E+00  0.15000E+01  0.00000E+00      0.76778E-01
 6 0.19135E+00  0.46195E+00  0.00000E+00      0.32250E+00
 7 0.47068E+00  0.98097E+00  0.00000E+00      0.98745E-01
 8 0.75000E+00  0.15000E+01  0.00000E+00      0.47607E-01
 9 0.35350E+00  0.35350E+00  0.00000E+00      0.31190E+00
10 0.64012E+00  0.64012E+00  0.00000E+00      0.13466E+00
11 0.92675E+00  0.92675E+00  0.00000E+00      0.56237E-01
12 0.12134E+01  0.12134E+01  0.00000E+00      0.19731E-01
13 0.15000E+01  0.15000E+01  0.00000E+00      0.00000E+00  *
14 0.46195E+00  0.19135E+00  0.00000E+00      0.31686E+00
15 0.98097E+00  0.47068E+00  0.00000E+00      0.81499E-01
16 0.15000E+01  0.75000E+00  0.00000E+00      0.00000E+00  *
17 0.50000E+00  0.00000E+00  0.00000E+00      0.30974E+00
18 0.75000E+00  0.00000E+00  0.00000E+00      0.18209E+00
19 0.10000E+01  0.00000E+00  0.00000E+00      0.97263E-01
20 0.12500E+01  0.00000E+00  0.00000E+00      0.42762E-01
21 0.15000E+01  0.00000E+00  0.00000E+00      0.00000E+00  *
```

```
----------STEP NUMBER (IPAS):    3

         INCREMENT                  (DPAS)= 0.30000E+06
         TOTAL LEVEL                (XPAS)= 0.90000E+06
         NUMBER OF ITERATIONS       (NITER)=       5
         METHOD NUMBER              (IMETH):       1
         TOLERANCE                  (EPSDL)= 0.10000E-01
         COEFFICIENT ALPHA          (OMEGA)= 0.10000E+01

ITERATION (ITER):  1 NORM (XNORM)= 0.18858E+00
ITERATION (ITER):  2 NORM (XNORM)= 0.17748E-16

NODES     X              Y              Z          DEGREES OF FREEDOM (* = PRESCRIBED)

 1  0.00000E+00   0.50000E+00   0.00000E+00    0.37698E+00
 2  0.00000E+00   0.75000E+00   0.00000E+00    0.24931E+00
 3  0.00000E+00   0.10000E+01   0.00000E+00    0.17107E+00
 4  0.00000E+00   0.12500E+01   0.00000E+00    0.13275E+00
 5  0.00000E+00   0.15000E+01   0.00000E+00    0.12164E+00
 6  0.19135E+00   0.46195E+00   0.00000E+00    0.37872E+00
 7  0.47068E+00   0.98097E+00   0.00000E+00    0.14264E+00
 8  0.75000E+00   0.15000E+01   0.00000E+00    0.77971E-01
 9  0.35350E+00   0.35350E+00   0.00000E+00    0.36602E+00
10  0.64012E+00   0.64012E+00   0.00000E+00    0.17911E+00
11  0.92675E+00   0.92675E+00   0.00000E+00    0.84453E-01
12  0.12134E+01   0.12134E+01   0.00000E+00    0.32900E-01
13  0.15000E+01   0.15000E+01   0.00000E+00    0.00000E+00 *
14  0.46195E+00   0.19135E+00   0.00000E+00    0.36786E+00
15  0.98097E+00   0.47068E+00   0.00000E+00    0.11206E+00
16  0.15000E+01   0.75000E+00   0.00000E+00    0.00000E+00 *
17  0.50000E+00   0.00000E+00   0.00000E+00    0.35918E+00
18  0.75000E+00   0.00000E+00   0.00000E+00    0.22601E+00
19  0.10000E+01   0.00000E+00   0.00000E+00    0.12893E+00
20  0.12500E+01   0.00000E+00   0.00000E+00    0.59047E-01
21  0.15000E+01   0.00000E+00   0.00000E+00    0.00000E+00 *
```

Figure 6.28 (*Contd.*)

464

```
----------STEP NUMBER (IPAS):    4

          INCREMENT                    (DPAS)= 0.30000E+06
          TOTAL LEVEL                  (XPAS)= 0.12000E+07
          NUMBER OF ITERATIONS         (NITER)=      5
          METHOD NUMBER                (IMETH):      1
          TOLERANCE                    (EPSDL)= 0.10000E-01
          COEFFICIENT ALPHA            (OMEGA)= 0.10000E+01

ITERATION (ITER):  1  NORM (XNORM)= 0.12672E+00
ITERATION (ITER):  2  NORM (XNORM)= 0.88471E-17

NODES      X            Y            Z         DEGREES OF FREEDOM (* = PRESCRIBED)

 1  0.00000E+00  0.50000E+00  0.00000E+00   0.41956E+00
 2  0.00000E+00  0.75000E+00  0.00000E+00   0.29097E+00
 3  0.00000E+00  0.10000E+01  0.00000E+00   0.21183E+00
 4  0.00000E+00  0.12500E+01  0.00000E+00   0.17232E+00
 5  0.00000E+00  0.15000E+01  0.00000E+00   0.16090E+00
 6  0.19135E+00  0.46195E+00  0.00000E+00   0.41958E+00
 7  0.47068E+00  0.98097E+00  0.00000E+00   0.17838E+00
 8  0.75000E+00  0.15000E+01  0.00000E+00   0.10517E+00
 9  0.35350E+00  0.35350E+00  0.00000E+00   0.40448E+00
10  0.64012E+00  0.64012E+00  0.00000E+00   0.21246E+00
11  0.92675E+00  0.92675E+00  0.00000E+00   0.10779E+00
12  0.12134E+01  0.12134E+01  0.00000E+00   0.44675E-01
13  0.15000E+01  0.15000E+01  0.00000E+00   0.00000E+00 *
14  0.46195E+00  0.19135E+00  0.00000E+00   0.40323E+00
15  0.98097E+00  0.47068E+00  0.00000E+00   0.13482E+00
16  0.15000E+01  0.75000E+00  0.00000E+00   0.00000E+00 *
17  0.50000E+00  0.00000E+00  0.00000E+00   0.39314E+00
18  0.75000E+00  0.00000E+00  0.00000E+00   0.25650E+00
19  0.10000E+01  0.00000E+00  0.00000E+00   0.15144E+00
20  0.12500E+01  0.00000E+00  0.00000E+00   0.70840E-01
21  0.15000E+01  0.00000E+00  0.00000E+00   0.00000E+00 *
```

```
--------STEP NUMBER (IPAS):    5

        INCREMENT                    (DPAS)= 0.30000E+06
        TOTAL LEVEL                  (XPAS)= 0.15000E+07
        NUMBER OF ITERATIONS         (NITER)=      5
        METHOD NUMBER                (IMETH):      1
        TOLERANCE                    (EPSDL)= 0.10000E-01
        COEFFICIENT ALPHA            (OMEGA)= 0.10000E+01

ITERATION (ITER):  1 NORM (XNORM)= 0.90549E-01
ITERATION (ITER):  2 NORM (XNORM)= 0.12718E-16

NODES   X           Y           Z           DEGREES OF FREEDOM (* = PRESCRIBED)

  1  0.00000E+00 0.50000E+00 0.00000E+00    0.45238E+00
  2  0.00000E+00 0.75000E+00 0.00000E+00    0.32353E+00
  3  0.00000E+00 0.10000E+01 0.00000E+00    0.24459E+00
  4  0.00000E+00 0.12500E+01 0.00000E+00    0.20489E+00
  5  0.00000E+00 0.15000E+01 0.00000E+00    0.19355E+00
  6  0.19135E+00 0.46195E+00 0.00000E+00    0.45083E+00
  7  0.47068E+00 0.98097E+00 0.00000E+00    0.20707E+00
  8  0.75000E+00 0.15000E+01 0.00000E+00    0.12797E+00
  9  0.35350E+00 0.35350E+00 0.00000E+00    0.43342E+00
 10  0.64012E+00 0.64012E+00 0.00000E+00    0.23810E+00
 11  0.92675E+00 0.92675E+00 0.00000E+00    0.12653E+00
 12  0.12134E+01 0.12134E+01 0.00000E+00    0.54483E-01
 13  0.15000E+01 0.15000E+01 0.00000E+00    0.00000E+00 *
 14  0.46195E+00 0.19135E+00 0.00000E+00    0.42938E+00
 15  0.98097E+00 0.47068E+00 0.00000E+00    0.15203E+00
 16  0.15000E+01 0.75000E+00 0.00000E+00    0.00000E+00 *
 17  0.50000E+00 0.00000E+00 0.00000E+00    0.41802E+00
 18  0.75000E+00 0.00000E+00 0.00000E+00    0.27886E+00
 19  0.10000E+01 0.00000E+00 0.00000E+00    0.16798E+00
 20  0.12500E+01 0.00000E+00 0.00000E+00    0.79523E-01
 21  0.15000E+01 0.00000E+00 0.00000E+00    0.00000E+00 *
```

Figure 6.28 (*Contd.*)

466

```
----------STEP NUMBER (IPAS):    6

         INCREMENT                    (DPAS)= 0.30000E+06
         TOTAL LEVEL                  (XPAS)= 0.18000E+07
         NUMBER OF ITERATIONS         (NITER)=        5
         METHOD NUMBER                (IMETH):        1
         TOLERANCE                    (EPSDL)= 0.10000E-01
         COEFFICIENT ALPHA            (OMEGA)= 0.10000E+01

ITERATION (ITER):  1 NORM (XNORM)= 0.66815E-01
ITERATION (ITER):  2 NORM (XNORM)= 0.11367E-16

NODES      X             Y             Z         DEGREES OF FREEDOM (* = PRESCRIBED)

  1  0.00000E+00  0.50000E+00  0.00000E+00   0.47809E+00
  2  0.00000E+00  0.75000E+00  0.00000E+00   0.34921E+00
  3  0.00000E+00  0.10000E+01  0.00000E+00   0.27077E+00
  4  0.00000E+00  0.12500E+01  0.00000E+00   0.23121E+00
  5  0.00000E+00  0.15000E+01  0.00000E+00   0.22005E+00
  6  0.19135E+00  0.46195E+00  0.00000E+00   0.47519E+00
  7  0.47068E+00  0.98097E+00  0.00000E+00   0.22994E+00
  8  0.75000E+00  0.15000E+01  0.00000E+00   0.14651E+00
  9  0.35350E+00  0.35350E+00  0.00000E+00   0.45575E+00
 10  0.64012E+00  0.64012E+00  0.00000E+00   0.25808E+00
 11  0.92675E+00  0.92675E+00  0.00000E+00   0.14143E+00
 12  0.12134E+01  0.12134E+01  0.00000E+00   0.62421E-01
 13  0.15000E+01  0.15000E+01  0.00000E+00   0.00000E+00 *
 14  0.46195E+00  0.19135E+00  0.00000E+00   0.44934E+00
 15  0.98097E+00  0.47068E+00  0.00000E+00   0.16528E+00
 16  0.15000E+01  0.75000E+00  0.00000E+00   0.00000E+00 *
 17  0.50000E+00  0.00000E+00  0.00000E+00   0.43691E+00
 18  0.75000E+00  0.00000E+00  0.00000E+00   0.29579E+00
 19  0.10000E+01  0.00000E+00  0.00000E+00   0.18050E+00
 20  0.12500E+01  0.00000E+00  0.00000E+00   0.86082E-01
 21  0.15000E+01  0.00000E+00  0.00000E+00   0.00000E+00 *
```

```
-------STEP NUMBER (IPAS):   7

        INCREMENT                          (DPAS)= 0.30000E+06
        TOTAL LEVEL                        (XPAS)= 0.21000E+07
        NUMBER OF ITERATIONS               (NITER)=        5
        METHOD NUMBER                      (IMETH):        1
        TOLERANCE                          (EPSDL)= 0.10000E-01
        COEFFICIENT ALPHA                  (OMEGA)= 0.10000E+01

ITERATION (ITER):  1 NORM (XNORM)= 0.50274E-01
ITERATION (ITER):  2 NORM (XNORM)= 0.16652E-16

NODES   X           Y           Z         DEGREES OF FREEDOM (* = PRESCRIBED)

 1  0.00000E+00 0.50000E+00 0.00000E+00   0.49835E+00
 2  0.00000E+00 0.75000E+00 0.00000E+00   0.36952E+00
 3  0.00000E+00 0.10000E+01 0.00000E+00   0.29162E+00
 4  0.00000E+00 0.12500E+01 0.00000E+00   0.25228E+00
 5  0.00000E+00 0.15000E+01 0.00000E+00   0.24131E+00
 6  0.19135E+00 0.46195E+00 0.00000E+00   0.49435E+00
 7  0.47068E+00 0.98097E+00 0.00000E+00   0.24813E+00
 8  0.75000E+00 0.15000E+01 0.00000E+00   0.16140E+00
 9  0.35350E+00 0.35350E+00 0.00000E+00   0.47321E+00
10  0.64012E+00 0.64012E+00 0.00000E+00   0.27377E+00
11  0.92675E+00 0.92675E+00 0.00000E+00   0.15324E+00
12  0.12134E+01 0.12134E+01 0.00000E+00   0.68770E-01
13  0.15000E+01 0.15000E+01 0.00000E+00   0.00000E+00 *
14  0.46195E+00 0.19135E+00 0.00000E+00   0.46483E+00
15  0.98097E+00 0.47068E+00 0.00000E+00   0.17560E+00
16  0.15000E+01 0.75000E+00 0.00000E+00   0.00000E+00 *
17  0.50000E+00 0.00000E+00 0.00000E+00   0.45151E+00
18  0.75000E+00 0.00000E+00 0.00000E+00   0.30888E+00
19  0.10000E+01 0.00000E+00 0.00000E+00   0.19016E+00
20  0.12500E+01 0.00000E+00 0.00000E+00   0.91133E-01
21  0.15000E+01 0.00000E+00 0.00000E+00   0.00000E+00 *
```

Figure 6.28 (*Contd.*)

```
----------STEP NUMBER (IPAS):   8

        INCREMENT                   (DPAS)= 0.30000E+06
        TOTAL LEVEL                 (XPAS)= 0.24000E+07
        NUMBER OF ITERATIONS        (NITER)=       5
        METHOD NUMBER               (IMETH):       1
        TOLERANCE                   (EPSDL)= 0.10000E-01
        COEFFICIENT ALPHA           (OMEGA)= 0.10000E+01

ITERATION (ITER):   1  NORM (XNORM)= 0.38323E-01
ITERATION (ITER):   2  NORM (XNORM)= 0.10329E-16

NODES    X              Y              Z         DEGREES OF FREEDOM (* = PRESCRIBED)

 1  0.00000E+00   0.50000E+00   0.00000E+00        0.51437E+00
 2  0.00000E+00   0.75000E+00   0.00000E+00        0.38561E+00
 3  0.00000E+00   0.10000E+01   0.00000E+00        0.30818E+00
 4  0.00000E+00   0.12500E+01   0.00000E+00        0.26907E+00
 5  0.00000E+00   0.15000E+01   0.00000E+00        0.25827E+00
 6  0.19135E+00   0.46195E+00   0.00000E+00        0.50947E+00
 7  0.47068E+00   0.98097E+00   0.00000E+00        0.26256E+00
 8  0.75000E+00   0.15000E+01   0.00000E+00        0.17326E+00
 9  0.35350E+00   0.35350E+00   0.00000E+00        0.48695E+00
10  0.64012E+00   0.64012E+00   0.00000E+00        0.28614E+00
11  0.92675E+00   0.92675E+00   0.00000E+00        0.16259E+00
12  0.12134E+01   0.12134E+01   0.00000E+00        0.73821E-01
13  0.15000E+01   0.15000E+01   0.00000E+00        0.00000E+00 *
14  0.46195E+00   0.19135E+00   0.00000E+00        0.47697E+00
15  0.98097E+00   0.47068E+00   0.00000E+00        0.18369E+00
16  0.15000E+01   0.75000E+00   0.00000E+00        0.00000E+00 *
17  0.50000E+00   0.00000E+00   0.00000E+00        0.46294E+00
18  0.75000E+00   0.00000E+00   0.00000E+00        0.31910E+00
19  0.10000E+01   0.00000E+00   0.00000E+00        0.19770E+00
20  0.12500E+01   0.00000E+00   0.00000E+00        0.95069E-01
21  0.15000E+01   0.00000E+00   0.00000E+00        0.00000E+00 *
```

```
----------STEP NUMBER (IPAS):      9

        INCREMENT                  (DPAS)= 0.30000E+06
        TOTAL LEVEL                (XPAS)= 0.27000E+07
        NUMBER OF ITERATIONS       (NITER)=      5
        METHOD NUMBER              (IMETH):      1
        TOLERANCE                  (EPSDL)= 0.10000E-01
        COEFFICIENT ALPHA          (OMEGA)= 0.10000E+01

ITERATION (ITER):  1  NORM (XNORM)= 0.29481E-01
ITERATION (ITER):  2  NORM (XNORM)= 0.12335E-16

NODES    X            Y            Z          DEGREES OF FREEDOM (* = PRESCRIBED)

  1  0.00000E+00  0.50000E+00  0.00000E+00    0.52704E+00
  2  0.00000E+00  0.75000E+00  0.00000E+00    0.39835E+00
  3  0.00000E+00  0.10000E+01  0.00000E+00    0.32132E+00
  4  0.00000E+00  0.12500E+01  0.00000E+00    0.28240E+00
  5  0.00000E+00  0.15000E+01  0.00000E+00    0.27174E+00
  6  0.19135E+00  0.46195E+00  0.00000E+00    0.52143E+00
  7  0.47068E+00  0.98097E+00  0.00000E+00    0.27400E+00
  8  0.75000E+00  0.15000E+01  0.00000E+00    0.18269E+00
  9  0.35350E+00  0.35350E+00  0.00000E+00    0.49778E+00
 10  0.64012E+00  0.64012E+00  0.00000E+00    0.29591E+00
 11  0.92675E+00  0.92675E+00  0.00000E+00    0.17000E+00
 12  0.12134E+01  0.12134E+01  0.00000E+00    0.77829E-01
 13  0.15000E+01  0.15000E+01  0.00000E+00    0.00000E+00 *
 14  0.46195E+00  0.19135E+00  0.00000E+00    0.48653E+00
 15  0.98097E+00  0.47068E+00  0.00000E+00    0.19007E+00
 16  0.15000E+01  0.75000E+00  0.00000E+00    0.00000E+00 *
 17  0.50000E+00  0.00000E+00  0.00000E+00    0.47192E+00
 18  0.75000E+00  0.00000E+00  0.00000E+00    0.32714E+00
 19  0.10000E+01  0.00000E+00  0.00000E+00    0.20362E+00
 20  0.12500E+01  0.00000E+00  0.00000E+00    0.98159E-01
 21  0.15000E+01  0.00000E+00  0.00000E+00    0.00000E+00 *
```

Figure 6.28 (*Contd.*)

```
----------STEP NUMBER (IPAS):   10

                INCREMENT                       (DPAS)= 0.30000E+06
                TOTAL LEVEL                     (XPAS)= 0.30000E+07
                NUMBER OF ITERATIONS            (NITER)=        5
                METHOD NUMBER                   (IMETH):        1
                TOLERANCE                       (EPSDL)= 0.10000E-01
                COEFFICIENT ALPHA               (OMEGA)= 0.10000E+01

ITERATION (ITER):  1  NORM (XNORM)= 0.22832E-01
ITERATION (ITER):  2  NORM (XNORM)= 0.15457E-16

NODES    X              Y              Z           DEGREES OF FREEDOM (* = PRESCRIBED)

 1  0.00000E+00  0.50000E+00  0.00000E+00         0.53707E+00
 2  0.00000E+00  0.75000E+00  0.00000E+00         0.40843E+00
 3  0.00000E+00  0.10000E+01  0.00000E+00         0.33173E+00
 4  0.00000E+00  0.12500E+01  0.00000E+00         0.29298E+00
 5  0.00000E+00  0.15000E+01  0.00000E+00         0.28244E+00
 6  0.19135E+00  0.46195E+00  0.00000E+00         0.53089E+00
 7  0.47068E+00  0.98097E+00  0.00000E+00         0.28307E+00
 8  0.75000E+00  0.15000E+01  0.00000E+00         0.19017E+00
 9  0.35350E+00  0.35350E+00  0.00000E+00         0.50635E+00
10  0.64012E+00  0.64012E+00  0.00000E+00         0.30364E+00
11  0.92675E+00  0.92675E+00  0.00000E+00         0.17587E+00
12  0.12134E+01  0.12134E+01  0.00000E+00         0.81006E-01
13  0.15000E+01  0.15000E+01  0.00000E+00        -0.00000E+00 *
14  0.46195E+00  0.19135E+00  0.00000E+00         0.49407E+00
15  0.98097E+00  0.47068E+00  0.00000E+00         0.19510E+00
16  0.15000E+01  0.75000E+00  0.00000E+00        -0.00000E+00 *
17  0.50000E+00  0.00000E+00  0.00000E+00         0.47901E+00
18  0.75000E+00  0.00000E+00  0.00000E+00         0.33348E+00
19  0.10000E+01  0.00000E+00  0.00000E+00         0.20829E+00
20  0.12500E+01  0.00000E+00  0.00000E+00         0.10059E+00
21  0.15000E+01  0.00000E+00  0.00000E+00        -0.00000E+00 *
```

```
----------STEP NUMBER (IPAS):   11

        INCREMENT                  (DPAS)= 0.30000E+06
        TOTAL LEVEL                (XPAS)= 0.33000E+07
        NUMBER OF ITERATIONS       (NITER)=      5
        METHOD NUMBER              (IMETH):      1
        TOLERANCE                  (EPSDL)= 0.10000E-01
        COEFFICIENT ALPHA          (OMEGA)= 0.10000E+01

ITERATION (ITER):  1  NORM (XNORM)= 0.17771E-01
ITERATION (ITER):  2  NORM (XNORM)= 0.10600E-16

NODES    X            Y            Z          DEGREES OF FREEDOM (* = PRESCRIBED)

 1  0.00000E+00  0.50000E+00  0.00000E+00       0.54500E+00
 2  0.00000E+00  0.75000E+00  0.00000E+00       0.41642E+00
 3  0.00000E+00  0.10000E+01  0.00000E+00       0.33998E+00
 4  0.00000E+00  0.12500E+01  0.00000E+00       0.30136E+00
 5  0.00000E+00  0.15000E+01  0.00000E+00       0.29091E+00
 6  0.19135E+00  0.46195E+00  0.00000E+00       0.53837E+00
 7  0.47068E+00  0.98097E+00  0.00000E+00       0.29025E+00
 8  0.75000E+00  0.15000E+01  0.00000E+00       0.19609E+00
 9  0.35350E+00  0.35350E+00  0.00000E+00       0.51312E+00
10  0.64012E+00  0.64012E+00  0.00000E+00       0.30976E+00
11  0.92675E+00  0.92675E+00  0.00000E+00       0.18051E+00
12  0.12134E+01  0.12134E+01  0.00000E+00       0.83523E-01
13  0.15000E+01  0.15000E+01  0.00000E+00       0.00000E+00 *
14  0.46195E+00  0.19135E+00  0.00000E+00       0.50004E+00
15  0.98097E+00  0.47068E+00  0.00000E+00       0.19908E+00
16  0.15000E+01  0.75000E+00  0.00000E+00       0.00000E+00 *
17  0.50000E+00  0.00000E+00  0.00000E+00       0.48461E+00
18  0.75000E+00  0.00000E+00  0.00000E+00       0.33848E+00
19  0.10000E+01  0.00000E+00  0.00000E+00       0.21198E+00
20  0.12500E+01  0.00000E+00  0.00000E+00       0.10252E+00
21  0.15000E+01  0.00000E+00  0.00000E+00       0.00000E+00 *
```

Figure 6.28 (*Contd.*)

---------STEP NUMBER (IPAS):   12

```
INCREMENT              (DPAS)= 0.30000E+06
TOTAL LEVEL            (XPAS)= 0.36000E+07
NUMBER OF ITERATIONS   (NITER)=       5
METHOD NUMBER          (IMETH):       1
TOLERANCE              (EPSDL)= 0.10000E-01
COEFFICIENT ALPHA      (OMEGA)= 0.10000E+01
```

ITERATION (ITER):   1 NORM (XNORM)= 0.13885E-01
ITERATION (ITER):   2 NORM (XNORM)= 0.23821E-16

NODES	X	Y	Z	DEGREES OF FREEDOM (* = PRESCRIBED)
1	0.00000E+00	0.50000E+00	0.00000E+00	0.55129E+00
2	0.00000E+00	0.75000E+00	0.00000E+00	0.42275E+00
3	0.00000E+00	0.10000E+01	0.00000E+00	0.34651E+00
4	0.00000E+00	0.12500E+01	0.00000E+00	0.30800E+00
5	0.00000E+00	0.15000E+01	0.00000E+00	0.29762E+00
6	0.19135E+00	0.46195E+00	0.00000E+00	0.54430E+00
7	0.47068E+00	0.98097E+00	0.00000E+00	0.29593E+00
8	0.75000E+00	0.15000E+01	0.00000E+00	0.20079E+00
9	0.35350E+00	0.35350E+00	0.00000E+00	0.51849E+00
10	0.64012E+00	0.64012E+00	0.00000E+00	0.31460E+00
11	0.92675E+00	0.92675E+00	0.00000E+00	0.18419E+00
12	0.12134E+01	0.12134E+01	0.00000E+00	0.85517E-01
13	0.15000E+01	0.15000E+01	0.00000E+00	0.00000E+00 *
14	0.46195E+00	0.19135E+00	0.00000E+00	0.50476E+00
15	0.98097E+00	0.47068E+00	0.00000E+00	0.20223E+00
16	0.15000E+01	0.75000E+00	0.00000E+00	0.00000E+00 *
17	0.50000E+00	0.00000E+00	0.00000E+00	0.48904E+00
18	0.75000E+00	0.00000E+00	0.00000E+00	0.34245E+00
19	0.10000E+01	0.00000E+00	0.00000E+00	0.21489E+00
20	0.12500E+01	0.00000E+00	0.00000E+00	0.10404E+00
21	0.15000E+01	0.00000E+00	0.00000E+00	0.00000E+00 *

```
--------STEP NUMBER (IPAS):    13

           INCREMENT                  (DPAS)= 0.30000E+06
           TOTAL LEVEL                (XPAS)= 0.39000E+07
           NUMBER OF ITERATIONS       (NITER)=         5
           METHOD NUMBER              (IMETH):         1
           TOLERANCE                  (EPSDL)= 0.10000E-01
           COEFFICIENT ALPHA          (OMEGA)= 0.10000E+01

ITERATION (ITER):  1 NORM (XNORM)= 0.10880E-01
ITERATION (ITER):  2 NORM (XNORM)= 0.93022E-17

NODES    X              Y              Z         DEGREES OF FREEDOM (* = PRESCRIBED)

 1  0.00000E+00  0.50000E+00  0.00000E+00    0.55627E+00
 2  0.00000E+00  0.75000E+00  0.00000E+00    0.42776E+00
 3  0.00000E+00  0.10000E+01  0.00000E+00    0.35169E+00
 4  0.00000E+00  0.12500E+01  0.00000E+00    0.31325E+00
 5  0.00000E+00  0.15000E+01  0.00000E+00    0.30294E+00
 6  0.19135E+00  0.46195E+00  0.00000E+00    0.54899E+00
 7  0.47068E+00  0.98097E+00  0.00000E+00    0.30044E+00
 8  0.75000E+00  0.15000E+01  0.00000E+00    0.20451E+00
 9  0.35350E+00  0.35350E+00  0.00000E+00    0.52273E+00
10  0.64012E+00  0.64012E+00  0.00000E+00    0.31843E+00
11  0.92675E+00  0.92675E+00  0.00000E+00    0.18711E+00
12  0.12134E+01  0.12134E+01  0.00000E+00    0.87096E-01
13  0.15000E+01  0.15000E+01  0.00000E+00    0.00000E+00  *
14  0.46195E+00  0.19135E+00  0.00000E+00    0.50849E+00
15  0.98097E+00  0.47068E+00  0.00000E+00    0.20473E+00
16  0.15000E+01  0.75000E+00  0.00000E+00    0.00000E+00  *
17  0.50000E+00  0.00000E+00  0.00000E+00    0.49254E+00
18  0.75000E+00  0.00000E+00  0.00000E+00    0.34558E+00
19  0.10000E+01  0.00000E+00  0.00000E+00    0.21720E+00
20  0.12500E+01  0.00000E+00  0.00000E+00    0.10524E+00
21  0.15000E+01  0.00000E+00  0.00000E+00    0.00000E+00  *
```

Figure 6.28 (*Contd.*)

474

```
---------STEP NUMBER (IPAS):    14

          INCREMENT             (DPAS)= 0.30000E+06
          TOTAL LEVEL           (XPAS)= 0.42000E+07
          NUMBER OF ITERATIONS  (NITER)=       5
          METHOD NUMBER         (IMETH):       1
          TOLERANCE             (EPSDL)= 0.10000E-01
          COEFFICIENT ALPHA     (OMEGA)= 0.10000E+01

ITERATION (ITER):  1  NORM (XNORM)= 0.85444E-02

NODES   X            Y            Z            DEGREES OF FREEDOM (* = PRESCRIBED)

 1  0.00000E+00  0.50000E+00  0.00000E+00      0.56021E+00
 2  0.00000E+00  0.75000E+00  0.00000E+00      0.43172E+00
 3  0.00000E+00  0.10000E+01  0.00000E+00      0.35578E+00
 4  0.00000E+00  0.12500E+01  0.00000E+00      0.31742E+00
 5  0.00000E+00  0.15000E+01  0.00000E+00      0.30715E+00
 6  0.19135E+00  0.46195E+00  0.00000E+00      0.55271E+00
 7  0.47068E+00  0.98097E+00  0.00000E+00      0.30400E+00
 8  0.75000E+00  0.15000E+01  0.00000E+00      0.20745E+00
 9  0.35350E+00  0.35350E+00  0.00000E+00      0.52609E+00
10  0.64012E+00  0.64012E+00  0.00000E+00      0.32146E+00
11  0.92675E+00  0.92675E+00  0.00000E+00      0.18941E+00
12  0.12134E+01  0.12134E+01  0.00000E+00      0.88347E-01
13  0.15000E+01  0.15000E+01  0.00000E+00      0.00000E+00 *
14  0.46195E+00  0.19135E+00  0.00000E+00      0.51145E+00
15  0.98097E+00  0.47068E+00  0.00000E+00      0.20670E+00
16  0.15000E+01  0.75000E+00  0.00000E+00      0.00000E+00 *
17  0.50000E+00  0.00000E+00  0.00000E+00      0.49532E+00
18  0.75000E+00  0.00000E+00  0.00000E+00      0.34806E+00
19  0.10000E+01  0.00000E+00  0.00000E+00      0.21903E+00
20  0.12500E+01  0.00000E+00  0.00000E+00      0.10619E+00
21  0.15000E+01  0.00000E+00  0.00000E+00      0.00000E+00 *
```

```
----------STEP NUMBER (IPAS):   15

          INCREMENT                    (DPAS)= 0.30000E+06
          TOTAL LEVEL                  (XPAS)= 0.45000E+07
          NUMBER OF ITERATIONS         (NITER)=       5
          METHOD NUMBER                (IMETH):       1
          TOLERANCE                    (EPSDL)= 0.10000E-01
          COEFFICIENT ALPHA            (OMEGA)= 0.10000E+01

ITERATION (ITER):  1  NORM (XNORM)= 0.67222E-02

NODES     X            Y            Z          DEGREES OF FREEDOM (* = PRESCRIBED)
  1  0.00000E+00  0.50000E+00  0.00000E+00     0.56333E+00
  2  0.00000E+00  0.75000E+00  0.00000E+00     0.43486E+00
  3  0.00000E+00  0.10000E+01  0.00000E+00     0.35903E+00
  4  0.00000E+00  0.12500E+01  0.00000E+00     0.32072E+00
  5  0.00000E+00  0.15000E+01  0.00000E+00     0.31049E+00
  6  0.19135E+00  0.46195E+00  0.00000E+00     0.55565E+00
  7  0.47068E+00  0.98097E+00  0.00000E+00     0.30682E+00
  8  0.75000E+00  0.15000E+01  0.00000E+00     0.20979E+00
  9  0.35350E+00  0.35350E+00  0.00000E+00     0.52875E+00
 10  0.64012E+00  0.64012E+00  0.00000E+00     0.32386E+00
 11  0.92675E+00  0.92675E+00  0.00000E+00     0.19124E+00
 12  0.12134E+01  0.12134E+01  0.00000E+00     0.89337E-01
 13  0.15000E+01  0.15000E+01  0.00000E+00     0.00000E+00 *
 14  0.46195E+00  0.19135E+00  0.00000E+00     0.51379E+00
 15  0.98097E+00  0.47068E+00  0.00000E+00     0.20826E+00
 16  0.15000E+01  0.75000E+00  0.00000E+00     0.00000E+00 *
 17  0.50000E+00  0.00000E+00  0.00000E+00     0.49751E+00
 18  0.75000E+00  0.00000E+00  0.00000E+00     0.35002E+00
 19  0.10000E+01  0.00000E+00  0.00000E+00     0.22047E+00
 20  0.12500E+01  0.00000E+00  0.00000E+00     0.10695E+00
 21  0.15000E+01  0.00000E+00  0.00000E+00     0.00000E+00 *
```

Figure 6.28 (*Cont d.*)

476

```
----------STEP NUMBER (IPAS):    16

            INCREMENT               (DPAS)= 0.30000E+06
            TOTAL LEVEL             (XPAS)= 0.48000E+07
            NUMBER OF ITERATIONS    (NITER)=        5
            METHOD NUMBER           (IMETH):        1
            TOLERANCE               (EPSDL)= 0.10000E-01
            COEFFICIENT ALPHA       (OMEGA)= 0.10000E+01

ITERATION (ITER):  1  NORM (XNORM)= 0.52958E-02

NODES    X             Y             Z          DEGREES OF FREEDOM (* = PRESCRIBED)

  1  0.00000E+00  0.50000E+00  0.00000E+00      0.56580E+00
  2  0.00000E+00  0.75000E+00  0.00000E+00      0.43735E+00
  3  0.00000E+00  0.10000E+01  0.00000E+00      0.36160E+00
  4  0.00000E+00  0.12500E+01  0.00000E+00      0.32333E+00
  5  0.00000E+00  0.15000E+01  0.00000E+00      0.31313E+00
  6  0.19135E+00  0.46195E+00  0.00000E+00      0.55798E+00
  7  0.47068E+00  0.98097E+00  0.00000E+00      0.30906E+00
  8  0.75000E+00  0.15000E+01  0.00000E+00      0.21163E+00
  9  0.35350E+00  0.35350E+00  0.00000E+00      0.53086E+00
 10  0.64012E+00  0.64012E+00  0.00000E+00      0.32577E+00
 11  0.92675E+00  0.92675E+00  0.00000E+00      0.19268E+00
 12  0.12134E+01  0.12134E+01  0.00000E+00      0.90121E-01
 13  0.15000E+01  0.15000E+01  0.00000E+00      0.00000E+00 *
 14  0.46195E+00  0.19135E+00  0.00000E+00      0.51564E+00
 15  0.98097E+00  0.47068E+00  0.00000E+00      0.20950E+00
 16  0.15000E+01  0.75000E+00  0.00000E+00      0.00000E+00 *
 17  0.50000E+00  0.00000E+00  0.00000E+00      0.49925E+00
 18  0.75000E+00  0.00000E+00  0.00000E+00      0.35158E+00
 19  0.10000E+01  0.00000E+00  0.00000E+00      0.22162E+00
 20  0.12500E+01  0.00000E+00  0.00000E+00      0.10754E+00
 21  0.15000E+01  0.00000E+00  0.00000E+00      0.00000E+00 *
```

```
------STEP NUMBER (IPAS):   17

         INCREMENT                  (DPAS)= 0.30000E+06
         TOTAL LEVEL                (XPAS)= 0.51000E+07
         NUMBER OF ITERATIONS       (NITER)=      5
         METHOD NUMBER              (IMETH):      1
         TOLERANCE                  (EPSDL)= 0.10000E-01
         COEFFICIENT ALPHA          (OMEGA)= 0.10000E+01

ITERATION (ITER):  1  NORM (XNORM)= 0.41766E-02

NODES    X            Y            Z          DEGREES OF FREEDOM (* = PRESCRIBED)

 1 0.00000E+00 0.50000E+00 0.00000E+00         0.56776E+00
 2 0.00000E+00 0.75000E+00 0.00000E+00         0.43932E+00
 3 0.00000E+00 0.10000E+01 0.00000E+00         0.36363E+00
 4 0.00000E+00 0.12500E+01 0.00000E+00         0.32540E+00
 5 0.00000E+00 0.15000E+01 0.00000E+00         0.31522E+00
 6 0.19135E+00 0.46195E+00 0.00000E+00         0.55982E+00
 7 0.47068E+00 0.98097E+00 0.00000E+00         0.31083E+00
 8 0.75000E+00 0.15000E+01 0.00000E+00         0.21310E+00
 9 0.35350E+00 0.35350E+00 0.00000E+00         0.53253E+00
10 0.64012E+00 0.64012E+00 0.00000E+00         0.32727E+00
11 0.92675E+00 0.92675E+00 0.00000E+00         0.19583E+00
12 0.12134E+01 0.12134E+01 0.00000E+00         0.90742E-01
13 0.15000E+01 0.15000E+01 0.00000E+00         0.00000E+00  *
14 0.46195E+00 0.19135E+00 0.00000E+00         0.51711E+00
15 0.98097E+00 0.47068E+00 0.00000E+00         0.21048E+00
16 0.15000E+01 0.75000E+00 0.00000E+00         0.00000E+00  *
17 0.50000E+00 0.00000E+00 0.00000E+00         0.50063E+00
18 0.75000E+00 0.00000E+00 0.00000E+00         0.35281E+00
19 0.10000E+01 0.00000E+00 0.00000E+00         0.22253E+00
20 0.12500E+01 0.00000E+00 0.00000E+00         0.10802E+00
21 0.15000E+01 0.00000E+00 0.00000E+00         0.00000E+00  *
```

Figure 6.28 (*Contd.*)

```
----------STEP NUMBER (IPAS):    18

             INCREMENT                  (DPAS)= 0.30000E+06
             TOTAL LEVEL                (XPAS)= 0.54000E+07
             NUMBER OF ITERATIONS       (NITER)=      5
             METHOD NUMBER              (IMETH):      1
             TOLERANCE                  (EPSDL)= 0.10000E-01
             COEFFICIENT ALPHA          (OMEGA)= 0.10000E+01

ITERATION (ITER):   1  NORM (XNORM)= 0.32967E-02

NODES    X            Y            Z         DEGREES OF FREEDOM (* = PRESCRIBED)

 1  0.00000E+00  0.50000E+00  0.00000E+00    0.56930E+00
 2  0.00000E+00  0.75000E+00  0.00000E+00    0.44088E+00
 3  0.00000E+00  0.10000E+01  0.00000E+00    0.36524E+00
 4  0.00000E+00  0.12500E+01  0.00000E+00    0.32703E+00
 5  0.00000E+00  0.15000E+01  0.00000E+00    0.31688E+00
 6  0.19135E+00  0.46195E+00  0.00000E+00    0.56128E+00
 7  0.47068E+00  0.98097E+00  0.00000E+00    0.31223E+00
 8  0.75000E+00  0.15000E+01  0.00000E+00    0.21425E+00
 9  0.35350E+00  0.35350E+00  0.00000E+00    0.53385E+00
10  0.64012E+00  0.64012E+00  0.00000E+00    0.32846E+00
11  0.92675E+00  0.92675E+00  0.00000E+00    0.19473E+00
12  0.12134E+01  0.12134E+01  0.00000E+00    0.91233E-01
13  0.15000E+01  0.15000E+01  0.00000E+00    0.00000E+00 *
14  0.46195E+00  0.19135E+00  0.00000E+00    0.51827E+00
15  0.98097E+00  0.47068E+00  0.00000E+00    0.21126E+00
16  0.15000E+01  0.75000E+00  0.00000E+00    0.00000E+00 *
17  0.50000E+00  0.00000E+00  0.00000E+00    0.50172E+00
18  0.75000E+00  0.00000E+00  0.00000E+00    0.35379E+00
19  0.10000E+01  0.00000E+00  0.00000E+00    0.22324E+00
20  0.12500E+01  0.00000E+00  0.00000E+00    0.10839E+00
21  0.15000E+01  0.00000E+00  0.00000E+00    0.00000E+00 *
```

```
------------STEP NUMBER (IPAS):   19

         INCREMENT                    (DPAS)= 0.30000E+06
         TOTAL LEVEL                  (XPAS)= 0.57000E+07
         NUMBER OF ITERATIONS         (NITER)=     5
         METHOD NUMBER                (IMETH):     1
         TOLERANCE                    (EPSDL)= 0.10000E-01
         COEFFICIENT ALPHA            (OMEGA)= 0.10000E+01

ITERATION (ITER):   1  NORM (XNORM)= 0.26039E-02

NODES    X              Y              Z          DEGREES OF FREEDOM (* = PRESCRIBED)

  1  0.00000E+00  0.50000E+00  0.00000E+00        0.57053E+00
  2  0.00000E+00  0.75000E+00  0.00000E+00        0.44211E+00
  3  0.00000E+00  0.10000E+01  0.00000E+00        0.36652E+00
  4  0.00000E+00  0.12500E+01  0.00000E+00        0.32833E+00
  5  0.00000E+00  0.15000E+01  0.00000E+00        0.31819E+00
  6  0.19135E+00  0.46195E+00  0.00000E+00        0.56244E+00
  7  0.47068E+00  0.98097E+00  0.00000E+00        0.31334E+00
  8  0.75000E+00  0.15000E+01  0.00000E+00        0.21517E+00
  9  0.35350E+00  0.35350E+00  0.00000E+00        0.53489E+00
 10  0.64012E+00  0.64012E+00  0.00000E+00        0.32941E+00
 11  0.92675E+00  0.92675E+00  0.00000E+00        0.19545E+00
 12  0.12134E+01  0.12134E+01  0.00000E+00        0.91622E-01
 13  0.15000E+01  0.15000E+01  0.00000E+00        0.00000E+00 *
 14  0.46195E+00  0.19135E+00  0.00000E+00        0.51919E+00
 15  0.98097E+00  0.47068E+00  0.00000E+00        0.21187E+00
 16  0.15000E+01  0.75000E+00  0.00000E+00        0.00000E+00 *
 17  0.50000E+00  0.00000E+00  0.00000E+00        0.50259E+00
 18  0.75000E+00  0.00000E+00  0.00000E+00        0.35456E+00
 19  0.10000E+01  0.00000E+00  0.00000E+00        0.22381E+00
 20  0.12500E+01  0.00000E+00  0.00000E+00        0.10869E+00
 21  0.15000E+01  0.00000E+00  0.00000E+00        0.00000E+00 *
```

Figure 6.28 (*Contd.*)

```
-----------STEP NUMBER (IPAS):    20

         INCREMENT                (DPAS)= 0.30000E+06
         TOTAL LEVEL              (XPAS)= 0.60000E+07
         NUMBER OF ITERATIONS     (NITER)=    5
         METHOD NUMBER            (IMETH):    1
         TOLERANCE                (EPSDL)= 0.10000E-01
         COEFFICIENT ALPHA        (OMEGA)= 0.10000E+01

ITERATION (ITER):  1  NORM (XNORM)= 0.20578E-02

NODES   X             Y             Z           DEGREES OF FREEDOM (* = PRESCRIBED)

 1  0.00000E+00  0.50000E+00  0.00000E+00    0.57150E+00
 2  0.00000E+00  0.75000E+00  0.00000E+00    0.44309E+00
 3  0.00000E+00  0.10000E+01  0.00000E+00    0.36753E+00
 4  0.00000E+00  0.12500E+01  0.00000E+00    0.32936E+00
 5  0.00000E+00  0.15000E+01  0.00000E+00    0.31923E+00
 6  0.19135E+00  0.46195E+00  0.00000E+00    0.56335E+00
 7  0.47068E+00  0.98097E+00  0.00000E+00    0.31422E+00
 8  0.75000E+00  0.15000E+01  0.00000E+00    0.21590E+00
 9  0.35350E+00  0.35350E+00  0.00000E+00    0.53572E+00
10  0.64012E+00  0.64012E+00  0.00000E+00    0.33016E+00
11  0.92675E+00  0.92675E+00  0.00000E+00    0.19602E+00
12  0.12134E+01  0.12134E+01  0.00000E+00    0.91931E-01
13  0.15000E+01  0.15000E+01  0.00000E+00    0.00000E+00  *
14  0.46195E+00  0.19135E+00  0.00000E+00    0.51992E+00
15  0.98097E+00  0.47068E+00  0.00000E+00    0.21236E+00
16  0.15000E+01  0.75000E+00  0.00000E+00    0.00000E+00  *
17  0.50000E+00  0.00000E+00  0.00000E+00    0.50327E+00
18  0.75000E+00  0.00000E+00  0.00000E+00    0.35517E+00
19  0.10000E+01  0.00000E+00  0.00000E+00    0.22426E+00
20  0.12500E+01  0.00000E+00  0.00000E+00    0.10892E+00
21  0.15000E+01  0.00000E+00  0.00000E+00    0.00000E+00  *

END OF PROBLEM,    409 UTILIZED REAL WORDS OVER   20000
```

Figure 6.28 Example of transient heat flow.

6.7.2 *Plane Stress Problem*

Consider the concrete elliptical arch shown below:

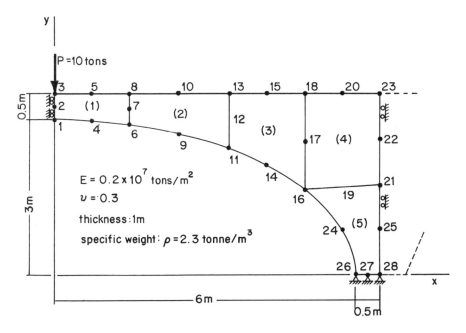

The loads are composed of the distributed dead weight and one concentrated force of 10 tons, as shown. Eight-node isoparametric elements of type 2 (with plane stress hypothesis) are used (paragraph 4.3.5).

Input records and results of MEF obtained by using blocks 'LINM' and 'VALP' are shown in Figures 6.29 and 6.30.

```
                    F.E.M.3.
               G.TOUZOT , G.DHATT

IMAGE OF DATA CARDS
====================

                                C O L U M N   N U M B E R
  CARD           1         2         3         4         5         6         7         8
 NUMBER  12345678901234567890123456789012345678901234567890123456789012345678901234567890
   1     COMT
   2     ELASTIC ANALYSIS OF AN ELLIPTIC HALF BRIDGE ARCH IN PLANE STRESS
   3
   4     COOR
   5         28    2
   6          3  0.00  3.50                    23        6.50     3.50     5
   7          5  0.75  3.50                    20        5.75     3.50     5
   8          2  0.00  3.25
   9          7  1.50  3.20
  10         12  3.50  2.97
  11         17  5.00  2.58
  12         19  5.75  1.70
  13          1  0.00  3.00
  14          4  0.75  2.98
  15          6  1.50  2.90
  16          9  2.50  2.73
  17         11  3.50  2.44
  18         14  4.25  2.12
  19         16  5.00  1.66
  20         24  5.75  0.86
  21         26  6.00  0.00
  22         27  6.25  0.00
  23         28  6.50  0.00
  24         25  6.50  0.87
  25         21  6.50  1.75
  26         22  6.50  2.62
  27         -1
  28     COND
  29     11   26   27   28
  30     10    1    2    3   25   21   22   23
  31     0000000000
  32     PREL
  33          1    4  2.0E6  0.3          0.0        2.3
  34         -1
  35     ELEM
  36          5    8    2    1    0
  37          1    4    5    0    1
  38
  39
  40
  --------------------------------------------------------------------------------------
  CARD    12345678901234567890123456789012345678901234567890123456789012345678901234567890
 NUMBER      1         2         3         4         5         6         7         8
                                C O L U M N   N U M B E R
```

```
                                COLUMN NUMBER
          1         2         3         4         5         6         7         8
 CARD    12345678901234567890123456789012345678901234567890123456789012345678901234567890
NUMBER   --------------------------------------------------------------------------------
  41     5         1    0    0    1    0    26   27   28   25   21   19   16   24
  42     -1
  43     SOLC
  44     1         0.00      -10.00
  45     3
  46     -1
  47     SOLR
  48     LINM 1
  49
  50     STOP
         --------------------------------------------------------------------------------
 CARD    12345678901234567890123456789012345678901234567890123456789012345678901234567890
NUMBER            1         2         3         4         5         6         7         8
                                COLUMN NUMBER

                              END OF DATA
```

Figure 6.29 (*Contd.*)

COMMENTS
=========

ELASTIC ANALYSIS OF AN ELLIPTIC HALF BRIDGE ARCH IN PLANE STRESS

INPUT OF NODES (M= 0)
=====================

```
            MAX. NUMBER OF NODES            (NNT)=  28
            MAX. NUMBER OF D.O.F. PER NODE  (NDLN)=  2
            DIMENSIONS OF THE PROBLEM       (NDIM)=  2
            COORDINATE SCALE FACTORS        (FAC)= 0.10000E+01  0.10000E+01  0.10000E+01
            WORKSPACE IN REAL WORDS         (NVA)=  20000
```

INPUT OF BOUNDARY CONDITIONS (M= 0)
===================================

BOUNDARY CONDITIONS CARDS

```
>>>>>1100000000  0.00000E+00  0.00000E+00  0.00000E+00  0.00000E+00  0.00000E+00  0.00000E+00  0.00000E+00
>>>>>      26    27    28     0           0           0           0           0           0           0
>>>>>1000000000  0.00000E+00  0.00000E+00  0.00000E+00  0.00000E+00  0.00000E+00  0.00000E+00  0.00000E+00
>>>>>       1     2     3    25    21    22    23     0           0           0
>>>>>0000000000  0.00000E+00  0.00000E+00  0.00000E+00  0.00000E+00  0.00000E+00  0.00000E+00  0.00000E+00
```

```
            TOTAL NUMBER OF NODES                         (NNT)=  28
            TOTAL NUMBER OF D.O.F.                        (NDLT)= 56
            NUMBER OF EQUATIONS TO BE SOLVED              (NEQ)=  43
            NUMBER OF PRESCRIBED NON ZERO D.O.F.          (NCLNZ)= 0
            NUMBER OF PRESCRIBED ZERO D.O.F.              (NCLZ)= 13
            TOTAL NUMBER OF PRESCRIBED D.O.F.             (NCLT)= 13
```

NODAL COORDINATES ARRAY

NO	D.L.	X	Y	Z	EQUATION NUMBER	(NEQ)
1	2	0.00000E+00	0.30000E+01	0.00000E+00	-7	1
2	2	0.00000E+00	0.32500E+01	0.00000E+00	-8	2
3	2	0.00000E+00	0.35000E+01	0.00000E+00	-9	3
4	2	0.75000E+00	0.29800E+01	0.00000E+00	4	5
5	2	0.75000E+00	0.35000E+01	0.00000E+00	6	7
6	2	0.15000E+01	0.29000E+01	0.00000E+00	8	9
7	2	0.15000E+01	0.32000E+01	0.00000E+00	10	11
8	2	0.16250E+01	0.35000E+01	0.00000E+00	12	13
9	2	0.25000E+01	0.27300E+01	0.00000E+00	14	15
10	2	0.24167E+01	0.35000E+01	0.00000E+00	16	17
11	2	0.35000E+01	0.24400E+01	0.00000E+00	18	19
12	2	0.35000E+01	0.29700E+01	0.00000E+00	20	21
13	2	0.35000E+01	0.35000E+01	0.00000E+00	22	23
14	2	0.32500E+01	0.21200E+01	0.00000E+00	24	25
15	2	0.40833E+01	0.35000E+01	0.00000E+00	26	27
16	2	0.50000E+01	0.16600E+01	0.00000E+00	28	29
17	2	0.50000E+01	0.25800E+01	0.00000E+00	30	31
18	2	0.48750E+01	0.35000E+01	0.00000E+00	32	33
19	2	0.57500E+01	0.17000E+01	0.00000E+00	34	35
20	2	0.57500E+01	0.35000E+01	0.00000E+00	36	37
21	2	0.65000E+01	0.17500E+01	0.00000E+00	-11	38
22	2	0.65000E+01	0.26200E+01	0.00000E+00	-12	39
23	2	0.65000E+01	0.35000E+01	0.00000E+00	-13	40
24	2	0.57500E+01	0.86000E+00	0.00000E+00	41	42
25	2	0.65000E+01	0.87000E+00	0.00000E+00	-10	43
26	2	0.60000E+01	0.00000E+00	0.00000E+00	-1	-2
27	2	0.62500E+01	0.00000E+00	0.00000E+00	-3	-4
28	2	0.65000E+01	0.00000E+00	0.00000E+00	-5	-6

Figure 6.29 (*Contd.*)

```
INPUT OF ELEMENT PROPERTIES (M= 0)
==================================

        NUMBER OF GROUPS OF PROPERTIES      (NGPE)=   1
        NUMBER OF PROPERTIES PER GROUP      (NPRE)=   4

CARDS OF ELEMENT PROPERTIES

>>>>>   1  0.20000E+07  0.30000E+00  0.00000E+00  0.23000E+01
>>>>>  -1  0.00000E+00  0.00000E+00  0.00000E+00  0.00000E+00

INPUT OF ELEMENTS (M= 0)
========================

        MAX. NUMBER OF ELEMENTS             (NELT)=   5
        MAX. NUMBER OF NODES PER ELEMENT    (NNEL)=   8
        DEFAULT ELEMENT TYPE                (NTPE)=   2
        NUMBER OF GROUPS OF ELEMENTS        (NGRE)=   1
        INDEX FOR NON SYMMETRIC PROBLEM     (NSYM)=   0
        INDEX FOR IDENTICAL ELEMENTS        (NIDENT)= 0

ELEMENT:  1 TYPE: 2 N.P.: 8 D.O.F.: 16 N. PROP:  0 EL. PROP:  4 GROUP: 0
   CONNECTIVITY (NE)      1    4    6    7    8    5    3    2
ELEMENT:  2 TYPE: 2 N.P.: 8 D.O.F.: 16 N. PROP:  0 EL. PROP:  4 GROUP: 0
   CONNECTIVITY (NE)      6    9   11   12   13   10    8    7
ELEMENT:  3 TYPE: 2 N.P.: 8 D.O.F.: 16 N. PROP:  0 EL. PROP:  4 GROUP: 0
   CONNECTIVITY (NE)     11   14   16   17   18   15   13   12
ELEMENT:  4 TYPE: 2 N.P.: 8 D.O.F.: 16 N. PROP:  0 EL. PROP:  4 GROUP: 0
   CONNECTIVITY (NE)     16   19   21   22   23   20   18   17
ELEMENT:  5 TYPE: 2 N.P.: 8 D.O.F.: 16 N. PROP:  0 EL. PROP:  4 GROUP: 0
   CONNECTIVITY (NE)     26   27   28   25   21   16    9   24

        MEAN BAND HEIGHT=   9.1 MAXIMUM=  15
        LENGTH OF A TRIANGLE IN KG          (NKG)=  393
        NUMBER OF INTEGRATION POINTS        (NPG)=   45
```

INPUT OF CONCENTRADED LOADS (M= 0)
==

CARDS OF NODAL LOADS

>>>>>   1 0.00000E+00 -0.10000E+02
>>>>>   3   0   0   0   0   0   0   0   0   0   0   0
>>>>> -1 0.00000E+00 0.00000E+00

ASSEMBLING OF DISTRIBUTED LOADS (M= 0)
==

ASSEMBLING AND LINEAR SOLUTION (M= 0)
==
       INDEX FOR RESIDUAL COMPUTATION   (NRES)=   1
       ENERGY   (ENERG)= 0.15641E-01

       ABSOLUTE VALUE OF MINIMUM PIVOT   = 0.33159E+06 EQUATION: 7
              ALGEBRAIC VALUE= 0.33159E+06 EQUATION: 7
              DETERMINANT   = 0.39532E+02 * 10 **   270

MAX. RESIDUAL VALUE= 0.19895E-12 EQUATION   13

Figure 6.29 (*Contd.*)

SOLUTION

NODES	X	Y	Z	DEGREES OF FREEDOM (* = PRESCRIBED)	
1	0.00000E+00	0.30000E+01	0.00000E+00	0.00000E+00 *	-0.11440E-02
2	0.00000E+00	0.32500E+01	0.00000E+00	0.00000E+00 *	-0.11456E-02
3	0.00000E+00	0.35000E+01	0.00000E+00	0.00000E+00 *	-0.11428E-02
4	0.75000E+00	0.29800E+01	0.00000E+00	0.48677E-04	-0.10173E-02
5	0.75000E+00	0.35000E+01	0.00000E+00	-0.85677E-04	-0.10116E-02
6	0.15000E+01	0.29000E+01	0.00000E+00	0.78758E-04	-0.76959E-03
7	0.15000E+01	0.32000E+01	0.00000E+00	-0.25684E-04	-0.76475E-03
8	0.16250E+01	0.35000E+01	0.00000E+00	-0.13203E-03	-0.72133E-03
9	0.25000E+01	0.27300E+01	0.00000E+00	0.75766E-04	-0.43504E-03
10	0.24167E+01	0.35000E+01	0.00000E+00	-0.12565E-03	-0.45424E-03
11	0.35000E+01	0.24400E+01	0.00000E+00	0.67096E-04	-0.21661E-03
12	0.35000E+01	0.29700E+01	0.00000E+00	-0.74605E-05	-0.20884E-03
13	0.32500E+01	0.35000E+01	0.00000E+00	-0.10106E-03	-0.26136E-03
14	0.42500E+01	0.21200E+01	0.00000E+00	0.47532E-04	-0.11537E-03
15	0.40833E+01	0.35000E+01	0.00000E+00	-0.68853E-04	-0.13360E-03
16	0.50000E+01	0.16600E+01	0.00000E+00	0.24909E-04	-0.54279E-04
17	0.50000E+01	0.25800E+01	0.00000E+00	-0.77484E-05	-0.59181E-04
18	0.48750E+01	0.35000E+01	0.00000E+00	-0.41166E-04	-0.66645E-04
19	0.57500E+01	0.17000E+01	0.00000E+00	0.10861E-04	-0.34608E-04
20	0.57500E+01	0.35000E+01	0.00000E+00	-0.20207E-04	-0.35330E-04
21	0.65000E+01	0.17500E+01	0.00000E+00	0.00000E+00 *	-0.26278E-04
22	0.65000E+01	0.26200E+01	0.00000E+00	0.00000E+00 *	-0.24819E-04
23	0.65000E+01	0.35000E+01	0.00000E+00	0.00000E+00 *	-0.28280E-04
24	0.57500E+01	0.86000E+00	0.00000E+00	0.19781E-05	-0.26376E-04
25	0.65000E+01	0.87000E+00	0.00000E+00	0.00000E+00 *	-0.19766E-04
26	0.60000E+01	0.00000E+00	0.00000E+00	0.00000E+00 *	0.00000E+00 *
27	0.62500E+01	0.00000E+00	0.00000E+00	0.00000E+00 *	0.00000E+00 *
28	0.65000E+01	0.00000E+00	0.00000E+00	0.00000E+00 *	0.00000E+00 *

STRESSES IN ELEMENT 1

P.G.	X	Y	EPSX	EPSY	GAMXY	SIGX / SIG1	SIGY / SIG2	TAUXY / TAUMAX	TETA
1	0.16501E+00	0.30570E+01	0.50204E-04	-0.19642E-04	0.54953E-04	0.97387E+02 0.11202E+03	-0.10067E+02 -0.24703E+02	0.42272E+02 0.68363E+02	19.1
2	0.15655E+00	0.32504E+01	-0.29925E-04	0.30349E-06	0.45342E-04	-0.65568E+02 -0.39706E+00	-0.19064E+02 -0.84235E+02	0.34879E+02 0.41919E+02	61.8
3	0.15655E+00	0.34437E+01	-0.11227E-03	0.20509E-04	0.50859E-04	-0.23322E+03 -0.21709E+02	-0.28946E+02 -0.24045E+03	0.39122E+02 0.10937E+03	79.5
4	0.73750E+00	0.30386E+01	0.22213E-04	-0.64565E-05	-0.13663E-05	0.44562E+02 0.44587E+02	0.45554E+00 0.43051E+00	-0.10510E+01 -0.22078E+02	-1.4
5	0.71875E+00	0.32400E+01	-0.25735E-04	0.10875E-04	-0.41179E-05	-0.49390E+02 -0.71102E+01	0.69326E+01 -0.49568E+02	-0.31676E+01 0.28339E+02	-86.8
6	0.73750E+00	0.34414E+01	-0.69797E-04	0.26667E-04	0.10082E-04	-0.13582E+03 0.12993E+02	0.12589E+02 -0.13622E+03	0.77553E+01 0.74607E+02	87.0
7	0.13184E+01	0.29883E+01	-0.14472E-04	0.24319E-04	0.36014E-04	-0.15772E+02 0.54784E+02	0.43907E+02 -0.26650E+02	0.27703E+02 0.40717E+02	68.6
8	0.13184E+01	0.32116E+01	-0.26598E-04	0.54277E-05	0.34284E-04	-0.54879E+02 0.58451E+01	-0.56082E+01 -0.66332E+02	0.26372E+02 0.36088E+02	66.5
9	0.13850E+01	0.34350E+01	-0.36593E-04	-0.10515E-04	0.36637E-04	-0.87358E+02 -0.32706E+02	-0.47238E+02 -0.10189E+03	0.28182E+02 0.34593E+02	62.7

Figure 6.29 (*Contd.*)

STRESSES IN ELEMENT   2
P.G.   X   Y   EPSX   EPSY   GAMXY   SIGX   SIGY   TAUXY   TETA
                                                                            SIG1   SIG2   TAUMAX

1 0.17172E+01 0.29429E+01 -0.38983E-04 0.10309E-04 0.34476E-04 -0.78879E+02 -0.30453E+01 0.26520E+02   72.5
                                                                    0.53089E+01 -0.87233E+02 0.46271E+02   66.9
2 0.17212E+01 0.31861E+01 -0.18897E-04 0.44171E-05 0.24287E-04 -0.38620E+02 -0.27519E+01 0.18682E+02
                                                                    0.52112E+01 -0.46583E+02 0.25897E+02   45.3
3 0.17749E+01 0.34292E+01 -0.30142E-05 -0.27797E-05 0.19267E-04 -0.84574E+01 -0.80966E+01 0.14821E+02
                                                                    0.65450E+01 -0.23099E+02 0.14822E+02   69.9
4 0.25031E+01 0.28168E+01 -0.49804E-04 -0.79131E-05 0.35385E-04 -0.11468E+03 -0.50229E+02 0.27220E+02
                                                                   -0.40271E+02 -0.12463E+03 0.42181E+02   75.5
5 0.24896E+01 0.31150E+01 -0.21234E-04 0.52067E-05 0.14675E-04 -0.43236E+02 -0.25573E+01 0.11288E+02
                                                                    0.36516E+00 -0.46158E+02 0.23262E+02   86.5
6 0.24386E+01 0.34132E+01 0.86355E-05 0.19183E-04 0.13133E-05 0.31627E+02 0.47853E+02 0.10102E+01
                                                                    0.47916E+02 0.31564E+02 0.81759E+01   71.0
7 0.32918E+01 0.26268E+01 -0.45736E-04 0.13102E-04 0.45959E-04 -0.91881E+02 -0.13611E+01 0.35553E+02
                                                                    0.10810E+02 -0.10405E+03 0.57431E+02   66.8
8 0.32704E+01 0.30079E+01 -0.15496E-04 0.82194E-06 0.17127E-04 -0.33515E+02 -0.84106E+01 0.13175E+02
                                                                   -0.27655E+01 -0.39160E+02 0.18197E+02   7.7
9 0.31244E+01 0.33891E+01 0.21894E-04 -0.91426E-05 0.85331E-05 0.42090E+02 -0.56583E+01 0.65639E+01
                                                                    0.42975E+02 -0.65442E+01 0.24760E+02

| STRESSES IN ELEMENT | | 3 | | | | SIGX | SIGY | TAUXY | TETA |
P.G.	X	Y	EPSX	EPSY	GAMXY	SIG1	SIG2	TAUMAX	
1	0.36906E+01	0.25063E+01	-0.50552E-04	0.22342E-04	0.49581E-04	-0.96372E+02	0.15772E+02	0.38139E+02	72.9
						0.27514E+02	-0.10811E+03	0.67814E+02	
2	0.36732E+01	0.29400E+01	-0.12359E-04	-0.29005E-05	0.60445E-05	-0.29075E+02	-0.14524E+02	0.46496E+01	73.7
						-0.13165E+02	-0.30434E+02	0.86347E+01	
3	0.35143E+01	0.33738E+01	0.29214E-04	-0.20784E-04	0.11104E-04	0.50502E+02	-0.26418E+02	0.85417E+01	6.3
						0.51439E+02	-0.27355E+02	0.39397E+02	
4	0.42687E+01	0.22755E+01	-0.43049E-04	0.12248E-04	0.57440E-04	-0.86557E+02	-0.14655E+01	0.44185E+02	67.0
						0.17330E+02	-0.10533E+03	0.61332E+02	
5	0.42604E+01	0.28100E+01	-0.12171E-04	-0.12015E-05	0.13653E-04	-0.27541E+02	-0.10665E+02	0.10502E+02	64.4
						-0.56311E+01	-0.32575E+02	0.13472E+02	
6	0.41396E+01	0.33445E+01	0.26343E-04	-0.11199E-04	0.84516E-05	0.50514E+02	-0.72430E+01	0.65015E+01	6.3
						0.51237E+02	-0.79658E+01	0.29601E+02	
7	0.48440E+01	0.19702E+01	-0.31920E-04	-0.16373E-05	0.42812E-04	-0.71234E+02	-0.24645E+02	0.32932E+02	62.6
						-0.76011E+01	-0.88277E+02	0.40338E+02	
8	0.48351E+01	0.26380E+01	-0.90960E-05	-0.32866E-05	0.20509E-04	-0.22158E+02	-0.13221E+02	0.15776E+02	80.3
						-0.12950E+02	-0.22428E+02	0.47391E+01	
9	0.47427E+01	0.33057E+01	0.23960E-04	-0.34094E-05	-0.93369E-05	0.50412E+02	0.83047E+01	-0.71822E+01	-9.4
						0.51603E+02	0.71134E+01	0.22245E+02	

Figure 6.29 (*Contd.*)

STRESSES IN ELEMENT   4

P.G.	X	Y	EPSX	EPSY	GAMXY	SIGX SIG1	SIGY SIG2	TAUXY TAUMAX	TETA
1	0.51816E+01	0.18744E+01	-0.18281E-04	-0.38572E-05	0.26007E-04	-0.42722E+02 -0.87505E+01	-0.20531E+02 -0.54503E+02	0.20005E+02 0.22876E+02	59.5
2	0.51816E+01	0.25835E+01	-0.53909E-05	-0.28796E-05	0.70215E-05	-0.13747E+02 -0.60787E+01	-0.98832E+01 -0.17551E+02	0.54012E+01 0.57362E+01	54.8
3	0.51150E+01	0.32933E+01	0.16035E-04	-0.12475E-05	-0.11558E-04	0.34420E+02 0.37119E+02	0.78311E+01 0.51320E+01	-0.88910E+01 0.15994E+02	-16.9
4	0.57625E+01	0.19019E+01	-0.15848E-04	0.17604E-05	0.17059E-04	-0.33670E+02 -0.12663E+01	-0.65801E+01 -0.38984E+02	0.13122E+02 0.18859E+02	68.0
5	0.57813E+01	0.25975E+01	-0.47053E-05	-0.40096E-06	0.56588E-05	-0.10606E+02 -0.18256E+01	-0.39836E+01 -0.12764E+02	0.43529E+01 0.54691E+01	63.6
6	0.57625E+01	0.32961E+01	0.17411E-04	-0.23222E-05	-0.60956E-05	0.36727E+02 0.37435E+02	0.63538E+01 0.56464E+01	-0.46889E+01 0.15894E+02	-8.6
7	0.63350E+01	0.19347E+01	-0.13309E-04	0.36681E-05	0.60716E-05	-0.26831E+02 -0.96875E-01	-0.71319E+00 -0.27641E+02	0.46705E+01 0.13869E+02	80.2
8	0.63434E+01	0.26145E+01	-0.40499E-05	-0.36218E-06	0.10913E-05	-0.91396E+01 -0.33446E+01	-0.34662E+01 -0.92612E+01	0.83947E+00 0.29583E+01	81.8
9	0.63434E+01	0.32996E+01	0.19064E-04	-0.43792E-05	-0.50739E-05	0.39011E+02 0.39428E+02	0.29447E+01 0.25272E+01	-0.39030E+01 0.18451E+02	-6.1

```
STRESSES IN ELEMENT   5
 P.G.       X            Y            EPSX           EPSY          GAMXY           SIGX           SIGY         TAUXY     TETA
                                                                                   SIG1           SIG2         TAUMAX

  1  0.60451E+01  0.19842E+00  0.15686E-05 -0.27606E-04 -0.66658E-07 -0.14754E+02 -0.59639E+02 -0.51275E-01   -0.1
                                                                    -0.14754E+02 -0.59639E+02  0.22442E+02    33.9
  2  0.58345E+01  0.86013E+00 -0.59590E-05 -0.18340E-04  0.30279E-04 -0.25189E+02 -0.44236E+02  0.23292E+02
                                                                    -0.95492E+01 -0.59876E+02  0.25163E+02    55.3
  3  0.53578E+01  0.14906E+01 -0.17882E-04 -0.16946E-05  0.43111E-04 -0.40418E+02 -0.15515E+02  0.33162E+02
                                                                     0.74568E+01 -0.63389E+02  0.35423E+02    -2.8
  4  0.62436E+01  0.19659E+00  0.26761E-05 -0.25115E-04 -0.27145E-05 -0.10678E+02 -0.53434E+02 -0.20881E+01
                                                                    -0.10576E+02 -0.53536E+02  0.21480E+02    22.2
  5  0.61250E+01  0.86250E+00 -0.27121E-05 -0.17696E-04  0.14640E-04 -0.17628E+02 -0.40681E+02 -0.11262E+02
                                                                    -0.13040E+02 -0.45269E+02  0.16115E+02    53.8
  6  0.58564E+01  0.15134E+01 -0.14356E-04 -0.61158E-05  0.25957E-04 -0.35584E+02 -0.22907E+02  0.19967E+02
                                                                    -0.82268E+01 -0.50194E+02  0.20949E+02    -4.2
  7  0.64422E+01  0.19544E+00  0.37803E-05 -0.26063E-04 -0.43832E-05 -0.87761E+01 -0.54789E+02 -0.33717E+01
                                                                    -0.86298E+01 -0.55035E+02  0.23203E+02    -2.2
  8  0.64155E+01  0.86787E+00  0.62431E-06 -0.15796E-04 -0.12372E-05 -0.90426E+01 -0.34304E+02 -0.95171E+00
                                                                    -0.90068E+01 -0.34340E+02  0.12666E+02    64.6
  9  0.63549E+01  0.15416E+01 -0.10641E-04 -0.46963E-05  0.73090E-05 -0.26484E+02 -0.17338E+02  0.56223E+01
                                                                    -0.14664E+02 -0.29158E+02  0.72474E+01
```

Figure 6.29 (*Contd.*)

EQUILIBRIUM RESIDUALS AND REACTIONS

NODES	X	Y	Z	DEGREES OF FREEDOM (* = PRESCRIBED)	
1	0.00000E+00	0.30000E+01	0.00000E+00	-0.20842E+02 *	0.69014E-13
2	0.00000E+00	0.32500E+01	0.00000E+00	0.19449E+02 *	0.44506E-12
3	0.00000E+00	0.35000E+01	0.00000E+00	0.30737E+02 *	0.41300E-13
4	0.75000E+00	0.29800E+01	0.00000E+00	-0.25174E-13	0.50543E-13
5	0.75000E+00	0.35000E+01	0.00000E+00	-0.14433E-14	-0.39899E-13
6	0.15000E+01	0.29000E+01	0.00000E+00	0.26201E-13	0.33307E-13
7	0.15000E+01	0.32000E+01	0.00000E+00	-0.26645E-14	-0.27534E-12
8	0.16250E+01	0.35000E+01	0.00000E+00	-0.22538E-13	-0.49738E-13
9	0.25000E+01	0.27300E+01	0.00000E+00	0.81810E-14	-0.17653E-13
10	0.24167E+01	0.35000E+01	0.00000E+00	0.61201E-14	-0.22260E-13
11	0.35000E+01	0.24400E+01	0.00000E+00	-0.14655E-13	0.21316E-13
12	0.35000E+01	0.29700E+01	0.00000E+00	0.29310E-13	-0.36859E-13
13	0.32500E+01	0.35000E+01	0.00000E+00	-0.25979E-13	-0.18430E-13
14	0.42500E+01	0.21200E+01	0.00000E+00	-0.36776E-15	-0.90206E-14
15	0.40833E+01	0.35000E+01	0.00000E+00	-0.25327E-14	-0.47740E-14
16	0.50000E+01	0.16600E+01	0.00000E+00	-0.11102E-14	0.15543E-14
17	0.50000E+01	0.25800E+01	0.00000E+00	-0.31086E-14	0.57732E-14
18	0.48750E+01	0.35000E+01	0.00000E+00	0.62172E-14	0.98810E-14
19	0.57500E+01	0.17000E+01	0.00000E+00	0.88818E-15	-0.55511E-14
20	0.57500E+01	0.35000E+01	0.00000E+00	-0.22482E-14	-0.18874E-14
21	0.65000E+01	0.17500E+01	0.00000E+00	-0.18816E+02 *	0.32196E-14
22	0.65000E+01	0.26200E+01	0.00000E+00	0.17578E+01 *	0.27756E-15
23	0.65000E+01	0.35000E+01	0.00000E+00	0.11144E+02 *	-0.22204E-15
24	0.57500E+01	0.86000E+00	0.00000E+00	0.20296E-14	0.14988E-14
25	0.65000E+01	0.87000E+00	0.00000E+00	-0.24541E+02 *	-0.20817E-14
26	0.60000E+01	0.00000E+00	0.00000E+00	0.80242E+01 *	0.75938E+01 *
27	0.62500E+01	0.00000E+00	0.00000E+00	-0.98394E+01 *	0.19985E+02 *
28	0.65000E+01	0.00000E+00	0.00000E+00	0.29277E+01 *	0.20392E+01 *

END OF PROBLEM,   874 UTILIZED REAL WORDS OVER   20000

Figure 6.29 Plane elasticity (equilibrium) problem.

```
                      F.E.M.3.
                G.TOUZOT , G.DHATT

IMAGE OF DATA CARDS
===================
                                  C O L U M N   N U M B E R
   CARD              1         2         3         4         5         6         7         8
  NUMBER    1234567890123456789012345678901234567890123456789012345678901234567890123456789012345678
  -------
     1      COMT
     2           ELASTIC ANALYSIS OF AN ELLIPTIC HALF BRIDGE ARCH IN PLANE STRESS
     3
     4      COOR
     5        28    2    2
     6         3       0.00                              23        6.50      3.50              5
     7         5       0.75                              20        5.75      3.50              5
     8         2       0.00
     9         7       1.50
    10        12       3.50
    11        17       5.00
    12        19       5.75
    13         1       0.00
    14         4       0.75
    15         6       1.50
    16         9       2.50
    17        11       3.50
    18        14       4.25
    19        16       5.00
    20        24       5.75
    21        26       6.00
    22        27       6.25
    23        28       6.50
    24        25       6.50
    25        21       6.50
    26        22       6.50
    27        -1
    28      COND
    29        11
    30        26   27   28
    31        10    1    2    3   25   21   22   23
    32      0000000000
    33
    34      PREL
    35         1    4
    36         1       2.0E6      0.3       0.0             2.3
    37        -1
    38      ELEM
    39         5    8    2    1    0
    40         1    4    5    0    1
  -------
              1234567890123456789012345678901234567890123456789012345678901234567890123456789012345678
   CARD              1         2         3         4         5         6         7         8
  NUMBER                           C O L U M N   N U M B E R
```

Figure 6.30 (Contd.)

```
CARD                         C O L U M N   N U M B E R
NUMBER       1         2         3         4         5         6         7         8
         12345678901234567890123456789012345678901234567890123456789012345678901234567890
---------------------------------------------------------------------------------------
   41    5         1         0    1    0    26  27  28   25   21  19  16  24
   42    -1
   43    SOLC
   44    1        0.00       -10.00
   45    3
   46    -1
   47    SOLR
   48    VALP   20 0.001    0.0         5    0   12   1.D-12
   49    STOP
   50
---------------------------------------------------------------------------------------
CARD     12345678901234567890123456789012345678901234567890123456789012345678901234567890
NUMBER       1         2         3         4         5         6         7         8
                             C O L U M N   N U M B E R
```

E N D   O F   D A T A

COMMENTS
=========

ELASTIC ANALYSIS OF AN ELLIPTIC HALF BRIDGE ARCH IN PLANE STRESS

INPUT OF NODES (M= 0)
=====================

```
        MAX. NUMBER OF NODES                (NNT)=       28
        MAX. NUMBER OF D.O.F. PER NODE      (NDLN)=       2
        DIMENSIONS OF THE PROBLEM           (NDIM)=       2
        COORDINATE SCALE FACTORS            (FAC)= 0.10000E+01 0.10000E+01 0.10000E+01
        WORKSPACE IN REAL WORDS             (NVA)=    20000
```

INPUT OF BOUNDARY CONDITIONS (M= 0)
===================================

BOUNDARY CONDITIONS CARDS

```
>>>>>1100000000 0.00000E+00 0.00000E+00 0.00000E+00 0.00000E+00 0.00000E+00 0.00000E+00
>>>>>      26  27  28     0     0     0     0     0     0     0     0     0
>>>>>1000000000 0.00000E+00 0.00000E+00 0.00000E+00 0.00000E+00 0.00000E+00 0.00000E+00
>>>>>       1   2   3    25   21  22  23     0     0     0     0     0     0
>>>>>0000000000 0.00000E+00 0.00000E+00 0.00000E+00 0.00000E+00 0.00000E+00 0.00000E+00
```

```
        TOTAL NUMBER OF NODES                       (NNT)=   28
        TOTAL NUMBER OF D.O.F.                     (NDLT)=   56
        NUMBER OF EQUATIONS TO BE SOLVED            (NEQ)=   43
        NUMBER OF PRESCRIBED NON ZERO D.O.F.      (NCLNZ)=    0
        MUMBER OF PRESCRIBED ZERO D.O.F.           (NCLZ)=   13
        TOTAL NUMBER OF PRESCRIBED D.O.F.          (NCLT)=   13
```

Figure 6.30 (*Contd.*)

NODAL COORDINATES ARRAY

NO	D.L.	X	Y	Z	EQUATION NUMBER	(NEQ)
1	2	0.00000E+00	0.30000E+01	0.00000E+00	-7	1
2	2	0.00000E+00	0.32500E+01	0.00000E+00	-8	2
3	2	0.00000E+00	0.35000E+01	0.00000E+00	-9	3
4	2	0.75000E+00	0.29800E+01	0.00000E+00	4	5
5	2	0.75000E+00	0.35000E+01	0.00000E+00	6	7
6	2	0.15000E+01	0.29000E+01	0.00000E+00	8	9
7	2	0.15000E+01	0.32000E+01	0.00000E+00	10	11
8	2	0.16250E+01	0.35000E+01	0.00000E+00	12	13
9	2	0.25000E+01	0.27300E+01	0.00000E+00	14	15
10	2	0.24167E+01	0.35000E+01	0.00000E+00	16	17
11	2	0.35000E+01	0.24400E+01	0.00000E+00	18	19
12	2	0.35000E+01	0.29700E+01	0.00000E+00	20	21
13	2	0.32500E+01	0.35000E+01	0.00000E+00	22	23
14	2	0.42500E+01	0.21200E+01	0.00000E+00	24	25
15	2	0.40833E+01	0.35000E+01	0.00000E+00	26	27
16	2	0.50000E+01	0.16600E+01	0.00000E+00	28	29
17	2	0.50000E+01	0.25800E+01	0.00000E+00	30	31
18	2	0.48750E+01	0.35000E+01	0.00000E+00	32	33
19	2	0.57500E+01	0.17000E+01	0.00000E+00	34	35
20	2	0.57500E+01	0.35000E+01	0.00000E+00	36	37
21	2	0.65000E+01	0.17500E+01	0.00000E+00	-11	38
22	2	0.65000E+01	0.26200E+01	0.00000E+00	-12	39
23	2	0.65000E+01	0.35000E+01	0.00000E+00	-13	40
24	2	0.57500E+01	0.86000E+00	0.00000E+00	41	42
25	2	0.65000E+01	0.87000E+00	0.00000E+00	-10	43
26	2	0.60000E+01	0.00000E+00	0.00000E+00	-1	-2
27	2	0.62500E+01	0.00000E+00	0.00000E+00	-3	-4
28	2	0.65000E+01	0.00000E+00	0.00000E+00	-5	-6

```
INPUT OF ELEMENT PROPERTIES (M= 0)
==================================

         NUMBER OF GROUPS OF PROPERTIES    (NGPE)=    1
         NUMBER OF PROPERTIES PER GROUP    (NPRE)=    4

CARDS OF ELEMENT PROPERTIES

>>>>>   1  0.20000E+07  0.30000E+00  0.00000E+00  0.23000E+01
>>>>>  -1  0.00000E+00  0.00000E+00  0.00000E+00  0.00000E+00

INPUT OF ELEMENTS (M= 0)
========================

         MAX. NUMBER OF ELEMENTS             (NELT)=    5
         MAX. NUMBER OF NODES PER ELEMENT    (NNEL)=    8
         DEFAULT ELEMENT TYPE                (NTPE)=    2
         NUMBER OF GROUPS OF ELEMENTS        (NGRE)=    1
         INDEX FOR NON SYMMETRIC PROBLEM     (NSYM)=    0
         INDEX FOR IDENTICAL ELEMENTS        (NIDENT)=  0

ELEMENT:  1 TYPE: 2 N.P.: 8 D.O.F.: 16 N. PROP:  0 EL. PROP:  4 GROUP:  0
  CONNECTIVITY (NE)    1    4    6    7    8    5    3    2
ELEMENT:  2 TYPE: 2 N.P.: 8 D.O.F.: 16 N. PROP:  0 EL. PROP:  4 GROUP:  0
  CONNECTIVITY (NE)    6    9   11   12   13   10    8    7
ELEMENT:  3 TYPE: 2 N.P.: 8 D.O.F.: 16 N. PROP:  0 EL. PROP:  4 GROUP:  0
  CONNECTIVITY (NE)   11   14   16   17   18   15   13   12
ELEMENT:  4 TYPE: 2 N.P.: 8 D.O.F.: 16 N. PROP:  0 EL. PROP:  4 GROUP:  0
  CONNECTIVITY (NE)   16   19   21   22   23   20   18   17
ELEMENT:  5 TYPE: 2 N.P.: 8 D.O.F.: 16 N. PROP:  0 EL. PROP:  4 GROUP:  0
  CONNECTIVITY (NE)   26   27   28   25   21   19   16   24

         MEAN BAND HEIGHT=    9.1 MAXIMUM=   15
         LENGTH OF A TRIANGLE IN KG          (NKG)=  393
         NUMBER OF INTEGRATION POINTS        (NPG)=   45
```

Figure 6.30 (*Contd.*)

INPUT OF CONCENTRADED LOADS (M= 0)
==================================

CARDS OF NODAL LOADS

>>>>>   1 0.00000E+00 -0.10000E+02
>>>>>   3   0   0   0   0   0   0   0   0   0   0
>>>>> -1 0.00000E+00 0.00000E+00

ASSEMBLING OF DISTRIBUTED LOADS (M= 0)
======================================

SUBSPACE ITERATION (M= 0)
=========================

```
        NUMBER OF DESIRED EIGENVALUES              (NVALP)=  3
        MAX. NUMBER OF ITERATIONS PERMITTED        (NITER)= 20
        INDEX FOR DIAGONAL MATRIX                  (NMDIAG)= 0
        CONVERGENCE TOLERANCE ON EIGENVALUES       (EPSLB)= 0.10000E-02
        SHIFT                                      (SHIFT)= 0.00000E+00
        SUBSPACE DIMENSION                         (NSS)=    5
        MAX. NUMBER OF ITERATION IN JACOBI         (NSWM)=  12
        CONVERGENCE TOLERANCE IN JACOBI            (TOLJAC)= 0.10000E-11
```

ITERATION 1 MAX. ERROR= 0.5E+06 EXACT EIGENVALUES: 0
ITERATION 2 MAX. ERROR= 0.4E+00 EXACT EIGENVALUES: 0
ITERATION 3 MAX. ERROR= 0.1E-01 EXACT EIGENVALUES: 3

. . . . CONVERGENCE IN 4 ITERATIONS

EIGENVALUE NO.   1 = 0.56152E+04

EIGENVECTOR:

NODES	X	Y	Z	DEGREES OF FREEDOM (* = PRESCRIBED)	
1	0.00000E+00	0.30000E+01	0.00000E+00	0.00000E+00 *	0.66375E+00
2	0.00000E+00	0.32500E+01	0.00000E+00	0.00000E+00 *	0.66172E+00
3	0.00000E+00	0.35000E+01	0.00000E+00	0.00000E+00 *	0.65867E+00
4	0.75000E+00	0.29800E+01	0.00000E+00	-0.19930E-01	0.60786E+00
5	0.75000E+00	0.35000E+01	0.00000E+00	0.45051E-01	0.60490E+00
6	0.15000E+01	0.29000E+01	0.00000E+00	-0.40117E-01	0.48147E+00
7	0.15000E+01	0.32000E+01	0.00000E+00	0.18612E-01	0.47890E+00
8	0.16250E+01	0.35000E+01	0.00000E+00	0.78526E-01	0.45397E+00
9	0.25000E+01	0.27300E+01	0.00000E+00	-0.43577E-01	0.28471E+00
10	0.24167E+01	0.35000E+01	0.00000E+00	0.80294E-01	0.29658E+00
11	0.35000E+01	0.24400E+01	0.00000E+00	-0.43399E-01	0.14162E+00
12	0.35000E+01	0.29700E+01	0.00000E+00	0.72541E-02	0.13794E+00
13	0.32500E+01	0.35000E+01	0.00000E+00	0.68736E-01	0.17225E+00
14	0.42500E+01	0.21200E+01	0.00000E+00	-0.32304E-01	0.72376E-01
15	0.40833E+01	0.35000E+01	0.00000E+00	0.48682E-01	0.85795E-01
16	0.50000E+01	0.16600E+01	0.00000E+00	-0.17548E-01	0.30240E-01
17	0.50000E+01	0.25800E+01	0.00000E+00	-0.50562E-02	0.33255E-01
18	0.48750E+01	0.35000E+01	0.00000E+00	0.29860E-01	0.39282E-01
19	0.57500E+01	0.17000E+01	0.00000E+00	-0.78157E-02	0.17301E-01
20	0.57500E+01	0.35000E+01	0.00000E+00	0.14539E-01	0.16647E-01
21	0.65000E+01	0.17500E+01	0.00000E+00	0.00000E+00 *	0.11601E-01
22	0.65000E+01	0.26200E+01	0.00000E+00	0.00000E+00 *	0.95616E-02
23	0.65000E+01	0.35000E+01	0.00000E+00	0.00000E+00 *	0.11585E-01
24	0.57500E+01	0.86000E+00	0.00000E+00	-0.19224E-02	0.13207E-01
25	0.65000E+01	0.87000E+00	0.00000E+00	0.00000E+00 *	0.98898E-02
26	0.60000E+01	0.00000E+00	0.00000E+00	0.00000E+00 *	0.00000E+00 *
27	0.62500E+01	0.00000E+00	0.00000E+00	0.00000E+00 *	0.00000E+00 *
28	0.65000E+01	0.00000E+00	0.00000E+00	0.00000E+00 *	0.00000E+00 *

EIGENVALUE NO. 2 = 0.37888E+05

Figure 6.30 (*Contd.*)

EIGENVECTOR:

NODES	X	Y	Z	DEGREES OF FREEDOM (* = PRESCRIBED)		
1	0.00000E+00	0.30000E+01	0.00000E+00	0.00000E+00 *	-0.43557E+00	
2	0.00000E+00	0.32500E+01	0.00000E+00	0.00000E+00 *	-0.43947E+00	
3	0.00000E+00	0.35000E+01	0.00000E+00	0.00000E+00 *	-0.43308E+00	
4	0.75000E+00	0.29800E+01	0.00000E+00	0.75478E-01	-0.30806E+00	
5	0.75000E+00	0.35000E+01	0.00000E+00	-0.66589E-01	-0.30728E+00	
6	0.15000E+01	0.29000E+01	0.00000E+00	0.12568E+00	-0.50310E-01	
7	0.15000E+01	0.32000E+01	0.00000E+00	0.19735E-01	-0.43091E-01	
8	0.16250E+01	0.35000E+01	0.00000E+00	-0.83712E-01	-0.50229E-02	
9	0.25000E+01	0.27300E+01	0.00000E+00	0.10301E+00	0.25248E+00	
10	0.24167E+01	0.35000E+01	0.00000E+00	-0.30655E-01	0.23703E+00	
11	0.35000E+01	0.24400E+01	0.00000E+00	0.43413E-01	0.31596E+00	
12	0.35000E+01	0.29700E+01	0.00000E+00	0.53209E-01	0.32955E+00	
13	0.32500E+01	0.35000E+01	0.00000E+00	0.32536E-01	0.31620E+00	
14	0.42500E+01	0.21200E+01	0.00000E+00	0.34673E-02	0.27207E+00	
15	0.40833E+01	0.35000E+01	0.00000E+00	0.74869E-01	0.29474E+00	
16	0.50000E+01	0.16600E+01	0.00000E+00	-0.12682E-01	0.20050E+00	
17	0.50000E+01	0.25800E+01	0.00000E+00	0.22806E-01	0.21003E+00	
18	0.48750E+01	0.35000E+01	0.00000E+00	0.77638E-01	0.23648E+00	
19	0.57500E+01	0.17000E+01	0.00000E+00	-0.23282E-02	0.14099E+00	
20	0.57500E+01	0.35000E+01	0.00000E+00	0.40232E-01	0.17473E+00	
21	0.65000E+01	0.17500E+01	0.00000E+00	0.00000E+00 *	0.11596E+00	
22	0.65000E+01	0.26200E+01	0.00000E+00	0.00000E+00 *	0.14199E+00	
23	0.65000E+01	0.35000E+01	0.00000E+00	0.00000E+00 *	0.15438E+00	
24	0.57500E+01	0.86000E+00	0.00000E+00	0.15066E-02	0.10386E+00	
25	0.65000E+01	0.87000E+00	0.00000E+00	0.00000E+00 *	0.71782E-01	
26	0.60000E+01	0.00000E+00	0.00000E+00	0.00000E+00 *	0.00000E+00 *	
27	0.62500E+01	0.00000E+00	0.00000E+00	0.00000E+00 *	0.00000E+00 *	
28	0.65000E+01	0.00000E+00	0.00000E+00	0.00000E+00 *	0.00000E+00 *	

EIGENVALUE NO.   3 = 0.10014E+06

EIGENVECTOR:

NODES	X	Y	Z	DEGREES OF FREEDOM (* = PRESCRIBED)		
1	0.00000E+00	0.30000E+01	0.00000E+00	0.00000E+00 *	-0.41128E+00	
2	0.00000E+00	0.32500E+01	0.00000E+00	0.00000E+00 *	-0.42594E+00	
3	0.00000E+00	0.35000E+01	0.00000E+00	0.00000E+00 *	-0.41626E+00	
4	0.75000E+00	0.29800E+01	0.00000E+00	0.12574E+00	-0.22292E+00	
5	0.75000E+00	0.35000E+01	0.00000E+00	-0.69262E-01	-0.22541E+00	
6	0.15000E+01	0.29000E+01	0.00000E+00	0.17483E+00	0.10569E+00	
7	0.15000E+01	0.32000E+01	0.00000E+00	0.68934E-01	0.11778E+00	
8	0.16250E+01	0.35000E+01	0.00000E+00	-0.33533E-01	0.14924E+00	
9	0.25000E+01	0.27300E+01	0.00000E+00	0.89392E-01	0.32437E+00	
10	0.24167E+01	0.35000E+01	0.00000E+00	0.86100E-01	0.32239E+00	
11	0.35000E+01	0.24400E+01	0.00000E+00	-0.47209E-01	0.12451E+00	
12	0.35000E+01	0.29700E+01	0.00000E+00	0.81544E-01	0.14886E+00	
13	0.32500E+01	0.35000E+01	0.00000E+00	0.18077E+00	0.20020E+00	
14	0.42500E+01	0.21200E+01	0.00000E+00	-0.87810E-01	-0.78360E-01	
15	0.40833E+01	0.35000E+01	0.00000E+00	0.18541E+00	-0.14116E-01	
16	0.50000E+01	0.16600E+01	0.00000E+00	-0.75138E-01	-0.21317E+00	
17	0.50000E+01	0.25800E+01	0.00000E+00	-0.10474E+00	-0.21304E+00	
18	0.48750E+01	0.35000E+01	0.00000E+00	0.13544E+00	-0.17181E+00	
19	0.57500E+01	0.17000E+01	0.00000E+00	-0.41600E-01	-0.21638E+00	
20	0.57500E+01	0.35000E+01	0.00000E+00	0.62931E-01	-0.27285E+00	
21	0.65000E+01	0.17500E+01	0.00000E+00	0.00000E+00 *	-0.22656E+00	
22	0.65000E+01	0.26200E+01	0.00000E+00	0.00000E+00 *	-0.28436E+00	
23	0.65000E+01	0.35000E+01	0.00000E+00	0.00000E+00 *	-0.29085E+00	
24	0.57500E+01	0.86000E+00	0.00000E+00	-0.27562E-01	-0.16051E+00	
25	0.65000E+01	0.87000E+00	0.00000E+00	0.00000E+00 *	-0.11610E+00	
26	0.60000E+01	0.00000E+00	0.00000E+00	0.00000E+00 *	0.00000E+00 *	
27	0.62500E+01	0.00000E+00	0.00000E+00	0.00000E+00 *	0.00000E+00 *	
28	0.65000E+01	0.00000E+00	0.00000E+00	0.00000E+00 *	0.00000E+00 *	

END OF PROBLEM, 1676 UTILIZED REAL WORDS OVER 20000

Figure 6.30 Natural frequencies of an elasticity problem.

Chute Panet, October 1979,
the day of the first snow
Monterey, California
August 22, 1983.

INDEX

Approximate functions, 67, 69
   algebraic form of, 30–32
   construction of, 14, 15, 17
   properties of, 32–36
Approximate integration, 258–264
Approximations, 69–76
   based on reference element, 30–36
   differential equation, 10
   four-model, 12–14
   linear, 17
   methods of, 18
   nodal, 9–14, 30, 32–33, 73
   non-nodal, 157
   numerical example, 71–75
   of functions, 156–157
   of physical quantity, 10
   with finite elements, 14–17
Arrays, 369, 372, 374, 376, 388
Assembly of three one-dimensional
      elements, 221
Assembly subroutine, 216
Assembly techniques, 169, 214–219

Back-substitution, 270, 274
BBMEF program, 358-363
Bidirectional formula, 252
Bidirectional method, 255
Bilinear elements, 102–103
Boundary conditions, 11, 131, 135, 136,
      137, 141–144, 148, 156, 157,
      227–230, 237, 388, 448
Boundary integral method, 142
Boundary value problems, 132
Buckling load of structures, 333–334

Cauchy condition, 136
Central finite difference method, 328
Chain rule, 43
Cholesky decomposition, 279

Coding techniques, 356–503
Collocation by sub-domains, 159–160
Collocation method, 159
Collocation points, 158
Column matrix, 75
Complementary functionals, 154
Complete cubic elements, 95, 98, 105,
      112–113
Complete polynomial for three-noded
      triangle, 35–36
Complete polynomial interpolation
      functions, 34–36
Complete quadratic elements, 103, 112,
      115
Computation cost, 364
Computation time, 364
Computer codes, common features of,
      356–357
Computer programming, 1
Concrete elliptical arch, 481
Connectivity matrix, 27
Connectivity table, 236
Continuity, 104, 109, 156
   inter-element, 33
   within element, 33
Continuity conditions, 173, 174
Continuity requirements, 34
Continuous functions, 34, 71
Continuous problems, 135
Continuous rectangular elements, 108–109
Continuous systems, 130, 131, 133
Continuum mechanics, 1, 130
Contour integral for four-node element,
      183
Convergence, 340
   to exact result, 173–174
Convergence criteria, 318
Convergence tests, 293, 295
Coordinate systems, 89–91, 101–102, 110

Core and out-of-core storage, 370–376
Corrector formulas, 317
Crout decomposition, 279
Cubic elements, 83, 85, 107, 242, 256–257
Curvilinear, 96–97, 106–107, 113, 119

Damping matrix, 133
Data storage, 364
Decomposition algorithms, 280
Decomposition solution, 280
Deflation process, 269
Differential equations, 131
   approximation solution, 10
Differential operators, transformation of, 42–47
Direct integration of second order systems, 324–329
Direct integration formulas, 254, 259, 260, 261
Dirichlet boundary conditions, 136
Discrete systems, 130, 131, 133
Discretization procedures, 129
Duhamel integration, 331

Eigenvalue problems, 132, 136
Eigenvalues, 331, 336, 337, 430
   computations of, 340–349
   intermediary, 341
   largest, 341
   separation of, 338, 339
   shifting of, 339
   smallest, 340, 348
   successive, 341
   see also Matrix eigenvalue problems
Eigenvectors, 331, 334–337, 430
Elastic springs, 213
Elasticity problems, 199
Element connectivities, 27–30
Element integral forms, 170, 172
Element mass matrix, 173
Element matrices
   assembly by expansion, 207–212
   classical forms of, 185
   computational techniques for, 185
   control subroutes for computation of, 192–193
   matrix organization for numerical computations of, 191–192
Element residual function, 173
Element vectors, 208
Elements
   boundary between, 20
   classical, 20, 25–27, 182

fifth order, 86
general, 87–89
geometrical definition, 15, 17, 18
hexahedronal, 184
list of, 78–80
location table, 214
modifications of, 123–124
partition of domain into, 18–20
reference, 21–24
using complete polynomial of order r, 94
various types of, 77–128
with higher degrees of inter-element continuity, 84–87
with n equidistant nodes, 84
see also under specific types
Engineering sciences, 1
Equations
   formulation of system, 227
   generation of, 129
   global system of, 227–235
   properties of system of, 165–166
   transformation of, 129
Equilibrium problems, 132, 135
Euclidian norm, 294
Execution functional blocks, 401–440
   organization of, 402
Explicit Euler's method, 307–308
Explicit predictor formulas, 317

Factorized matrix, 278
Fifth order elements, 100
Finite difference method, 129
Finite element discretization process, 1
Finite element method
   applications, 1
   definition, 169–173
   historical evolution, 2–3
   introduction to, 1–8
   main purpose of, 130
   state of the art, 2–3
   teaching methods, 3
FORTRAN, 50, 64, 169, 284, 343, 367
Fourier series, 159
Full non-symmetrical matrix, 222
Full symmetrical matrix, 222
Functional analysis, 2
Functional blocks, 356–357, 366, 367, 370, 376–440
   reading data, 382–401
Functionals, 145–155
   additional, 149–155

associated with integral formulation, 146–148
discretization of, 163–165
first variation, 145–146
formulation of Poisson's equation, 146–148
second variation, 146
Functions, approximation of. *See* Approximation

Galerkin integral forms, 170
Galerkin method, 160–162
Gauss method, 243–248, 251, 252
Gauss–Jacobi formula, 256
Gauss–Radau method, 255, 256
Gauss–Seidel method, 269
Gaussian elimination, 269–274
  matrix formulation of, 276–277
Gaussian elimination algorithm, 280
Gaussian elimination computer code, 274
Gaussian elimination reformulation, 275–276
Generalized variables, 38
Geometrical transformations, 36
Global discretized form, assembly of, 207–214
Global load vector, 212
Global matrices, 219–227, 418
Global matrices assembly, 406
Global residual function, 173
Global stiffness matrix, 212
Global tangent matrix, 298
Global vectors, 208

Heat conduction, 447–480
Helmholtz equation, 136
Hermite elements, 98–101, 113, 120
  higher order, 107–109
  with one non-nodal degree of library, 88
Hermite polynomials, 36, 84, 85, 181
Heterogeneous elements, 121–123
Hexahedrons, 114–120
Hierarchical elements, 124–126
Houbolt's method, 325

Implicit Euler's method, 311–312
Incomplete cubic elements, 96, 99, 106, 117
Incomplete quadratic elements, 104, 115
Incremental method, 300–301
Independent variables, change of 301–302
Infinite elements, 127–128
Initial value problems, 132

Integral equations, 139
Integral forms, 139, 149
  additional 143–145
  discretization, 155–168, 177–185
Integral transformations, 140–145
Integration formulas, 243
  for exact integration, 259
Integration methods for first order systems, 307
Integration points, choice of number of, 264–265
Integro-differential equations, 131
Interpolation functions, 31–33, 36, 50, 77, 100, 109, 125, 128, 250
  construction of, 74
  properties of, 73
Inverse iteration algorithm, 341
Inverse iteration method, 340
Isoparametric concept, 2
Isoparametric elements, 77, 102, 106, 448, 481
Isotropic conductivity coefficient, 190

Jacobi method, 349
Jacobi rotation method, 342, 343
Jacobian matrix, 24, 43, 44, 45, 49, 53–64, 77, 81, 88, 92, 101, 102, 111, 181, 242, 263
  ill-conditioned, 48–49

Lagrange–Hermite element, 88
Lagrange multipliers, 130, 149–155
Lagrange polynomial, 36, 84, 106, 248
Lagrangian elements, 81–84, 92–98
  higher order, 103–107, 112–120
Laplace equation, 254, 259, 448
  numerical patch test for, 175
  variational patch test for, 176
Laplacian operator, 133
LDU decomposition, 278
Least squares method, 162–163
Legendre polynomial, 244, 245
Line elements, 80–89
Linear constraints
  at element level, 234
  between variables, 232–239
Linear elements, 91, 111
Linear equations, solution of systems of, 269–275
Linear systems, 305
Location table, 217

Mass matrix, 132

Mathematical models, 1, 9, 131
Matrix algebra, 169
Matrix eigenvalue problems, 333–352
Matrix factorization, 275–282
Matrix formulation, 169–239
Matrix segmentation, 226
Maximum norm, 294
MEF program, 366–376
 applications of, 447–503
 input values for, 440–447
Merging process, 172
Mixed functionals, 152–154
Mixed integral form, 144–145
Modal superposition
 for first order system, 321–323
 for second order system, 330–332
Modularity, 365–366
Multidimensional regions, 128
Multipurpose programs, 363–366

Navier–Stokes equations, 137, 139, 291–292
Neuman condition, 136
Newmark's method, 327, 328
Newton–Cotes method, 248–252
Newton–Raphson algorithm, 296
Newton–Raphson method, 297–300, 314, 316, 325, 424
Nodal coordinates, 27–30
Nodal variables, 38
Non-compatible elements, 97
Non-linear differential operator, 179–180
Non-linear equations, 138
Non-linear problems, 291, 304, 309, 313–316, 424
Non-linear spring, 295, 296, 297, 301
Non-linear systems, 291–306
Non-stationary problems of first and second order, 334
Non-symmetrical band matrix, 223
Non-symmetrical skyline matrix, 223
Numerical analysis, 1
Numerical integration, 240–269
 for four-node trapezoidal element, 263
 in one dimension, 243–252
 in three dimensions, 256
 in two dimensions, 252–256
Numerical integration code, 265–269
Numerical procedures, 240–355
Numerical tests, 174

Orthogonality conditions, 337
Over-relaxation factor, 293

Partial differential equations, 1, 131, 133, 135, 148
Patch tests, 174–177
Penalty parameter, 130
Physical systems, 129, 131, 134
Plane elasticity, 199
Plane stress problems, 481–503
Poisson's equation, 135, 157
 collocation by sub-domain for, 160
 comparison of results from various methods of solution, 165
 complementary functional for, 154
 explicit computations for, 185
 functional formulation for, 146–148
 generalized functional, 151
 integral form of, 139, 181
 matrix expression for element integral form, 171
 matrix formulation method, 235–239
 mixed functional for, 153
 mixed integral form for, 144–145
 over square domain, 159
 second variation of functional corresponding to, 147
 solution by BBMEF program, 358–363
 solution by Galerkin method, 161–162
 solution by least squares method, 163
 solution by Ritz method, 164
 supplementary form for, 144
 weak integral form for, 141–143
Polynomial basis
 choice of, 36–38, 39
 complete and incomplete, 37
Ponderation function, 248
Predictor–corrector formulas, 316–319
Prisms, 120
Problem data base, 370–376
Problem description, 364
Problem size, 364
Problem types, 363–364
Programming norms, 367–370
Propagation problems, 132, 137
Pseudo-dynamic memory management, 367

Quadratic elements, 82, 93, 107
Quadrilateral elements, 101–109, 122–124, 126, 252, 262
Quasi-harmonic problems, 193

Rainfall estimation, 71
Ramp function, 331

Rayleigh quotient, 338, 339, 340, 348
Reactions, 230
Recurrence formula, 311
Reference elements, 23, 25, 26, 30, 64, 73, 85, 89, 242–243
Ritz method, 163–165, 348
Ritz subspace, 349
Ritz vectors, 348
Row pivoting, 273
Runge–Kutta formula, 320
Runge–Kutta type explicit methods, 319–320

Semi-implicit Euler's method, 314–316
Shape functions, 39, 41
Simultaneous equations of the first order, 306
Skyline envelope, 224
Skyline matrix
    in compact storage, 282–291
    non-symmetrical, 223
    out of core, 226
    stored out of core by blocks, 287
Skyline profile, 225
Skyline storage, 225
    for non-symmetrical matrix, 225
    Spectral decomposition, 336
Square elements, 242
Stability, 308–309, 315
Stability conditions, 310
Stability criteria, 313
Stationarity conditions, 150, 152, 153, 154, 163, 348
Stationarity principle, 148–149
Storage strategies, 222–227
Structural mechanics
    assembly for elements of, 212–213
Structural vibrations, 333
Sturm sequence, 339
Subspace iteration method, 349, 430
Substitution method, 292–297, 295, 316, 325
    algorithm for, 293
Superparametric elements, 126–127
Surface integration
    for eight-node element, 184
    in three dimensions, 183
Symmetrical banded matrix, 223
Symmetrical matrix, 273
    with skyline profile, 225

Symmetry, 222
Taylor series, 65, 298
Tetrahedronal elements, 243, 257
Tetrahedrons, 110–113
Time-dependent problems, 305–332, 430
Transformation
    of contour integrations, 182
    of derivatives, 181
    of differential operators, 42–47
    of domain of integrations, 182
    of integral, 44
    of matrix, 181
    of nodal variables, 181
Transformation functions, 77
Transformation matrices, 342
Transformations
    variable, 230–232
Trapezoidal elements, 263
Triangular elements, 23, 89–101, 208, 236, 242, 255
    explicit computations for, 185
Triangular matrix, 273, 277
Triangular system, 274
Triangularization
    of matrix of coefficients, 269
    of non-symmetric system, 272
Truncation errors, 34, 64–70, 81, 92, 325
    evaluation of, 68–70
    reduction of, 70

Upper triangular system, 274

Variables, 369, 376
Variation operator, 145
Variational formulation of engineering problems, 129–168
Variational method, 175–177
Vector norms, 293
Vectors, 75

Weak integral forms, 141–143
Weighted residual method, 2, 130, 138–139, 144, 145, 149
Weighting coefficients, 248
Weighting functions, 142, 145, 147, 149, 155, 157, 159, 160, 170
    choice of, 158–163
Wilson's method, 327–328